T0173111

OCR GCSE (9–1)

DESIGN & TECHNOLOGY

Andy Knight
Chris Rowe
Sharon McCarthy
Jennifer Tilley
Chris Walker

HODDER
EDUCATION
AN HACHETTE UK COMPANY

This resource is endorsed by OCR for use with the GCSE (9–1) in Design and Technology specification J310. In order to gain OCR endorsement, this resource has undergone an independent quality check. Any references to assessment and/or assessment preparation are the publisher's interpretation of the specification requirements and are not endorsed by OCR. OCR recommends that a range of teaching and learning resources are used in preparing learners for assessment. OCR has not paid for the production of this resource, nor does OCR receive any royalties from its sale. For more information about the endorsement process, please visit the OCR website, www.ocr.org.uk.

Although every effort has been made to ensure that website addresses are correct at time of going to press, Hodder Education cannot be held responsible for the content of any website mentioned in this book. It is sometimes possible to find a relocated web page by typing in the address of the home page for a website in the URL window of your browser.

Orders: please contact Hachette UK Distribution, Hely Hutchinson Centre, Milton Road, Didcot, Oxfordshire, OX11 7HH. Telephone: +44 (0)1235 827827. Email education@hachette.co.uk Lines are open from 9 a.m. to 5 p.m., Monday to Friday. You can also order through our website: www.hoddereducation.co.uk

ISBN: 9781510401136

© Andy Knight, Chris Rowe, Sharon McCarthy, Jennifer Tilley, Chris Walker 2017

First published in 2017 by
Hodder Education,
An Hachette UK Company
Carmelite House
50 Victoria Embankment
London EC4Y 0DZ

www.hoddereducation.co.uk

Impression number   10 9 8 7 6 5

Year   2023

All rights reserved. Apart from any use permitted under UK copyright law, no part of this publication may be reproduced or transmitted in any form or by any means, electronic or mechanical, including photocopying and recording, or held within any information storage and retrieval system, without permission in writing from the publisher or under licence from the Copyright Licensing Agency Limited. Further details of such licences (for reprographic reproduction) may be obtained from the Copyright Licensing Agency Limited, www.cla.co.uk

Cover photo © Brain light/Alamy Stock Photo

Illustrations by DC Graphic Design Limited

Typeset in Myriad Pro Light 10/12pt by Integra Software Services Pvt. Ltd., Pondicherry, India

Printed and bound by CPI Group (UK) Ltd, Croydon, CR0 4YY

A catalogue record for this title is available from the British Library.

CONTENTS

INTRODUCTION TO OCR GCSE (9–1) DESIGN AND TECHNOLOGY

How to use this book

This book has been written to help you master the skills, knowledge and understanding you need for OCR GCSE (9-1) Design and Technology.

Throughout the course you will develop your ability to explore design opportunities and stakeholder needs, wants and values, and to develop and effectively communicate realistic design proposals that meet these requirements. You will develop your decision making skills and the ability to critique and refine your ideas throughout the process. You will also gain a broad knowledge of materials, components, technologies and practical skills and will learn how to use these safely to develop high quality, imaginative and functional prototypes.

The course content is divided into 'core' and 'in-depth' principles of design and technology:
- The **'core'** principles offer a broad set of principles that **all** learners must know. These principles are covered in Section 1 of the textbook.
- You are also required to demonstrate your **'in-depth'** knowledge, understanding and design development skills in relation to **one or more** of the areas of learning you have chosen to work with and have an interest in. These 'in-depth' principles are covered in Section 2 of the textbook, and are divided into separate chapters for each of the main categories for in-depth learning:
  - papers and boards (Chapter 8)
  - natural and manufactured timber (Chapter 9)
  - ferrous and non-ferrous metals (Chapter 10)
  - thermo and thermosetting polymers (Chapter 11)
  - woven, non-woven and knitted textiles (Chapter 12)
  - design engineering (Chapter 13).

The first five in-depth areas relate directly to material categories. Those who choose to follow the design engineering area are likely to be more interested in electronic and mechanical systems and control.

The content of the OCR specification is divided into eight topic areas:
1. Identifying requirements
2. Learning from existing products and practice
3. Implications of wider issues
4. Design thinking and communication
5. Material considerations
6. Technical understanding
7. Manufacturing processes and techniques
8. Viability of design solutions

Each of these topics is explored through an enquiry approach. The questions that form this enquiry approach are included at the start of each section within the textbook.

Summary of assessment

The topic areas listed above will be assessed in the Principles of Design and Technology written paper that you will take at the end of the course. (You will find more information to help you prepare for this written paper in Chapter 15 of this textbook.) They will also be assessed in the Iterative Design Challenge non-exam assessment (NEA) that you will carry out in the final year of the course. (You will find information to help you with how to approach the NEA in Chapter 14 of this textbook.)

The table below summarises how you will be assessed for OCR GCSE (9-1) Design and Technology.

Component	Assessment type	Time	Marks available	% of qualification
Principles of Design and Technology	Written paper	2 hours	100 marks	50%
Iterative Design Challenge	Non-exam assessment	Approx. 40 hours	100 marks	50%

Features of this book

Throughout each chapter you will find a range of features to support your learning.

LEARNING OUTCOMES

Learning outcomes are included at the start of each topic within a chapter and tell you what you should know and understand by the end of the topic.

KEY POINT

Key point boxes highlight and summarise important points.

ACTIVITY

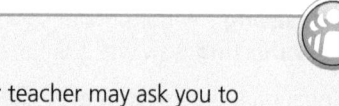

Short activities are included throughout each chapter. Your teacher may ask you to complete these to help you develop your knowledge and understanding of a topic.

STRETCH AND CHALLENGE

These activities will help you to develop your knowledge and understanding of a topic further. They may ask you to complete further research or to consider some of the more challenging aspects of the course.

KEY TERMS

All of the important terms you need to know and understand are defined.

PRACTICE QUESTIONS

You will find these at the end of each section. Practice questions are designed to check your knowledge and understanding of the content in each section. Please bear in mind that these are the work of the authors and do not necessarily reflect the type of questions you will face in the examination.

What is iterative design?

In simple terms, iterative design means the creation and refining of a design by using a repeating process of prototyping, analysing and testing.

Each time the process is repeated and refined, this is referred to as an 'iteration'. Every cycle of prototyping, analysing and testing of a design informs and reveals further refinements that are necessary or possible, which in turn leads to a new iteration of the design. The first prototype is the first iteration of the design; the second prototype is the second iteration, and so on. Repeating the cycle produces a more refined and improved iteration of the design each time.

The iterative design process allows the designer to explore different design avenues and try out potential ideas so that they can discover design problems early in the design process and make changes and refinements that lead to improved iterations of their original design. It is more effective than a linear design process (for example, **Situation → Brief → Analysis → Design ideas → Final design → Planning → Making → Evaluation**), in which a designer might not examine how well a product actually meets the needs of the stakeholder until it is completed, at which stage it is likely to be too late to change anything or improve it.

The iterative design process encourages you to be more creative and to take 'design risks'. These may or may not be successful, but they are necessary in order to discover whether ideas will work or not. These 'failures' will be rewarded as they are valuable steps to finding new avenues of exploration and ways forward in the refining process. It means your ability to undertake a process of design thinking is assessed, rather than just focusing on ensuring you produce the most perfect outcome.

Iterative designing using Explore, Create, Evaluate

Underpinning the iterative design process in GCSE Design and Technology are the three main interconnected stages: Explore, Create and Evaluate.

You will need to demonstrate your knowledge, understanding and skills through:
- **exploring** needs, requirements and opportunities
- **creating** solutions that resolve those needs and requirements
- **evaluating** how well solutions meet the needs.

Repeating this cycle will create multiple design iterations that show how a design has evolved to meet the stakeholder's needs.

Management of the cycle is central to the whole iterative design process, which can be successful only when the three stages are managed effectively. Decisions such as which stakeholder's needs are most important, which design ideas are worth exploring or pursuing further, what tests need to be done, which materials could be used, etc., are integral to the iterative design process.

Core principles of design and technology

This section looks at the core principles of design and technology that you must know and understand in order to make informed decisions as a designer.

All students will need knowledge and understanding of the principles considered in this section.

The section explores the following questions outlined in the OCR specification:

1.1 How can exploring the context a design solution is intended for inform decisions and outcomes?
1.2 Why is usability an important consideration when designing prototypes?
2.1 What are the opportunities and constraints that influence design and making requirements?
2.2 How do developments in design technology influence design decisions and practice?
3.1 What are the impacts of new and emerging technologies when developing design solutions?
3.2 How do designers choose appropriate sources of energy to make products and power systems?
3.3 What wider implications can have an influence on the processes of designing and making?
4.1 How can design solutions be communicated to demonstrate their suitability? (The content also relates to in-depth areas, therefore there will be some differences dependent on the design approach you are taking.)
4.2 How do designers source information and thinking when problem solving?
5.1 What are the main categories of materials available to designers when developing design solutions?
5.2 Why is it important to consider the characteristics and properties of materials and/or system components when designing?
6.3 How do we introduce controlled movement to products and systems?
6.4 How do electronic systems provide functionality to products and processes?
7.6 How do new and emerging technologies have an impact on production techniques and systems?

These questions are considered in the following chapters:

Chapter 1 Identifying requirements

Chapter 2 Learning from existing products and practice

Chapter 3 Implications of wider issues

Chapter 4 Design thinking and communication

Chapter 5 Material considerations

Chapter 6 Mechanical devices and electronic systems

Chapter 7 New and emerging technologies

CHAPTER 1
Identifying requirements

Different people have different needs and wants – what is desirable for and helpful to one person may cause problems and difficulties for someone else. All new designs of products and systems should ideally make things easier, quicker or cheaper for a user, but as designers we also need to consider the impact on the environment and other users. For example, the product might have an impact on nature, waste energy or cause wasteful by-products. Social, moral and cultural issues are also important considerations when designing products and systems. Economic factors also need to be considered.

The smartphone, for example, is a relatively recent invention. Smartphones have met people's needs by enabling social interaction and access to information – people can keep in touch, use social media, shop, listen to music, read the news or check the weather forecast while on the move. Smartphones are also useful in emergencies. There are, however, downsides to any new technology. Using smartphones inappropriately, when driving for example, increases the risk of accidents. Loud phone conversations in public places can be irritating. Although we interact through our devices, this can mean we interact less in person, which can affect our social confidence. It's hard to switch off from the 'connected' world. Smartphones are also seen as 'fashion items' – people will buy a new phone even though their old one is still working, and this can waste resources and have a negative effect on the environment.

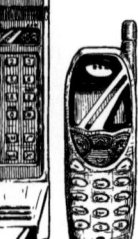

Figure 1.1 **The evolution of the mobile phone**

Exploring the context of a design solution

LEARNING OUTCOMES

By the end of this section you should know about and understand how exploring the context a design solution is intended for can inform decisions and outcome, including:
→ where and how the product or system is used
→ primary user and wider stakeholder requirements
→ social, cultural, moral and economic factors, and how they affect design decisions.

Basic human needs include food, shelter, clothing and safety, and some designs are intended to meet these. However, most people also want products in all areas of their lives, for example, for self-expression and to make day to day life easier and more comfortable.

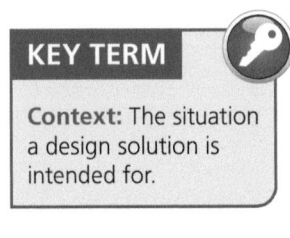

KEY TERM

Context: The situation a design solution is intended for.

The situation or **context** for which a design solution is intended has an effect on many decisions that will be made; it is important to consider the surroundings (place) in which the product will be used, and its users. Purpose (what the product needs to do) and price can also be important factors.

When considering the context for which a design solution is intended, a useful starting point could be to create a concept map considering the general influences such as 'people, place, purpose and price' or 'who, why, what, when, where and how?'

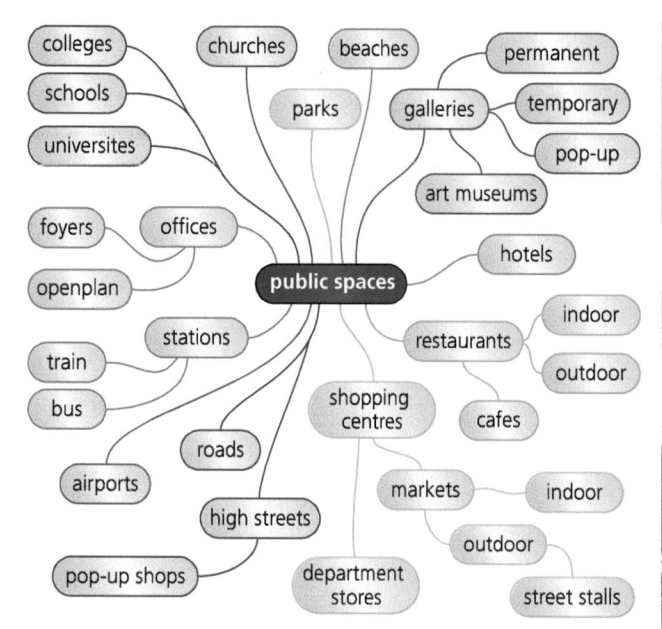

Figure 1.2 **A concept map exploring concepts of public space**

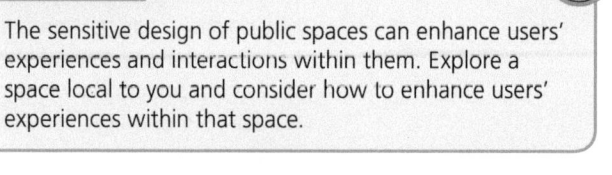

ACTIVITY

The sensitive design of public spaces can enhance users' experiences and interactions within them. Explore a space local to you and consider how to enhance users' experiences within that space.

ACTIVITY

Choose another context. This could be an OCR set context or your own context.

Complete a mind map of all the possible ideas you can think of related to this context, then share your ideas with others in your group. You will have thought of different approaches and discussion might spark further ideas. Think as widely and creatively as possible.

Figure 1.3 **Examples of indoor and outdoor public spaces**

Where and how the product or system is used

Where, when and how a product/system is used can affect many design decisions. For example, an outdoor product will need to be suitable for extremes of weather, while a product for a public place will need to be durable and possibly vandal-proof.

The primary user and wider stakeholders

Most products and systems are used or maintained by humans. The primary user is the most obvious user to consider, but any product will have other **stakeholders** with an interest in it. For example, a company will have an interest in the chairs it provides in its office in order to ensure its workers (the **primary users**) can work efficiently and comfortably. The Chartered Institute of Ergonomics and **Human Factors** (CIEHF) and the Health and Safety Executive (HSE) have an interest too, as wider stakeholders, to provide guidance for the design of office furniture to minimise back pain and RSI. A stakeholder is any person, group or organisation with an interest in a product or system.

KEY TERMS

Stakeholder: A person, group or organisation with an interest in a product/system, for example parents/schools when designing products for children.

Primary user: The person or group of people who will use a product or system.

Human factors: The scientific discipline concerned with the interaction between humans and elements of a product and/or system; also called ergonomics.

In industry, designers consider all stakeholders who will have an interest in the product they are designing, not just the person or people who will use the products, or the person buying the product. When designing (for example, when completing your NEA), always try to consider the wider stakeholders that have an interest in the designs you are developing, as well as considering how the primary user will interact with the product. For example:

- ask your primary user what they feel are important things to consider. You can observe the design context and use similar or alternative products;
- talk to other stakeholders who have an interest in the product/system (wider stakeholders);
- get ongoing feedback from the primary user and other stakeholders at the developmental stage;
- complete user-testing with designs and with prototypes, allowing for further iterations along the way to an optimum solution.

Figure 1.4 **The primary users**

Whenever you interact with stakeholders you will need to keep a record of your questions and any feedback they provide. You could do this by taking photographs or videos.

Before approaching stakeholders you may complete a **task analysis** – thinking about 'who, why, what, where, when and how' will enable you to develop a list of questions and to plan what you need to find out from any stakeholders you interact with. Perhaps the easiest way to analyse the task is to create a concept map (see Figure 1.2).

Figure 1.5 **Example of task analysis**

Figure 1.6 **Housing design has changed due to the use of new building materials and technologies to light and heat our homes and conserve energy**

Social, cultural, moral and economic considerations

Making, using and disposing of products at the end of their life can contribute to pollution and use up resources. Some products genuinely improve the quality of life for the users, others might be good in one society or **culture**, but not so good in a different one. Some products are not really needed at all.

New products can change the way we work and live. Traditional ways of doing things are sometimes best, but in other situations improved products can build and improve on past designs. Some products encourage social interaction, such as the mobile phone or a musical instrument. But these products can also make us more isolated.

Every person has a right to basic freedoms – enough to eat, safety, care (especially the young and elderly), a place to live. Most successful products need to be able to be used by a wide variety of people, the elderly, left and right handers and those with disability; we call this inclusive design.

We also need to consider economic issues, as making, using and disposing of products will have an impact on the economy and create or affect jobs. For example, many modern products are manufactured by computer controlled systems (CAM), resulting in the loss of jobs for skilled workers in factories, but some new jobs are created. Many products that we buy are manufactured by people who are badly paid and who work in poor conditions, sometimes abroad.

Culture relates to the ideas and activities of groups of people. It is about the way that people behave and relate to one another; and how they live, work and spend their leisure time. It is also about beliefs and aspirations. When we design products or systems we always need to look beyond our own experiences. We live in a culturally rich and diverse country, and as designers we have a responsibility to respect and understand the cultural beliefs and differences of others in order to avoid giving offence.

Cultures change with time, particularly with the influences of new ideas, new technologies and the availability – or scarcity – of resources. In an increasingly technological world, cultures can also change to the point of being lost altogether. For example, traditional skills and trades are lost because technology changes and old skills are no longer needed.

We now see similar products and building designs all over the world, as a result of **globalisation**. Marketing strategies by large companies have ensured that there are outlets for their products all over the world. The big brand names in fashion, fast foods, drinks, supermarkets, DIY stores, cars and motorbikes can be seen in many parts of the world, and this product globalisation creates cultural change and taps into new markets.

However, there is also an increasing desire to preserve cultural identity, traditions and language and this can influence designing. For example, traditional forms of dress may re-emerge where once such forms might have been abandoned.

Figure 1.7 **Traditional forms of dress have changed with the influence of new materials, fashions, mass production techniques and the globalisation of manufacturing industries**

KEY TERMS

Culture: The ideas and activities of groups of people; the way that people behave and relate to one another; the beliefs and aspirations of a group of people.

Globalisation: Businesses and organisations operating globally and developing international influence.

5

Similarly, after having tried all sorts of high-tech materials, some architects are now exploring the use of ancient materials and techniques such as straw, clay and rammed earth, which have proved their resistance. As we can see in the countless structures still standing these days, traditional materials do not produce waste, they are easily recyclable and they are obtained from renewable sources. These are all qualities that are important for a **sustainable economic future**.

Although more and more new technologies are being discovered through the sharing of ideas and international collaborations, many designers and manufacturers are looking to past traditions and methods to develop ideas.

> **KEY TERM**
>
> **Sustainable economic growth:** Development that aims to satisfy the economic needs of humans while sustaining natural resources (such as materials) and the environment for future generations.

Figure 1.8 How the Korean *hanbok* dress was transformed by Karl Lagerfeld in 2015

Figure 1.9 A modern mud house by Austrian architect Martin Rauch, compared with a traditional mud hut in a Maasai village in Kenya

It is also important to research and understand cultural issues when designing. For example, colours have different meanings in different cultures. In Western cultures, white symbolises purity, peace and cleanliness; brides traditionally wear white dresses at their weddings. But in China, Korea and some other Asian countries, white represents death, mourning and bad luck, and is traditionally worn at funerals. Islamic art is different from Western art; within some Islamic traditions, using images of people is not permitted in religious art and geometric patterns are used to create an impression of continuous repetition that is believed by some to represent the infinite nature of God.

Usability when designing prototypes

> **LEARNING OUTCOMES**
>
> By the end of this section you should know and understand why usability is an important consideration when designing prototypes
> → the impact of a solution on a user's lifestyle
> → the ease of use and inclusivity of design solutions
> → ergonomic considerations and anthropometric data to support ease of use
> → the importance of aesthetics

Figure 1.10 Human users should be at the centre of any design activity

Designing any product involves consideration of many factors. Production, marketing, materials and cost are all important, but in recent years designers have realised that the human users must be a central part of the process. If a manufacturer makes a product that is uncomfortable or unsafe, or that people find difficult to use, the product will fail. Full consideration of **usability** is therefore very important, and designers need to consider the characteristics and needs of a product's users.

Observation is a key design method that can be used to identify the problems that can arise when people interact with products, services and environments. Always design with the user in mind.

The impact of a solution on a user's lifestyle

Any new product has the potential to change the way we live and work, products can change the way we communicate with others, use our leisure time and make day to day tasks easier. With the user in mind, it is important to understand how the design of a product and its ease of use can impact our lives: a badly designed product could lead to discomfort and even injury.

Ease of use and inclusivity

Sometimes a product is easy and straightforward to use – think of a toothbrush, for example. There are clear interfaces between your hand and the handle and between the bristles and your teeth. Other products are more complex to use, such as a smartphone or a car; for example, driving a car involves understanding the controls.

The ease of use of a toothbrush design is mainly concerned with physical sizes and shapes of people and how well they grip or move muscles. Driving a car or using a smartphone is more complicated and involves the brain and a level of understanding. This relationship between people and the objects they use is referred to as **ergonomics**. (Ergonomics is explored in more detail later in this section.) By considering all these factors, a designer can improve the usability of any design/system.

> **KEY TERMS** 🔑
>
> **Usability:** How easy a product is to use, how clear and obvious the functions are.
>
> **Ergonomics:** The study of how we use and interact with a product or system.

Figure 1.11 A toothbrush is simple to use, while car dashboard controls are often complex

Inclusive design

KEY TERMS

Inclusive design: Designing for the widest possible audience.

Anthropometrics: The study of the sizes of the human body.

Inclusive design aims to remove the barriers that create undue effort and separation. It enables everyone to participate equally, confidently and independently in everyday activities. This does not mean you have to design for all 7 billion people on Earth, but you should aim to exclude as few as possible.

Design has traditionally catered only for the perceived majority of so-called normal users. But what is 'normal'? Most of us will suffer at least temporary disability at some point in our lives; and we will all grow old. We are also variously tall, short, fat, male or female. All of these factors will have an impact on how we use products. In order to ensure everyone participates fully and enjoys equality, we need to move away from thinking about 'regular' and 'specialised' design and create inclusive designs that are not difficult for anyone to use.

Ergonomic considerations and anthropometric data

Ergonomics

Ergonomics is about understanding the interactions between people and the things they do, the objects they use, and the environments they work, travel and play in. Sometimes described as the 'fit' between humans and the products they use, it is not just about size, it is about how easy it is to understand how to use a product, through awareness of the other senses that users draw on when interacting with it.

If a good fit is achieved then people are more comfortable, they can do things more quickly and easily, and they make fewer mistakes. So when we talk about 'fit', we don't just mean the physical fit of a person, we are concerned with how easy it is to understand how to use a product, through awareness of the other senses that users draw on when interacting with it.

Dimensions (mm)	Age Range 5–9 Combined (Percentiles)			Age Range 13–18 Combined (Percentiles)			Age Range 19–65 Men (Percentiles)			Women (Percentiles)		
	5%	50%	95%	5%	50%	95%	5%	50%	95%	5%	50%	95%
1 Height	1058	1264	1483	1470	1685	1857	1630	1745	1860	1510	1620	1730
2 Eye level	895	1055	1180	1456	1570	1740	1520	1640	1760	1410	1515	1620
3 Shoulder height	843	1014	1198	1184	1352	1525	1340	1445	1550	1240	1330	1420
4 Elbow height	610	720	805	945	1005	1170	1020	1100	1180	950	1020	1090
5 Hip height	496	619	754	734	855	990	850	935	1020	750	820	890
6 Knuckle height (fist grip height)	375	480	565	690	720	815	700	765	830	670	720	770
7 Fingertip height	298	390	470	420	620	695	600	675	730	560	620	680
8 Vertical reach (standing position)	1241	1521	1820	1758	2033	2220	1950	2100	2250	1810	1940	2070
9 Forward grip reach (standing)	442	531	640	594	689	809	720	790	860	660	725	790

Figure 1.12 **Anthropometric data for standing**

KEY POINTS

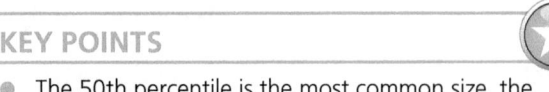

- The 50th percentile is the most common size, the average.
- The 5th percentile indicates that five per cent of people (or one person in 20) is smaller than this size.
- The 95th percentile indicates that five per cent of people (or one person in 20) is larger than this size.
- Very few people are extremely large or very small.
- Be careful not to mix up ergonomics and anthropometrics: an easy way to remember is 'anthro' is man in Greek and metrics should help you remember measurements.

Anthropometrics

Anthropometrics are people measurements. Anthropometric data comes in the form of charts and tables. It may provide specific sizes, such as finger lengths and hand spans, but it also offers average group sizes for people of different age ranges and genders. Other sizes to consider are height, reach, grip and sight lines.

The use of percentiles is an important aspect of anthropometrics. The sizes of the human body given in anthropometric data are usually presented in tables, and normally include the 5th percentile (the smallest 5 per cent), the 50th percentile (the average) and the 95th percentile (the largest 5 per cent). Look at the anthropometric data below and notice the different percentile measurements.

Applying anthropometric data

The percentile you follow from anthropometric data tables will depend on what is being designed and who will be using it. Is the product for all potential users, or just the ones of above or below average dimensions? If you pick the right percentile, 95 per cent of people will be able to use your design. For example, a doorway needs to be high enough to allow the 95th percentile to pass through.

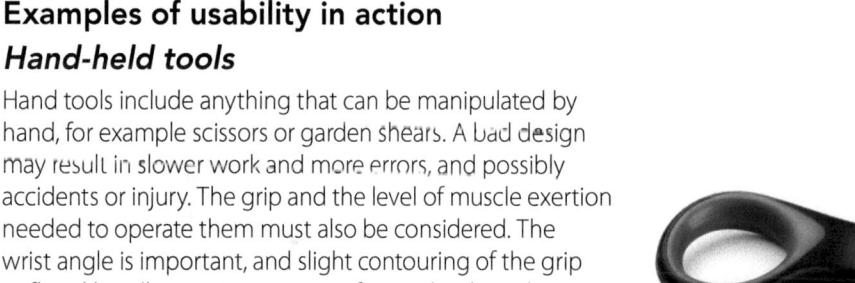

STRETCH AND CHALLENGE

Consider which percentile you would need to use for the following and explain why:
- The size of a handle for an upright vacuum cleaner.
- Ventilation slots in the casing of an electric motor on a lawnmower.
- The diameter of a screw-top water bottle.
- The zip on a jacket.

Examples of usability in action
Hand-held tools

Hand tools include anything that can be manipulated by hand, for example scissors or garden shears. A bad design may result in slower work and more errors, and possibly accidents or injury. The grip and the level of muscle exertion needed to operate them must also be considered. The wrist angle is important, and slight contouring of the grip or flared handles can increase comfort and reduce slippage in sweaty hands. Longer handles distribute the forces more evenly across the fingers.

When designing hand-held tools, smooth handles that require wrist rotation should be avoided because of the increased risk of slippage and rotational wrist damage. Padding handles helps to reduce the force needed to grip the tool. The effect of tool weight is also a factor. Use of rests, supports, two-hand grips and so on can all help to decrease the effort required to use heavy tools.

Figure 1.13 **Consider the design and usability of hand tools**

Try to design tools and hand-held items for operation with both hands. When only right-handed users are considered, left-handers may be at an increased risk of injury.

Design of seating

Approximately 50 per cent of people in the developed world suffer some form of back complaint, which is often related to poor seat design. How we sit and what we sit on affects the health of the spine. There is no single ideal sitting posture, and we can't design a chair for the best way to sit. What we need is a variety of chairs that allow us to sit in a variety of postures that relate to the context they are to be used in.

Optimum seat height depends on the user and context the seating is used in, but anthropometric data is applied, so that a minimum height should be 380 mm based on the 5th percentile in women (see below). Seat contouring and cushioning can help to distribute weight over a larger area and promote good posture. The correct seat angle helps users to maintain good contact with the backrest. For most purposes a five to ten degree angle is recommended. Professional designers create full-size mock-ups to test comfort; testing using real people is always recommended. Aircraft and car designers build full-scale mock-ups for use with real people, in order to test the **psychological factors**: not only physical sizes but also how people respond to and how they feel about a prototype.

KEY TERM

Psychological factors: Such things as mental reaction time, various acquired meanings associated with certain colours (e.g. red often means danger), the capabilities and limitations of short-term memory.

Aesthetic considerations

KEY TERMS

Aesthetically pleasing: Beautiful to the senses.

Aesthetics: Factors concerned with the appreciation of beauty – this can include how something looks, sounds, feels, tastes and smells.

Products and systems have to be designed to suit the users' needs. They must work well, be easy to use and improve our lives. But they should also ideally be **aesthetically pleasing**. **Aesthetics** includes all of the senses: sight, hearing, touch, taste and smell. A person could be drawn to a product because of one or a combination of these aesthetic factors – for example, because of the visual impact of the product or as a result of its shape, colour and texture or feel.

Products can be evaluated aesthetically using various measures aside from beauty. We might speak of objects as having a high-tech aesthetic, or a rugged aesthetic, or a calming aesthetic. Aesthetic judgement is personal (it is subjective), but visual design elements can be combined in certain ways to evoke a common emotional response. Designers combine design elements such as form, texture, scale, colour and symmetry in different ways to spark different emotions.

The Bauhaus, a design school founded by Walter Gropius in Germany in 1919, for example, was very influential in shaping an understanding of design and taste. Design was considered crucial and integral to the production process rather than as merely a visual add-on.

'Form follows function' was a phrase often used to counter the prevailing view that beauty was achieved by including additional features and decoration that didn't improve a product's function. Architects and industrial designers in the twentieth century were beginning to show that the form or shape of an object should be based on its intended purpose or function. In essence, form and function must be balanced and the form (aesthetics) should always communicate with function.

Figure 1.14 **Barcelona Chair, Mies van der Rohe (1929)**

Figure 1.15 **Juicy Salif, lemon squeezer, Philippe Starck (1990)**

Examples of products that demonstrate a sensitive consideration of both aesthetics and function

Colour

Colour is important in product design. It elicits responses by stimulating emotions, and can be used to excite and persuade. Colours can also have negative associations, however.

A designer must be aware of how people respond to colour and colour combinations. For example, in 1998 Apple broke with tradition by introducing iMac computers in a wide range of colours. They realised that home computers did not need to look like the usual office machine and that customers wanted a more visually interesting and appealing design. Two years earlier, in 1996, the media had pronounced Apple all but dead. The company had lost $878 million in 1997, but under the guidance of Steve Jobs it earned $414 million in 1998 – its first profit in three years. This was the beginning of the iProduct revolution.

The success of the iMac was down to its simplicity in both form and colour – a great example of the form-follows-function principle. After the release of the iMac, multi-coloured translucent plastic housing became such a common aesthetic in the consumer products industry that Apple had to move on and iterate, dropping the bright array of colours from the product line with the release of the flat-panel iMac in 2002. Other companies followed and now most consumer electronics devices use brushed aluminium, frosty white or glossy black – the colours of more recent iMac iterations.

KEY POINT

Research suggests that 70 per cent of consumer purchasing decisions are made in-store. Catching the consumer's attention and conveying information effectively are critical to successful sales. Careful use of colour can catch a consumer's eye.

Figure 1.16 The original iMac (1998) and the iMac 2016 iteration

Colours and typical meanings

Table 1.1 **Colours and their typical meanings**

Red		Aggressive, passion, strong and heavy, danger, socialism, heat
Blue		Comfort, loyalty, for boys, sea, sky, peace and tranquillity, conservativism, cold
Yellow		Caution, spring and brightness, joy, cowardice, sunlight
Green		Money, health, jealousy, greed, food, nature
Brown		Nature, aged and eccentric, rustic, soil and earth, heaviness
Orange		Warmth, excitement and energy, religion, fire, gaudiness
Pink		Soft, healthy, childlike and feminine, gratitude, sympathy
Purple		Royalty, sophistication and religion, creativity, wisdom
Black		Dramatic, classy and serious, modern, evil, mourning
Grey		Business, cold and distinctive, humility, neutrality
White		Clean, pure and simple, innocent, elegant, peace

The colour wheel

The first colour wheel was produced by Sir Isaac Newton in 1666. All colours are derived from the primary colours of red, yellow and blue. The primary colours mix to produce the secondary colours (green, orange and purple), which sit between the primary colours on the wheel.

Colour harmony is achieved when colours combine to create order, balance and a sense of pleasure – for example, colours next to each other on the colour wheel create visual harmony, as in a red, orange and yellow sunset. Complementary colours are opposite each other on the colour wheel, for example red and green. Opposing colours create maximum contrast and they can work well together, as seen in nature; sometimes designers choose colours to create contrast and impact rather than harmony.

Figure 1.17 **Colour wheel showing primary and secondary colours**   Figure 1.18 **Red and green work well together in nature**

Proportion and symmetry

KEY TERMS

Proportion: The relative size and scale of the various elements in a design.

Symmetry: When elements are arranged in the same way on both sides of an axis or when rotated around a point.

Asymmetry: The absence of symmetry of any kind.

Proportion refers to the relative size and scale of the parts in a design. It is the relationship between objects or between parts of a whole. This means that we need often to consider proportion in terms of the context or user in order to determine proportions.

Figure 1.19 **Symmetry creates aesthetically pleasing results**

Symmetry creates balance, and balance in design in turn creates harmony, order and aesthetically pleasing results. Symmetry is found everywhere in nature, and this is probably why we find it so beautiful. Symmetry can be reflective (a mirror image) or rotational (turned around a central axis). Too much symmetry, however, can be boring. **Asymmetry** is a break in symmetry that, when used effectively, can make things more interesting from a design point of view.

Mathematics in aesthetics

There is much debate over whether the **golden ratio** in mathematics can be linked to aesthetics. It is believed by many that it can be used to create pleasing, natural-looking compositions in your design work, and that images featuring the golden ratio can be interpreted by the eye faster than any other.

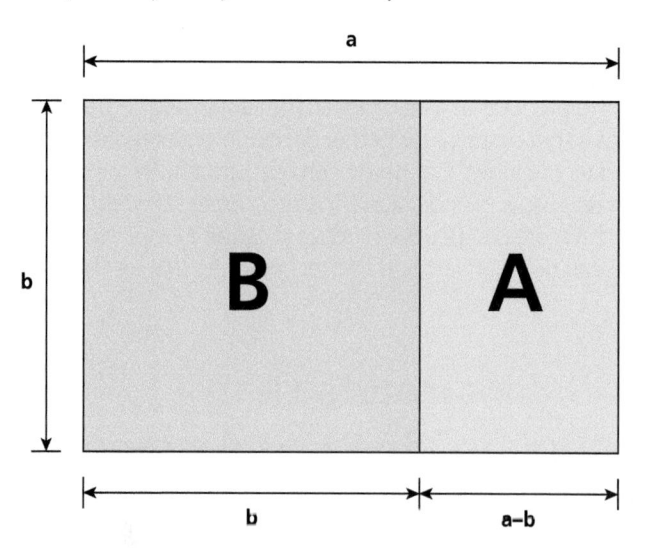

KEY TERM

Golden ratio: A common mathematical ratio found in nature that can be used to create pleasing, natural-looking compositions in your design work; also known as the Golden Mean, the Golden Section, or by the Greek letter Phi.

Figure 1.20 Approximately equal to a 1:1.61 ratio, the golden ratio can be illustrated using a golden rectangle (a large rectangle consisting of a square with sides equal in length to the shortest side of the rectangle, and a smaller rectangle)

The principle of the golden ratio is believed to have guided artists and architects over the centuries, from artist Leonardo da Vinci (1452–1519) to architect Le Corbusier (1887–1965). For example, Gaudi is said to have used it in his designs for the facades of the Sagrada Familia in Barcelona.

It also appears in some patterns in nature, including the spiral arrangement of plants. You can easily incorporate the golden ratio into design work – for example when designing a system interface you can make sure the content and sidebars follow the 1:1.61 ratio.

STRETCH AND CHALLENGE

Compare two different products: one that uses the golden ration in its design and another that doesn't. Which design looks best? Do you think the golden ratio improves the design?

Fashion, trends and taste

What we consider to be aesthetically pleasing can change over time, based on the influence of fashions and trends. It is also highly individual and based on our own tastes. Taste depends on the individual. In product design we often refer to 'good taste' or 'bad taste', but it is linked to aesthetics and personal appreciation of beauty. Product designers will research the likes and dislikes of their market thoroughly to ensure their creations will appeal to targeted groups, and will acknowledge that taste changes quickly, depending on factors such as peer pressure, fashion, trends and celebrity endorsement. More information on fashions, trends and taste can be found in Chapter 2.

Successful product design involves learning from other designs with similar features. Professional designers analyse existing products as a key stage in the design and development of new products, or as a starting point for further development or redesign. We can learn from the materials and technologies that have been used, from how the product was made and how easy it is to use and understand. In the commercial world this can result in the development of a more competitive product. It can also help identify features of successful products that can be improved, and find technologies that can be used in different ways.

Opportunities and constraints that influence design and making requirements

LEARNING OUTCOMES

By the end of this section you should know and understand how to explore and critique existing designs, systems and products to identify features and methods, including:
→ how to identify materials, components and processes that have been used
→ how fashion, trends, taste and/or style have influenced existing products
→ the influence of marketing and branding
→ the impact on society
→ the impact on usability
→ the impact on the environment
→ life cycle assessment
→ the work of past and present professionals and companies in the area of Design and Technology

Learning from existing solutions is one of the most valuable sources of information for designers.

Exploring and critiquing existing designs, systems and products

When exploring existing products, you need to understand the choices the designer and manufacturer made in order to ensure that the product is fit for purpose and use. This can include the following:
- Function
- Materials and components
- Methods of construction and manufacture
- Ergonomic and anthropometric considerations
- Aesthetics, fashion and style
- How the product impacts the environment during its use
- Ease of recycling
- How the product has been influenced by the work of past and present designers.

When analysing existing products, try to examine actual products rather than images. This helps you get a true 'feel' for the product and establish its strengths and weaknesses. If you aren't the user of the product, observe the product in use by its primary users.

The study of existing products and contexts can develop understanding of other topics included in this course, such as materials considerations, manufacturing processes, technical understanding, ergonomics, and inclusive design and wider issues, such as learning why historical and current professionals have approached design in different ways.

When approaching the analysis of existing products you should identify the features of the product and ask yourself questions about them, for example:

- Start by analysing the aesthetics (appearance) of the product – its colour, shape, proportions and texture. Does it have any special finishes or protective components? Is it obvious how the product functions?
- What is the function of the product? Does it function well, and how is this function achieved? Is the product successful? Where is the product intended to be used and by whom (context)?
- Is it mass or batch produced, and what manufacturing methods have been used? What materials have been used? Are their properties suitable for this use? Can they easily be recycled at the end of the product's life? Does the product use energy, and if so how is this created? Is the product easy to maintain, and can parts be replaced?

Materials, components and processes

It is useful to question why a designer has chosen particular materials, components or processes. Their choices will be dependent upon the context for which their design solution is intended, and are likely to be based on the following considerations:

- Where and how is the product to be used?
- What are the needs of the primary user and wider stakeholders?
- What are the social, cultural, moral and economic factors that influenced the decisions about which materials, processes and components to use?

Analysing existing products that are made from multiple materials and relevant to the context and users they are designed for can help you to develop understanding of the properties and uses of materials, and why they have been chosen. For any given context, existing products can be studied in depth to develop understanding of the physical and working properties of specific materials and/or system components, and of the use of standard components such as clips, fasteners and bindings; hinges, brackets and screws; and rivets, hinges, caps, fasteners and bolts. Analysing products also develops an awareness of the processes that can be used to ensure the structural integrity of a product (for example triangulation, plastic webbing and reinforcing). You can identify the methods used to shape, fabricate, construct and assemble items, such as perforating and folding, steaming and pressing, bending, casting and vacuum forming. You can also identify the methods used for manufacturing at different scales of production, for example one-off, bespoke production, batch production and mass production.

Figure 2.1 **Examples of products from different material categories can be used to develop knowledge of materials, properties and constructional techniques, and understanding of structural integrity**

MILK CARTON
/PACKAGE/

More information on materials, components and processes can be found in Section 2 of this textbook.

STRETCH AND CHALLENGE

Select a product, and list the materials and components that have been used to make it. Identify why the designer selected each one.
You should consider:
● cost
● properties
● aesthetics

Disassembly

Disassembling (taking apart) a piece of machinery or a product can be a good way to examine it in detail, to see how many materials and components it is made from and how they are assembled and constructed. Consider each part, and also what might happen to it at the end of its useful life.

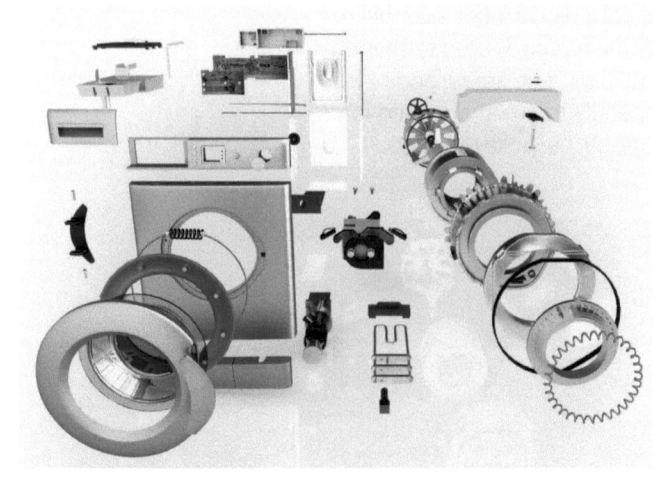

ACTIVITY

Select a product from your chosen in-depth areas of learning and disassemble it if possible.
● Take photos of the product once separated into all component parts.
● What materials have been used and why?
● What methods of construction and manufacture have been used and why?
● What ergonomic and anthropometric measurements have been considered?
● What strengths and weaknesses can you list relating to the product?
● How easy is the product to disassemble, and could parts be recycled?

Figure 2.2 **Disassembling a piece of machinery can be a good way to examine it in detail**

Fashion, trends, taste and/or style

Throughout design history there have been distinctive styles that can be linked to a period in time. From the Victorian period to postmodernism, many continue to influence designers today.

- Victorian 1830s–1890s
- Art nouveau 1890 1905
- De Stijl 1914–31
- Bauhaus 1919–35
- Streamlining 1930–1950
- Scandinavian design 1935–present
- Art deco 1925–39
- Organic design 1930–present
- Pop art 1960s
- Memphis 1980–90
- Minimalist design 1960s–present

Designers are able to emulate styles of the past or follow new trends, and manufacturers produce these products using modern materials and manufacturing processes. In the twenty-first century, fashion is very fast-moving and can be influenced by any number of factors. Recent trends across various areas of design include eco design and **ethical** approaches to design, for example. Many companies employ trend forecasters to predict what will be in fashion many seasons ahead and what will influence the market. These forecasters can predict the mood, behaviour and buying habits of consumers all over the world. **Twenty-first century design** is developing with the ability to reflect on the successes of the past, but be involved in the inventions of the future through the advancement of new technologies, global endeavour and changes in ways of thinking.

> ### STRETCH AND CHALLENGE
>
> Choose one of the styles listed above and research in detail the materials associated with the era. How did these influence shape and form?

Marketing and branding

Marketing and branding are well established tools that influence customer decisions. The aim of all advertising and marketing activity is to influence potential customers, to persuade them to buy the advertiser's product and therefore allow a company to make money. For marketing to be successful in the first place, there needs to be a demand for the product. It is therefore vital to understand who the target market is and to design the marketing to draw them in.

The manufacturing industry must take into account customer requirements in order to develop new and improved products with a competitive edge. New products are usually developed either because of **market pull**, where consumers demand a particular type of product, or by **technology push**, where new materials and/or technologies lead to innovative products that are released on to the market. New materials and technologies can cause products to become obsolete; sometimes companies plan for products to become obsolete, which is known as planned obsolescence and is covered in chapter 3.

Successful marketing usually includes promoting a brand image and developing new markets for the product. A 'brand' can be a name or a logo, and some brands can be incredibly influential on the decision to buy. If a particular brand is 'in fashion' it can be extremely profitable for the company. The product quality may be comparable to a similar non-branded item, but the addition of a brand or logo can influence the buying decision.

Figure 2.3 One example where branding is a prime influencing factor is within the sportswear market; sportswear is now worn by many people in their daily life, and not just for sports

KEY TERMS

Ethical: Correct, good or honourable. Aim for an ethical approach when you are designing products.

Market pull: A need for a product that arises from customers or market research to solve a need, or to compete with a product launched by another manufacturer.

Marketing: The business of promoting and selling a product; can include advertising and promotion, and market research.

Technology push: When research and development of new technology drives new product development, e.g. touchscreen and fingerprint technology in smartphones.

Twenty-first century design: Current and forward-thinking design that considers the evolving practice of design to shape behavior for a preferable future.

17

Impact on society

When analysing products, you need to consider their impact on society and people. People in different countries and cultures will have different feelings towards products (think of the earlier example of differing interpretations of colours).

It is not only cultural differences – society is continually changing, and people's tastes and fashions reflect this. One example is the growing popularity of social media. Young consumers have grown used to mobile phones and computers, and the younger generation prefers to use digital technology to connect with people and shop online, while older people may prefer to stick to traditional methods or find it harder to adapt to new technologies.

Products can have negative as well as positive impacts on society. For example, increasing use of digital products is making us more sedentary, contributing to increasing levels of obesity in society.

Impact on usability

When analysing products, consider too how usability, ergonomics and appearance have been incorporated into the design, while still maintaining function.

Identify the ergonomic features in a design and consider how anthropometric data was used. As an example, in a garlic crusher, the length of the lever and width of the grip and the force needed to crush a garlic clove have all been taken into account. The edges will be rounded for comfort, and the designer will have considered the size of users' hands (anthropometric data) in order to improve usability and ergonomics. How do you think it could be redesigned?

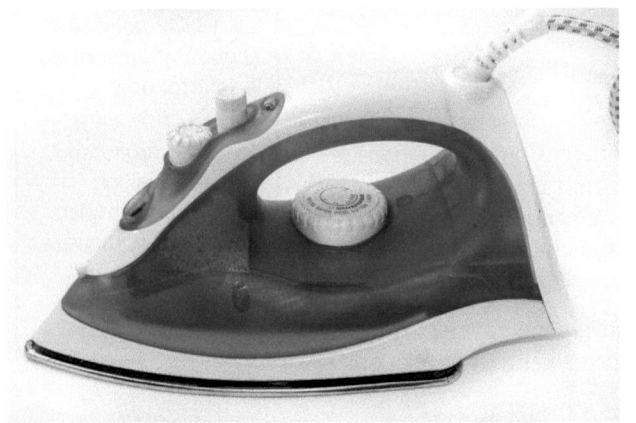

Figure 2.4 **Think about how advances in materials and technology have affected some of the domestic products we use. Since Hotpoint's first electrical iron was introduced in 1905, much of the design has stayed the same, but modern irons have more safety features, thermostats and steam functions and are more lightweight and ergonomic, which all improve usability.**

Figure 2.5 **Think about the development of the vacuum cleaner, from carpet sweepers to the first electrical models in the early 1900s, to today's Dyson cord-free bagless models, and even robotic versions that make household chores easier for us all**

STRETCH AND CHALLENGE

Choose a mechanical or electrical domestic product and research how it has changed over the years as materials, manufacturing techniques and technology have been developed.

The environment and lifecycle assessment

One of the main issues relating to product development is the impact on the environment through the consumption of energy and materials. In the past, keeping development costs down and product prices low tended to be more important than considering the impact on the environment when using materials and manufacturing products. This, combined with the influence of advertising, encouraging us to buy more products and increasing demand for new products, has resulted in products flooding the market, encouraging a **throwaway society**/junk culture. This in turn affects the environment as a result of both the production and final disposal of products.

In recent years, however, people have realised that the impact on the environment of certain manufacturing processes can be irreversible, and that certain materials will run out if we continue to use them. This has resulted in consumers and companies seeking ways to be environmentally friendly, and is supported by the introduction of laws that force companies to recycle or dispose of products in a way that is more environmentally friendly at the end of their life.

Many designers today encourage the use of environmentally friendly and sustainable methods of manufacture and materials; and some designers are contantly looking to new material developments that are environmentally friendly. Products that can easily be disassembled at the end of their life are being developed, as we realise the importance of moving towards a **circular economy,** where all waste can be put back into the system and reused or recycled. This topic is covered fully in Chapter 3.

When disassembling products, think about how easy it is to take the product apart and recycle the components. Could fittings be redesigned to make disassembly easier? Could the product be made from fewer parts and materials, and be better for the environment? Can it be rethought?

The work of past and present professionals and companies in the area of Design and Technology

There are many examples of designers past and present who have led the development of new technologies, and use of materials and aesthetics in products. These designers are often influenced by a particular style or movement that might be current or historical. When you look at Braun's products by Dieter Rams you can see how they have influenced current designs by Jonny Ive at Apple, in the use of minimal colour and simplicity; in fashion we can see the influence of Coco Chanel in dress and accessory design by Oscar de la Renta and Donna Karan. As we advance further into the twenty-first century, designers are becoming more involved in collaborations with scientists and technologists to incorporate new technological developments into products, ensuring their knowledge and understanding of design offers a greater potential for new products and technologies to succeed.

KEY TERMS

Circular economy: An alternative to a traditional linear economy (manufacture, use, dispose), in which we use resources for as long as possible then reuse and regenerate products and materials; a cradle-to-cradle model.

Throwaway society: A society influenced by consumerism and excessive consumption of products.

ACTIVITY

Choose a design style or designer and research the main influences in their style. Produce an image board and identify the key similarities in style.

How developments in design and technology influence decisions

LEARNING OUTCOMES

By the end of this section you should know and understand how new and emerging technologies influence and inform design decisions, including both contemporary and potential future scenarios from perspectives such as:
→ ethics
→ the environment
→ product enhancement.

Critical evaluation of how new and emerging technologies influence and inform design decisions

KEY TERM

Emerging technologies: New technologies that are currently being developed, or will be developed within the next five to ten years.

When examining products and developing them to make use of the latest technologies, whether in the design, materials or manufacturing, it is crucial to be able to evaluate the wider impact of new and **emerging technologies**. There will always be arguments for and against adopting them; new technologies can have all kinds of impacts on people and society (both positive and negative), as well as giving rise to economic and environmental issues.

Not every product that uses new technologies is immediately successful. One very famous example is the Sinclair C5, launched in 1985 as a small one-person battery-assisted vehicle. Sir Clive Sinclair himself said in 2005 that the C5 'was early for what it was. People reacted negatively and the press didn't help. It was too low down and people felt insecure, hence it got bad press.' Professor Stuart Cole of the University of South Wales believes the C5 suffered as a result of the design of the roads and attitudes of the time: 'In the days before unleaded petrol, your face would have been at the height of every exhaust pipe, and drivers weren't used to having to consider slower-moving cyclists. But with more cycle lanes, better education, and workplaces providing showers, etc., the world now is much more geared up for people looking for alternatives to the car, and hopefully will become even more so in the future.'

Another example is Google Glass. In 2013 Google allowed several thousand people to pay to test their prototype. It was met with hype and celebrity endorsement, with tech magazines calling it a great invention. Users reported bugs, however, and privacy concerns were raised, eventually leading Google to stop production of the product, but they are still looking at future iterations.

It is good to remember that companies that create products that fail learn from them. Before the iPad, Apple released its prototype with the Apple Newton in 1993. This was a personal digital assistant that was too expensive; the average consumer had no use for a personal digital assistant in the early nineties. Apple discontinued the Newton in 1998.

Figure 2.6 **The Sinclair C5, Google Glass and Apple Newton: perhaps all these products have helped inspire future products or will return in another form in future**

Ethics, the environment and product enhancement

Ethics

With the use of any new technologies and the development of new products, the impact on the environment will result in ethical decisions; how companies manage waste and use resources needs consideration. New products can put pressure on people to keep up with the latest gadgets as well as creating waste. Often the mining of minerals and metals needed for electronic gadgets can cause political problems in countries as well as affecting the landscape and habitat.

The environment

Waste generated by obsolete products can create environmental issues as can the extraction and processing of raw materials into products; this topic is dealt with in depth in chapter 3.

Product enhancement

Many companies enhance products regularly. An enhancement is a change or upgrade that increases a product's capabilities. Communication and software companies regularly provide software or hardware enhancements, which can sometimes cause other products to become obsolete. This can have an effect on the environment. Planned obsolescence is covered in detail in Chapter 3.

Meeting customer requirements increasingly requires constant upgrading of products with new technologies. New technology, however, must be managed. This is especially true for electronic systems where technology life is often very short. To the customer this means that improved performance from upgraded and new technologies is more easily available and affordable, but companies must think about compatibility with older technologies and this can sometimes hinder development.

ACTIVITY

1 What are the latest technological developments within your chosen area of in-depth learning?
2 How has this development benefited the function of a product of your choice?
3 Discuss the advantages and disadvantages of using new technologies in products.
4 Discuss examples where new technology or an upgrade has caused an existing product to become obsolete.

New technology is constantly emerging, so that existing technology either becomes outdated or needs to be refined and improved. Significant changes in technology are bound to occur that are likely to have a major impact on our lifestyle, industries and the environment. The challenge is to identify these and decide what will actually be a game-changer, and what will simply be a brief fad.

Designers also need to consider the implications of other wider issues that arise when designing and making products, including choosing appropriate energy sources to make products and power systems, as well as other wider environmental, social and economic issues that can have an influence on the processes of designing and making.

The impacts of new and emerging technologies

LEARNING OUTCOMES

By the end of this section you should know and understand the impacts of new and emerging technologies when developing design solutions within different contexts on:
→ industry and enterprise, such as the circular economy,
→ people, in relation to lifestyle, culture and society
→ other environment
→ sustainability.

All new technology will have an impact on our lives, and while it provides many benefits there are also downsides. We need to consider ethical and moral issues that may arise, for example, and assess the likely impact on business, industry, individuals and the environment we live in.

In order to understand the wider implications that emerging technologies might have, consider some recent technological advances and their impact.

For example, internet search engines bring access to all kinds of information, from local restaurant reviews to international news. The internet also provides effective means of communication, including email and instant messaging. It has enabled globalisation, with companies able to make transactions with clients and suppliers worldwide. We can track the weather, or access millions of books, journals and other materials, with a click, facilitating learning. Sometimes, though, the internet can have negative effects. People can become addicted to using social networks, be victims of cyberbullying, computer viruses or even online fraud. Illegally downloading copyrighted material affects the music industry in particular, threatening artists' livelihoods.

Perhaps less obvious is the effect of the internet on our environment. Online shopping results in fuel use for deliveries, and therefore air pollution, although it does also reduce short journeys made by individuals. It has made goods easier to get hold of, however, meaning that we might upgrade products more often. Choices like buying an e-book instead of a printed book, watching a film online rather than on DVD, or organising a work meeting to take place on Skype rather than flying from New York to London to meet face to face, all reduce environmental impact. However, Melbourne-based research centre CEET estimated that the telecommunications industry as a whole emitted over 800 million tonnes of carbon dioxide in 2013, and that the energy demands of the internet could double by 2020.

CEET reports that the internet now accounts for almost 2 per cent of the world's energy consumption, which means that if the internet were a country it would rank as the fifth largest for energy consumption.

Artificial intelligence

Artificial intelligence is considered by some to be a threat to employment. Will new technologies – from industrial robots to advances in machine learning – lead to mass unemployment? Artificial intelligence agents are already involved in every aspect of our lives, however – they keep our inboxes free of spam, they help us make our web transactions, they fly our planes and they may soon drive our cars. Robots have already revolutionised manufacturing, too, as demonstrated by modern car assembly. Work continues to develop robots that can function independently and safely in a normal environment.

Biometrics

The most obvious use of biometrics is in our passports, where they have allowed the use of automated passport checks. Elsewhere, the use of biometrics is becoming more common – for example, fingerprint scanners are used for a variety of purposes including recording arrival and departure times at places of work.

Virtual reality

Current virtual reality is impressive, and is increasingly used for training purposes. It is very expensive but even so it allows training that would otherwise be impossible, for example allowing pilots to practise emergency routines.

Drones

Drones are already used in numerous industries, from retail to manufacturing. This is currently still new technology, but the use of unmanned flying objects could change the world in which we live. Drones have the potential to carry out courier roles, resulting in job losses, but there will also be a need for more technically skilled people to manage an efficient drone delivery network.

Figure 3.1 New technology: drones

Impact on the environment

Designing products to last has become increasingly important in the twenty-first century. We all want the latest technology and designs, however. You might find it funny when you come across a product that your parents or grandparents use that seems really old-fashioned, like a vacuum cleaner with a bag, a chunky TV that looks like it belongs in a museum, or a CD player. If these items still function, though, why do we need new ones? As companies introduce technologies, many old products become obsolete and it might get harder to find the vacuum cleaner bags, buy CDs as easily, or have access to as many features or be able to link a product to a new digital system.

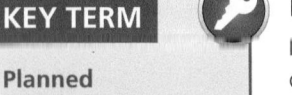

KEY TERM

Planned obsolescence: The business practice of deliberately outdating an item before the end of its useful life.

Planned obsolescence

Planned obsolescence is deliberately making an item out of date by stopping its supply or service support and by introducing a newer model or version. This is to encourage consumers to upgrade, and is common in the computer industry, for example.

Washing machines and white goods are also designed with planned obsolescence in mind, meaning they will last only a few years. Two or three years after purchase, a washing machine might need only minor, inexpensive repairs. After four or five years, however, the vital parts will begin to wear out, the company won't provide replacements and a new machine will be needed.

Another example is a mobile phone. Mobile phones are often designed with only current technology in mind, despite the manufacturers' knowledge of future technological developments. For instance, a mobile phone might have connectors and chargers that fit current products, such as headphones and computers. Eventually the 'old' USB connections will be upgraded and this will make the product obsolete. The customer will therefore need a new phone, even though their old one still works. Sometimes these changes are phased in and companies will have the changes planned years ahead.

Sometimes planned obsolescence can be a positive thing – new products such as medical syringes and disposal razors can avoid the spread of infections. While a partly disposable toothbrush is unavoidable, you can reduce up to 93 per cent of toothbrush waste by choosing a reuseable handle with replaceable heads.

What can you do as a consumer?

We can all contribute towards saving resources and energy by:
- conserving energy wherever possible: turning down central heating, sharing travel, walking or using a bike
- making decisions on whether powered products and systems are necessary; are all gadgets actually useful?
- using energy-efficient products: refrigerators, freezers, washing machines, cars with good fuel consumption
- using appliances efficiently: filling dishwashers, boiling just enough water in kettles
- choosing reuseable products over disposable ones
- buying products that can be upgraded or repaired
- choosing products with minimal packaging, reducing the waste that you create.

Sustainability

Sustainability refers to the concept of meeting the needs of the present without compromising the ability of future generations to meet their own needs. It applies to economic development, environment, food production, energy and social organisation. Making products sustainable means considering the long-term effects that using technology and materials to create products will have.

We must design with sustainability in mind, which means doing the following:
- Choosing non-toxic, sustainable or recycled materials that don't require as much energy to process.
- Manufacturing and producing products using less energy.
- Making products fuel- and material-efficient.
- Producing products that are long-lasting and better-functioning so there is less replacement and use of products.
- Designing products that can be recycled when their use is done (disassembly).
- Developing products that are profitable but also offer income to producers.
- Considering the impacts of a design on all of the stakeholders involved in its development.

You can use the following questions to assess sustainability:

- **Rethink:** How can it do the job better? Is there another way of doing it altogether? Is it energy efficient? Has it been designed for disassembly?
- **Reuse:** Which parts can I use again? How easy is it to take apart? Has it another valuable use without recycling or discarding it?
- **Recycle:** How easy is it to take apart? Can the materials be recycled? How much energy would it take to reprocess materials?
- **Repair:** Which parts are likely to fail or wear? How easy is it to replace parts?
- **Reduce:** Which parts are not needed? Do we need as much material? Can we simplify the product?
- **Refuse:** Is it really necessary? Is it going to last? Is it fairtrade? Is it too unfashionable to be trendy and too costly to be stylish? Is there something ethically wrong? Is it made from material that is scarce?

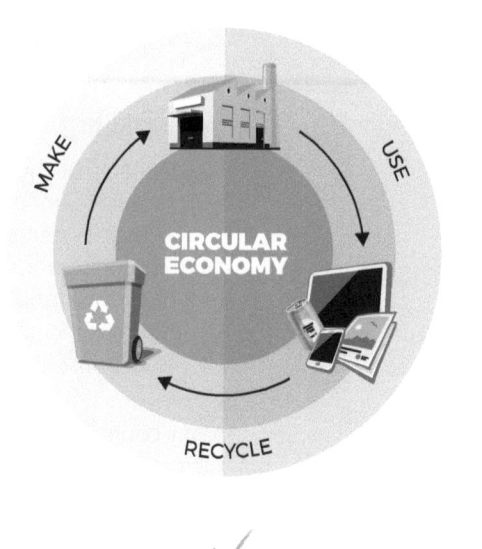

Circular economy

A circular economy is an alternative to a traditional linear economy (make, use, dispose), in which we keep resources in use for as long as possible, extract the maximum value from them while they are in use, then recover and regenerate parts and materials at the end of their life.

Cradle to cradle

In cradle-to-cradle production (as opposed to cradle-to-grave), all material inputs and outputs are seen either as technical or biological nutrients. Technical nutrients can be recycled or reused with no loss of quality, and biological nutrients can be composted or consumed.

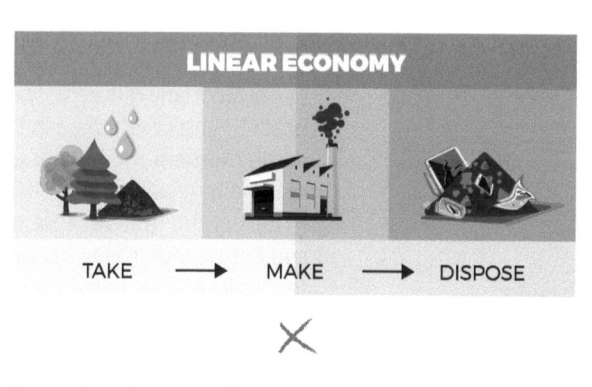

Figure 3.2 Circular vs linear economy

The following are some examples of this approach in action:

- Sports brand Adidas and environmental initiative Parley for the Oceans released trainers with uppers made using recycled plastic recovered from the sea. Parley also partnered with G-Star RAW to produce collections of denim garments made from recycled plastics removed from the oceans.
- Lauffenmuhle invented a textile yarn designed for workwear such as uniforms, which uses a blend of cellulosic fibres derived from FSC-certified wood (Forest Stewardship Council sustainably sourced wood) and biodegradable synthetic polymers.
- Bill Stumpf and Don Chadwick designed the Aeron Chair to be sparing of natural resources, durable and repairable, and constructed for ease of disassembly and recycling.

STRETCH AND CHALLENGE

Find your own examples of cradle-to-cradle production and the circular economy.

Figure 3.3 The Aeron Chair

Choosing appropriate sources of energy

LEARNING OBJECTIVES

By the end of this section you should know and understand:
→ How designers choose appropriate sources of energy to make products and power systems
→ The generation of electricity and how energy is stored and transferred.
→ The appropriate use in products and systems of renewable and non-renewable sources including fossil fuels, nuclear fuel, bio-fuel, wind, hydro-electricity, tidal and solar.

Many countries already have huge energy demands and demand for energy continues to increase, particularly as developing countries emerge as rapidly growing industrial powers with plans for further expansion. Meeting these demands poses genuine concerns over the use of existing resources and the environmental effects of using existing technologies to generate energy.

The three fossil fuels – oil, natural gas and coal – are **finite** – the deposits that exist cannot be replenished when they are used. With further high-level use, all are in danger of running out.

There are many different opinions and calculations about this, and none really agree on the exact timing of when each fossil fuel could run out. How fast we are using each fossil fuel at the moment might change in the future. If we start switching to alternative fuel sources that are renewable rather than non-renewable and if we make more use of recycled plastic materials, the reserves that we have will obviously last longer. However, while new energy sources provide many benefits and opportunities, they may also introduce their own technological and environmental challenges.

How energy is stored and transferred

We obtain energy for the generation of electricity from many different energy sources, including renewable and non-renewable sources. These different energy sources store and transfer energy in different ways to produce electricity, and are discussed in more detail below. There are different ways in which energy can be stored, including:

● **Kinetic energy:** the energy in a moving object is called kinetic energy. Dynamos and wind-up mechanisms transfer potential energy (stored until it is released) into kinetic energy.
● **Thermal energy:** thermal energy is energy that comes from heat. The sun, radiators and fires give off thermal energy.
● **Chemical energy:** chemical energies are available in many different forms and are stored in fuels that we burn to release thermal energy. Batteries also use chemical energy.

Renewable and non-renewable sources of energy

Non-renewable sources of energy

Non-renewable energy sources come out of the ground as liquids, gases, and solids.

Coal, crude oil, and natural gas are all considered fossil fuels because they were formed from the buried remains of plants and animals that lived millions of years ago.

Compression over time fossilised the remains of these plants and animals, creating carbon-rich fuel sources that store chemical energy. In power stations these fossil fuels are burnt to generate heat (thermal energy), which is used to heat water and generate steam, that turns turbines and generators (kinetic energy) to generate electricity (electrical energy).

KEY TERM

Non-renewable energy: Sources come out of the ground as liquids, gases and solids and cannot be quickly replenished.

Uranium ore, a solid, is mined and converted to a fuel used at nuclear power plants. Uranium is not a fossil fuel, but it is classified as a non-renewable fuel. It generates heat (thermal energy) through the process of nuclear fission, which is then used to heat water, generate steam and turn turbines and generators to generate electricity.

Table 3.1 **Non-renewable sources of electricity**

Method	How it is used to generate electricity
Nuclear	Nuclear fission generates heat, which heats water to generate steam, steam turns turbines, turbines turn generators, electricity is distributed.
Gas/coal/oil	Fuel is burnt to generate heat, which heats water to generate steam, steam turns turbines, turbines turn generators, electricity is distributed.

We are approaching a worldwide crisis in terms of oil supply, and there is a global movement to reduce the amount of fossil fuels that we use.

Using fossil fuels to generate energy also has a significant environmental impact. Burning fuels produces waste products, including various gases such as sulphur dioxide, nitrogen oxide and other volatile organic compounds, which can have a harmful effect on the environment. The burning of fossil fuels creates carbon dioxide, which many scientists agree is a significant contributory factor to global warming.

Renewable sources

There are plenty of alternative energy sources that are renewable, or in constant supply. **Renewable energy** sources such as solar and wind can be replenished naturally in a short period of time. Most people are now realising that a massive switch to these types of power source will be required in the future.

Table 3.2 **Renewable sources of electricity**

Method	How it is used to generate electricity
Hydroelectric	Dam is used to trap water, the water released turns turbines, turbines turn generators, electricity is distributed.
Wind	Blades are designed to catch wind, blades turn turbines using gears, turbines turn generators, electricity is distributed.
Solar photovoltaic	Photovoltaic cells convert light to electricity.
Tidal barrages	Barrage built across river estuary, turbines turn as tide enters (and when tide leaves), turbines turn generators, electricity is distributed.
Wave	Motion of waves forces air up cylinder to turn turbines, turbines turn generators, electricity is distributed.
Geothermal	Cold water is pumped underground through heated rocks, steam turns turbines, turbines turn generators, electricity is distributed.
Biomass	Fuel (wood, sugar cane, etc.) is burnt to generate heat, which heats water to generate steam, steam turns turbines, turbines turn generators, electricity is distributed.

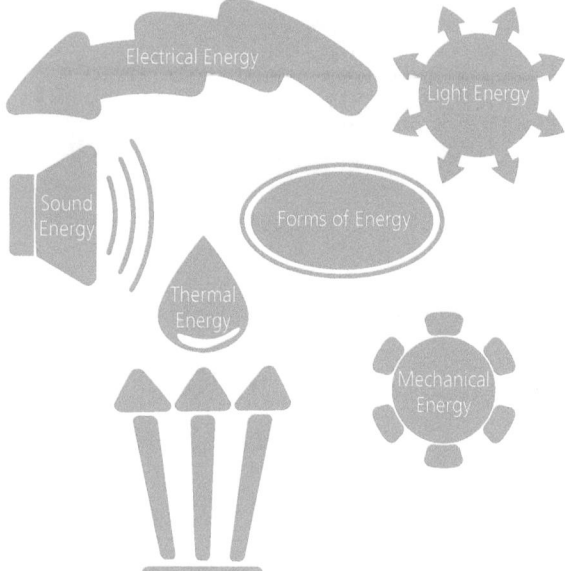

Figure 3.4 Forms of energy

KEY TERM

Renewable energy: Sources can be replenished naturally in a short period of time.

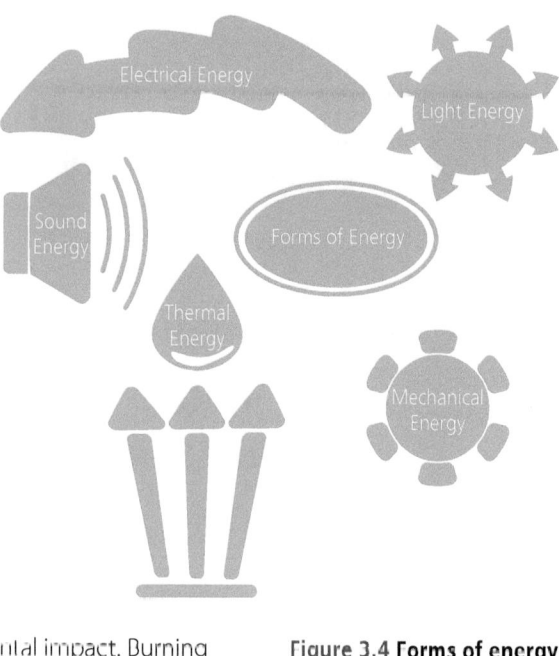

Figure 3.5 Renewable energy

Figure 3.6 **Batteries come in a variety of shapes and sizes**

Batteries

A battery is a self-contained, chemical power pack that can produce a limited amount of electrical energy wherever it's needed. Unlike normal electricity, which flows to your home through wires that start off in a power plant, a battery slowly converts chemicals packed inside it into electrical energy, typically released over a period of days, weeks, months or even years. Although we get through billions of them every year and they have a big environmental impact, we couldn't live our modern lives without batteries.

Wider implications that can have an influence on the processes of designing and making

LEARNING OBJECTIVES

By the end of this section you should know and understand the wider environmental, social and economic influences that can have an influence on the processes of designing and making, including:
→ environmental initiatives
→ fair trade
→ social and ethical awareness
→ global sustainable development

As well as the impacts of new and emerging technologies and choosing appropriate sources of energy, there are other wider issues a designer must consider when designing and making products. These include the environmental, social and economic factors that can have an influence on the designing and making process.

Environmental initiatives

Figure 3.7 **SOKO, Kenya**

Earlier in this chapter we have already considered the environmental implications of wasting resources and of the continued use of non-renewable energy when designing and making products.

There are many environmental initiatives in place that aim to limit the impact that the design and making processes have on the environment. We have already explored the concept of the circular economy, in which we use resources for as long as possible then reuse and regenerate products and materials. We have considered the use of renewable sources of energy as alternatives to fossil fuels.

Fashion companies, for example, realise they have a responsibility, and are committed to reducing their impact on the planet. For example, ASOS have worked with eco-friendly brands and global initiatives to manufacture clothing, accessories and beauty products that fit within their criteria for sustainability. They have set up a fair-trade clothing label ASOS Made In Kenya, made in partnership with SOKO Kenya.

Fair trade

Fair trade is about establishing better prices, working conditions and terms of trade for farmers and workers.

Many supermarkets and department stores now stock fair trade goods and ingredients, such as tea, sugar, coffee, rice, dried fruit and chocolate. These products have been made with fairtrade standards in mind.

Social and ethical awareness

Products are made by real people – so there are moral implications for all of us when we buy things. We need to consider the conditions of those involved in designing and making the products we buy. As designers, we also need to consider the social and ethical issues associated with our designs and their impact.

The cotton industry, for example, has a huge ethical and environmental impact, as it struggles with issues around child labour and welfare of workers. Despite some positive examples from cotton producing plants that promote good working practices, organisations like ActionAid work alongside their partners, such as the Labour Education Foundation, to continue improving working conditions, health, education and childcare.

ACTIVITY

Try to find examples of Fairtrade products and examples of companies' ethical policies next time you are shopping on the high street or online.

Figure 3.8 Cotton pickers and cotton plants ready for harvest

What can you do?

You can make a difference by being aware of sustainable and ethical issues and making informed choices when you design or buy products or by sourcing ethically produced materials. For example, buying clothes made with Fairtrade cotton helps low-paid cotton farmers around the world. The price of cotton has dropped in the past 30 years even though the cost of producing it has risen, and that means farmers in places like India, Kyrgyzstan and West Africa are struggling to survive. Buying Fairtrade products ensures that farmers receive a fair and stable price for their cotton, as consumers favouring these products means designers and manufacturers will ensure both awareness of issues and working standards are raised.

Global sustainable development

Global sustainable development is, as defined by the Brundtland Commission, development that 'meets the needs of the present generation without compromising the ability of future generations to meet their own needs'. When discussing sustainability, we must first consider that the product development and production process requires energy to source and process materials, which can cause pollution and affect climate change, damaging people's health. The materials and energy we use can, therefore, have an impact on resources and the environment causing consequences for future generations. As manufacturing has become increasingly globalised, governments worldwide have recognised the need for collective action in order to minimise its consequences.

The Paris Agreement made in 2016 was the first Global Sustainable Development Agreement. It focuses on reducing greenhouse gases and emissions but also includes wider global goals. The choice of materials, technologies and the way in which we use and dispose of products has a significant impact on achieving these goals, and technological research, design and development is a key driving force behind it. Closer to home, the UK Sustainable Development Strategy recognises the need for a new, more environmentally sound approach to development in terms of energy production, transport and waste management. Designers and consumers can all do small things to make a difference towards these goals; consider the approaches and steps identified earlier in this chapter.

ACTIVITY

Research the issues described in this section, and provide some examples of industries (for example, coffee, clothing, electronics) in which individuals or groups are disadvantaged by unfair working conditions or trading practices.

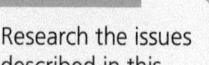

Communication is very important in any design process. Professional design teams and designers often need to pitch ideas to clients and stakeholders, and the ability to present ideas and design thinking clearly can be the difference between success and failure of a product. Designers use a range of graphical techniques to communicate their ideas and the constructional and technical considerations related to a design solution.

When thinking about problems, designers often rely on different design approaches and sources of information. Technological innovation and invention give designers the opportunity to collaborate on complex projects. A design team might include individuals who have training and experience in a variety of disciplines, such as electrical engineering, industrial design, fashion and architecture; they all bring specialist knowledge to a project. Good collaboration can be key to a team's effectiveness in reaching a successful outcome; the final result can often be more successful than that of an individual designer undertaking the same problem.

Communication of design solutions

LEARNING OUTCOMES

By the end of this section you should know and understand:
→ How design solutions can be communicated to demonstrate their suitability
→ The use of graphical techniques to communicate ideas, modifications, constructional and technical considerations, using clear 2D and 3D sketches with notes, sketch modelling, exploded drawings and mathematical modelling

The ability to represent design concepts using effective visual methods enables communication between designers and stakeholders. Designers often find that sketching, modelling and other visual techniques is an efficient way of developing ideas and communicating instructions and technical details.

2D and 3D sketching

A sketch can express more than words and provide greater understanding of an idea. Sketches are an easy and fast way to communicate an idea to others.

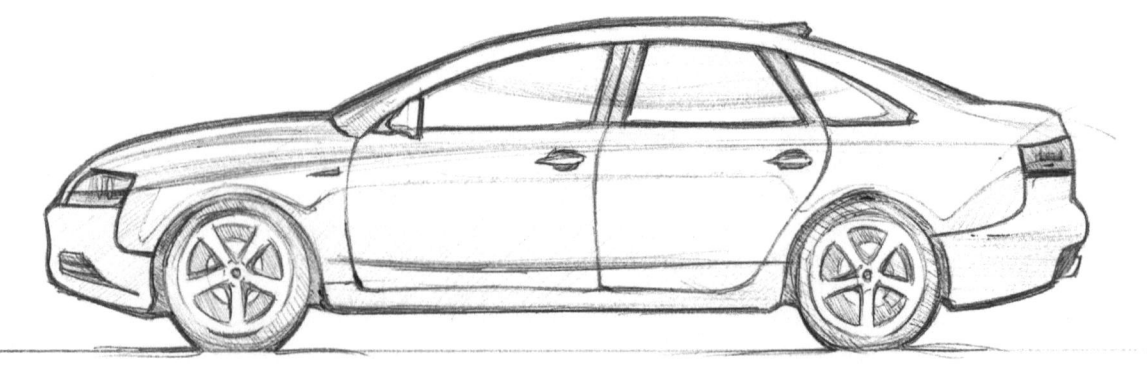

Figure 4.1 **Sketches are produced by designers of all disciplines**

Graphical techniques
Different types of sketches

There are many instances in which sketching can be used effectively during the designing and making process:

- Sometimes sketching is used right at the start of a design task. The main purpose is to facilitate understanding and analysis of the problems and contexts. In this type of sketch there can be more writing than sketching (for example, mind maps or thumbnail sketches). These sketches are not always about shape and form – they can be rough and lack detail but still enable an exploration of concepts. They are more about understanding and considering what the stakeholders want, who the product is for and what it needs to do, and about examining where the product will be used and analysing the context.
- First initial ideas are often communicated through sketches. Sometimes these are still very rough and don't always make sense for anyone who isn't directly involved in the project. They are a quick way for designers to get their initial ideas on paper and explore solutions. The idea is to grasp the overall idea and its principles rather than the details.
- Sketching can also be used at a later stage to help explain the design concept, a product's function, structure and form or a specific detail of a design. These types of sketches communicate a design in a clear and neutral manner, focusing more on explaining the idea rather than on selling it. They need to be readable to people other than those involved in the design process and they are often produced to get stakeholder feedback.
- Sketches can also be used to try to sell the design concept; this can involve sketching a product in use and creating a story board although some designers prefer to use a CAD program or more formal illustrations at this stage of the design process rather than sketching.

Sketches can be two dimensional (**2D**) or three dimensional (**3D**). 2D sketches look flat on the page and will only show two dimensions of a design, although different line thicknesses, colour and texture can be added to them to help to communicate ideas more clearly. Sketches can be produced in pencil or pen, fine liners are a commonly used medium. All students and designers develop individual techniques for sketching ideas quickly and effectively.

Perspective drawing

Perspective drawing is a type of 3D sketching technique that allows a designer to create believable sketches. Correct perspective can convey the proportions of an object and will make the object seem 'natural'.

There are many types of perspective drawing. The most commonly used are one-, two- and three-point perspective.

- One-point perspective is where lines converge to one point, this is known as a vanishing point.
- Two-point perspective is mostly used by product designers to create believable sketches faster than with the three-point perspective. It uses two vanishing points rather than one.
- Three-point perspective is the exact translation of the real-life situation. Three-point perspective is mostly used in architecture.
- Moving the vanishing point in perspective can change the appearance of a perspective drawing. For this reason, it's rarely used in engineering but product designers find it useful to communicate ideas to stakeholders and clients.

KEY TERMS

2D sketching: 'flat' drawings that only show two dimensions.

3D sketching: drawings in three dimensions that show depth.

Oblique drawing: a simple 3D sketching technique with one face of the item square on and other angled lines to give depth.

Perspective drawing: a 3D sketching technique that shows objects in proportion.

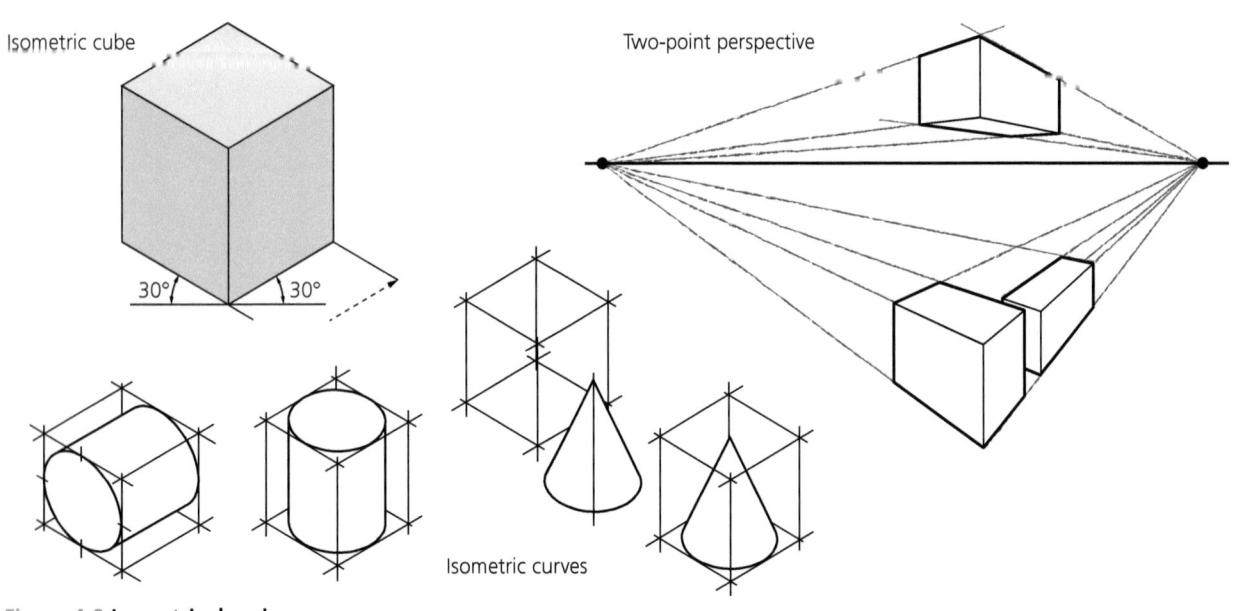

Figure 4.2 **Isometric drawing.**

KEY TERM

Isometric drawing: a 3D technique in which parallel lines at 30-degree angles are applied to the sides of the drawn object.

Isometric drawing

Isometric drawing is another way of presenting designs/drawings in three dimensions. In order for a design to appear three dimensional, a 30-degree angle is applied to its sides. The shapes above have been drawn in isometric projection. Designs drawn in this way can be drawn precisely using drawing equipment. An isometric grid can be used as an underlay, to ensure that lines are drawn in proportion. Many designers, however, use 'freehand' sketching in isometric projection to produce useful quick sketches or to put thoughts down on paper. Isometric projection is used to produce exploded views; these can help designers and engineers work out how parts of a product fit together to form a whole.

Oblique drawing

Oblique drawing is a simple 3D sketching technique that shows one face of the item square on and using real proportions, with other angled lines to give depth not necessarily in perspective. Isometric and perspective are more commonly used by designers for this reason.

Circles and ellipses

A compass is often used to draw circles. However, circles can be constructed in isometric perspective by first drawing a box and then constructing the diagonals and connecting the middles of the opposing sides. The shape create is, in fact, an ellipse.

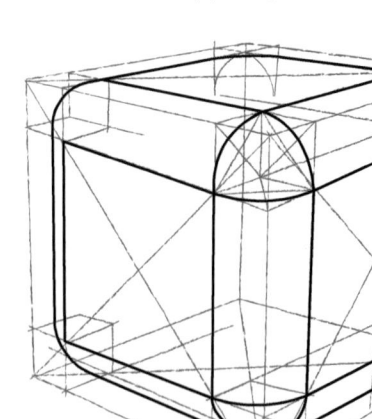

Figure 4.3 **Circles are a common feature in many designs. Use squares to create boxes on the corners of your shape then construct arches in these cubes to form circles or ellipses.**

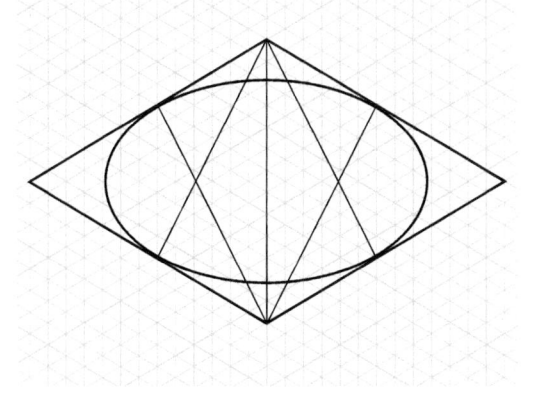

Figure 4.4 **Sketching ellipses and circles**

Details and notes

Adding details to 3D sketches at the later stages of product development provides essential information for stakeholders and helps a designer to give a more realistic view of their product. This can include, for example, information on the overall size of an object and how it will be used (i.e. user needs, the context and scenario of use, functional and material choices, possibilities for manufacture, etc.). Design is an **iterative** and reflective activity, in which design solutions are assessed and evaluated against the understanding of the design problem. Considering how the user will interact with and use the product can be done through storyboard/scenario sketches. (These can be helpful when presenting ideas to stakeholders.)

Working drawings

Working drawings contain all the information needed to make the design, including:

- dimensions
- details of components
- materials
- assembly instructions.

Working drawings are normally 2D **orthographic projections**, with a plan, front and side view, and often a sectional view. For some products a 3D technical drawing can be more appropriate, or a sectional or exploded view. Although traditionally these drawings were produced by draftsmen on drawing boards using T-squares, today's designers, architects and engineers tend to use CAD packages.

Sketch modelling

Sketch models are simple physical models made of soft, low-cost, easy-to-work materials such as cardboard, Styrofoam, foam board or calico. They are usually used to explore or create initial ideas and can provide 2D and 3D models to physically test with users and other stakeholders. Later models may become more and more accurate and focused on specific details.

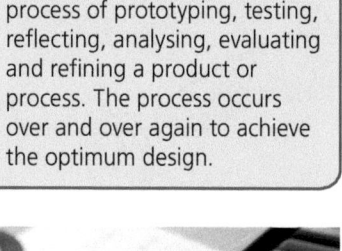

> **KEY TERM**
>
> **Iterative design:** A design process based on a cyclic process of prototyping, testing, reflecting, analysing, evaluating and refining a product or process. The process occurs over and over again to achieve the optimum design.

Figure 4.5 Detailed sketches can be produced by hand or using CAD

> **KEY TERMS**
>
> **Orthographic projections:** 2D sketches that show different views (for example, plan, front, side and sectional views).
>
> **Sketch models:** Quick models, often of just parts of a design, made from easy-to-work and low-cost materials such as cardboard or foam.

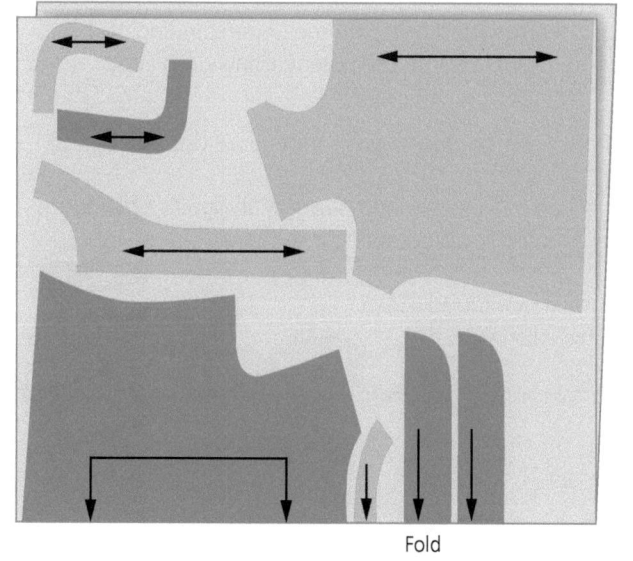

Figure 4.6 A lay plan including sectional view

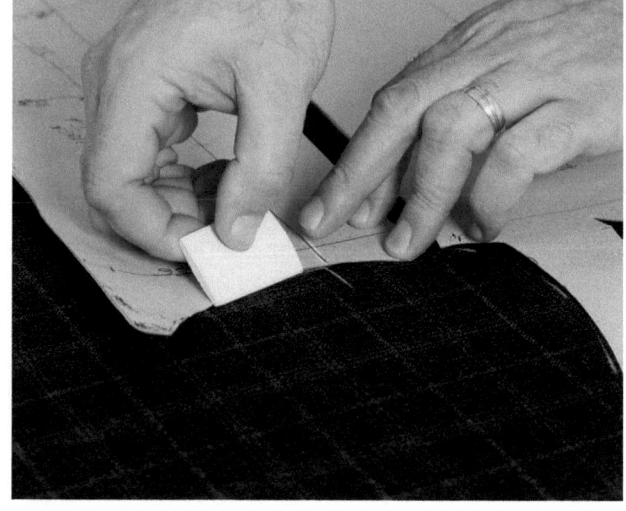

Figure 4.7 Sketch modelling examples

Exploded drawings

Accurate exploded views can take a long time to construct. This kind of 3D drawing is useful to use when explaining the layout of components and parts in relation to each other, the method of assembly and the general layout. They help engineers and designers visualise how the parts fit together and allow them to identify any potential construction problems.

Almost all exploded views are drawn in isometric perspective. An exploded view is created by constructing a box using any isometric perspective grid. The different parts that make up the product are then defined with some simple construction lines which are then 'copied' in the direction of assembly. Parts that are placed in the top surface of the product must be copied following the vertical lines. Parts that are connected on to the side of the product are copied following the horizontal directions.

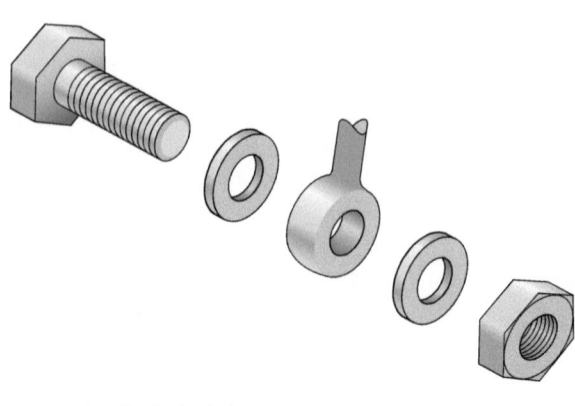

Figure 4.8 **Exploded view**

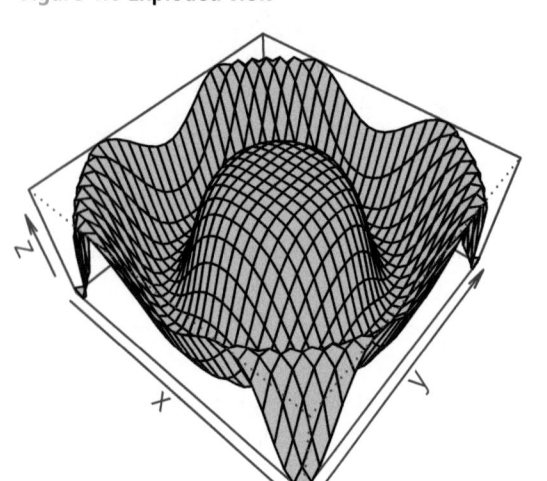

Figure 4.9 **Mathematical modelling surface mesh**

Mathematical modelling

Mathematical modelling is the representation of a real situation, but rather than a physical model it uses mathematical concepts and language. For example, mathematical modelling and computerised simulation software can be used to test circuits and mechanical devices without the need to physically build them.

As no physical components are used, this is a cost-effective process that means money isn't wasted on expensive parts. Modelling is also quick and can therefore speed up the production process. When designing structures such as buildings or moving parts in products, mathematical modelling can be used to predict stresses on components so that, if necessary, parts can be strengthened before physical prototypes are built. Used with simulation tools, these models can lead to rapid prototyping, software testing and verification.

Flowcharts

Flowcharts are used in designing to document and help with the understanding of simple processes. They might also help designers consider potential flaws and add alternative features.

There are many different types of flowcharts; the most common types of boxes used in a flowchart are:

- Terminal – this signals the start or end of a process, and is usually shown as a diamond.
- Process – this explains an activity or a step in a process, and is shown by a rectangular box.
- Input or output – this is often shown by a rounded box.
- Decision – this is usually shown by a diamond.

Other commonly used diagrams in engineering and design are block diagrams, which can give graphical representations of processes and orders. If they depict time, they move across the paper right to left or from top to bottom.

KEY TERM

Mathematical modelling: the representation of a real situation, but using mathematical concepts and language.

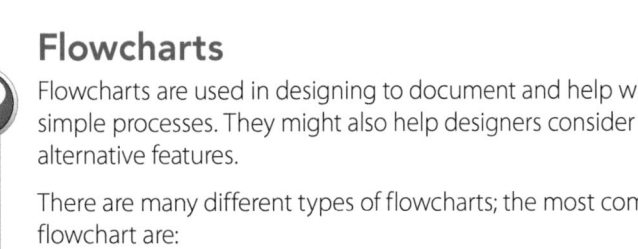

Figure 4.10 **3D example of flowchart symbols**

Schematic diagrams

Schematic diagrams are used to show the arrangement of components in electrical and mechanical systems. They are used to indicate the relative points of interconnection of the components rather than the actual positions of the components within a system as a whole.

Sources of information and thinking when problem-solving

> **LEARNING OUTCOMES**
>
> By the end of this section you should know and understand:
> - → How designers source information and the thinking used when problem solving
> - → Different design approaches, including user-centred design and systems thinking
> - → The importance of collaboration to gain specialist knowledge from across subject areas when delivering solutions in design and manufacturing industries

When considering how to approach a design problem, designers often make use of a wide range of sources of information.

- You can collect information first hand using questionnaires, by interviewing people, through **focus groups** and by carrying out surveys, and from first-hand research and observations.
- You can also get information from looking at solutions to similar problems, other products with similar features, and other products used in the context you are designing for.
- You can look to nature and incorporate its approaches to problems into design solutions (this is called **biomimicry**).
- You should consider the sizes of items that your product or system will need to function with or the space it needs to function within.
- You can look at data from magazines, reference books and government agencies. This could take the form of data about materials' properties or components' characteristics, or be tables of anthropometric data, information on forces, strengths to push and pull.
- These sources of information are used by designers to inspire new ideas or different ways of looking at a problem. They can help a designer to avoid **design fixation** (focusing on their own initial ideas and not considering alternative approaches) and to help them to create more innovative design solutions.

When gathering research and learning from the work of others, designers record and document this exploration as it occurs, so that they can refer back to it throughout the designing and making process. You will use a similar process when working on your NEA.

> **KEY TERMS**
>
> **Focus group:** An organised discussion led by a moderator, where a group of people are asked about their views and experiences, perceptions of and attitudes towards a product, brand, service, idea, advertisement or packaging.
>
> **User-centred design (UCD):** Sometimes called 'human-centred design', user-centred design is a design strategy, or design approach, with the aim of making products and systems useable. It focuses on the user interface and how the user interacts with and relates to the product.

User-centred design

User-centred design (UCD) is based on the understanding of users, the tasks they do and the environments in which they live and work. Users are involved at every stage throughout design and development of products. Design is driven and refined by user evaluation and feedback, and user-centred design considers the whole user experience. The process is iterative.

Some methods commonly used in UCD are:

- **Focus groups**, where an invited group of intended users share their thoughts, feelings, attitudes and ideas.
- **Usability testing**, which evaluates products by collecting data from people as they try out a product.
- **Participatory design** does not just ask users for their opinions on designs, but actively involves them in the design and decision-making processes at every stage.
- **Interviews** are usually carried out with one participant at a time.
- **Questionnaires** are used to ask users for their responses to a set of questions.

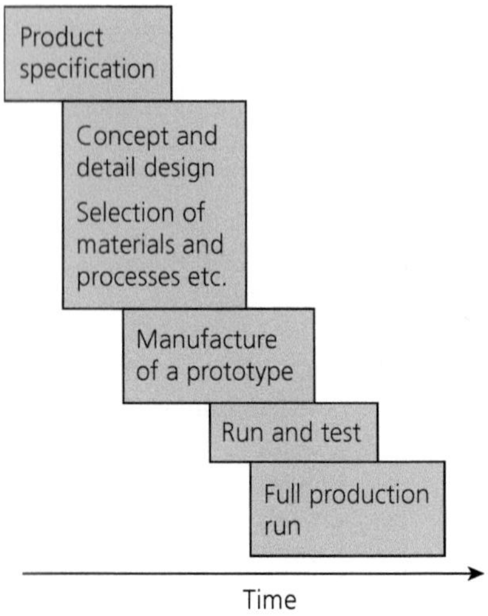

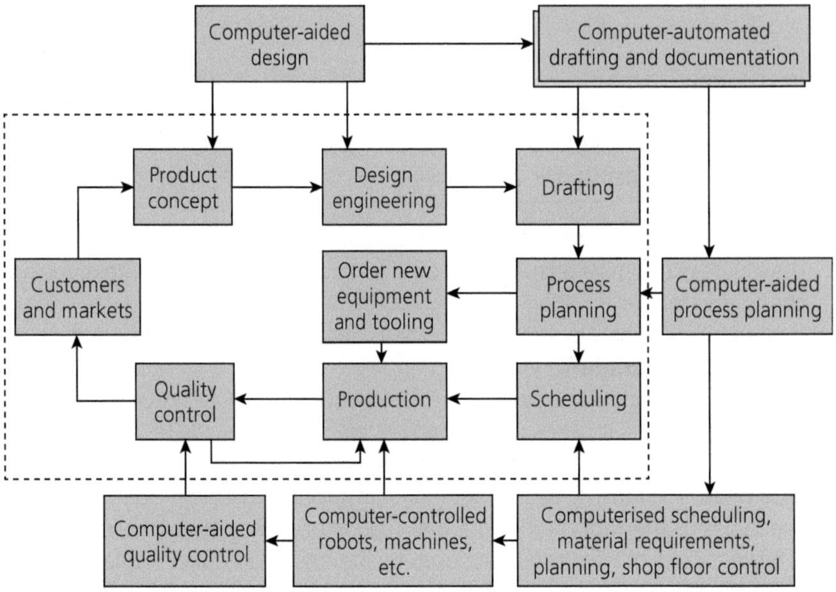

Figure 4.11 **Block diagrams showing a concurrent engineering process and how elements of the process overlap**

KEY TERM

Systems thinking: The understanding of a product or component as part of a larger system of other products and systems. In the iterative design process, consideration of the role of all components and sub-systems of the product or system, including the user experience and the marketing of the object being designed, ensures all aspects of the product are given the required attention to detail.

Systems thinking

A product is not just a product, it is actually a whole service. A camera, for example, is thought of as a product, but its real value is the service it offers to its owner: providing memories. Similarly, music players provide a service – the enjoyment of listening. Smartphones offer communication and interaction.

When thinking about a product, a designer should not only consider the product itself, but the whole experience associated with it – the opening of the packaging, the first and continued use of the product, how it is updated and maintained, and how it is disposed of, updated or exchanged at the end of its life. If you think of every product as a service, the point of any successful product will always be to offer great experiences to the user.

Applying **systems thinking** to designs means considering the whole of a problem rather than focusing one aspect alone, and helps a designer to look at all possible solutions to provide the best service to the user.

The Apple iPod, for example is a good example of systems thinking: not only is the iPod well-designed (it looks good and is easy to hold and use), but the process of purchasing, downloading a song to a computer and then to the iPod, or upgrading its software, is also a smooth and easy process. The 'Genius Bar' (the experts who offer free in-store advice to Apple customers) improves the customer service experience; and Apple also offers a recycling scheme to encourage customers to upgrade to newer models. In designing the iPod (and indeed all of their products) Apple have considered the entire experience of the product, from start to finish.

Collaboration

Design can be seen as a team sport, with more than one person seeking to find solutions with a mutual objective of improvement. Effective **collaboration** between designers, developers, users and other stakeholders can be the difference between success and failure.

Collaboration, however, adds complexity to a design project. Different collaboration styles are needed to ensure a project's success, but some stages may require more control in order to make the project deliverable. Therefore, a collaborative project can move through different types of collaboration as it follows the iterative design process.

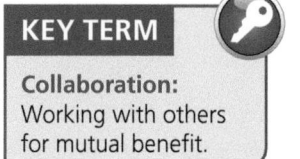

KEY TERM

Collaboration: Working with others for mutual benefit.

CHAPTER 5
Material considerations

This chapter looks at the main categories of materials available to designers when developing design solutions. It provides an overview and introduces the design considerations when selecting materials. Specific categories of materials are looked at in more detail in Chapters 8–12.

LEARNING OUTCOMES

By the end of this section you should know about and understand:
→ that products are predominantly made from multiple materials
→ about the main categories of materials
→ the developments in materials, including modern and smart materials, composite materials and technical textiles.

KEY TERMS

gsm: Grams per square metre. Used to classify the weights of paper and card.

Micron: One-thousandth of a millimetre. Used to classify the thickness of paper and card.

5.1 Categories of materials
Paper and boards

Paper and board are widely used by designers for a range of purposes, from the sketching, drawing and planning of ideas through to the modelling and prototyping of design solutions.

Papers and boards come in a wide range of different thicknesses, sizes and types, which are looked at in this section and also covered in more detail in Chapter 8.

In Europe the thickness of paper is known as its 'weight' and this is measured in grams per square metre, often abbreviated as g/m2 or **gsm**. This is the weight in grams of a single sheet of paper measuring 1 × 1m in size.

Common paper weights used by designers are:
- 80gsm
- 90gsm
- 100gsm
- 120gsm
- 130gsm
- 150gsm
- 170gsm.

A weight greater than 170gsm is classified as a board rather than a paper.

Boards are usually classified by thickness as well as by weight. This is because, depending on the type of board, different sheets may be the same weight but different thicknesses. For example, a sheet of corrugated cardboard and a sheet of mounting board may both be the same thickness but will weigh different amounts.

The thickness of board is measured in **microns**; a micron is one-thousandth of a millimetre.

Figure 5.1 Newspapers and magazines are a common use of paper

Paper

Paper is a familiar material that we come into contact with every day. Newspapers, magazines, comics, bus tickets, receipts and even toilet paper are common examples of different types of paper. Common paper types include:
- Layout paper
- Copier paper
- Cartridge paper
- Bleed proof paper
- Sugar paper.

For more information on different paper types, see Section 8.1.

Card and cardboard

Card and cardboard are thicker than paper, and are also commonly encountered in everyday life.

Card

Thin card is slightly thicker than paper, around 180 to 300gsm in weight. Like paper it is available in a wide range of colours, sizes and finishes, including metallic and holographic shades. Thin card is easy to fold, cut and print on, making it ideal for greetings cards, paperback book covers and so on, as well as for simple modelling applications.

Cardboard

Cardboard is available in many different sizes and surface finishes, with thickness from around 300 microns upwards. Cardboard is widely used for the packaging of many different products – for example it is used for cereal boxes, tissue boxes, sandwich packets, etc. – because it is relatively inexpensive and can be cut, folded and printed on to easily. Cardboard can be used to model design ideas and is often used to make templates for parts and pieces of products, which once correct can then be made from metal or other more resistant materials.

Corrugated cardboard

Corrugated cardboard is a strong but lightweight type of card that is made from two layers of card with another, fluted sheet in between. It is available in thicknesses ranging from 3mm (3000 microns) upwards. The fluted construction makes it very stiff and difficult to bend or fold, especially when folding across the flutes. Because of the spaces between the two layers created by the fluted sheet, corrugated card can absorb knocks and bumps. This makes it ideal for packaging fragile or delicate items that need protection during transportation. It is also widely used as packaging for takeaway foods, such as pizza boxes, as the fluted construction gives it good heat-insulating properties compared to normal cardboard.

Figure 5.2 Corrugated card board is stiff and often used as packaging for takeaway food

Double wall corrugated card is also available, which is twice as thick as corrugated card and gives extra strength and damage resistance.

For more information on cardboard, see Section 8.1.

Board sheets

Mounting board is a rigid type of card with a thickness of around 1.4mm (1400 microns) and a smooth surface. It is available in different colours but white and black are the most commonly available and used colours. Mounting board is often used for picture framing mounts and architectural modelling.

For more information on board sheets, see Section 8.1.

Laminated layers

Laminated layers include various other materials that come in sheet form, like paper and cardboard, and can be used in similar ways.

Foam board

Foam board is a lightweight board that is made up of polystyrene foam sandwiched between two pieces of thin card or paper. It has a smooth surface and is available in a range of colours, sheet sizes and thicknesses, with 5mm (5000 microns) being the most common. Foam board is a very lightweight but rigid material and is ideal for modelling and point-of-sale displays. It is easy to cut and can be easily folded with the correct technique.

Styrofoam

Figure 5.3 **Styrofoam™ is strong, lightweight, water-resistant and a good heat insulator.**

Styrofoam is a tradename for expanded polystyrene foam. It is available in a wide range of sizes and thicknesses but can be identified by its blue colour. Styrofoam has a structure of uniformly small, closed cells that make it easy to cut, shape and sand to a smooth finish. Styrofoam is strong and lightweight, as well as being water-resistant and having good heat-insulation properties. It is used as a wall insulation material in caravans, boats and lorries but it is also ideal for creating three-dimensional models and moulds for vacuum formed or glass fibre products.

Corriflute

Corriflute is an extruded corrugated plastic sheet similar in structure and thickness to corrugated cardboard. It is made from a high-impact polypropylene resin and available in a wide range of colours and sheet sizes. Corriflute is rigid and lightweight as well as being extremely waterproof. It is easy to cut but can be difficult to fold, especially across the flutes. Corriflute is often used for outside signs such as estate agents' signs on houses, and signs outside car forecourts or shops. It is also used for plastic containers, packaging, point-of-sale displays and for modelling purposes.

For more information on laminated layers, see Section 8.1.

Figure 5.4 **Corriflute is often used for outdoor signs.**

> ### ACTIVITY
>
> Copy and complete the following sentences using the words below.
>
> | cartridge | foam | weight | blue | microns | gsm | receipts | shape |
> |---|---|---|---|---|---|---|---|
>
> Newspapers, magazines, bus tickets, _____ and toilet paper are examples of different uses of paper. Common paper types include layout paper, copier paper, _____ paper, bleed proof paper and sugar paper.
>
> The thickness of paper is known as its _____ and this is measured in grams per square metre (g/m2 or ___).
>
> Paper with a weight greater than 170 gsm is classified as a board. The thickness of board is measured in _____. Examples of board are cardboard, corrugated cardboard and ____ board.
>
> Styrofoam is ____ in colour. It is ideal for creating 3D models because it is easy to cut, _____ and sand to a smooth finish.

Natural and manufactured timber

For thousands of years, trees have provided us with wood, a material used to make a wide variety of products. Wood is a natural material and timber is the general name given to wood materials once they have been processed into useable forms such as planks and strips. There are three main types of timber: hardwoods, softwoods and manufactured boards.

Figure 5.5 **Trees are cut down into logs**

Figure 5.6 **Logs are cut into timber**

Hardwoods

Hardwoods come from broad-leafed, deciduous trees that lose their leaves over winter, such as oak, birch and teak. These trees grow slowly and as a result the timber obtained from them tends to be dense, hard and heavy. Timbers come in many different colours and are generally used for high-quality items such as furniture.

Figure 5.7 **An oak tree**

Figure 5.8 **An oak chair**

Softwoods

Softwoods come from conifers – evergreen trees that keep their needles all year round, like pine, cedar and spruce. They grow faster than hardwood trees and the wood is usually lighter in colour. Softwoods are cheaper than hardwoods and are usually used in the building industry for roof, wall and door frames.

> **KEY POINT**
>
> The terms 'hardwood' and 'softwood' are used to describe the type of tree the wood came from. It does not necessarily mean that the wood itself is hard or soft. For example, balsa wood (a very soft and lightweight wood) is actually classified as a hardwood.

Figure 5.9 **A pine tree**

Figure 5.10 **A roof frame on a house**

Manufactured boards

Manufactured boards are sheets of timber made by gluing either wood fibres or wood layers together. This is a good way of making large flat boards that are stable and easy to work with, as natural timber can twist and warp. Examples of manufactured boards are MDF, plywood, and chipboard or block board.

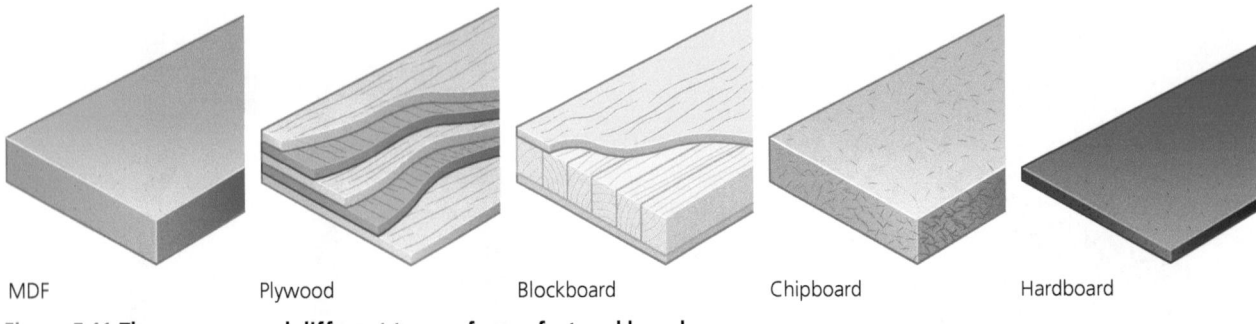

MDF   Plywood   Blockboard   Chipboard   Hardboard

Figure 5.11 There are several different types of manufactured boards.

Figure 5.12 A table made from MDF

More in-depth information on timber can be found in Chapter 9.

Metals

Metal is made by extracting metal ores from rocks in the Earth's crust by mining. The metal ore is then processed and refined to create a more useable material with improved properties.

There are two main categories or types of metal:
- **Ferrous metals** – metals that contain iron.
- **Non-ferrous metals** – metals that do not contain iron.

> ### KEY TERMS
>
> **Ferrous metal:**
> A metal that contains iron.
>
> **Non-ferrous metal:**
> A metal that does not contain iron.

Ferrous metals contain iron and will corrode quickly and easily because of their iron content unless they are treated with a suitable surface coating such as paint, oil or wax. The majority of ferrous metals are also magnetic so will be attracted to a magnetic force. Non-ferrous metals are much more resistant to corrosion and many are significantly better electrical conductors than ferrous metals. Generally non-ferrous metals are also more expensive than ferrous metals.

Both types of metals are available in a wide variety of shapes and sizes.

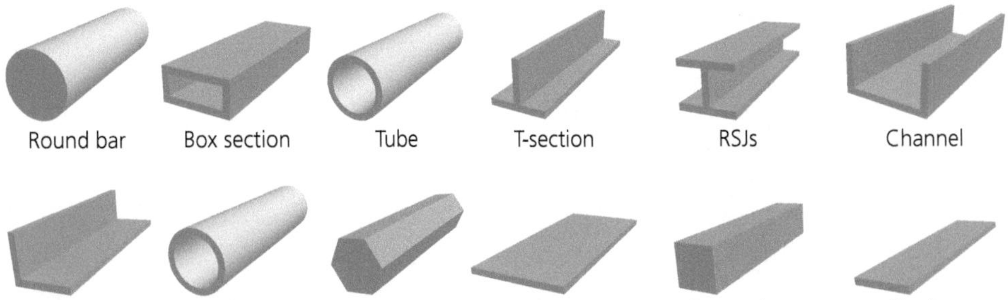

Round bar   Box section   Tube   T-section   RSJs   Channel

Angle   Pipe   Hexagon   Sheet   Square bar   Flat bar

Figure 5.13 Standard metal forms

Ferrous metals

Ferrous metals include:
- Mild steel
- Carbon steel
- Stainless steel
- Cast iron
- Wrought iron.

For more information on ferrous metals, see Section 10.1.

Non-ferrous metals

Non-ferrous metals include:
- Aluminium
- Copper
- Tin.

For more information on non-ferrous metals, see Section 10.1.

Figure 5.14 **Mild steel is widely used for building and engineering – for example, in steel joists and girders.**

Figure 5.16 **Aluminium has a wide range of uses, including ladders.**

Figure 5.15 **Stainless steel is resistant to corrosion and is used for cutlery.**

Figure 5.17 **Tin is often used for food containers, or 'tin' cans.**

KEY TERM

Alloy: A metal made by combining two or more metals to give greater strength or resistance to corrosion.

Alloys

An **alloy** is a metal that is mixed or combined with other substances to make it stronger, harder, lighter, or better in some other way.

Alloys include:
- Brass
- Bronze
- Pewter
- Lead/tin solder.

For more information on alloys, see Section 10.1.

Figure 5.18 **Brass is often used in decorative products such as this door knocker.**

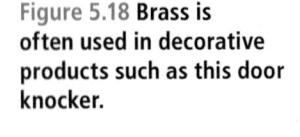

Figure 5.19 **Solder is used for electrical connections on printed circuit boards.**

ACTIVITY

Complete the table by adding the missing information.

Category	Definition	Example 1	Example 2	Example 3
Ferrous metals		Mild steel		
	Metals that do not contain iron	Copper		
Alloys		Brass		

Figure 5.20 **This Viking drinking horn is made from a naturally occurring polymer.**

Thermo and thermosetting polymers

A polymer is a very large, chain-like molecule made up of monomers, which are small molecules. Polymers can occur naturally or be manufactured. Examples of naturally occurring polymers include silk, wool, hair and even animal horn.

Manufactured polymers are commonly referred to as plastics and are derived from petroleum oil. Different types include nylon, polyethylene and acrylic. Rubber items are also a type of polymer. There are two families of polymers – thermo and thermosetting.

Thermo polymers

PET, HDPE, PVC, LDPE, PS, PP, ABS, acrylic and TPE are all examples of thermo polymers. They soften when they are heated and can be moulded into shape. They harden again once they have cooled. This can be repeated many times, which means thermo polymers can be recycled. When reheated, these plastics will try to return to their original shape. This is called **polymer memory**.

KEY TERM

Polymer memory: The ability of thermo polymers to return to their original state after reheating.

Figure 5.21 **These products are made from man-made polymer.**

Figure 5.22 **Drinks coaster made from recycled HDPE**

Thermosetting polymers

Silicone, polyester resin and epoxy resin are types of thermosetting polymer. These types of polymer undergo a chemical change when heated to become hard. Once they have 'set' they cannot be reheated and remoulded and so they cannot be recycled.

Figure 5.23 **Silicone cake moulds**

More in-depth information on polymers can be found in Chapter 11.

KEY POINT

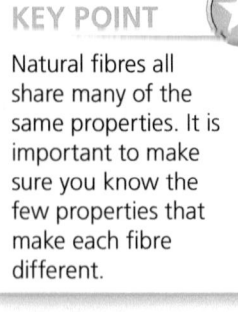

Natural fibres all share many of the same properties. It is important to make sure you know the few properties that make each fibre different.

Textile fibres and fabrics

Fibres are the starting point for all textile products. They are tiny hair-like structures that are spun (twisted) together to make yarns. These yarns are then woven or knitted together to create fabric.

Natural fibres

Nature provides us with a huge variety of fibres, which are found in both plants and animals. These are called natural fibres. Examples of natural fibres include cotton, wool and silk

Figure 5.24 **Fibres are spun together to make yarns and then woven or knitted together to create fabric.**

Figure 5.25 **Seed casings of the cotton plant**

Figure 5.26 **Silk worms wrap themselves in silk fibre to form a cocoon.**

Figure 5.27 **Wool fleece is sheared from a sheep.**

Synthetic fibres

Synthetic fibres (polyester, acrylic and nylon) are man-made and come from a range of sources including coal, oil, minerals and other petrochemicals. These fibres are mostly non-biodegradable and are therefore not sustainable. They can be engineered to give them a range of useful properties including flame resistance, crease resistance and stain resistance.

Mixed/blended fibres

Fibres can be mixed or blended together to improve the quality, aesthetics, function or cost of the final fabric.

Fibres are mixed by adding yarns of different fibres together during the production process. For example, a small percentage of elastane yarns may be mixed with cotton yarns during the weaving process to add strength to the weave and provide some elasticity. Fibres may be mixed for aesthetic reasons, such as mixing fibres of varying lustre or colour, giving a variable effect. The mixing of fibres happens after yarns are spun.

Blending of fibres is commonplace within the textiles industry. Fibres are blown together before they are spun into yarns. Popular blends include polyester cotton; this blend of polyester and cotton fibres is often used to produce shirts, and gives the finished fabric strength, breathability, absorbency and crease resistance. The blend also lowers the overall cost of the fabric.

Woven, non-woven and knitted fabrics
Woven fabrics

Woven fabrics are produced on manual or automatic looms. A woven fabric consists of warp and weft yarns. The warp yarns run vertically and the weft yarns are woven horizontally in an under/over configuration.

Different types of weave are produced for various uses. There is more on different types of weave in Section 12.2.

Figure 5.28 **Oil is used to make synthetic fibres.**

Figure 5.29 **Polyester cotton is a popular blended fibre.**

Figure 5.30 **An automatic loom**

Non-woven fabrics

Non-woven fabrics lack the strength of woven or knitted fabrics and are usually used for decorative or disposable products. These fabrics fall into two categories: bonded and felted.

Bonded fabrics are manufactured by applying pressure and heat or adhesives to bond the fibres together and are often used in disposable textiles such as wet wipes, tea bags, surgical masks, dressings and nappies. These fabrics lose their strength and structure once wet, so they are usually only suitable for one use.

Felted fabrics are produced by applying heat, moisture and friction to fibres, which matt together. The most commonly used fibres are wool and acrylic. Felt is often used for decorative purposes, such as appliqué, and is historically applied to the surface of pool and snooker tables. Felt is also used for cushioning and insulating various products.

Figure 5.31 Disposable textiles such as surgical masks are made from bonded fabrics.

Figure 5.32 Felt is applied to the surface of snooker tables.

Knitted fabrics

Knitted fabrics are made up of rows of interlocking loops, also known as stitches. The most commonly used knits are weft and warp.

For more information on different types of knitted fabric, see Chapter 12.

Figure 5.33 Knitted fabrics are made up of rows of interlocking stitches.

Modern and smart materials

Modern materials

Modern materials are those that are continually being developed through the invention of new or improved processes. Some will have practical applications for your projects in Design and Technology, while others will be more suited to a specific commercial or industrial use. Below are some examples of modern materials.

Polymorph

Polymorph comes in the form of polymer granules. When heated to 60°C in warm water the granules melt and can be moulded into shape. You can reheat it again using warm water or a hairdryer.

Teflon

Teflon is mainly used as a non-stick coating on cookware, but it is also used in paints, fabrics, carpets and clothing to repel liquids.

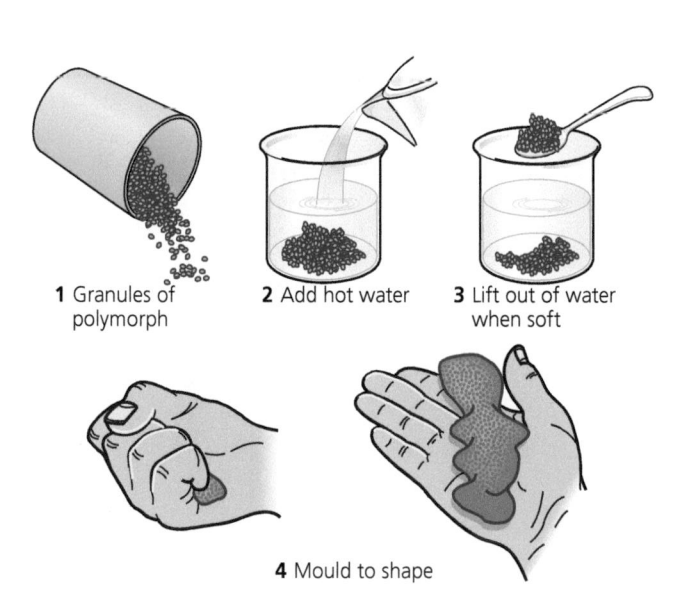

1 Granules of polymorph

2 Add hot water

3 Lift out of water when soft

4 Mould to shape

Figure 5.34 **The stages of using polymorph**

Lenticular plastic sheet

Lenticular plastic sheet is smooth on one side while the other side is made of small lenses. These lenses transform 2D images into a variety of visual illusions.

Flexiply

Flexiply is a form of plywood that is extremely flexible and can quite easily be bent into various shapes.

Figure 5.35 **Teflon-coated pans are non-stick**

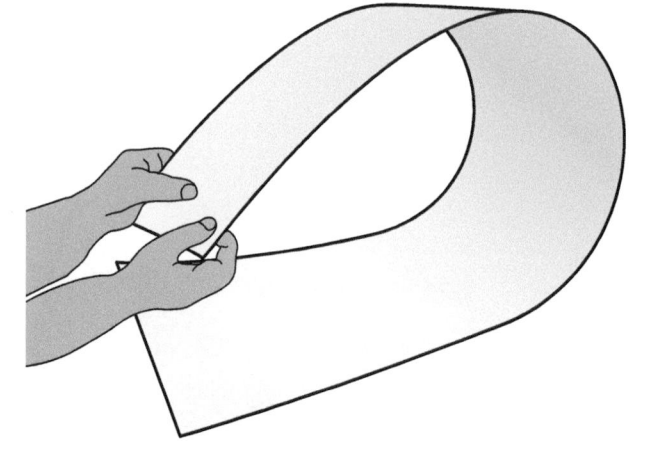

Figure 5.36 **Flexiply**

Precious metal clay

Precious metal clay (PMC) is made from 99 per cent silver or gold and 1 per cent clay. It can be shaped at room temperature then heated in a kiln to produce jewellery.

Figure 5.37 **An example of hand-made PMC jewellery**

Conductive polymers

Conductive polymers are plastic products that can conduct electricity.

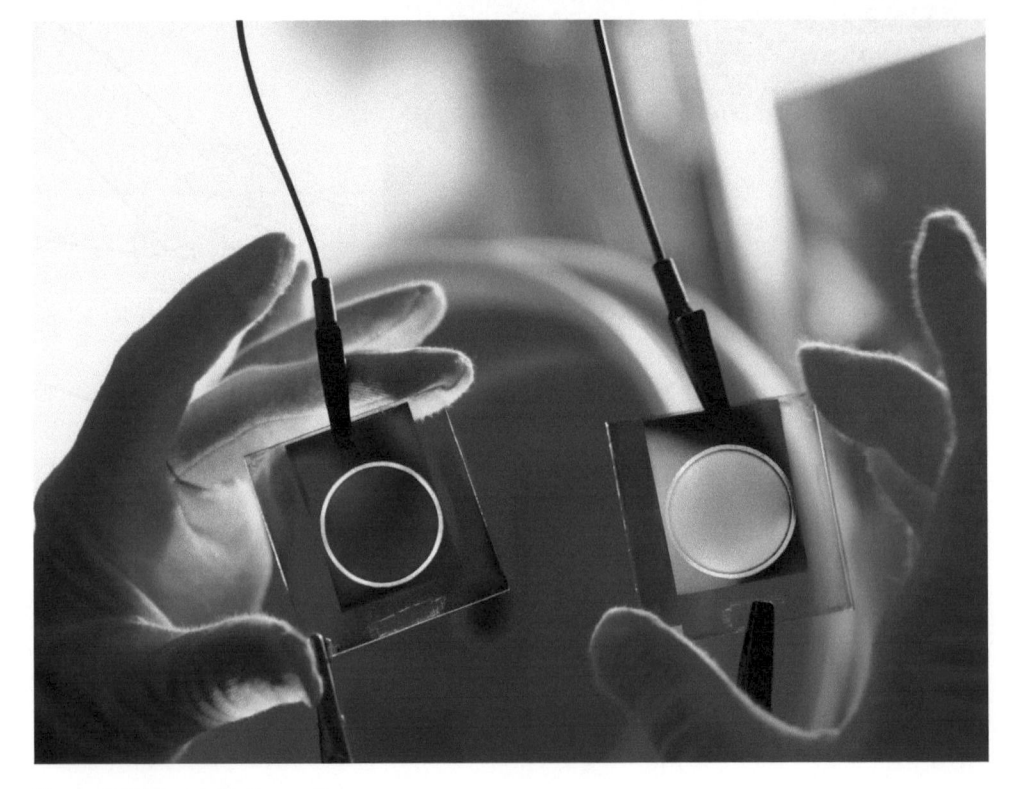

Figure 5.38 **A conductive polymer**

Smart materials

Smart materials respond to differences in temperature or light and change in some way. They are called smart because they sense conditions in their environment and respond to those conditions. Smart materials appear to 'think' and some have 'memory' as they revert back to their original state

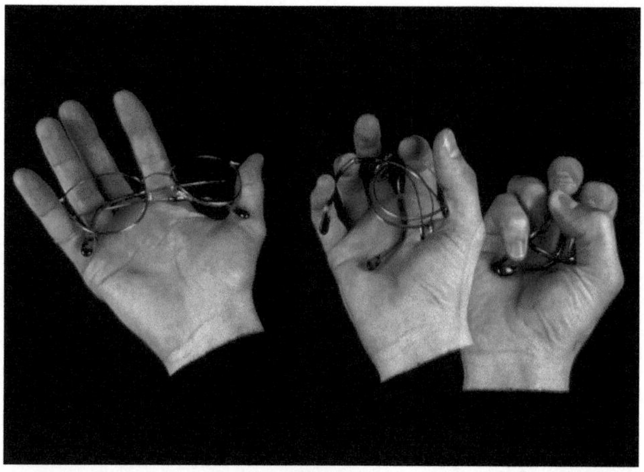

Shape-memory alloy

Shape-memory alloy (SMA) remembers its original shape when deformed and returns to it when heated. SMA can be used in many applications, such as spectacle frames that return to their original shape after being bent, and as a simple and effective way to move parts in machines, like cooling vents.

Figure 5.39 Shape-memory glasses

Shape-memory polymers

Shape-memory polymers can also be 'programmed' to remember their original shape when they are heated.

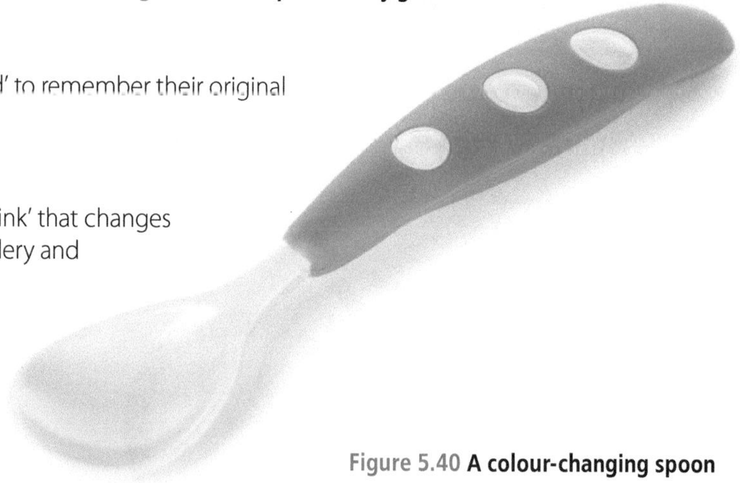

Thermochromic sheet

Thermochromic sheet is printed with liquid crystal 'ink' that changes colour above 27°C. It is used in children's toys, jewellery and temperature indicators.

Thermochromic pigments

Thermochromic pigments are often used on novelty mugs that reveal a design as hot water is poured into the mug. They can also be added to polymers to create plastics that react to heat, such as colour-changing drinks stirrers and baby-feeding spoons that warn you if food or drink is too hot.

Figure 5.40 A colour-changing spoon

Photochromic materials

Photochromic materials react to light. Spectacles that darken in bright sunlight use photochromic lenses.

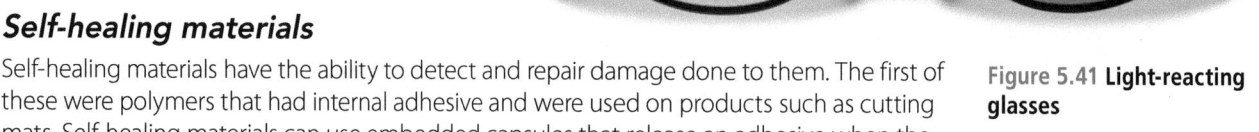

Self-healing materials

Self-healing materials have the ability to detect and repair damage done to them. The first of these were polymers that had internal adhesive and were used on products such as cutting mats. Self-healing materials can use embedded capsules that release an adhesive when the material cracks. BioConcrete heals itself using bacteria that react with any water that gets into it and produce limestone to fill any micro-cracks that appear.

Figure 5.41 Light-reacting glasses

Material science has developed rapidly over the past 20 years and will continue to move forward. It is difficult to keep up to date with current developments in this field and by the time you read this many more new materials will have already been developed and integrated into modern products. Try hard to keep abreast of current scientific progress as you study Design and Technology, by following the news on television and on recommended websites.

Composite materials

Composite materials are produced by bonding different materials to produce new materials with improved properties. Several composite materials have been around for many years and new composites are being developed. Composites are increasingly used in place of metals in machine tools.

Figure 5.42 **GRP boats**

Glass-reinforced plastic

Glass-reinforced plastic (GRP) is polyester resin reinforced with glass fibre strands. The glass fibre is available as woven fabric, matting or loose strands. It has all the properties of a polymer but is much stronger. It is used for large structural items such as boats and car bodies.

Carbon fibre

Carbon fibre is similar to GRP but uses carbon fibres instead. This makes the material even stronger and also lighter in weight. It is used for protective helmets, high-end bicycles and sporting equipment.

Figure 5.43 **A carbon fibre racing bike**

Figure 5.44 **A carbon fibre hockey stick**

Kevlar

Kevlar is similar to carbon fibre and has even stronger plastic woven in to it. It is even lighter and stronger than carbon fibre and is used for bulletproof vests.

Laminates

Laminates are materials made up of layers. Since the layers are usually of different materials, laminates are examples of composites. If a material is not strong or durable enough to survive by itself, you would combine it with one that is. For example, laminated glass has a thin layer of plastic sandwiched in the middle to make the glass stronger and tougher. A waterproof jacket is made from layers that stop wind and rain getting in but allow moisture vapour out.

Figure 5.45 **A Kevlar bulletproof vest**

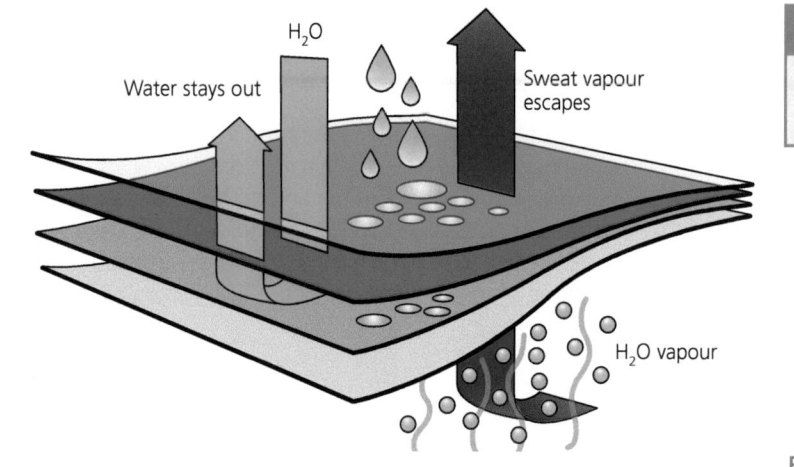

Figure 5.46 **Layers of fabric in a waterproof coat**

STRETCH AND CHALLENGE

Use the internet to research other examples of composite materials and their uses.

Technical textiles

Technical textiles are engineered specifically for their performance properties, not for their aesthetic value. The application of technical textiles in everyday products is rapidly growing and new technical textiles are constantly being developed by designers and engineers.

Table 5.1 **Common technical textiles and their applications**

Technical textile	Properties	Applications
Nomex	Heat- and flame-resistant	• Protective clothing for racing drivers, firemen, astronauts • Oven gloves • Fire-resistant insulation in buildings
Kevlar	Resistant to abrasion and damage by sharp or pointed objects	• Stab- and bullet-resistant vests for police and armed forces • Linings in motorcycle clothing and accessories • Car and motorcycle tyres • Protective gloves for butchers and fishmongers
Coolmax	Wicks water away from body, improves breathability	• Bedding • Sportswear • Uniforms • Underwear
Fastskin	Mimics the skin of a shark, giving a streamlining effect when underwater	• Competitive sportswear and swimwear

Figure 5.47 **Nomex is used to protect firefighters from flames and heat exposure.**

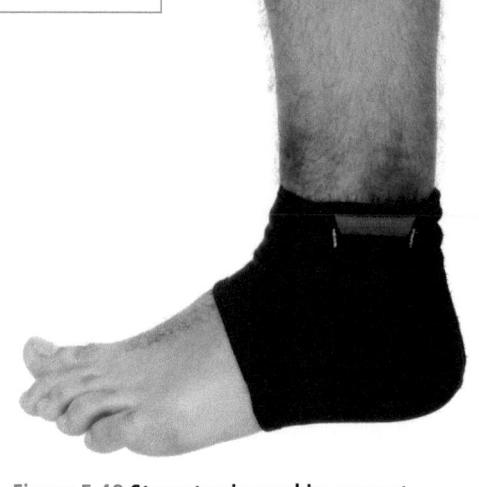

Figure 5.48 **Stomatex is used in support bandages as it allows the skin to breathe.**

53

5.3 Characteristic properties of the main categories of materials

When designing any product, the choice of materials is one of the most important decisions and will have a massive impact on the success of the final product. In order to choose the most appropriate material the designer will need to consider and understand the key properties of materials and make informed choices based on these characteristics. The same will apply when designing an electronic system; the designer must consider the wide range of components available and decide on the most suitable type for the application, based on certain characteristics it must have.

> **LEARNING OUTCOMES**
>
> By the end of this section you should know about and understand:
> → the characteristic properties of the main categories of materials, and why this makes them appropriate for different uses.

Density

Density is the mass of a material (its weight) divided by its volume (size). Materials such as metals are denser than materials such as softwood or Styrofoam.

For example, a block of steel and a block of softwood both measure 10cm long × 10cm wide × 10cm high. The steel block weighs 7.8kg. The softwood block weighs 0.65kg. By dividing the weight of the blocks by their volume we can find their densities:

Steel block $\dfrac{\text{Weight} = 7800\text{grams}}{\text{Volume } (10 \times 10 \times 10) = 1000\text{cm}^3} = 7.8\text{g/cm}^3$

The steel is much denser than the softwood.

Water has a density of 1g/cm³. Materials with a density less than water will float.

Strength

The strength of a material is its ability to withstand forces that try to bend or break it. Materials are often strong in one way but weak in another, depending on the types of forces that are applied.

Because there are different forces that can be applied to a material, there are different kinds of strength, related to how well a material can resist these forces:
- Compressive strength – how well the material can withstand compression forces (squashing forces).
- Tensile strength – how well the material can withstand tension forces (stretching forces).
- Yield strength – how well the material can withstand forces before it is permanently deformed in some way.
- Impact strength – how well the material can withstand sudden forces.

Many materials that have good strength characteristics against one type of force will respond poorly against another. For example composite materials, such as glass fibre and certain polymers, tend to have higher tensile strengths than compressive strengths. The tensile strength of timber is considerably higher than its compressive strength. Some materials such as metals tend to have tensile and compressive strengths that are very similar.

The strength of textiles and fabrics is sometimes referred to as 'tenacity'.

Hardness

The hardness of a material refers to how resistant it is to pressure from cutting, scratching or wear. Scratch hardness, indentation hardness and rebound hardness are different measures of hardness. Some materials (for example, metals) are much harder than others (for example, polymers and textiles). Metals such as high carbon steel are extremely hard materials.

Many hard materials are also brittle.

Durability

Durability is the ability of a material to withstand wear, pressure or damage. Some materials are more durable than others and are chosen for certain applications for their ability to withstand wear and damage.

Strength to weight ratio

Strength to weight ratio is a measure of a material's strength compared to its weight. Materials with high strength to weight ratios are used where a strong material is required but saving weight is also as important, such as aircraft. Carbon fibre, glass fibre and alloys such as titanium have extremely high strength to weight ratios. Most polymers have high strength to weight ratios especially if reinforced with fibres.

Stiffness

The stiffness of a material describes its rigidity – its ability to resist being deformed when a force is applied to it. Ceramics, glass and steel have good rigidity, whereas polymers such as foam and rubber are extremely flexible and have poor stiffness.

Elasticity

The elasticity of a material is its ability to return to its original shape when forces are applied to it that will make it bend or flex out of shape.

Different materials have different levels of elasticity. Some materials will not deform or will deform very little before breaking and therefore have a low elasticity. If a material has high elasticity it will return to its initial shape and size when the forces are removed.

A simple example of a material with high elasticity is elastic band. When you stretch an elastic band it is deformed out of shape, but returns to its original size once you let go.

Impact resistance

Impact resistance is the ability of a material to withstand a force or shock applied to it over a short period of time. Many polymers such as rubber and nylon have good impact resistance as well as softer metals such as mild steel.

Plasticity

Plasticity is the ability of a material to permanently change in shape when force is applied to it. Metals such as sheet steel are shaped by hammering or bending. This is an example of plasticity as the hammering and bending forces create permanent changes within the material itself.

ACTIVITY

Find or collect two or three different types of plastic carrier bags. Suspend each bag on a suitable hanger and gradually add heavy objects or weights into each one.

● Do the plastic handles snap or stretch?

The handles that stretch most have the highest ductility. The ones that snap without stretching have poor ductility.

Ductility and malleability

The ductility of a material is how easily it can be deformed or bent out of shape without snapping or breaking. The more ductile the material, the easier it is to stretch into a thin wire. Materials with poor ductility will break when stretched. Malleability refers to a material's ability to deform without breaking or snapping when hammered or rolled into a thin sheet.

Soft metals such as lead and polymers such as polypropylene have good ductility and malleability. Temperature and application of heat can alter both properties. Higher temperatures will increase a material's ductility and malleability, and lower temperatures can decrease them.

CLASS ACTIVITY

Take some thin strips of different woods, metals or plastics. Place one end in a bench vice and push the other end, making the strips bend. Start by bending each one a little and seeing if they return to their original shape, then continue pushing a little more and bending them further and further until they do not return to their original shape.

Which of the materials has the highest and lowest elasticity? Which has the highest and lowest elastic limit?

Health and safety: be very careful when bending materials that may splinter or shatter. Always wear appropriate protective equipment, including safety goggles to protect eyes and gloves and an apron to protect skin.

Brittleness

Brittleness is how easily a material will snap or break when bent or impacted. Very brittle materials will shatter or break up rather than bend and deform when forces are applied. Brittleness is the opposite of ductility and is also affected by heat. The lower the temperature, the less ductile and more brittle many materials will become. Many metals, especially cast iron and aluminium, are very brittle.

Corrosive resistance to chemicals and weather

Corrosive resistance refers to how susceptible a material is to degradation from elements such as oxygen, moisture and other chemicals. The most common form of corrosion is rust, which affects certain metals when they are exposed to oxygen and moisture. Ferrous metals (metals that contain iron) have a very low corrosion-resistance and are extremely susceptible to rust.

Corrosion also affects other materials, including timber and polymers. Some timbers such as oak and other dense hardwoods are resistant to corrosion (rot), whereas softwoods will rot within a few months if exposed to the elements, if they have not been treated with chemical inhibitors.

Many polymers degrade gradually over time if exposed to ultra-violet light, oxygen or chemicals such as chlorine that cause them to swell up, crack and/or break down. Natural and synthetic rubber products such as car or bicycle tyres are common polymers that will become brittle over time before cracking and splitting if left outside in the sun. Even harder plastics like uPVC windows and doors go brittle over time and need to be replaced every twenty or thirty years.

Figure 5.49 Rubber products like tyres become brittle over time.

Water resistance

Water resistance is the ability of a material to resist the ingress of water. Materials that do not absorb moisture or water are called waterproof. The majority of polymers such as rubber, PVC and polyurethane are water resistant and these are often used to coat other materials or products to make them water resistant. Products such as waterproof clothing, inflatable boats and footballs use natural or synthetic fabrics that are laminated or coated during manufacture by a waterproofing spray.

Absorbency

Absorbency is the ability of a material to absorb moisture. Many natural fabrics such as cotton, linen and wool have good absorbency. Most types of cardboard along with porous polymers such as foam also have good absorbency.

Flammability

Flammability is the ability of a substance to burn or ignite, causing fire. The vast majority of timbers, polymers, fabrics, papers and boards are flammable although some will ignite and burn more quickly and easily than others. Metals are not flammable and will not burn when exposed to a flame

Electrical conductivity

Electrical conductivity refers to how easy it is for electricity to flow through a material. Metals are good electrical conductors and have high electrical conductivity properties. Wood and rubber are poor electrical conductors and do not allow electricity to flow easily through them.

Thermal conductivity

Thermal conductivity refers to the way heat can be transferred through a material. Materials with good thermal conductivity will allow heat to be transferred through them relatively easily (and they will heat up). Metals often have good heat-conductivity properties and are therefore used to transfer heat in heating systems (metal radiators, etc.).

Thermal resistance is the ability of a material to resist the flow of heat. Materials with good thermal resistance will not let heat be transferred through them easily, and such materials are said to be good insulators. Generally, porous materials such as Styrofoam and softwoods are good insulators.

Figure 5.50 **A heat sink**

Thermal fabrics have excellent insulation properties and are used in many types of clothing to keep the wearer warm as well as products such as oven mitts, table pads and ironing board pads. Thermal fabrics are also used to keep cold things cold such as bottle jackets, cool bags etc. Acrylic and viscose are fabrics with good insulation properties.

ACTIVITY

Fill two cups with boiling water. Put a metal tablespoon in one and a wooden spoon in the other.

Leave for 1 minute and feel the end of the spoon sticking out of the cup.

The metal spoon will be much hotter than the wooden spoon as the metal conducts the heat and the wood insulates it.

Health and safety warning: be very careful with boiling water to avoid scalding.

Magnetic properties

A material with magnetic properties will emit forces that attract or repulse other materials. All materials have a certain amount of magnetism, but it is very low in most of them. Metals such as magnetite, iron, steel, nickel, cobalt and some alloys have high levels of magnetism.

Magnetite is a naturally occurring material that is magnetic. Other materials can be magnetised artificially by various methods. These are called ferro-magnetic materials. The magnetic properties of artificially magnetised materials fade over time, however, whereas magnetite is a permanently magnetic material and its magnetic property will not fade.

KEY POINTS

- Designers must consider the properties and characteristics of materials when designing.
- Different materials can have a range of different properties.
- A material's properties may change depending on things such as temperature or the shape and direction of the material.

STRETCH AND CHALLENGE

1   Complete the table below by adding three properties of the materials shown in the left-hand column.

Material	Property 1	Property 2	Property 3
Copper			
Softwood			
Nylon			
Foam rubber			
Glass			
Leather			

2   Add more materials of your own in the left hand column and fill in the three property columns.

CHAPTER 6
Mechanical devices and electronic systems

This chapter introduces some of the ways in which products and systems are 'brought to life' using movement and/or electronic control. Even very basic products such as a pair of garden secateurs or a door closer make use of a mechanical system, and domestic products we use daily such as kettles and toasters are controlled by electronic systems.

If you wish to further explore the use of mechanical or electronic systems, and would like to use such systems during your iterative project development, you will find the more in-depth technical information in Chapter 10 useful.

Figure 6.1 **Simple products that rely on mechanisms or electronics**

6.1 Controlled movement

LEARNING OUTCOMES

By the end of this section you should know about and understand how mechanical devices introduce controlled movement to products or systems, including:
→ an overview of the different types of motion
→ the effects of forces on the ease of movement
→ how mechanical devices are used to change the magnitude and direction of forces.

Movement and motion bring products to life. The movement might involve the entire product, for instance a vehicle, or just a part of the product, such as in an electric fan. The motion might be clearly visible, for example in the case of a food mixer, or it might take place inside the product and not be immediately visible from the outside, as in a computer printer. Sometimes, movement is a key part of the function of a product, such as a bicycle. In other products, movement adds to the functionality of the product to enhance it in some way; for example, a soft-close door mechanism is not essential to the function of a door, but it is a useful addition and stops the door slamming closed.

How mechanical systems produce different sorts of movement and types of motion

A **mechanism** is a series of parts that work together to control forces and motion in a desired way.

- A mechanism is an example of a **system**, and there is a huge variety of mechanisms that generate, control and change motion. Controlled motion is essential in many engineered products.
- A **force** is a push, a pull or a twist. Forces are measured in newtons (N).
- There are basically four different types of motion, and these are described below.

Rotary motion

This is motion that follows the path of a circle. Rotary motion is very common and can easily be seen in the rotation of wheels. The output shaft of an electric motor also moves in rotary motion. Rotary motion can be measured by counting the number of revolutions in a set period of time – revolutions per minute (rpm) is a common measurement of rotary speed.

Linear motion

This is motion in a straight line and is commonly seen – for example, a vehicle travelling in a straight line, or items travelling on a conveyor belt. The speed at which objects travel in a straight line is measured by dividing the distance they travel by the time taken:

$$\text{Speed} = \frac{\text{distance}}{\text{time}}$$

Speed is often measured in units of metres per second (m/s) or kilometres per hour (km/h).

KEY TERMS

Mechanism: A series of parts that work together to control forces and motion.

System: The general name for a set of mechanical or electronic parts that work together to produce a desired output.

Force: A push, a pull or a twist.

Figure 6.2 Wheels are an example of rotary motion.

STRETCH AND CHALLENGE

A conveyor must move items through a distance of 5.7m in 18s. Calculate the required speed of the conveyor.

$$\text{Speed} = \frac{\text{distance}}{\text{time}}$$

$$\text{Speed} = \frac{5.7}{18}$$

$$= 0.32 \text{ms}^{-1}$$

Figure 6.3 Items on a conveyor belt move with linear motion.

Oscillating motion

Oscillating motion is similar to circular motion, but the rotation moves back and forth in a circular path. Oscillating motion is quite common but you need to look carefully to spot it; a good example is the oscillating head of an electric toothbrush. The rate of oscillating motion is measured in oscillations per second or per minute.

Reciprocating motion

This is back-and-forth motion in a straight line, for example the blade on an electric jigsaw. Oscillating and reciprocating motions are similar in that they are both measured in oscillations per second or per minute.

Figure 6.4 **An electric toothbrush head uses oscillating motion.**

Figure 6.5 **A jigsaw blade moves with reciprocating motion.**

ACTIVITY

1. Take five products from around your home that exhibit movement (for example, a computer mouse, hole punch, toaster, electric drill, electric toothbrush, or food mixer). For each product state the kinds of motion involved when they are being used – there may be several different types of motion going on at once.
   Sometimes it is not immediately obvious which type of motion is involved. For example, when automatic doors slide open and closed you might describe their motion as reciprocating,; but if the same doors are locked open all day would 'linear motion' be a better description?
2. How would you describe the motion of the following items:
   a) A stapler? (Is it reciprocating or oscillating?)
   b) The handles of a pair of scissors?

Many mechanisms change one kind of motion to a different kind. They can also change the speed and direction of the motion, as well as the size and direction of forces. Examples can be found throughout this chapter.

The effect of forces on ease of movement

Mechanisms that control and change motion have an **input** and an **output**. On a pair of scissors, for example, the input motion is the movement of the handles, and the output is the movement of the blades. Mechanisms such as this can either:
- reduce the distance moved but increase the force being exerted, or
- increase the distance moved but reduce the force being exerted.

There is always a trade-off between force and distance moved – one can increase, but only if the other decreases. This means that mechanisms can be used to amplify (make bigger) forces so that very heavy objects can be lifted by one person, as in the example of a car jack, but the trade-off is that the person must move the input lever much further than the car gets lifted.

KEY TERMS

Input: The type of motion put in to a mechanism.

Output: The type of motion a mechanism produces.

Levers

The simplest example of a mechanism that controls and changes motion is a **lever**, which consists of a rigid bar that pivots on a **fulcrum**. The input force is often called the **effort** and the output force is called the **load**.

A car brake pedal is an example of a lever. The effort force applied by the driver's foot creates a load force, which is used to apply the car's brakes. The forces are represented by arrows that show the direction in which the forces act.

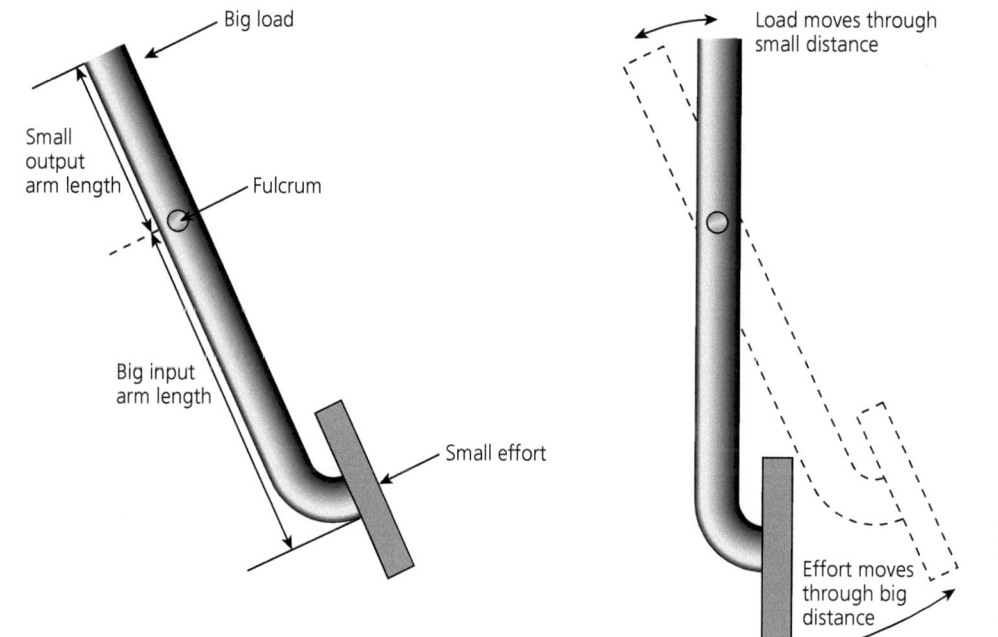

Figure 6.6 **A brake pedal is a lever.**

The diagram shows that, because of the position of the fulcrum, the effort moves through a larger distance than the load. Remember that the lever must obey the trade-off principle described above, so if the load moves through a smaller distance than the effort, then the load will be a larger force than the effort. This means that the brake pedal has effectively amplified the force applied by the driver's foot. This leads to the idea of 'leverage'.

The distance between the fulcrum and the force is called the **arm length**. The larger the arm length, the larger the distance the force must move, therefore the smaller the force.

ACTIVITY

Find five examples of levers from around your home. These could be complete products (for example, kitchen tongs, can opener, wheelbarrow, stapler or nail clippers), or parts within a product. Identify the fulcrum and the points where the effort and load are applied. Slightly harder products to analyse could include a door handle, spanner and tap (lever tap and rotary tap).

KEY TERMS

Arm length: The distance between the force being exerted and the fulcrum.

Effort: The input force.

Fulcrum: The pivot around which a lever turns.

Lever: A rigid bar that turns around a fulcrum.

Load: The output force.

KEY POINT

bigger arm length = bigger distance moved = smaller force

- So, for a lever to amplify a force, effort needs to be smaller than load.
- Therefore, input arm length must be greater than output arm length.

The arm length is in inverse proportion to the force. In other words, if the input arm is twice as long as the output arm, the effort will be half as big as the load.

Not all levers are used to amplify forces; in some applications, the load will intentionally be less than the effort.

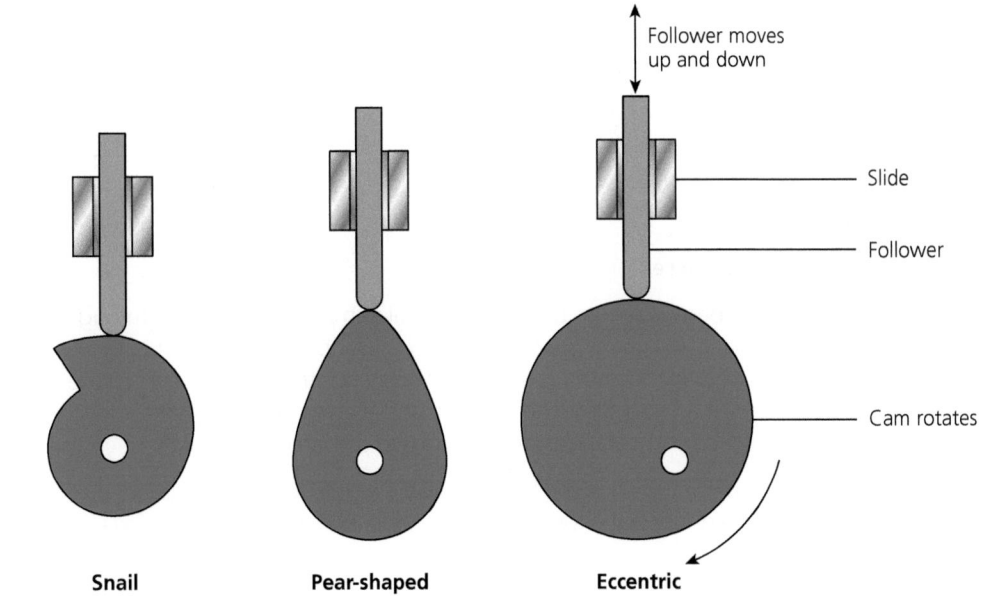

KEY TERMS

Cam and follower: A mechanism to convert rotary motion into reciprocating motion.

Pinion: A small driver gear (smaller than the driven gear).

Shaft: A rod that transfers the rotation through a mechanism.

Simple gear train: A pair of gears consisting of a driver gear and a driven gear.

Spur gear: A wheel with teeth around its edge.

Torque: A turning or twisting force.

How mechanical devices change the magnitude and direction of forces

Mechanisms that control and change rotary motion also have an input and an output. The basic mechanical trade-off principle still applies, but instead of force we refer to **torque**. Torque is a turning or twisting force. Rotary mechanical systems can either:

- reduce rotary speed but increase torque, or
- increase rotary speed but reduce torque.

Remember, if the output speed is faster than the input speed then the torque will be less; one can increase but only at the expense of the other.

Cams

A **cam and follower** is a mechanism to convert rotary motion into reciprocating motion. A cam is a specially shaped wheel and the follower rests on the edge of the cam. As the cam rotates, the follower moves up and down. Cams are used extensively in machinery and engines to produce a desired motion.

The profile (shape) of the cam determines the motion of the follower throughout one rotation cycle. With a little thought, you should be able to predict the type of reciprocating motion produced by a particular shape profile of cam, such as those shown in the diagram below.

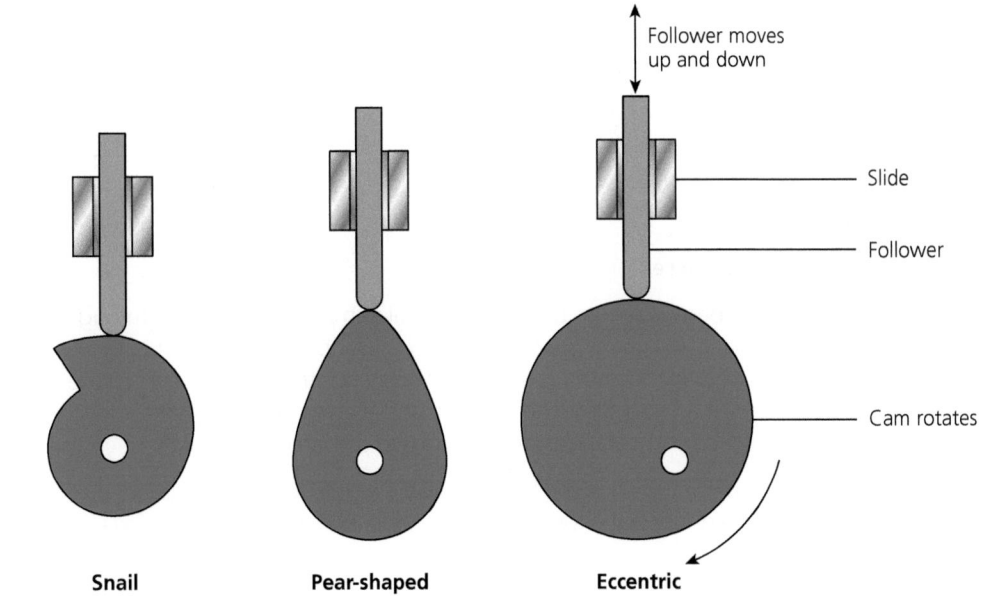

Follower moves up and down

Slide

Follower

Cam rotates

Snail   Pear-shaped   Eccentric

Figure 6.7 **Three types of cam**

Gears

A **spur gear** is a wheel with teeth around its edge. These teeth are designed to mesh with (link into) the teeth on another spur gear, which is usually of a different size – this pair of gears is called a **simple gear train** and consists of the driver gear (the input) and the driven gear (the output). If the driver gear is the smaller gear, it is sometimes called a **pinion**.

Gears are mounted on **shafts**, which carry the rotation to a different part of the mechanism.

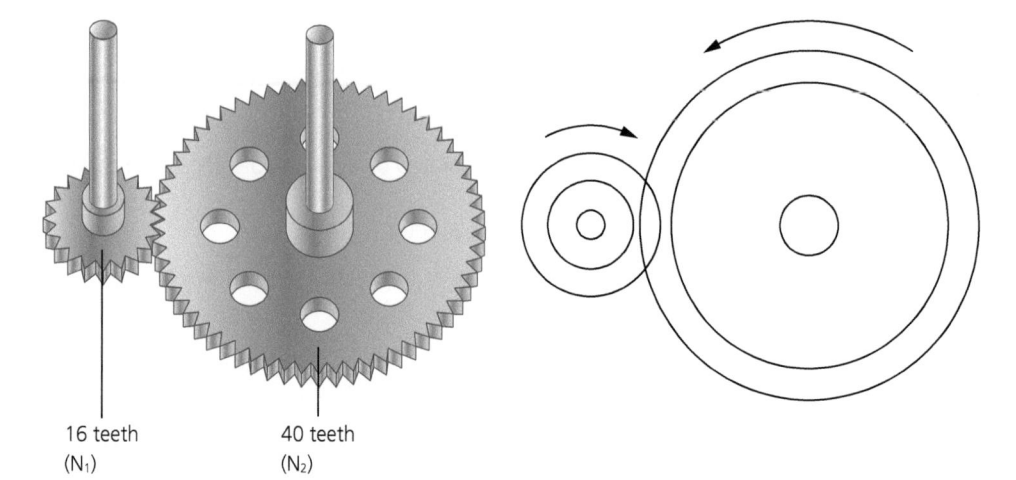

16 teeth
(N₁)

40 teeth
(N₂)

Figure 6.8 **A simple gear train**

If the two gears are different sizes they will rotate at different speeds. There are two simple rules to remember:

- The smaller gear will rotate faster than the larger gear.
- The gears will rotate in opposite directions.

A gearbox is a mechanical system that usually contains several simple gear trains working together to achieve a very large speed reduction. The photograph shows an electric motor combined with a gearbox, which reduces the speed of the motor by a ratio of 50:1.

Figure 6.9 **A motor with gearbox attached**

Gear systems usually require some form of **lubrication** to reduce friction and to prevent the teeth wearing away too quickly. Depending on the application, oil, grease, or a PTFE or graphite lubricant can be used.

KEY POINTS

- The smaller gear rotates quickly; the larger gear rotates slowly. Therefore, for a simple gear train to reduce the input rotational speed, the driven gear must be larger than the driver gear.
- The number of teeth on the gear is in inverse proportion to the speed it rotates. In other words, a gear with twice as many teeth will rotate at half the speed.
- Not all gear trains are used to reduce rotational speed; in some applications, the output speed will intentionally be higher than the input speed.

KEY TERM

Lubrication:
A substance applied to reduce friction between moving parts.

65

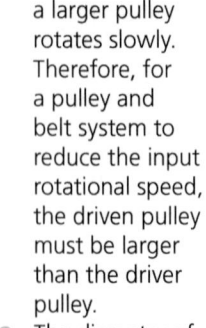

KEY POINTS

- A smaller pulley rotates quickly; a larger pulley rotates slowly. Therefore, for a pulley and belt system to reduce the input rotational speed, the driven pulley must be larger than the driver pulley.
- The diameter of the pulley is in inverse proportion to the speed it rotates. In other words, a pulley with twice the diameter will rotate at half the speed.

KEY TERMS

Pulley and belt drive: A method of transferring rotary motion between two shafts.

Linkage: A component used to direct forces and movement to where they are needed.

Pulleys and belts

Spur gears provide one method of transferring rotary motion between two shafts. An alternative is to use a **pulley and belt** drive, which behaves in a similar way to a simple gear train but with the following differences:

- The input shaft and the output shaft can be separated by a greater distance than can be achieved with spur gears.
- The input and output shafts rotate in the same direction.

A pulley and belt drive can provide speed reduction or increase in a similar way to the simple gear train.

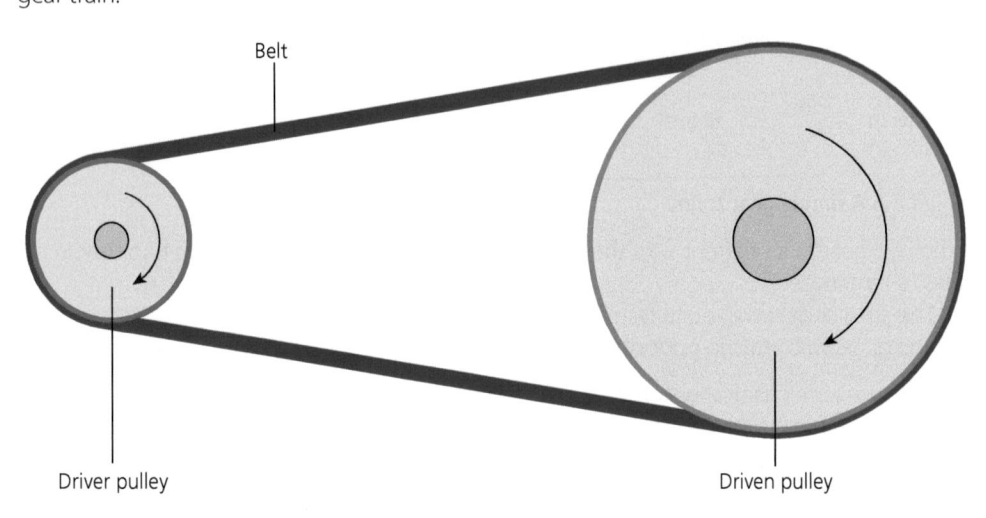

Figure 6.10 **Transferring rotary motion with pulleys and a belt**

Levers and linkages

A **linkage** is a component used to direct forces and movement to where they are needed. A linkage will often change the direction of motion, and it might also be used to convert between different types of motion. A simple linkage is often just a specially shaped lever. Some examples of linkages are shown in the diagram.

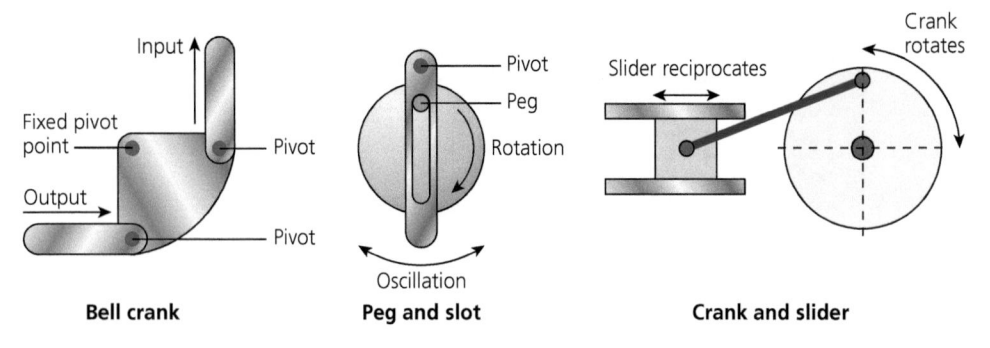

Figure 6.11 **Examples of linkages**

ACTIVITY

The best way to get a 'feel' for mechanical systems is to see them in operation. Completing at least one of the following activities will help you to explore mechanical systems further.

1. Try to find working examples of linkages and other mechanisms in the equipment you use regularly, e.g. on a bike, on garden tools or in the workshop equipment. Watch them in operation.
2. Visit an industrial museum to watch engines in operation is always worthwhile.
3. Find a free animation or video on the internet that shows linkages or a mechanism in operation. (This may allow you to observe the effect of changing dimensions of levers, sizes of gears and pulleys etc. There are numerous online videos of working mechanisms which you can study,
4. If your school has mechanical simulation software, build and run a virtual mechanical system.

6.2 Electronic systems

LEARNING OUTCOMES

By the end of this section you should know about and understand how simple electronic systems provide control functions, including:
→ switches and sensors, to respond to a variety of input signals
→ devices to produce a variety of outputs
→ how programmable components can be embedded into products in order to enhance and customise their operation.

Electronics permeate our lives in so many ways. It is hard to imagine going through a day without using an electronic device at some point: a smartphone, tablet, computer, TV or radio. The car we travel in is bursting with electronic systems to control the engine, to ensure our safety and to help with navigation. The heating and ventilation in the buildings where we work and live are monitored by electronics, not to mention security systems, CCTV, and so on.

The combination of electronic and mechanical systems creates powerful and useful design solutions.

KEY TERMS

Subsystem: A section with a specific role within a system.

System diagram: A diagram of the interconnections and flow of signals in an electronic system.

How electronic systems provide functionality in products

An electronic system can be broken down into smaller parts called **subsystems**, and these in turn can be classified into inputs, processes and outputs. A **system diagram** is drawn to show how the subsystems are interconnected.

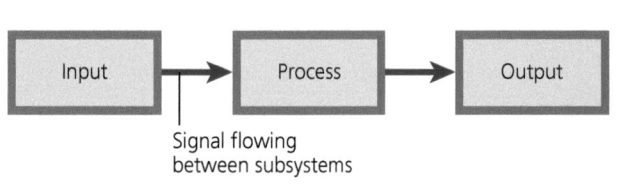

Figure 6.12 **A generic electronic system.**

Figure 6.13 **Push buttons are frequently used as inputs.**

Figure 6.14 **Identify the inputs and outputs on a microwave oven.**

Inputs consist of **sensors**, which are the 'eyes and ears' of an electronic system. They allow the system to monitor and measure a range of **physical quantities** such as light level, temperature, weight, etc. A push button is a very commonly used input component that senses when it has been pressed. A sensor produces an electrical **signal**, which is sent into the process subsystem.

The process subsystem will usually be a programmable **microcontroller**, and these are explained below.

The output components of an electronic system produce a physical output in response to signals from the process subsystem. Outputs can produce light, sound, motion, etc. A visual display screen is an output device increasingly found on modern products.

KEY TERMS

Embedding: Customising a microcontroller to be permanently placed within a product.

Microcontroller: A programmable electronic component that adds functionality to a product.

Physical quantity: Something that can be measured, e.g. light, temperature, speed.

Program: A set of instructions to tell a microcontroller how to carry out a task.

Sensor: A component that produces a signal in response to a specific physical quantity.

Signal: An electrical voltage that is used to represent information.

Embedding programmable components into products

Programmable microcontrollers are electronic components that can add incredible functionality to a product or a system. Microcontrollers take information from sensors and other inputs and then process this information in order to control a variety of outputs, which may include lights, sounds, motion or visual displays. Microcontrollers all start off blank, but the designer then writes a **program** that is downloaded into the microcontroller's memory. The program is a set of instructions that tells the microcontroller how to carry out a task that is specific to a product. This process of customising a generic microcontroller to a specific application and placing it permanently within the product is called **embedding**.

Microcontrollers are increasingly being used in products, and their small size, versatility and relative ease of use make them an attractive new technology. Even the most basic microcontroller is an extremely powerful miniature computer, capable of processing signals and performing calculations at astonishing speed. In many basic products the full potential of the microcontroller is never used. In a kettle, for example, the microcontroller simply needs to switch on the heating element until it senses that the water has reached boiling point, then to switch the heater off. Nonetheless, it is usually more cost effective to use a microcontroller for this simple task than to design a bespoke electronic system from scratch, and designers now realise that, once a microcontroller has been selected for a design, there is the potential to add advanced features that can enhance the product's functionality and increase the usefulness for the user.

Figure 6.15 **A smart fridge can sense what is stored inside it and re-order stock when it is needed**

For example, a kettle could have a 'keep warm' feature added, or it could heat the water to different temperatures the user has programmed or sound an alert when the water is ready. Some 'smart kettles' are now connected via WiFi to a network, allowing the user to control the kettle from their smartphone, and enabling the kettle to communicate with other appliances should it need to. These enhanced features are relatively straightforward to implement using microcontrollers, which is why microcontrollers are generally the first choice in modern electronic product design.

Several different kinds of microcontroller are available, each one offering benefits for different applications. Some typical microcontrollers that are suitable for GCSE projects are shown in the photograph.

Figure 6.16 A selection of microcontrollers typically used in GCSE projects

There is a growing interest in wearable technology such as intelligent watches, smart glasses, fitness/activity monitors, health monitors and jewellery. Such items are becoming highly desirable and electronics manufacturers are working with the fashion industry to create ergonomic and fashionable wearable technology.

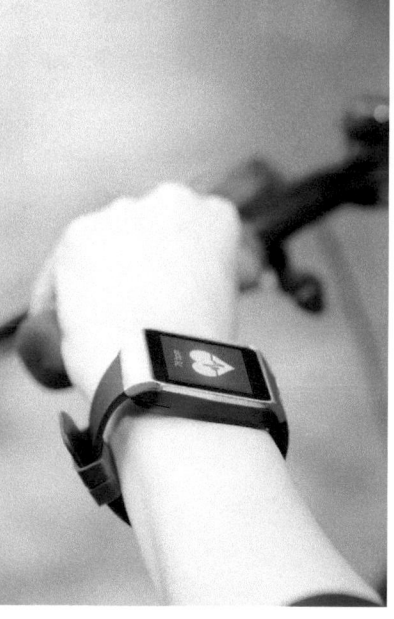
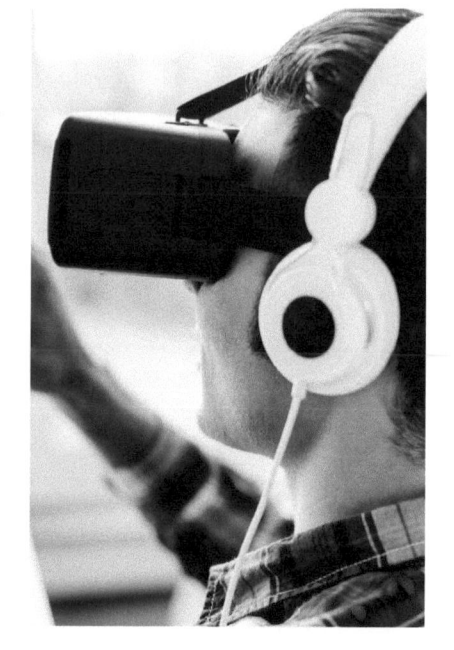

Figure 6.17 Examples of wearable technology

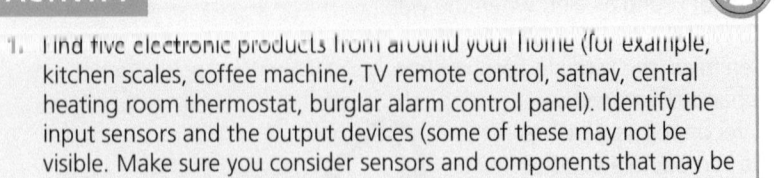

ACTIVITY

1. Find five electronic products from around your home (for example, kitchen scales, coffee machine, TV remote control, satnav, central heating room thermostat, burglar alarm control panel). Identify the input sensors and the output devices (some of these may not be visible. Make sure you consider sensors and components that may be hidden inside the product).
2. How do electronic kitchen scales sense weight.
3. Where does the 'beep' sound from a microwave come from?.

STRETCH AND CHALLENGE

Consider a relatively straightforward product found in the home, for example a tap, a vacuum cleaner or a doorbell. Write down ways in which a microcontroller could enhance the product's functionality so that the user has a better experience interacting with the product.

Input sensors, switches and control devices

Different microcontrollers use slightly different methods for connecting sensors and other inputs. Some microcontrollers have a dedicated set of input/output components that can be plugged together like building blocks. There are, however, some standard components that the various microcontroller systems will all use.

Light sensor

The light-sensing component in an electronic system will often be a light dependent resistor (LDR).

An LDR is connected into the microcontroller circuit so it produces a signal that rises when the light level increases. Inside the microcontroller, the signal will be converted into a number; the larger the number, the brighter the light detected. The microcontroller can then process this number or make decisions depending on how bright the light is.

Infrared sensor

Several different types of infrared (IR) sensors are available. Infrared is the invisible radiant energy given off by warm objects. There are two types of sensor that make use of this radiant energy:

- A simple IR sensor will produce a signal when it detects a warm object within range. This can be used, for example, to detect the presence of a person's hands so that an automatic tap or a hand dryer can be turned on.
- A passive infrared (PIR) sensor is used to detect a moving warm object, usually a person walking across a room. These sensors are used in burglar alarm systems, or in automatic lighting or heating systems that switch on when a person walks nearby.

Another type of IR sensor consists of an IR transmitter and receiver in a single package. These devices reflect an IR beam off an object to measure how far away the object is, a process known as range-finding. Such sensors are used in robotic systems to allow the robot to sense the presence and distance of nearby objects, without needing to touch them.

Figure 6.18 **An LDR and various IR sensors**

Switch sensors

A switch is a very simple electronic component that can be either off or on. When a switch is acting as an input for a microcontroller it will produce either an off signal or an on signal. Most people are familiar with simple push button switches, but a variety of other types of useful switches are also available:

- Tilt switches – these produce a signal when the switch is upright, and no signal when the switch is inverted. They sense being tilted, which is useful for detecting if a potentially dangerous product such as a room heater has fallen over. A slight variation on this is the vibration switch, which produces a brief signal when it is moved; this can be used in a security system to sound an alarm if a valuable object is tampered with.
- Push-to-make switches – these produce a signal when the push button is pressed. A **momentary** switch turns off when the button is released; a **latching** switch stays on until the button is pressed a second time.
- Time delay switches – there are two types:
  - The first type is similar to a latching push-to-make switch but it only stays on for a period of time before automatically switching off.
  - The other type is a timer switch, which can be programmed to switch on and off at various times throughout the 24-hour day.

KEY TERMS

Latching: A switch that stays on (or off) after the button is released.

Momentary: A switch that stays on only while it is held pressed.

Figure 6.19 **Different types of switch**

ACTIVITY

1. Some electronic products contain a light sensor. A smart phone, for example, senses the light level in the room and adjusts the brightness of its display to suit. A dusk-to-dawn security lamp senses the light level and switches on when it gets dark. Find a product which contains a light sensor and identify the exact position of the sensor. Photograph your findings.
2. Passive infra-red (PIR) sensors are frequently used in burglar alarm systems to detect a person moving across a room. Describe a product which uses a PIR sensor for an alternative purpose.
3. Photograph six examples of switches used in products around your home. Use words to describe the type of switch, e.g. 'momentary push switch' or 'latching rocker switch' etc.

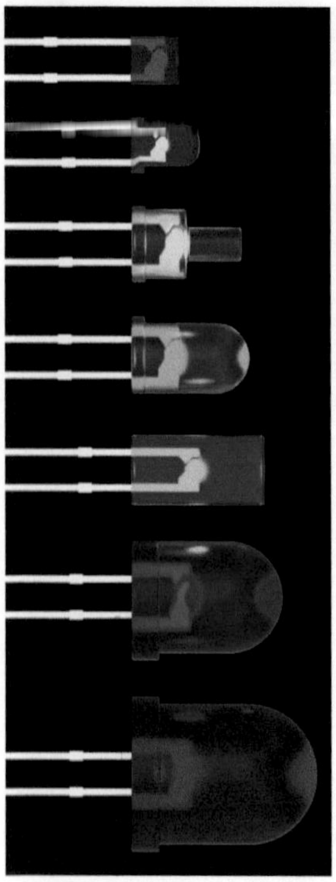

Figure 6.23 **A variety of LEDs**

Output devices

Light emitting diodes

Light emitting diode (LED) technology has progressed at an enormous rate and a very wide range of LEDs is now available. When specifying an LED for an application, a designer needs to consider:

- colour
- size (diameter)
- shape
- brightness
- normal or flashing.

The 'standard' size LED is 5mm diameter, and these are available in every visible colour as well as colours that are not visible, such as infrared and ultraviolet. Red-green-blue (RGB) LEDs can be made to light up any colour, including white.

Most LEDs are round but some other shapes are available, such as triangular or arrowhead. LED strips consist of several miniature LEDs on a flexible self-adhesive roll. LED strips are available as single-colour or RGB.

A standard brightness LED is suitable for use as an indicator on a control panel (e.g. to show that something is switched on). High brightness types are best for illumination applications or for attracting attention, though, e.g. as warning indicators. Some very high brightness LEDs can light up an entire room, and care needs to be taken not to look directly at these as they can damage the eyes.

For more information on how to use LEDs, refer to Section 10.2.

Speakers and buzzers

Microcontrollers can be programmed to produce musical notes and they can play tunes. Their ability to produce simple sounds is useful for creating alarm noises, and for attracting attention or for providing a simple 'click' feedback to indicate that a button has been pressed. Some microcontrollers can play back music files (such as MP3) or can record and play back voice files, thus allowing a product to speak information to the user, e.g. in a satnav. In order to convert the electronic signals into sound waves, a loudspeaker is needed.

A buzzer is not the same as a loudspeaker. A buzzer will produce a tone when it receives power, but that is all it can do; it cannot make other sounds and it cannot produce music or speech. Similarly, a loudspeaker will not emit a tone if it is simply connected to power; it can only reproduce the electrical sound waveform that it receives. So, if a designer wishes a microcontroller to emit sound from a loudspeaker then the microcontroller will need to be programmed to generate a suitable sound waveform.

Buzzers are used when it is only ever necessary to produce the same sound, such as in an alarm system. Some buzzers can be extremely loud and these are usually known as sirens.

Figure 6.20 **Loudspeaker, buzzer and siren**

Motors

An electric motor is a component that produces rotary motion when it receives power. Motors usually spin at a very high speed and they often have a gearbox attached to them to reduce the speed to a more useful and controllable level (see Section 3.1). Under the control of a microcontroller, a motor can be made to rotate forwards and in reverse, and its speed can be varied.

Electric motors are the most common component used to generate motion in products. Various mechanical systems may be used to convert the rotary motion into other types. Attaching an off-centred wheel directly to a motor shaft will cause the motor to produce a large amount of vibration when it spins, and this technique is used to create the 'rumble' in games console controllers, and the vibration in mobile phones when they are in 'silent mode'.

Figure 6.21 **Electric motors without and with a gearbox attached**

ACTIVITY

1. Find a product which uses different colour LEDs as indicators. Discuss how the choice of LED colour helps to convey information to the user.
2. Find a product which uses sound to catch the user's attention. Discuss how the sound has been designed to be attention-grabbing and able to be heard over background noise.

CHAPTER 7
New and emerging technologies

Designers and manufacturers have adapted and embraced new and emerging technologies in their designing and making activities. The past 40 years have seen a huge increase in the use of digital systems in design and technology, and new scientific technologies have led to advancements in materials that have changed the way in which products are produced and how they function.

7.1 The impact of new and emerging technologies on production techniques and systems

LEARNING OUTCOMES

By the end of this section you should know about and understand the benefits and implications of incorporating new and emerging technologies into production processes, including:
→ economies of scale
→ how disruptive technologies such as 3D printing and robotics are changing manufacturing.

Global communication systems have greatly increased the flow of digital information in text, visual and audio format. Large, detailed documents can be transferred instantly between sites, regardless of distance. The internet has enabled online research opportunities, and allows companies to make up-to-date checks on competitors and assess market trends. Marketing, advertising and sales opportunities have increased greatly as a result of the ability to communicate digitally on a global scale.

The traditional manufacturing industry is undergoing a digital transformation that has been accelerated by a range of technologies, including artificially intelligent robots, autonomous drones, sensors and 3D printing. Advanced technologies, for example **nanotechnology**, **cloud computing** and the **Internet of Things (IoT)** are changing the processes – reducing manufacturing production costs and improving manufacturing precision, speed, efficiency and flexibility.

Scientific developments mean that many new materials and composites with improved properties, such as insulation, conductivity and higher strength-to weight ratios, are now available to designers, such as the use of, self-healing polymers and even bio-materials that can be grown into products. .

While all of these new and emerging technologies have many benefits, there are also implications of incorporating them into our production processes. Advances in automation and robotics, for example, while improving speed and efficiency, may result in unemployment, with machines and software increasingly being used to perform jobs traditionally completed by humans. Increasing use of automation also has an impact on the environment and energy resources.

KEY TERMS

Cloud computing: A network of online servers that store and manage data.

Internet of Things (IoT): Where electronic devices connect within the existing internet infrastructure, to send and receive data without human intervention.

Nanotechnology: Technology on a microscopic scale.

Economies of scale

Economies of scale are the cost advantages that a manufacturer gains as a result of the size, output, or scale of their production. Traditionally, with larger-scale production the cost per unit of output tends to decrease because fixed costs such machinery are spread out over more units of output. Manufacture can often also become more efficient and create less waste with increasing scale, again leading to lower costs. Large companies benefit from bulk buying of materials and using standardised parts, and through lower-cost marketing and advertising.

Larger manufacturers, for example car manufacturers or white good manufacturers, also split up the different stages of the production process and have workers specialise in producing a certain part. For example, one worker might become highly specialised in the design of a car part; another in fitting a particular component. This requires less training and can result in more efficient production.

Many large companies produce components in large factories abroad might carry out final assembly closer to customers. Mass-producing component parts in Chinese or Malaysian plants can cut production costs. However, operating final assembly plants in places such as the United Kingdom or Germany might be preferable, as shipping the assembled product is then cheaper, subject to lower duty charges and some consumers prefer to buy from a company with manufacturing facilities in their own country.

New technologies mean that manufacturers can increasingly produce products and prototypes in smaller quantities but still benefiting from reduction in production costs normally associated with higher levels of production. For example, developments in computer-aided design, manufacture and engineering, technologies such as 3D printing and more flexible manufacturing systems mean that companies can now design, set-up and manufacture prototypes very quickly, allowing them to produce smaller batches as efficiently and cheaply as larger production runs. This means companies can produce bespoke products that meet the needs of individual clients much more easily and are able to react to fashions and trends. Producing products in smaller batches also reduces other costs such as the storage of final assembled products.

Disruptive technologies

This is how Harvard Business School professor Clayton M. Christensen coined the term 'disruptive technology' in his bestselling 1997 book:

> 'A disruptive technology is one that displaces an established technology and shakes up the industry or a ground-breaking product that creates a completely new industry.'

Disruptive technologies can allow smaller companies to take on large companies and create products that change the way we work and live, they often create new market demand and thinking. For example, the technology that created laptop computers, smart phones and tablets have allowed people to work from home and connect with others all over the world, they have changed the traditional office environment.

Some examples of disruptive technologies are provided below (although new developments are constantly emerging that will have an impact on manufacturing methods).

KEY TERM

Economies of scale: The cost advantages that manufacturers obtain due to the size, output or scale of their production.

Additive manufacturing

Additive manufacturing is the term used to describe technologies that translate 3D designs into solid, physical forms. Software is used to 'slice up' 3D designs into a series of layers, which are then sent in sequence to a rapid prototyping system. They are then built up layer by layer in a number of different ways:

● **Stereo lithography** – a laser traces the shape of a layer onto a bath of liquid resin which is cured, this process is repeated layer by layer until a final object is created.
● **Laser sintering** – this works in a similar way to stereo lithography. The laser traces the shape onto a powder, which could be polymer, metal or ceramic, this becomes solid and the process is repeated layer by layer until the object is completed.
● **3D printing** – 'prints' thermoplastic material (ABS or PLA are most common) layer by layer to build up a 3D shape. Complex shapes often require an additional support material to be printed to support the object while it sets. 3D printed products can be painted or electroplated for a higher quality finish.

3D printing technology has developed over the last 30 years. It started out as a rapid prototyping tool, but can now be used to print a wide range of materials, including polymers, paper, ceramics, metals, super alloys, wool and bio-materials, meaning that it is increasingly able to be used for final bespoke products. It is likely that further developments in additive manufacturing will continue to create many new applications, which will continue to have a significant impact on traditional manufacturing methods.

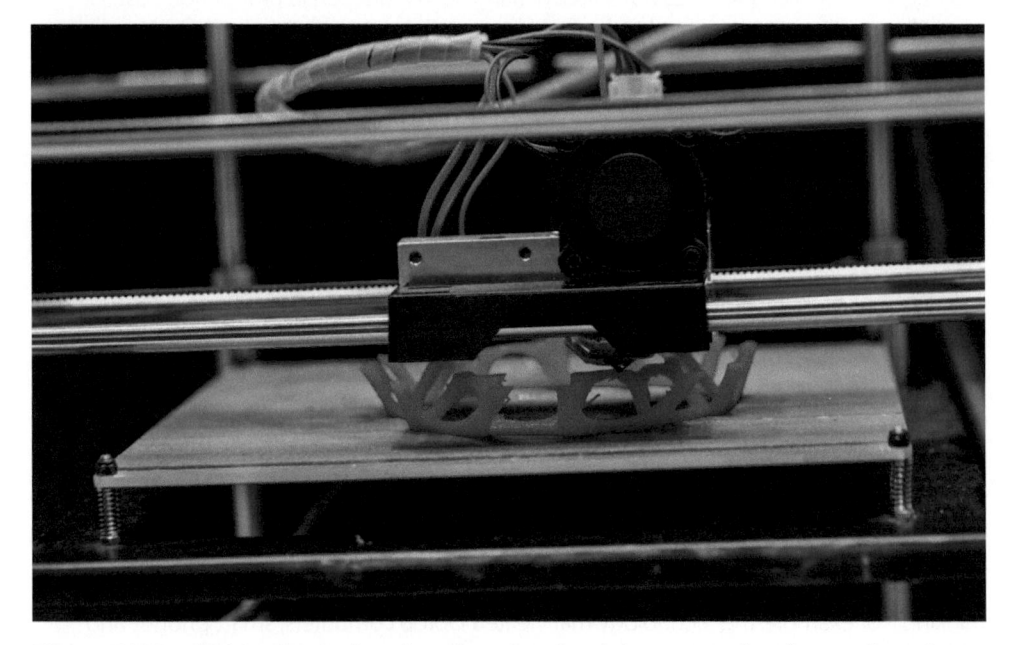

Figure 7.1 3D printing prototyping of products in polymer is commonplace in manufacturing.

Advanced robotics

Industrial robots are used extensively in automotive manufacturing, welding, material handling and painting. New and advanced robotics will increasingly be used in precision manufacturing and assembly, and in semiconductor fabrication.

The Internet of Things

The Internet of Things (IoT) allows electronic devices connected to each other to communicate with one another over the internet without human intervention. It enables connected devices to send and receive messages and notifications, for example it can be used to alert manufacturers to a defect in a product or production run. The application of IoT in manufacturing results in reduced downtime and fewer problems and defects with the final products.

Virtual reality

VR is being used to test new concepts in products allowing designers to assemble parts and design tooling in a virtual world before final manufacture. Ford and BAE are already adopting this process and its proving to cut costs and increase safety.

The development of emerging and disruptive technologies has had many benefits:
- Increased flexibility to produce complex products that are made from multiple materials and have multiple functions.
- Use of digital modelling and simulation tools to evaluate and plan manufacturing processes has reduced time to market and costs.
- Improved productivity and quality
- Sustainable manufacture as a result of improved use of material resources and reduced energy consumption.

Maker movement

Recent advances in digital design, engineering and simulation are already making a huge difference to manufacturing enterprises and small businesses. Affordable 3D printers, scanners and CAD software have made it possible for people to design and make high-tech products at home, enabling a maker movement. People are then able to share their ideas on the internet, as well as distributing their (often unique and customisable) items without involving 'middlemen', its lead to the creation of many maker communities.
- The website Etsy now has over one million sellers who have created products to be sold on the site.
- Other start-up companies like Kickstarter and Quirky allow makers to access resources and funding that can help turn an idea into a real life product.
- 'Makerspaces' such as Fab Lab in the UK are becoming more commonplace.
- Thousands of people attend Maker Faire each year, which takes place in many cities worldwide.

Figure 7.2 A Pebble smartwatch

A good example of the maker movement is the Pebble smart watch: it began as a prototype produced in a bedroom, and raised over $10 million on Kickstarter before it was launched commercially. The development of the Pebble smart watch led to increased interest and lead to research and developments in wearable technology by large manufacturers such as Samsung, Google and Apple.

Another example is when the British Atlas V rocket launched in 2015: astronaut Tim Peake uploaded code from small computers designed to enhance spacecraft sensors, satellite imaging, space measurements, data fusion and measurement of space radiation. These computers cost just a few dollars each and the code was written by school children in the UK as part of the Astro Pi project.

Further developments in technology are likely to see the maker movement develop even further in the years to come.

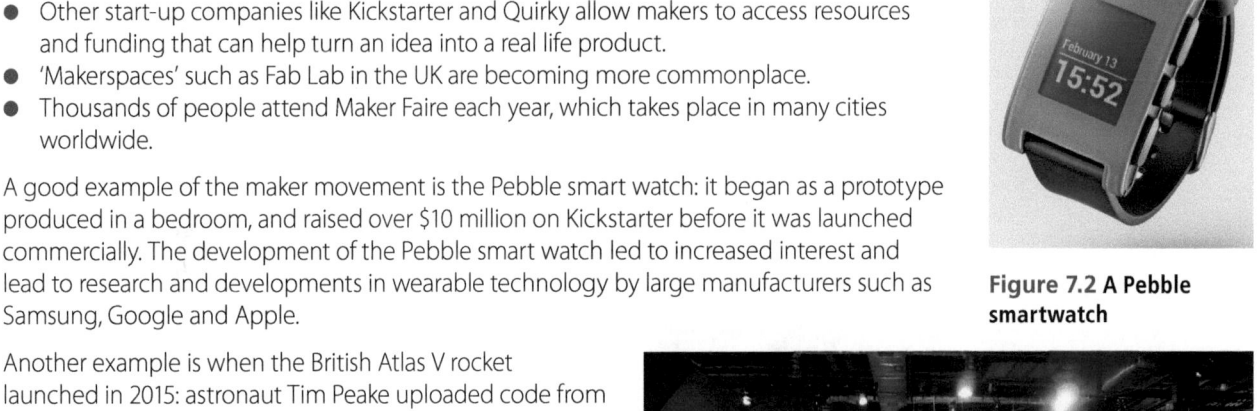

Figure 7.3 The maker movement Fab Lab space and Maker Faire UK

STRETCH AND CHALLENGE

Using two different materials from your material area, describe how rapid prototyping and disruptive technologies have been used in the development of products.

PRACTICE QUESTIONS: Core Technical Principles

1 Identify the primary users and wider stakeholders for the following products:
   - A children's tricycle
   - A bicycle alarm
   - A pair of running shoes

2 What issues would a designer need to consider when designing for a global market?

3 Discuss the impact of emerging technologies on product development.
   Use examples to support your discussion.

4 Describe how designers can use the Six Rs to design sustainable products.
   Use examples to support your discussion.

5 Give one example of a product that uses each of the following types of motion:
   - Linear
   - Reciprocating
   - Rotary
   - Oscillating

6 Give three benefits to a designer of using a microcontroller in electronic product design.

7 Identify and describe applications for:
   - Three types of input sensor
   - Three types of output device.

8 The figure below shows two pairs of sunglasses.

    a Explain one way that fashion and social trends have influenced and affected the design of sunglasses.

    b State two pieces of anthropometric data that would need to be considered when designing sunglasses.

    c Many products today are designed in a 'retro' style.
   Explain what is meant by the term 'retro'.

9 This figure shows a Digital Video Disc (DVD).

DVDs may soon become obsolete.

a Explain what is meant by the term 'obsolescence' and why products become obsolete.

b What impact does obsolescence have on the environment?

c This figure shows the lifecycle of a DVD. Fill in the missing stages to complete the diagram.

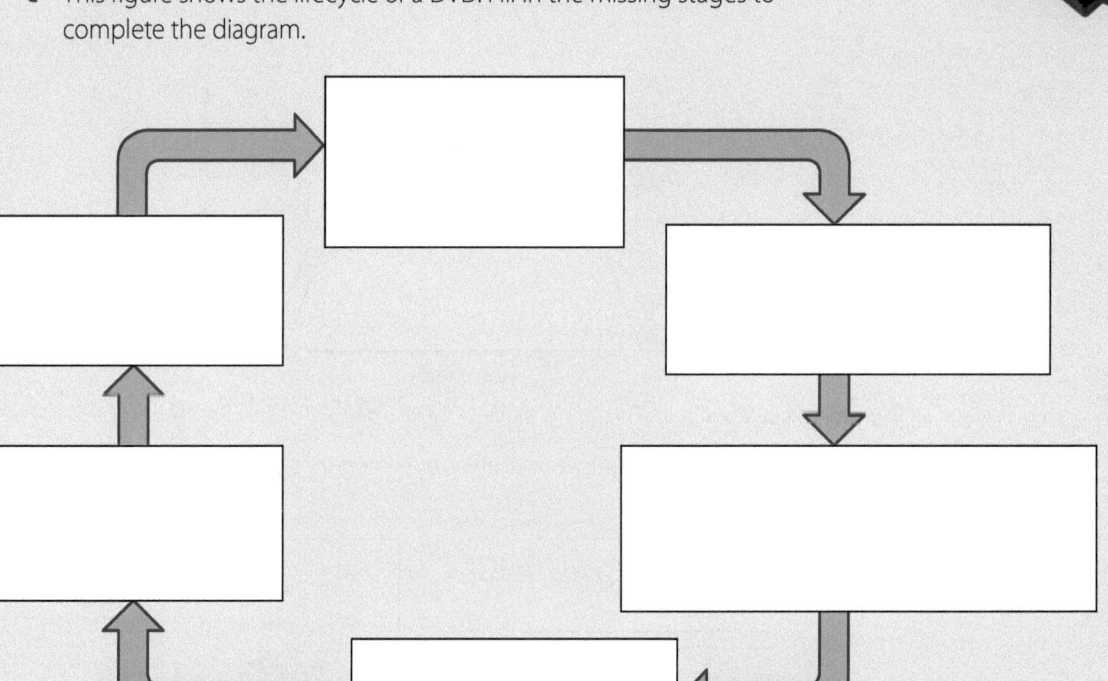

10 This figure shows a sandwich board, used to advertise outside shops and cafés.

The side legs are made from a hardwood timber.

a Name a hardwood timber that could be used for the side leg

b The main board is made from a man made from a manufactured board.
Name a manufactured board that could be used for the main board.

c Give two properties of manufactured boards that make them suitable for the main board.

d This figure shows an isosceles triangle that is formed from the floor when the board is fully opened.

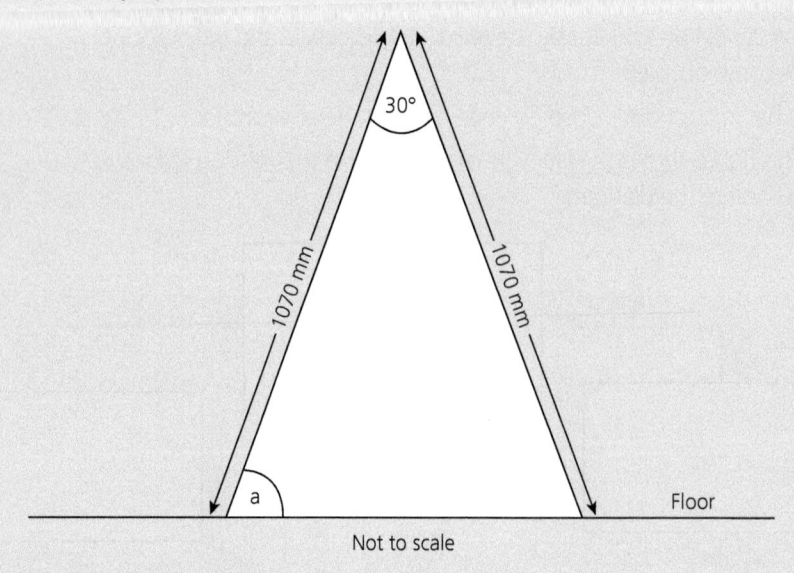

Calculate angle a.

11 a Match the fibres listed below to the correct category by adding lines to the diagram.

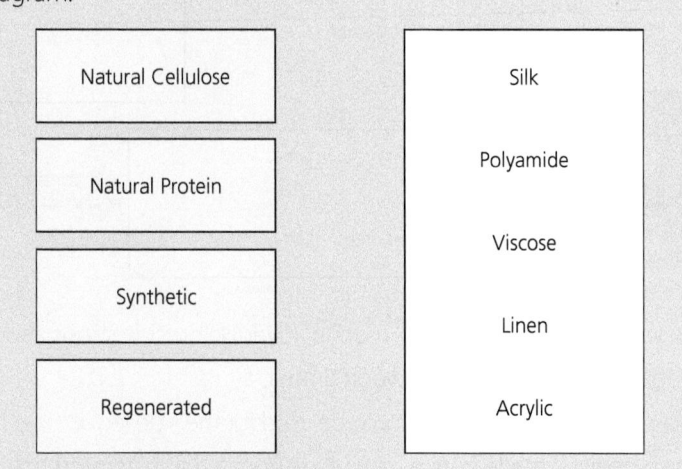

b Name the two types of fibre structure.

12 This figure shows a cardboard box that would be used for packaging.

a Calculate the **volume** of the box.

b Calculate the surface area of the box.

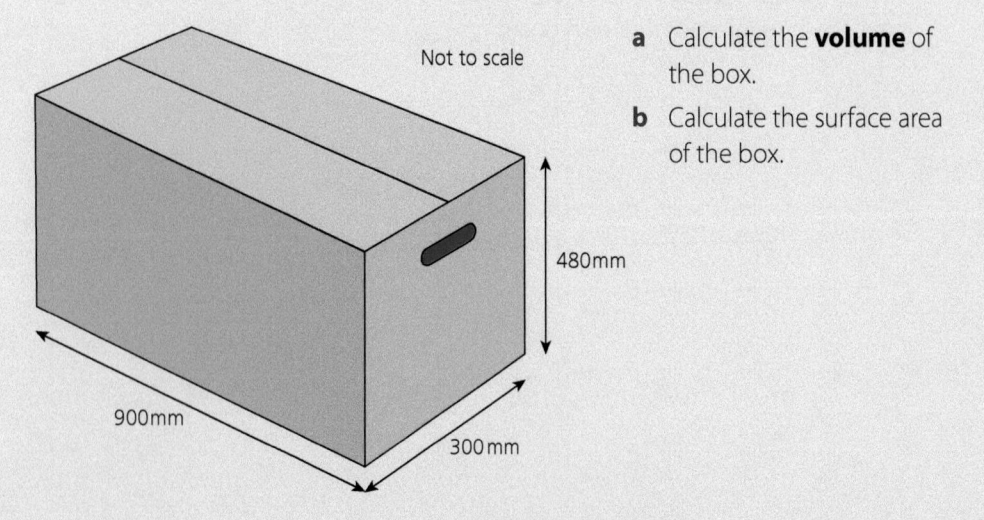

In-depth principles of design and technology

This section looks at the in-depth principles of design and technology that you must know about and understand in order to make informed decisions as a designer.

All students need knowledge and understanding of the principles considered in this section in relation to at least one material category or system. You should refer to the chapter(s) that is/are relevant to your chosen material(s).

The section explores the following questions outlined in the OCR specification:

5.1 What are the main categories of materials available to designers when developing design solutions?
5.2 What factors are important to consider when selecting materials and/or system components when designing?
5.3 Why is it important to understand the sources or origins of materials and/or system components?
5.4 Why is it important to know the different available forms of specific materials and/or systems components?
6.1 What gives a product structural integrity?
6.2 How can materials and products be finished for different purposes?
7.1 How can materials and processes be used to make iterative models?
7.2 How can materials be manipulated and joined in different ways in a workshop environment when making final prototypes?
7.3 How do designers and manufacturers ensure accuracy when making prototypes and products?
7.4 How do industry professionals use digital design tools when exploring and developing design ideas?
7.5 How do processes vary when manufacturing products to different scales of production?
8.1 How can cost and availability of specific materials and/or system components affect their selection when designing?

These questions are considered in the following chapters:

CHAPTER 8
Papers and boards

You learnt about the different types of papers and boards in Section 5.1. This chapter includes more in-depth information on paper and boards.

LEARNING OUTCOMES

By the end of this section you should know about and understand the physical and working properties of a range of papers and boards, including:
→ how easy they are to work with
→ how well they fulfil the required functions of products in different contexts.

8.1 Physical and working properties

Paper

Paper is a thin but extremely versatile material that is easy to work with. It is available in many different types, sizes, finishes and weights.

Virgin fibre paper

Paper is made from wood fibres called **cellulose**. Chemicals are used to break down or pulp the fibres by removing the lignin, which is the natural 'glue' that holds them together. Further chemicals can be added that bleach or colour the paper and give it a special texture. The pulp is then sprayed out in thin layers on to fine mesh, and compressed and dried out by running it through heated rollers. Paper made from 'new' wood fibres is called **virgin fibre paper**.

Recycled paper

Because of the environmental impact of making virgin fibre paper it is now becoming more expensive, and consequently **recycled paper** is now extremely common. Recycled paper is made by soaking and mixing waste paper in water to separate the fibres back into pulp. The pulp is then refined to remove any unwanted contaminants such as staples, plastic, etc., that may have been mixed in with the waste paper, then sprayed and compressed in the same way as virgin fibre paper.

Paper types

Layout paper

Layout paper is usually sold in A4- and A3-size pads for sketching and developing design ideas. It has a smooth surface for both pencil and pen work. The weight of layout paper is around 50gsm, which makes it thin enough to use to trace and copy parts of designs when developing ideas. Because of its light weight it has a low relative cost.

Copier paper

Copier paper is usually sold in A3- and A4-sized 'reams' – a ream is a pack of 500 sheets. It has a weight of approximately 80gsm and a smooth surface, making it ideal for most printers as well as for photocopying.

KEY TERMS

Cellulose: Wood fibres; an organic compound, structuraly important in plant life.

Recycled paper: Paper made from used paper products.

Virgin fibre paper: Paper made from 'new', unused wood fibres.

Cartridge paper

Cartridge paper is available in different weights between 80 and 140gsm, and is thicker and more expensive than layout or copier paper. It has a slightly textured surface and is creamier in colour. Cartridge paper is used for sketching, drawing and painting as it is an ideal surface for pencil, crayon, pastels, watercolour paints, inks and gouache. Because of this it is often used by artists.

Bleedproof paper

Bleedproof paper is available in similar thicknesses to cartridge paper but is smooth and bleached bright white. Bleedproof paper is ideal for drawing and sketching using marker pens as it stops the colours of the marker pens bleeding into the paper and mixing together or blurring the edges of lines.

Sugar paper

Sugar paper is usually sold in larger sheet sizes from A2 upwards (see Section 5.3), and comes in a wide variety of different colours. It is usually made from recycled paper and is around 100gsm in weight. It is relatively inexpensive and is commonly used for the mounting and displaying of work on noticeboards and walls rather than for drawing directly on to.

Table 8.1 **The properties, characteristics and uses of common paper types**

Common name	Weight (gsm)	Properties/working characteristics	Uses
Layout paper	50	Bright white, smooth, lightweight (thin) so slightly transparent and inexpensive	Sketching and developing design ideas; tracing parts of designs
Copier paper	80	Bright white, smooth, medium weight, widely available	Printing and photocopying
Cartridge paper	80–140	Textured surface with creamy colour	Drawing with pencil, crayon, pastels, watercolour paints, inks and gouache
Bleedproof paper	80–140	Bright white, smooth surface, stops marker 'bleed'	Drawing with marker pens
Sugar paper	100	Available in wide range of colours, inexpensive, rough surface	Mounting and display work

Boards

Card

Card is slightly thicker than paper and around 180 to 300gsm in weight. Like paper it is available in a wide range of colours, sizes and finishes, including metallic and holographic finishes. Thin card is easy to fold, cut and print on to, which makes it ideal for greetings cards, paperback book covers and so on, as well as for simple modelling.

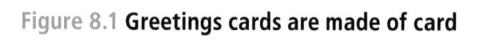

Figure 8.1 **Greetings cards are made of card**

Cardboard

Cardboard is available in many different sizes and surface finishes, with thickness from around 300 microns upwards. Cardboard is widely used for the packaging of many different products – for example it is used for cereal boxes, tissue boxes, sandwich packets and so on – because it is relatively inexpensive and can be cut, folded and printed on to easily. Cardboard can be used to model design ideas and is often used to make templates for parts and pieces of products, which once correct can then be made from metal or other materials.

Corrugated cardboard

Corrugated cardboard is a strong but **lightweight** type of card that is made from two layers of card with another, fluted sheet in between. It is available in thicknesses ranging from 3mm (3000 microns) upwards. The fluted construction makes it very stiff and difficult to bend or fold, especially when folding across the flutes. Because of the spaces between the two layers created by the fluted sheet, corrugated card can absorb knocks and bumps. This makes it ideal for packaging fragile or delicate items that need protection during transportation. It is also widely used as packaging for takeaway foods, such as pizza boxes, as the fluted construction gives it good heat-**insulating** properties compared to normal cardboard.

Double wall corrugated card is also available, which is twice as thick as corrugated card and gives extra strength and damage resistance.

Board sheets

Mounting board is a **rigid** type of card with a thickness of around 1.4mm (1400 microns) and a smooth surface. It is available in different colours, but white and black are the most commonly available and most used colours. Mounting board is often used for picture framing mounts and architectural modelling.

Figure 8.2 Cardboard packaging

Figure 8.3 Corrugated cardboard packaging

> **KEY TERMS**
>
> **Lightweight:** Weighs very little.
> **Rigid:** Difficult to bend.

Table 8.2 **The properties, characteristics and uses of common board types**

Common name	Thickness (microns)	Properties/working characteristics	Uses
Card	180–300	Available in a wide range of colours, sizes and finishes; easy to fold, cut and print on to	Greetings cards, paperback book covers, etc., as well as simple modelling
Cardboard	300 upwards	Available in a wide range of sizes and finishes; easy to fold, cut and print on to	General retail packaging such as food, toys, etc.; design modelling
Corrugated cardboard	3000 upwards	Lightweight yet strong; difficult to fold; good heat insulator	Pizza boxes, shoeboxes, larger product packaging, e.g. electrical goods
Mounting board	1400	Smooth; rigid; good fade-resistance	Borders and mounts for picture frames

Laminated layers

Foam board

Foam board is a lightweight board that is made up of polystyrene foam sandwiched between two pieces of thin card or paper. It has a smooth surface and is available in a range of colours, sheet sizes and thicknesses, with 5mm (5000 microns) being the most common. Foam board is a very lightweight but rigid material and is ideal for modelling and point-of-sale displays. It is easy to cut and can easily be folded with the correct technique.

Styrofoam

Styrofoam is a trade name for a type of expanded polystyrene foam. It is available in a wide range of sizes and thicknesses, and can be identified by its blue colour. Styrofoam has a structure of uniformly small, closed cells that make it easy to cut, shape and sand to a smooth finish. Styrofoam is strong and lightweight, as well as being water-resistant and having good heat-insulation properties. It is used as a wall insulation material in caravans, boats and lorries, but it is also ideal for creating three-dimensional models and moulds for vacuum formed or glass fibre products.

Corriflute

Corriflute is an extruded corrugated polymer sheet similar in structure and thickness to corrugated cardboard. It is made from a high-impact polypropylene resin and is available in a wide range of colours and sheet sizes. Corriflute is rigid and lightweight as well as being extremely waterproof. It is easy to cut but can be difficult to fold, especially across the flutes. Corriflute is often used for outside signs such as estate agents' signs on houses, and signs outside car forecourts or shops. It is also used for containers, packaging and point-of-sale displays and for modelling purposes.

PVC foam sheet

PVC foam sheet is a lightweight material and has good insulation properties, similar to Styrofoam, but is more robust and can be printed on to easily. PVC foam sheet is available in a range of different colours and is easy to cut and join to other materials. It is resistant to water and many chemicals so is often used for outdoor displays and signs.

Table 8.3 **The properties, characteristics and uses of common laminated layers**

Common name	Properties/working characteristics	Uses
Foam board	Smooth, rigid, very lightweight and easy to cut	Point-of-sale displays, ceiling-hung signs in supermarkets, architectural modelling
Styrofoam	Light blue colour, easy to cut, sand and shape, water-resistant, good heat and sound insulator	3D moulds for vacuum forming and GRP, wall insulation in caravans, boats, etc.
Corriflute	Wide range of colours, waterproof, easy to cut, rigid, lightweight	Outdoor signs, packaging and modelling
PVC foam sheet	Lightweight, good insulation properties, easy to print on, water-resistant	Outside displays and signs

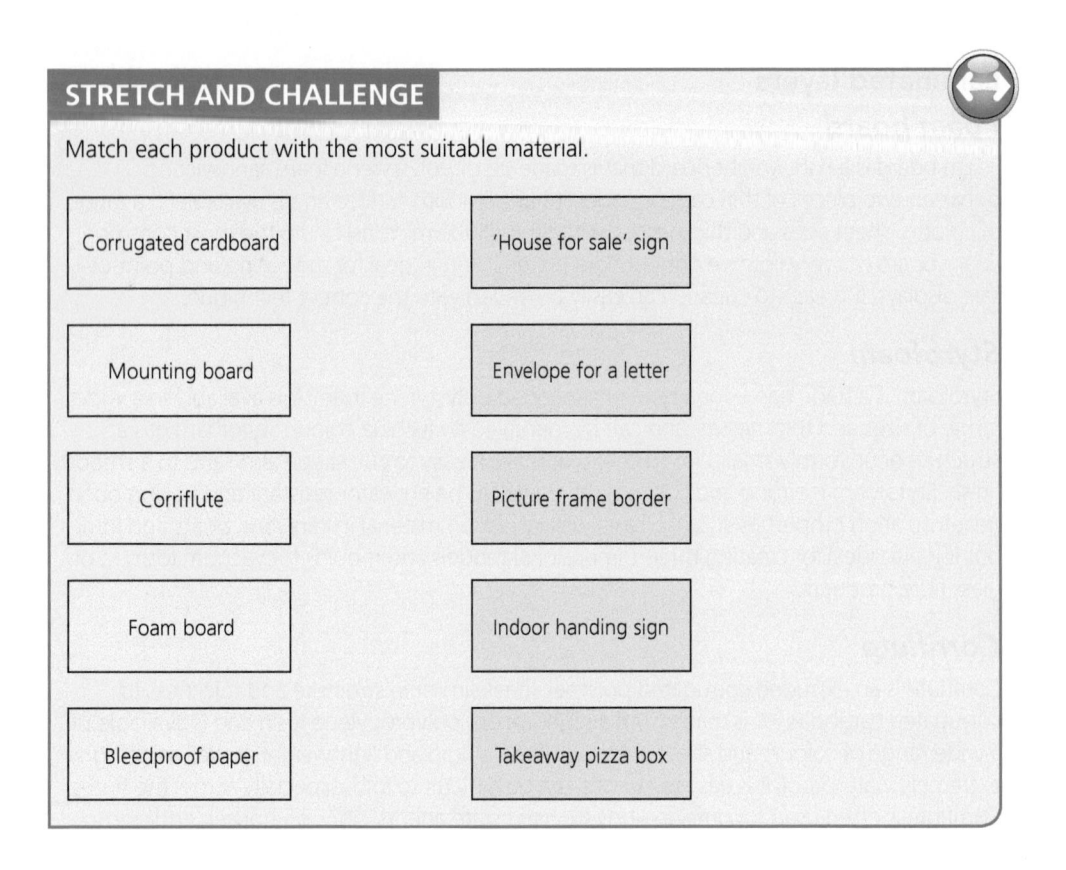

Match each product with the most suitable material.

Corrugated cardboard	'House for sale' sign
Mounting board	Envelope for a letter
Corriflute	Picture frame border
Foam board	Indoor handing sign
Bleedproof paper	Takeaway pizza box

8.2 Sources and origins

LEARNING OUTCOMES

By the end of this section you should know about and understand:
→ the sources and origins of paper, boards and laminated boards
→ the processes used to extract paper and board into a useable form
→ the ecological, social and ethical issues associated with processing paper and board
→ the lifecycle of papers and boards
→ recycling, reuse and disposal of paper and board.

Sources

Paper was first invented in China around AD100. The first paper was made from fibres of tree bark mixed in water. This mixture, known as pulp, was then drained and spread out on a bamboo-framed matting before being pressed down into a thin layer and dried out in the sun.

It was later discovered that lignin, which is the natural glue that holds the wood's fibres together, could be broken down more easily if plants with long cellulose fibres were used. This meant the fibres could be made into a finer pulp, which in turn made better quality paper.

The paper-making process

The methods used to make paper today are essentially the same or very similar to the original method, but mechanical methods are now used to separate the wood fibres.

The mechanical pulping process

Using large tanks known as pulpers, the raw wood chippings are soaked in up to a hundred times their weight of water and then pulverised with large steel rotor blades. The finished pulp is then pumped into a paper machine, where it is sprayed on to large sheets of thin mesh and pressed through a series of rollers into paper.

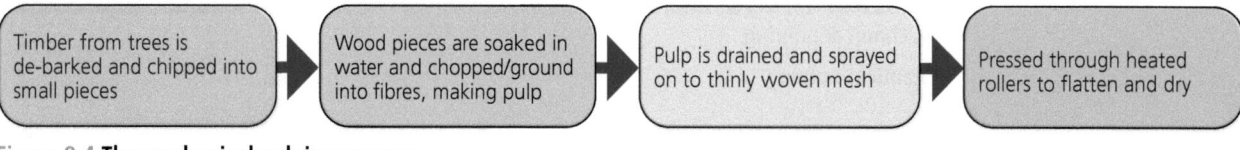

Figure 8.4 The mechanical pulping process

Mechanically pulped paper is suitable for paper products that use 'bulk' grades of paper such as newspaper and toilet tissue. Because of the damage done to the fibres in the grinding process, however, it has a low strength. It will tear relatively easily and disintegrate quickly when wet. The mechanical process also means that a large amount of lignin remains in the pulp mixture, which can lead to the paper yellowing over time or when exposed to bright light.

The chemical pulping process

The chemical pulping process uses chemicals such as caustic soda and sodium sulphate to help break down the wood pieces and chemically remove the lignin, rather than removing it by mechanical grinding. This separates the fibres without damaging them and creates a much stronger pulp. Other fibres such as cotton and linen are then added to the pulp mixture to improve or change the texture, and this is called blending. Chemicals such as bleaching agents, dyes and fillers are then also added to give the paper a specific colour or property. The pulp is then dried, pressed and formed into rolls of paper in the same way as before. Chemically pulped paper is generally of a much higher quality than mechanically pulped paper, and can be bleached or coloured to a much brighter finish. Specialist coatings can also be added to the paper once it has been formed into sheets, such as glossy or shiny finishes (see Section 5.5).

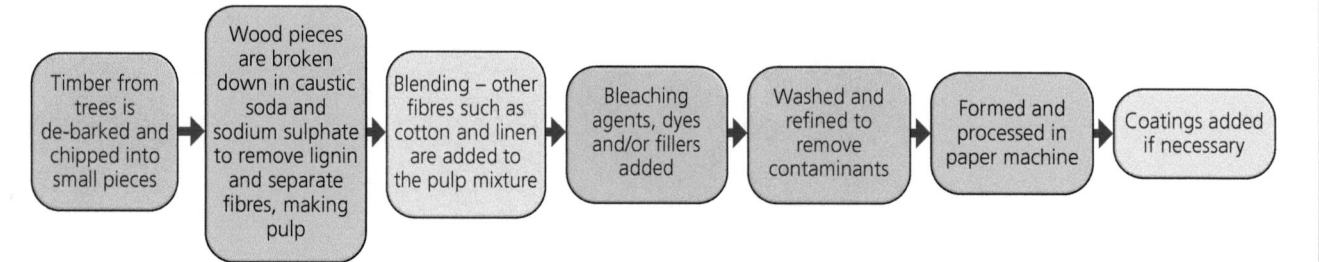

Figure 8.5 The chemical pulping process

Recycled paper

Because the main ingredient of paper is wood, it requires a large number of trees to make it. It takes around twelve mature trees to make one tonne of newspaper, and around twenty-four trees to make one tonne of copier paper. Recycled paper was created to reduce the number of trees needed to make paper and therefore to lessen the environmental impact of paper production. Using recycled paper instead of virgin wood means that fewer trees are needed to make the same amount of paper.

Paper cannot be recycled indefinitely as the fibres get shorter and weaker every time they are recycled. After around five or six times, the fibres usually become too short and weak to be of use and will not pulp adequately. To maintain the quality and strength of recycled paper, many manufacturers use a mixture of recycled paper and new virgin wood chippings to make the pulp. The ratio is usually between 55 and 80 per cent recycled paper and between 20 and 45 per cent virgin wood chippings, depending on the type of pulping process used. This cuts down the number of trees needed by about 60 to 80 per cent.

Table 8.4 **The percentages of recycled and virgin paper in mechanical and chemical pulping**

Pulping process	Percentage recycled paper	Percentage virgin
Mechanical pulping	55%	45%
Chemical pulping	77%	23%

Card

Cardboard is made in a number of different ways. One method is by sandwiching and pasting multiple layers of paper together to make a thicker paper. Another method is to press the layers of wet pulp used for making paper together into a thicker layer. By layering sheets of paper or pulp together, card of various thicknesses can be produced.

Corrugated card

Corrugated card was first produced in the 1850s. It was originally used as packaging for products to be shipped, to protect them from damage. Due to the thicker and more cushioned construction of the cardboard, it quickly replaced normal thick card. Its rigid yet lightweight properties also made it ideal for cardboard boxes. Corrugated cardboard is made by passing paper through a corrugation machine, which has three layers. The centre layer is treated with high pressure steam to heat-soften the fibres. The paper is pressed into the required thickness and shape by crimping it to give it a wavy shape. The two outer layers of paper are then glued on each side of the wavy centre layer. After the corrugator has heated, glued and pressed the paper together to form the corrugated cardboard, it is cut into large pieces or 'blanks', which then go to other machines for printing, cutting and gluing together.

Double-wall corrugated card is made in the same way but has an additional wavy and flat layer added to make it even more rigid and give extra protection.

Figure 8.6 **A typical corrugator is around 90 metres long – roughly the length of a football field**

Foam board and Styrofoam

Foam board and Styrofoam are slightly different types of expanded polystyrene foam. Foam board uses foam sandwiched between two outer layers of paper or thin card. Expanded polystyrene foam (Styrofoam) is made from small pellets of polymer (around 1mm in diameter) that are heated to around 200°C, allowing the gas in the polymer to escape and air to enter. The air entering the polymer expands the polymer to up to forty times its original size. The enlarged balls are then fused together using heat to create large blocks that can be cut into sheets or other required shapes.

For more information on the sources and origins of foam board and polystyrene foam, see Section 8.6.

Corriflute

Corriflute is a tradename for corrugated polypropylene. It is similar to corrugated card in appearance but is manufactured in one piece by extrusion, instead of being constructed in layers. To make Corriflute, the molten polypropylene is drawn through a former that moulds the polymer into the corrugated shape required.

Metal former

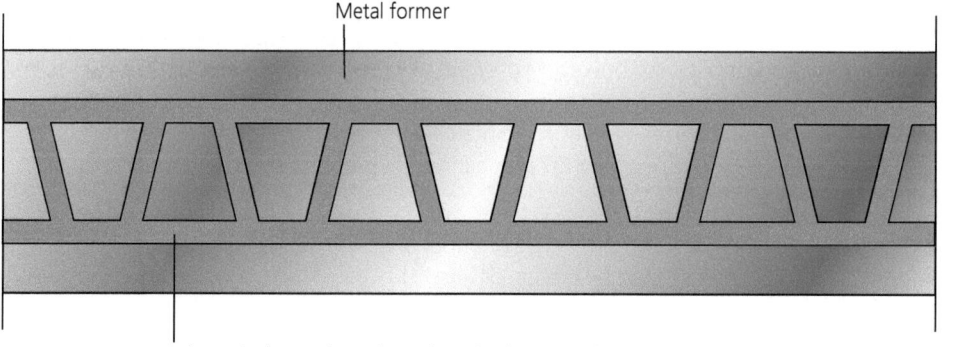

Polymer is drawn through voids in the former to form the required shapes

Figure 8.7 **Section through Corriflute**

For more information on polypropylene, see Section 8.1.

Ecological, social and ethical issues

The two main materials used in the manufacture of paper and card are wood pulp and recycled paper. The use of trees for wood pulp requires the harvest of many hectares of forestland. Most trees, however, come from forests managed by the Forest Stewardship Council (FSC) or the paper manufacturer themselves. This ensures that one or more trees is planted for every one felled, and ensures a continuing supply of trees for the supply of wood.

Paper manufacturing can be a source of both air and water pollution. Modern paper production, however, produces significantly less pollution than was previously the case. The use of recycled paper reduces the demand for fresh pulp, which in turn lowers the amount of energy needed to produce the pulp, along with the amount of air and water pollution produced during manufacture. Recycled paper produces 73 per cent less air pollution than paper made from raw materials, but bleaching agents, dyes and additives are still used in the manufacturing process. Hydrogen peroxide and sodium hydrosulphite, which are the most common bleaching agents, can be harmful to the environment. Lignin, however, which is a waste product extracted during the pulping process, can be burnt as a fuel oil substitute in some manufacturing plants, and the burning of bark and other residues can often supply power or steam to local firms or energy suppliers.

A large quantity of water is also used in the paper-making process, but this can be reused or returned to the water source.

It is estimated that every adult male in the UK generates around one tonne of waste each year. Around 40 per cent of this is waste paper.

The ecological, social and ethical issues associated with processing the Polymer elements in laminated layers such as foam board, Styrofoam, PVC foam and Corriflute are covered in Section 2.1.

Lifecycle

Paper and cardboard do not stay in the environment for a long time after they are discarded. Because of their fibrous construction and susceptibility to moisture they decompose quickly compared to all other materials. Paper products take only two to four weeks to decompose, and cardboard around two months. Laminated layers, however, take much longer to decompose as they contain polymers, which can resist chemical attacks. Styrofoam can take around fifty years to decompose and PVC can take up to 500 years. Also, because paper and cardboard are made almost entirely of natural materials (wood) they do not release any harmful chemicals during decomposition. Meanwhile, laminated layers that contain Styrofoam or PVC can release toxic chemicals.

Today, the majority of paper and cardboard products we use are recycled. Currently around 70 per cent of paper worldwide is recycled back into new paper. Paper fibres can be recycled only a certain number of times, however, so eventually they reach the end of their useful life and then are often burned as a fuel source instead of being left to decompose. Although paper is by far the most commonly recycled material and accounts for more than half of all recycled materials, some papers and cards cannot be recycled and these do end up in landfill.

Food packaging is the most common type of paper and card that is not recycled. This is because it is often contaminated by grease, oil and other liquids that are absorbed into the fibres of the paper. When this occurs, the paper cannot be recycled as the fibres cannot be separated from the grease and oils, which get into the pulp mixture during the recycling process. Later in the process, the oil separates and causes problems with bleaching and washing the paper. When the water is squeezed out of the pulp as the sheets of paper go through the paper machine, the oil residue creates dark marks and holes that make the paper very low quality or in some cases unsuitable for any use.

Figure 8.8 **Pizza boxes are a classic example of how grease can contaminate paper and card. Oil from the cheese and grease from the meat run down into the cardboard meaning it cannot be recycled. In the UK, approximately one million pizzas are delivered in corrugated cardboard boxes every day.**

Figure 8.9 **Composting bins can be bought for a few pounds at most garden centres, hardware shops or DIY outlets.**

In cases where paper or card cannot be recycled, they do not need to go into landfill. An alternative is for them to be composted using a composting bin. Food-soaked paper and cardboard can be composted with no ill effects. Composting is a natural process that not only reduces the amount of paper put into landfill but also benefits the environment by creating new soil that is rich in minerals.

Composting occurs when two different types of ingredients – called 'greens' and 'browns' – are mixed together.
- Brown ingredients are items such as cardboard, paper, small twigs, eggshells, items made from wood, etc.
- Green ingredients are things such as grass cuttings, vegetable peelings, leaves, old flowers or plants, teabags, etc.

Green ingredients rot quickly, releasing important nitrogen and providing moisture. Brown ingredients rot more slowly, adding fibre and carbon to the mixture and allowing air pockets to form, which are necessary for composting to occur.

Laminated boards

Laminated boards such as Styrofoam, PVC foam and Corriflute are much harder to recycle because they are made of polymers that need to be sorted, cleaned and chipped before being melted down into useable granules. The recycling process can also use lots of energy, which does further environmental damage. Foam board has two outers of paper that need to be separated from the foam core before it can be recycled, which can be a difficult and time-consuming process. For this reason foam board usually ends up in landfill.

For more information on the lifecycle and recycling of polymers, see Section 8.2.

STRETCH AND CHALLENGE

Use the internet to find out what percentage of waste is recycled in your area. You can find this information on your local council website.

8.3 Commonly available forms

LEARNING OUTCOMES

By the end of this section you should know about and understand:
→ that paper and boards are available in a range of stock forms
→ the relationship between each size
→ the most common sheet sizes and their uses.

Paper and board sizes

Paper and boards are available in standard-sized sheets ranging from A10, which is approximately the size of a postage stamp, through to 4A0, which is larger than a kingsize bed sheet. The most common sizes used by designers are between A6 and A0.

Table 8.5 **Each sheet size is twice the size of the one before, e.g. A3 is twice the size of A4**

Size	A10	A9	A8	A7	A6	A5	A4	A3	A2	A1	A0	2A0	4A0
Length (mm)	37	52	74	105	148	210	297	420	594	841	1189	1682	2378
Width (mm)	26	37	52	74	105	148	210	297	420	594	841	1189	1682

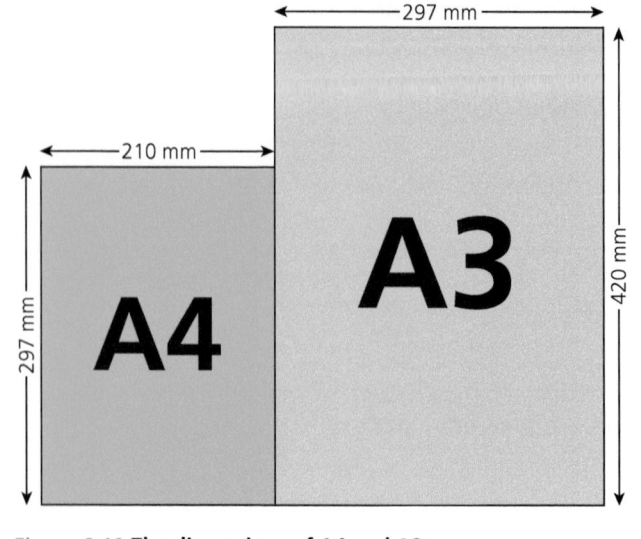

Figure 8.10 **The dimensions of A4 and A3**

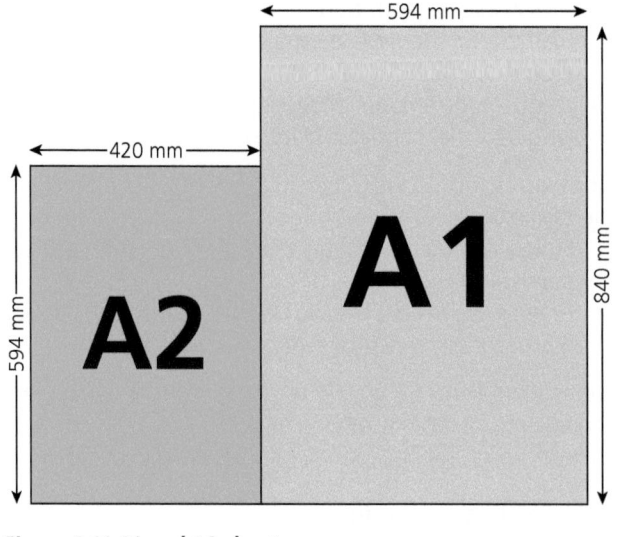

Figure 8.11 **A1 and A2 sheets**

If you fold a sheet of paper in half it then becomes the next size below – for example, an A1 sheet folded in half becomes A2 size.

Laminated board sizes

Foam board

Like paper and cardboard, foam board is available in standard sheet sizes ranging from A4 to A0. Many suppliers also stock standard imperial sizes up to 8 × 4ft (2440 × 1220mm), which is a standard commercial size for many other materials such as MDF, plywood, plasterboard, and so on.

Foam board is available in thicknesses of 3mm, 5mm or 10mm. The standard colours available are white, black or black/grey (one side black, one side grey).

Corriflute

Corriflute is available in a range of sizes, but typical sheet sizes are:
- 450 × 600mm
- 600 × 900mm
- 900 × 1200mm

It is also available from specialist suppliers in standard imperial sizes:
- 8 × 4ft (2440 × 1220mm)
- 8 × 6ft (2440 × 1830mm)

Corriflute is available in thicknesses of 2mm to 10mm and comes in a range of different colours.

PVC foam

PVC foam is available in standard paper sizes and larger sizes, typically:
- 10 × 4ft (3050 × 1220mm)
- 10 × 5ft (3050 × 1530mm)
- 10 × 7ft (3050 × 2030mm)

It is available in thicknesses of 1mm, 2mm, 3mm, 4mm, 5mm, 6mm, 8mm, 10mm, 13mm, 15mm, 19mm and 25mm and in a wide range of standard and special designer colours with gloss or matt finishes.

Styrofoam

Styrofoam is available in sheet form or in blocks. Sheet thicknesses range from 5mm up to 165mm in increments of 5mm or 10mm. Thicknesses above 165mm are available, but these are considered a block rather than a sheet.

Sheet sizes range from 600 × 300mm to over 3 × 1.5m depending on the supplier.

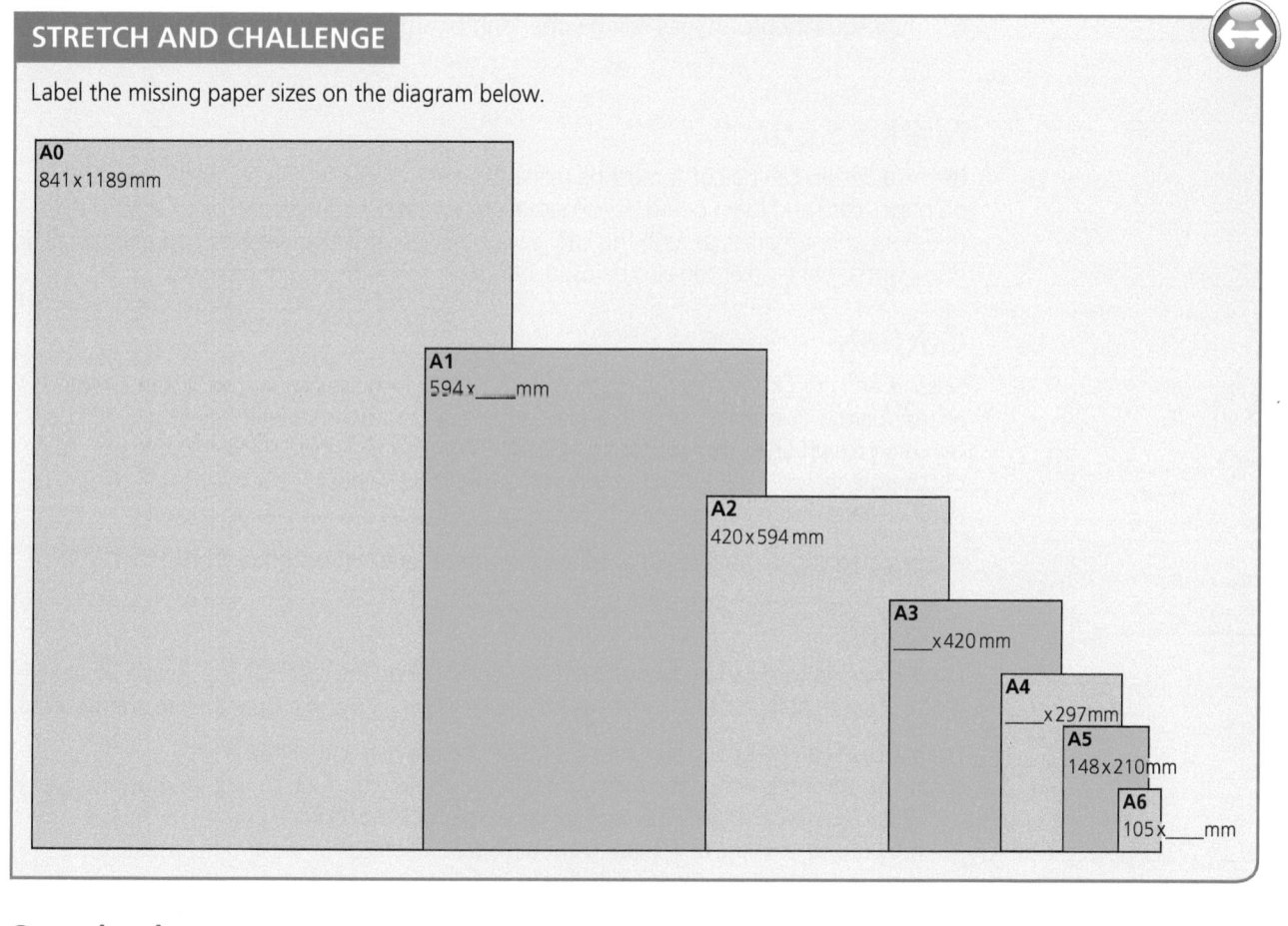

STRETCH AND CHALLENGE

Label the missing paper sizes on the diagram below.

A0
841 × 1189 mm

A1
594 × ____ mm

A2
420 × 594 mm

A3
____ × 420 mm

A4
____ × 297 mm

A5
148 × 210 mm

A6
105 × ____ mm

Standard components

When producing paper and board products, a range of standard components can be used to join materials together, either temporarily or permanently.

Component	Uses
Paper clip	Temporary fastening together of a few sheets of paper. More can be added at any time.
Bulldog clip	More secure method of holding larger number of sheets together temporarily
Filing strip	Strong, temporary method of securing large number of sheets together along one edge. Requires 2 holes to be punched.
Treasury tag	Temporary method allowing large number of sheets to be held loosely together. Requires 1 hole to be punched.
Split pin	Often used for pop up cars and cardboard mechanisms.
Staple	Common method of temporarily joining a few sheets together. More secure than a paper clip. Large staples often used on heavy duty corrugated cardboard boxes
Clic rivet	Temporary method for joining corrugated cardboard and foam board together.
Comb binders	Semi permanent method of binding documents with multiple pages along one edge. Requires a special binding machine which punches the sheet and attaches the binder.
Binder spines	Simple temporary method of binding documents with multiple pages along one edge.

8.4 Manipulating and joining

> **LEARNING OUTCOMES**
>
> By the end of this section you should know about and understand:
> → a range of specialist techniques used to shape, fabricate, construct and assemble high-quality prototypes using paper and board in a workshop.

Marking out

Before a design can be cut, it must be marked out. Pencil or pen can be used to mark out on paper, card and foam board. Styrofoam, Corriflute and PVC foam can be marked with a thin permanent marker, though this can leave a mark on the material. A chinagraph pencil or non-permanent marker can also be used, but the lines can be smudged easily.

Cutting

Scissors and craft knives are the most obvious and best choices for cutting paper and card. Foam board, PVC foam and Corriflute can also be cut using a craft knife. Styrofoam can be cut using a craft knife, but only up to a thickness of around 10mm. Thicker Styrofoam can be cut using a serrated-edged knife (such as a bread knife), bandsaw, hacksaw blade or hot wire cutter. Final shaping can be done by sanding and smoothing with abrasive paper.

A laser cutter can be used to cut any 2D shape out of card, PVC foam, foam board or Corriflute.

Folding

Paper and thin card can be folded easily by hand. Scoring the material first using a blunt knife blade or other dull pointed object will help fold thicker card and ensure a clean, sharp crease.

Foam board can be folded by cutting through the foam in one of two ways:
● Hinge cutting is where the foam board is cut partway through so that the bottom layer of card acts as a hinge, allowing the card to be folded backwards.
● Vee cutting is where a 'V'-shaped cut is made in the foam board and the material removed. This allows the foam board to be folded inwards and gives a clean, tidy fold.

PVC foam cannot be folded unless it is cut partway through in a similar way to foam board. Similarly, Corriflute is not easily folded but can be done by cutting a section of material away from the top layer between the flutes, allowing the material to be folded backwards.

Joining

As with all materials there are several permanent and temporary ways of joining paper and board together.

Adhesives

There is a range of different adhesives available for papers and boards. Different glues are suitable for different products and purposes.

Paper and thin card

Glue sticks are an easy, clean, safe method of gluing paper together. They are ideal for young children and are fairly long-lasting. They create a bond that is only strong enough for paper and very thin card but this can weaken and unstick over time. Spray mount (spray adhesive) comes in an aerosol can and is a quick way to mount drawings on to backing paper. It is commonly used to mount photographs. Spray adhesive gives a thin even coat that dries quickly.

Corrugated cardboard

PVA glue is ideal for joining thicker card and corrugated card. Drying times can be between a few minutes and two to three hours, but only a thin layer is required for a strong, long-lasting bond. Superglue and hot glue guns can also be used to join boards and corrugated card when a fast-setting bond is required.

Foam board and Styrofoam

PVA glue is also ideal for joining foam board and Styrofoam. Masking tape or dressmakers pins can be used to hold the materials together while the glue dries. Hot glue guns and solvent-based adhesives such as cyanoacrylate (Superglue) must not be used on foam board and Styrofoam as the solvent will melt the foam.

PVC foam and Corriflute

Contact adhesive such as glue stick and solvent-based adhesives such as Superglue and polystyrene cement are the best methods of joining Corriflute and PVC foam. Epoxy resins can also be used, providing the surfaces are roughened up with abrasive to give the glue a 'key'. Epoxy resins must be mixed and applied in a well ventilated area and the wearing of protective gloves is advised.

For more information on joining laminated boards using adhesive, please refer to Section 8.7.

Table 8.6 **Common adhesives and their properties**

Adhesive	Material	Properties
Glue sticks	Paper and thin card	• Easy to apply, inexpensive and mess-free • Can unstick over time, pieces can break off, making surface lumpy • Quick-setting
Spray glues	Paper and thin card	• Give a light even coating • Can be permanent or temporary • Quick-setting • Can be messy
PVA glue	Thicker card, corrugated cardboard, foam board, Styrofoam	• Dries clear • Inexpensive • Can be watered down
Hot glue guns	Thick card, Corriflute	• Quick-setting • Can burn material and user • Cool melt versions available • Coloured glue sticks available
Cyanoacrylate glue (Superglue)	Corriflute, PVC foam	• Quick-setting • Very strong bond • Short shelf life (approx. one year)
Polystyrene cement	Corriflute, PVC foam	• Dries quickly • Clear finish
Contact adhesive	Corriflute, PVC foam	• Must be applied to both surfaces • Needs to partly set before joining • Instant bond • Does not require clamping
Epoxy resin	Corriflute, PVC foam	• Needs to be mixed with hardener • Very strong bond • Gives off strong fumes

8.5 Structural integrity

LEARNING OUTCOMES
By the end of this section you should know about and understand:
➜ how and why papers and boards can be reinforced to withstand external forces and stresses
➜ the processes that can be used to ensure the structural integrity of a product.

The structural integrity of a product or object is its ability to hold together under a load, including its own weight, without bending or breaking. If an object has structural integrity it will perform its function without failing for as long as it was designed to do so.

If the object does not hold together because it has been stressed more than it can withstand, this is known as structural failure. This could refer to just a part of the object's structure or the whole thing.

As a thin sheet, material paper can appear to have little structural integrity. The structural integrity of paper can be greatly increased in various ways, however.

Folding

Folding a sheet of paper or thin card turns it from a flat sheet that will bend and flex easily into a structure. The fold creates a rigid section within the paper that can support its own weight and additional weight.

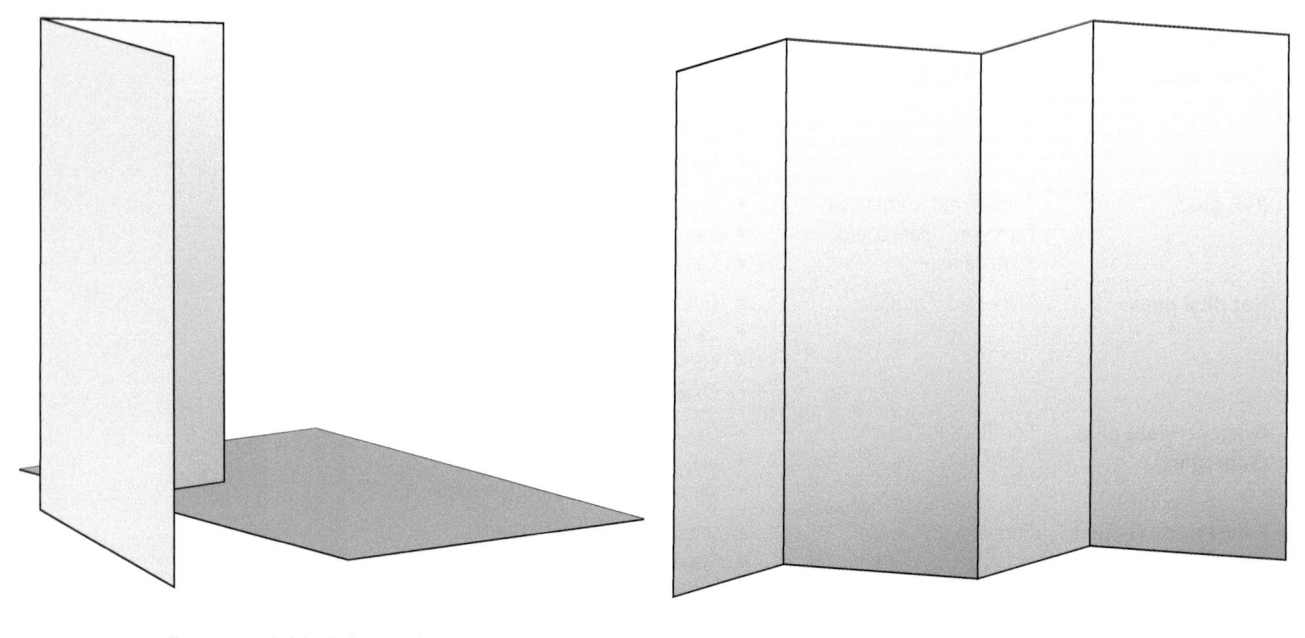

Figure 8.12 **A flat and a folded sheet of paper**

Figure 8.13 **Paper folded multiple times**

By folding the paper a number of times the structural integrity can be further increased.

Curving and bending

As well as folding, paper and card can be bent into curved shapes such as a cylinder, a shape that has inherent strength and structural stability.

The wavy centre section of corrugated card is what gives it its rigidity and strength. A standard cardboard box with four vertical-folded corners will be able to withstand and hold the weight of an average-sized human. The structural integrity provided by the box shape means that many boxes can be stacked on top of one another without collapsing. The triangulated shapes in the centre of Corriflute work in the same way.

The structural integrity of cardboard can also be increased by shaping it into a structure. Cardboard egg boxes are an excellent example of this. When the card is manufactured, the pulp is sprayed on to a dimpled mould instead of a flat mesh sheet, similar to the papier mâché process. The shape of the carton not only supports each individual egg and isolates it from the other eggs, but the structure created also helps to absorb shock and protect the eggs against breakage. This type of structure is called a monocoque, meaning 'single structural skin'. Monocoque structures are commonly used in boat hull design, and monocoque chassis are used on many modern vehicles.

Structural integrity can also be increased by reinforcing one material with another. Foam board is made up of a thin layer of foam sandwiched between two layers of paper or thin card. On their own, the sheets of paper and foam would have very little structural integrity or strength. If stood on their edge they would collapse or fall over. By joining the materials together, however, they gain structural integrity. The foam board inside allows the outer paper layers to stand on their edge and not bend as easily. The paper outer layers spread any load across the foam instead of it being concentrated in one place.

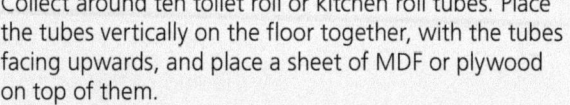

ACTIVITY

Collect around ten toilet roll or kitchen roll tubes. Place the tubes vertically on the floor together, with the tubes facing upwards, and place a sheet of MDF or plywood on top of them.

Gently stand on the structure, and it should take your weight. Remove two of the central toilet/kitchen rolls, and repeat the exercise. Progressively remove more tubes and repeat the tests until the structure collapses.

ACTIVITY

Cut down a toilet roll so that it is the same length as its diameter. Try to squash the tube one way then the other.

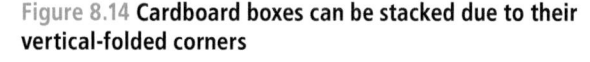
Figure 8.14 **Cardboard boxes can be stacked due to their vertical-folded corners**

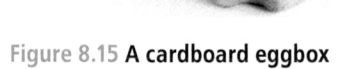
Figure 8.15 **A cardboard eggbox**

8.6 Making iterative models

LEARNING OUTCOMES

By the end of this section you should know about and understand:
→ the processes and techniques used to produce early models to support iterative design.

Processes and techniques

Paper and boards can be used in a variety of useful and interesting ways to help you explore your early design ideas and to support iterative designing in school.

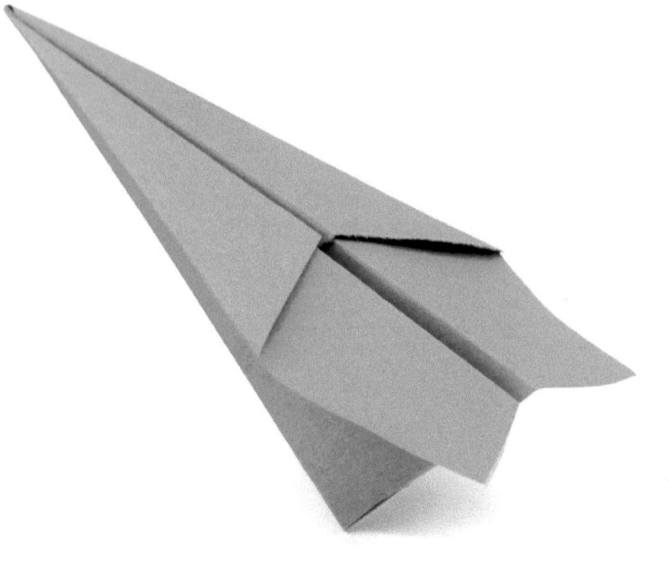

Paper and thin card are extremely versatile materials for model-making as they can be easily folded, cut, bent, printed on and stuck together. Paper modelling originates in Japan – origami is the art of paper modelling without the use of glue. Everyday folded objects such as paper aeroplanes and paper hats, which we have all made or seen, are examples of simple origami.

Paper and card modelling can use printed parts or patterns, which can be cut out and assembled by slotting or gluing. Paper and card can be modelled into all manner of forms and are frequently used for modelling vehicles such as cars, lorries, boats and spacecraft, as well as animals, furniture, buildings and bridges. The internet provides examples of many downloadable models with realistic textures, from simple shapes up to complex life-size models with extremely fine details and thousands of pieces to assemble.

Figure 8.16 A paper aeroplane

The most common methods of assembling a 3D shape using paper or card involves the use of nets. Nets can be drawn or printed on to paper or card and then scored, folded, cut out and joined to form the shape required.

STRETCH AND CHALLENGE

Draw a net and use it to construct the following shapes:
- A cube 50 × 50 × 50mm
- A cuboid 60 × 40 × 30mm
- An equilateral triangular prism (triangle sides 50 × 60mm high)
- A square-based pyramid 60 × 60 × 60mm

Paper and card models

Paper and card are extremely useful for making templates or patterns for items that will be made from sheet metal, polymer or textiles. By first making a template in paper or card, designers can easily trim and shape the pattern to exactly the right size and fold or bend it to the required shape. If the template is not correct and requires modification or completely redoing, it is much easier and cheaper to do this in card than in metal or polymer. Once it is correct it can be laid out on to the sheet material and drawn around so that the 'real' item can be made to the exact size.

Models from other materials

Foam board is an extremely popular material for making architectural models. The thickness of the material means it can be stood up and joined to form walls, floors and roofs. Windows, doors and other apertures are easy to cut out and the surface can be scored, indented, printed or drawn on to in order to achieve realistic architectural effects.

Figure 8.17 **A card template laid out on sheet metal**

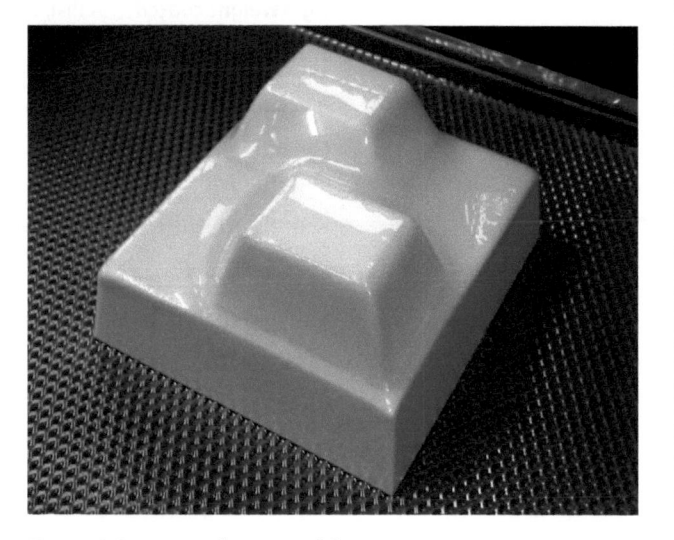

Figure 8.18 **An architectural model made using foam board**

Styrofoam is an excellent modelling material due to its ability to be shaped and sanded easily. It can also be used to form lightweight building blocks, which can be joined together to form larger models. Styrofoam is often used in the modelling process as a mould for vacuum forming or GRP.

For more information on the use of Styrofoam, vacuum forming and polymer laminated layers (Corriflute and PVC foam) for modelling, see Section 8.6.

Figure 8.19 **A Styrofoam mould**

8.7 Finishes

LEARNING OUTCOMES

By the end of this section you should know about and understand:
→ the different finishes that can be applied to paper, board and laminated layers
→ the processes used for finishing paper, boards and laminated layers for specific purposes.

Paper

The properties and aesthetics of paper and board can be altered and improved by applying different surface treatments. A range of different coatings can be applied to improve the opacity, lightness, surface smoothness, lustre and colour-absorption of paper.

Some types of coating are done during the making process while the paper is in the paper-making machine. Other types are applied separately after the paper has been produced.

Cast coatings are applied after the paper has been produced. One or both sides of the paper are coated with china clay, chalk, starch, latex or other chemicals. The wet coated paper is then pressed or rolled against a polished hot metal drum, creating a finish so smooth, reflective and shiny that it can seem almost mirror-like. Cast-coated paper also holds ink well and produces sharp, bright images when printed on.

Supercalendering

Supercalendering gives paper an even smoother surface. The paper passes through a calender or supercalender, which is a series of rollers with alternately hard and soft surfaces. The pressure on the paper creates a smooth and thin paper with a very high lustre surface. Supercalendered paper is primarily used for glossy magazines and high-quality colour printing.

Types of paper finish

Table 8.7 **Characteristics and uses of different paper finishes**

Paper finish type	Characteristics	Uses
Cast-coated paper	Provides the highest gloss surface of all coated papers and boards	Labels, covers, cartons and cards
Lightweight coated	A thin, coated paper, which can be as light as 40gsm	Magazines, brochures and catalogues
Silk or silk matt-finished paper	Smooth, matt surface; high readability and high image quality	Product booklets and brochures
Calendered or glossy paper	Glazed shiny surface; can be coated and/or uncoated	Colour printing
Machine-finished paper	Smooth on both sides; no additional coatings applied after leaving the paper-making machine	Booklets and brochures
Machine coated	Coating applied while it is still on the paper machine	All types of coloured print
Matt-finished paper	Slightly rough surface prevents light from being reflected; can be either coated or uncoated	Art prints and other high quality print work

Card and board finishes

Varnish

Varnish coatings are a thin coating of matt, silk or gloss varnish that can be applied to paper or card products to enhance their look and feel. The coating also adds additional protection and makes the paper or card last longer. Varnished card has a high shine finish and feels like plastic to the touch.

Spot UV varnish is a special varnish that is applied to the printing surface, then cured or hardened by ultraviolet (UV) light during the printing process. Spot varnish is only used on certain areas of the paper or card. The varnish makes the coated area shinier and clearer than the surrounding uncoated areas, making parts stand out.

Hot foil application

Hot foil application or foil blocking is used to produce metallic finishes such as gold or silver on card and paper products. Multi-coloured and holographic foils are also available. It is often used for lettering on invitations and business cards.

The hot foiling process uses a magnesium or copper die shaped to the required foil design that is heated up. The foil is fed between the card and the die, which is then pressed against the card, releasing the foil on to the card where the die touches. Hot foil application is currently the only method of achieving a metallic glossy finish.

Embossing and debossing

Embossing and debossing actually give paper and card a three-dimensional image that can be seen and felt. Embossing creates a raised area on the paper or card that stands out slightly. Debossing has the opposite effect and creates a sunken or lowered area. Using both techniques can create stunning visual effects.

Figure 8.20 An invitation with foil lettering

The embossing and debossing process uses two metal formers (a male and a female) in the shape of the design required. These fit together perfectly. The card is placed between the two formers and heat and pressure are applied. This squeezes and deforms the fibres of the material into the shape of the formers.

The embossing process

1 Card is placed between the male and female dies.
2 Heat and pressure are applied.

Card retains shape when removed from the dies.

Figure 8.21 **A business card with embossing and debossing**

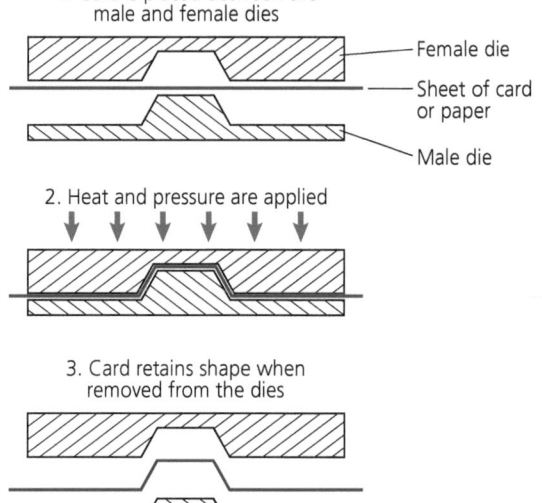

1. Card is placed between the male and female dies

— Female die
— Sheet of card or paper
— Male die

2. Heat and pressure are applied

3. Card retains shape when removed from the dies

Laminating

The coating processes applied to paper described at the start of this section are sometimes called laminates, but they use spray-on methods that are done during or straight after the paper-making process.

The other form of laminating, described below, is usually done to finished documents such as menus, posters, signs, identity badges and other printed documents. It is done in order to:

- improve their strength and resistance to bending, creasing or ripping
- waterproof the document, allowing it to be wiped clean and preventing it smudging or going soggy
- improve the appearance by making the document shiny
- increase the lifespan of the printed document.

Laminating involves applying a film of clear plastic between 1.2 and 1.8mm thick to one or both sides of paper or thin card. There are three methods of laminating a document:

- Pouch lamination
- Thermal or hot lamination
- Cold lamination.

Figure 8.22 A laminating machine and laminated document

Pouch lamination

Pouch lamination uses thin clear plastic pouches that are available to fit standard paper sizes such as A3 and A4. The pouches are coated on the inside with a thin layer of heat-activated glue. The document is placed inside the film pouch, which is slightly larger than the paper, before being fed through a laminating machine. Inside the machine the pouch is heated and pressed between rollers. The heat activates the glue, which seals the pouch together as it is pressed through the rollers, encasing the document inside it.

Pouch laminators can do only one document at a time so are ideal for small volume items. Laminating pouches and laminating machines are reasonably cheap to purchase and readily available in schools.

Thermal or hot lamination

Where large volumes of laminated documents are required, commercial lamination methods are required. Thermal lamination uses rolls of thin heat-sensitive polymer film. There are different types of film available with different properties:

- **Polypropylene film** has good resistance to water, chemicals and cracking, excellent transparency, and can be recycled. Commonly used for write on/wipe off products such as calendars, labels, etc.
- **PET (polyester) film** has all the benefits of polypropylene film with improved gloss, scratch-resistance and durability. It is commonly used for book covers and documents that have undergone foil application.
- **Nylon film** is considerably more expensive than other film due to the high temperatures needed for it to laminate, but it has a high chemical-, mechanical- and abrasion-resistance. It will also absorb moisture in a similar way to the paper, which prevents the documents curling, and is commonly used for soft covers such as on paperback books.

As well as clear laminating films, metallic colours such as gold and silver are available. Holographic films and iridescent films that shift and change colour as they catch the light can be used to give interesting effects to the printed documents being laminated.

Cold laminating

Cold laminating is done when only one side of the paper or card is to be coated. Cold laminating does not use heat, and many cold lamination machines are inexpensive hand-operated machines that do not require any power. The lack of heat makes them ideal for documents that might be damaged by heat, such as photographs.

Cold lamination uses film that has a thin coating of pressure-sensitive adhesive applied to one side. The film is placed over the document and passed through the machine's rollers, which press down and smooth out the adhesive. This sticks the film firmly to the document. It can then be trimmed to size if necessary.

Cold lamination is mainly used in the sign-making industry. By adjusting the pressure applied by the rollers, cold lamination can be used on other graphic products such as foam board and PVC foam. It is also used in other areas of industry, such as coating sheets of glass or as a protective coating for stainless steel, aluminium and acrylic.

Other laminated layers such as Corriflute and PVC foam, which are polymers, can have additional surface finishes applied to them. For more information, see Chapter 11, Polymers.

8.8 Using digital design tools

> **LEARNING OUTCOMES**
>
> By the end of this section you should know about and understand:
> → the use of 2D and 3D digital technology and tools to present, model, design and manufacture solutions in paper and board.

Use of 2D and 3D digital technology and tools
Digital technologies: CAD and CAM

Computer-aided design (**CAD**) is a digital design tool that uses computer technology to produce designs instead of, or in addition to, drawing by hand. CAD allows the designer to produce detailed working drawings of designs to an accuracy of a thousandth of a millimetre. Drafting on CAD can be done in two dimensions, for example on paper, or in three dimensions. CAD also enables the designer to explore design ideas and visualise concepts through fully rendered 3D images, which can be rotated and viewed from any angle. Many CAD programs can also simulate how a design will behave in use and perform in the real world.

> **KEY TERMS**
>
> **CAD:** Computer-aided design.
> **CAM:** Computer-aided manufacture.

There are many different types of CAD program available, from simple programs that can be downloaded or used online free of charge through to industry-standard packages. Many schools have CAD packages that are simple to use and can create detailed working drawings of final products in isometric and orthographic projections. While different CAD packages will vary in the way they work, the majority will have similar tools and commands, and will usually have built-in tutorials to teach you how to use them.

An important advantage of CAD is that it allows drawings of products to be created and turned into real products through the use of computer-aided manufacture (**CAM**). A wide range of CAM machines is available that can create products from paper and boards. Some of these are 2D machines that can cut, score and engrave on different sheet materials, and others can mill and form 3D shapes out of Styrofoam.

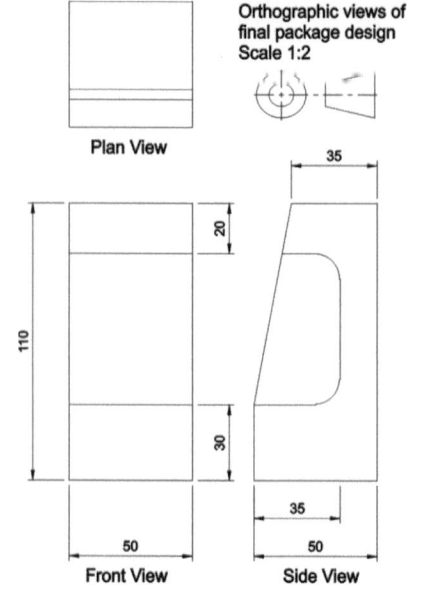

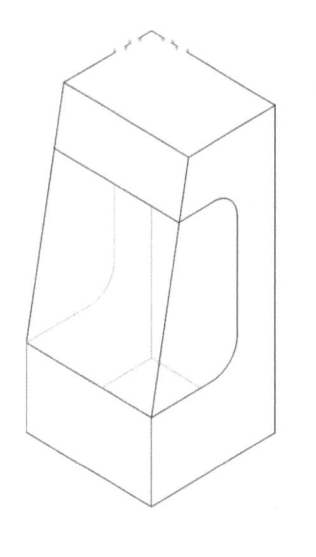

Orthographic views of
final package design
Scale 1:2

Plan View

35

20

110

30

50

Front View

35

50

Side View

Figure 8.23 **An orthographic in Techsoft**

Once a drawing of a design has been created on CAD, it can be output in various ways. The most common method of outputting a design is by printing it on to paper. There is a vast array of printers available and most households with a computer also have a printer. Smaller printers that can print sizes up to A4 are relatively inexpensive. Larger printers up to and over A0 in size are used in industry for engineering drawings, house plans and maps.

Vinyl cutters

After printers, the simplest CAM machines that many schools have available are vinyl cutters. These are 2D machines that can transfer line drawings on CAD on to self-adhesive vinyl. Vinyl cutters have rollers that move the vinyl sheet backwards and forwards on one axis and a blade that moves horizontally back and forth on the opposite axis. The blade cuts through the top surface of the vinyl but leaves the backing sheet intact so that the finished design can be peeled off and applied to a sign, window or vehicle, for example.

Vinyl cutters are available in a variety of different sizes, from small A4 size up to commercial sizes of over A0.

Figure 8.24 **A vinyl cut logo**

Laser cutters are usually used for cutting sheet materials, but they can also cut a wide range of other materials. Most schools have laser cutters that can cut materials such as paper, cardboard, foam board, PVC foam, polystyrene sheet and acrylic. More expensive machines are available that can cut plywood, MDF, hardwood and thin sheet metals.

Laser cutters have a flat 'bed' that the sheet material is laid on to. The laser then moves across the sheet along the x- and y-axes, cutting all the way through or part way into the material. The speed of the laser and distance from the bed can be altered and set to suit the specific material being cut.

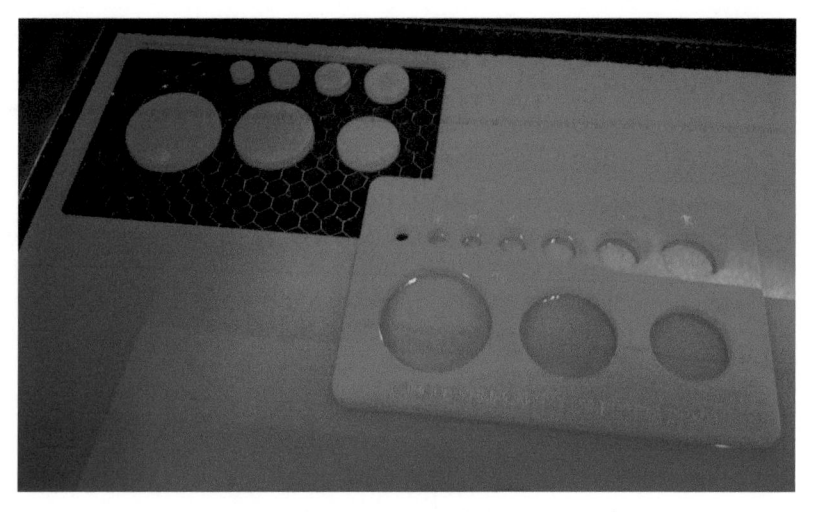

Figure 8.25 **A laser cutter and a laser-cut item**

Laser cutters are available with bed sizes from approximately A3 up to industrial machines with beds 3 metres in length. Different machines cut the work in different ways – by melting, burning, vaporising or blowing away material with a jet of gas, depending on the material being cut. Laser cutters are much quicker and more accurate than cutting by hand and can cut out intricate designs that would be extremely difficult or impossible to do by hand.

Milling machines

Computer numeric controlled (**CNC**) milling machines use a rotating cutter to remove material from the surface of a workpiece. Milling machines are mainly used for the shaping of blocks of aluminium and other metals, but can also be used in exactly the same way for shaping and milling designs in Styrofoam. For more information on milling machines, see Section 8.7.

KEY TERM

CNC: Computer numeric control.

KEY POINT

A printer is a digital design tool that can be used with paper and board when designing products.

Figure 8.26 **A CNC milling machine milling some Styrofoam**

8.9 Manufacturing methods and scales of production

LEARNING OUTCOMES

By the end of this section you should know about and understand:
→ the methods used for manufacturing at different scales of production
→ manufacturing processes used for larger scales of production
→ methods of ensuring accuracy and efficiency when manufacturing at larger scales.

Scales of production

Paper and board products can be made using different methods depending on the quantities required.

One-off production

One-off production is the making of a single product (known as a 'one-off'). A handmade birthday card is an example of a one-off product. One-off products are labour intensive and time consuming to produce. Some one-off products are entirely handmade, whereas others use automated processes. One-off products are often produced for a specific client's requirements or needs. Often designers create a one-off prototype of their design to show what it will look like or how it will function.

Batch production

Batch production is a system of production in which a limited number of items is produced in one go. Batch production enables the production of similar items with slight variations, such as changes to text or colour.

Digital printing

For paper and card products, digital printing is the most cost-effective batch production method as it allows variations to be made easily, such as enlargement and reduction, cropping, rotating, and so on. Digital printing is ideal for personalised items such as party invitations or company name badges, as the main design stays the same but the details can be changed for each copy printed.

Digital printers are inexpensive to buy, readily available and many people have them for household use. The cost per sheet of using a digital printer is high compared to other types of commercial printing, but because there are no set up costs and only a limited number of print-outs are needed, they are the best option for small print runs.

Figure 8.27 **A digital printer**

Screen printing

Screen printing is another method of printing that is suitable for batch production. Screen printing is used to create repeating patterns or designs, such as on wallpaper or fabrics.

Screen printing uses a porous fabric mesh screen stretched over a wooden frame. The screen is the exact size of the pattern or design required. The lightest colour is printed first, by masking off the parts of the screen that are not in that colour on the design. This creates a stencil of the coloured parts to be printed. The frame is then laid on to the paper or fabric, a small quantity of ink is added and a squeegee is used to spread the ink evenly over the whole of the screen. The ink is pushed through the mesh by the squeegee in the unmasked areas and on to the paper or fabric below. The screen is then moved to the next position and the process is repeated.

Once the lightest colour has been printed across the whole area required, the screen is washed and masked up for the next colour. The process is repeated until all the colours have been printed.

The design to be printed

Wooden frame

Mesh screen

Figure 8.28 **The screen is made to the same size as the design.**

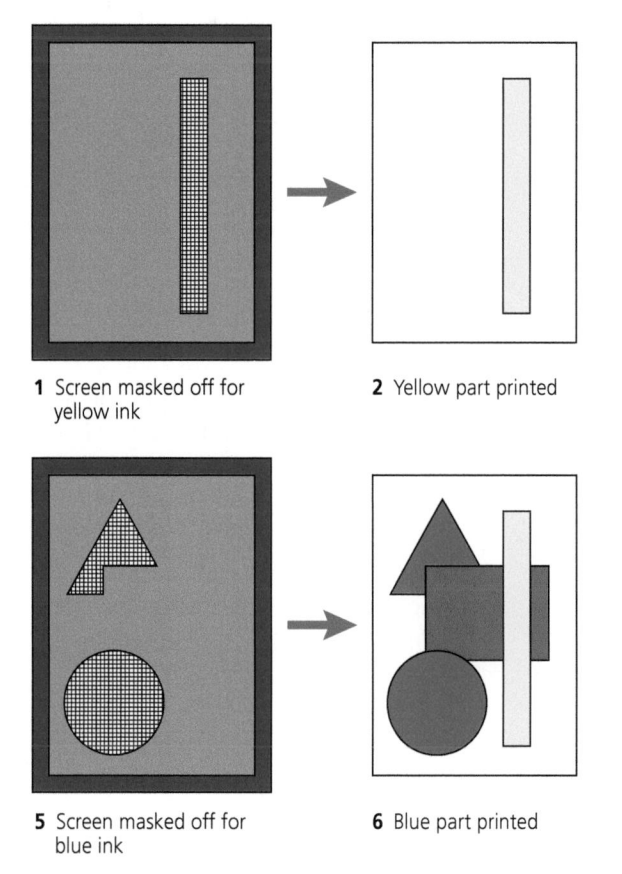

1 Screen masked off for yellow ink

2 Yellow part printed

5 Screen masked off for blue ink

6 Blue part printed

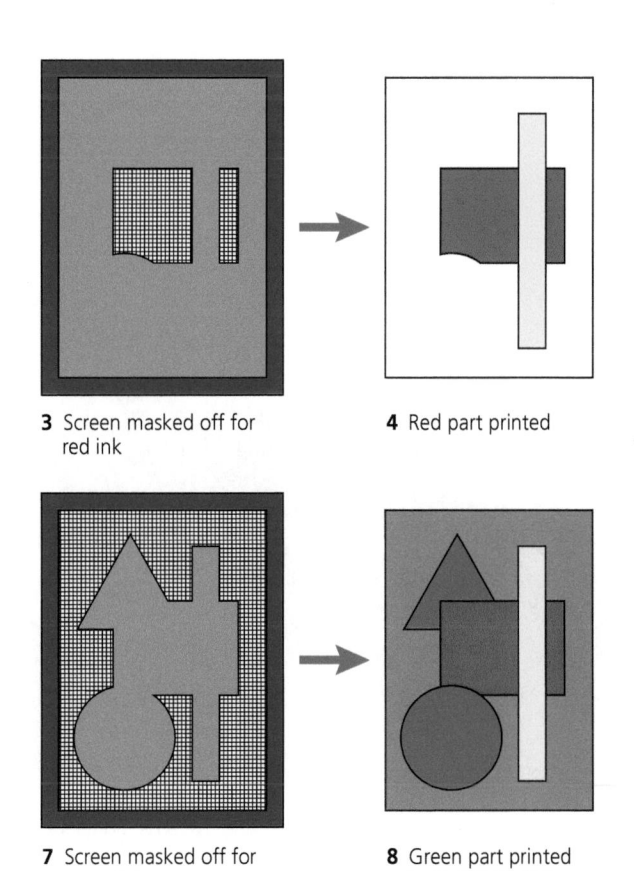

3 Screen masked off for red ink

4 Red part printed

7 Screen masked off for green ink

8 Green part printed to complete design

Figure 8.29 **The screen printing process**

Stencils and templates

Stencils and templates can also be used for batch production of paper and board products. If multiple numbers of the same part are required, the shape can be drawn out once on card and cut out to create a cardboard template. This template can then be laid on to the material and drawn around, then moved and drawn around again and again until the required number is reached. Using a template ensures that every piece is exactly the same and saves the time of drawing each part out individually.

To reduce waste, the template pieces should be tessellated – arranged in such a way that the material is used to its maximum capacity and there is only minimal waste.

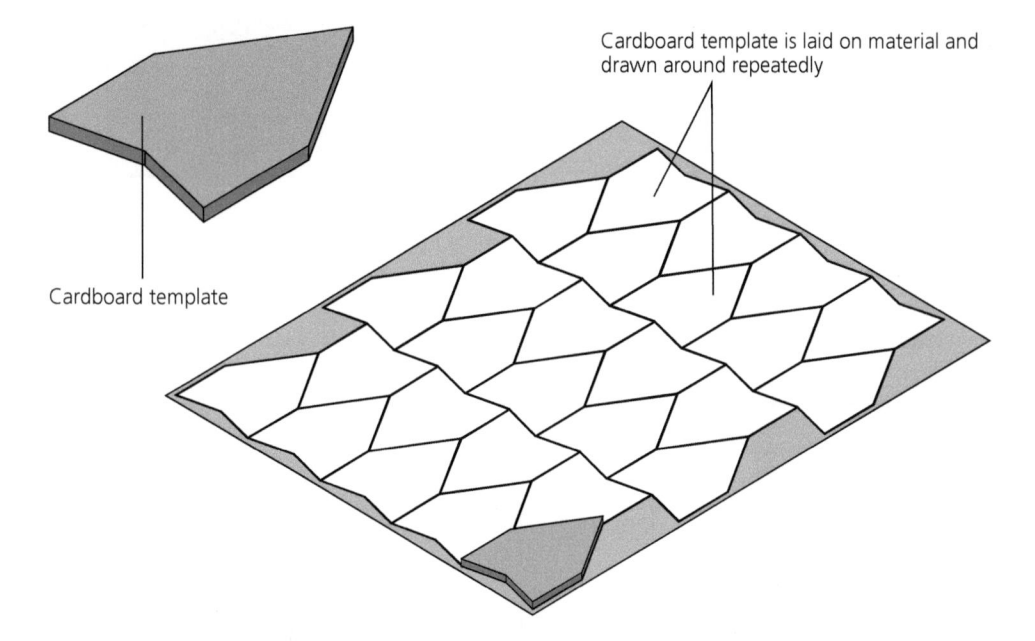

Cardboard template is laid on material and drawn around repeatedly

Cardboard template

Figure 8.30 **An example of tessellation**

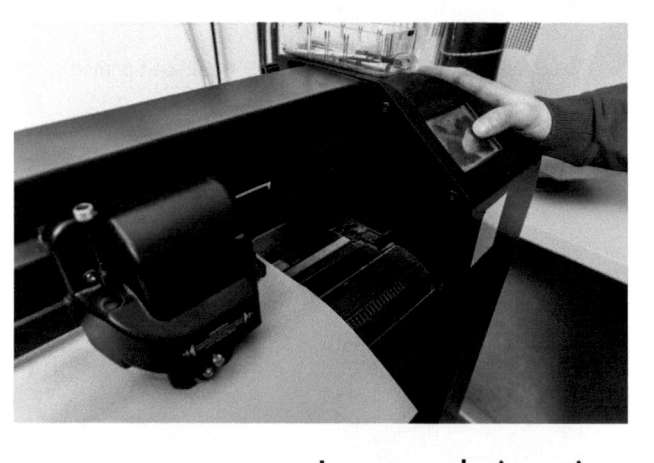

Figure 8.31 **A vinyl cutter**

Vinyl cutting

Vinyl cutting is frequently used in sign making and display work. There are many different sizes available, from A4 up to A0. Most schools have a vinyl cutter of at least A4 size. The most common uses are for creating logos, designs and text to stick on to packaging, point-of-sale displays and so on, where they cannot be printed.

Vinyl-cut lettering and images are often used on the sides of cars, buses and lorries. The advantages of using vinyl graphics instead of painting or sign writing on to the vehicle is that they can be removed relatively easily by applying a little heat to soften the vinyl and peeling it off.

Large-scale (mass) production

Mass production is when a very large number (possibly thousands or millions) of a product is produced. Newspapers and magazines are examples of mass produced paper and board products.

Offset lithography

When large numbers of graphic products are required, commercial printing methods are the best option. Offset lithography is one of the most common forms of commercial printing used today. The process uses four ink colours: cyan, magenta, yellow and black (called the 'key'). These colours are often abbreviated as CMYK. Just these four colours can be overlaid to create others, for example printing cyan on top of yellow creates green.

The initial setting of the print run is the most expensive part of the process, as it requires an image setter to produce films for each of the four colours of the artwork. These are then used to produce a set of printing plates.

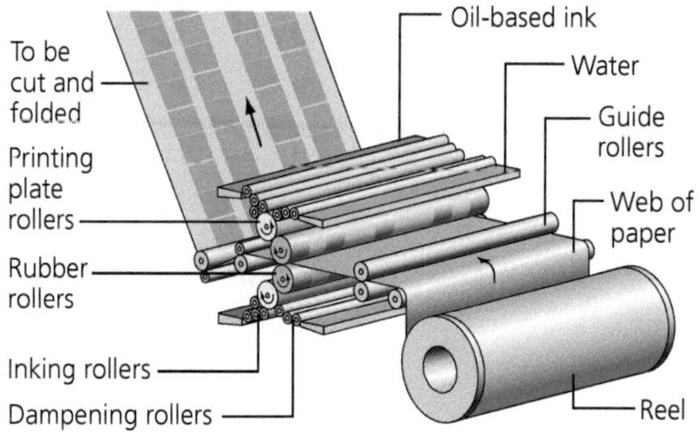

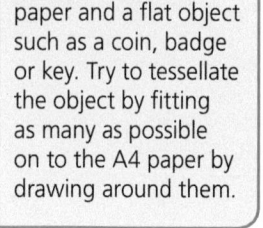

Figure 8.32 Offset lithography printing

The principal of offset lithography is that oil and water do not mix (oil will separate from water). The image on the plate is set to attract the ink but repel water, whereas the non-image parts of the plate attract moisture and repel the ink. The plate is kept wet during the printing process so that the ink will stick only to the areas required.

Die cutting

Die cutting is used when you wish to cut many identically shaped products, such as in mass production. The process works in a very similar way to a pastry cutter: a sharp, shaped blade (the die) is lowered on to the material to be cut and a press is applied to force through the material.

A similar method is also used to crease paper and card products, only a blunt or rounded blade rather than a sharp blade is used. The material is not cut all the way through, but a crease is left. This is used when creating products that need to be folded such as boxes and surface developments.

Flexography

Flexography is another common type of mass-production printing process. Flexography uses water-based inks rather than oil-based, which allows a wider range of inks to be used. The water-based inks dry much faster so the process is much quicker, and although the print quality is not as good, the costs are significantly lower. It is largely used for printing on packaging such as corrugated cardboard boxes, cartons, sweet wrappers and plastic carrier bags, where the quality of the print is less important.

As in offset lithography, the four colours – CMYK – are printed one at a time. A flexible printing plate is mounted on to the plate cylinder, and the material to be printed is pressed against it by the impression cylinder. Ink is transferred to the printing plate from an ink pan via two rollers (a fountain roller and an anilox roller).

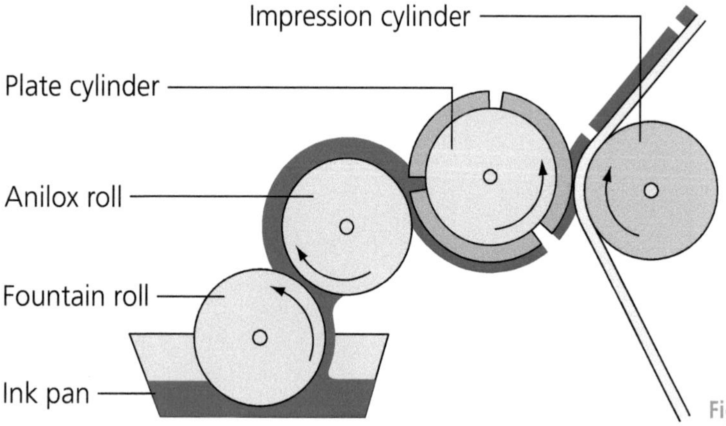

Figure 8.33 Flexography printing

ACTIVITY

Use a magnifying glass to look closely at the quality of printing on some different products. Compare the print quality of the illustrations and text in this book to the printing on items such as corrugated cardboard, drink cartons, and so on.

STRETCH AND CHALLENGE

Take a sheet of A4 paper and a flat object such as a coin, badge or key. Try to tessellate the object by fitting as many as possible on to the A4 paper by drawing around them.

8.10 Cost and availability

LEARNING OUTCOMES

By the end of this section you should know about and understand:
→ how the cost and availability of specific papers and boards can affect decisions when designing.
→ how to calculate the quantities, cast and sizes of materials required in a design or product.

Cost and availability considerations

- **Paper and cardboard** are among the most inexpensive and widely available materials. Normal A4 white paper that we use every day for sketching, printing and writing on costs around 1p per sheet, while larger sizes such as A1 and A0 cost around 10p per sheet.
- **Cardboard and glossy paper types** are more expensive but are still widely available in high street stationery shops, and they are inexpensive compared to other materials such as woods, metals and plastics.
- **Foam board** is widely available in sizes from A4 to A1. It is kept in stock by art and hobby shops. Prices range from approximately 50p for an A4 sheet up to around £2.50 for A1. The thicker the sheets are, the more expensive they become.
- **PVC foam** is also widely available from specialist suppliers but can also be bought on the internet under various trade names. Small sheets (A4 size) cost around £2 per sheet, and this ranges up to large (2400 × 1200mm) sheets that can cost around £50.
- **Styrofoam** is used in many industries as an insulation material and is available from many DIY stores and on the internet. A standard-sized sheet of 25mm-thick Styrofoam (2400 × 600mm) costs around £7.

As with most other materials, the cost reduces the more you purchase. Therefore when designing a product and planning the manufacture of a model or prototype, it is essential that you carefully consider and calculate the amount of the material you will need. If, for example, you need just over three A1 sheets of foam board, it may be cheaper to buy seven sheets of A2 than four sheets of A1.

Another consideration to take into account is that the cost per mm^2 of all sheet materials will generally decrease as the sheet size increases. In many cases it may be more economical to buy a large size sheet and cut it into smaller sheets yourself.

Example

A designer needs seven pieces of A4-size foam board to make a prototype product.

- One A4 sheet of 25mm foam board costs £0.45. Seven sheets @ £0.45 = £3.15 total.
- A pack of ten A4 sheets costs £3.00.
- A sheet of A1-size foam board costs £2.75.

The cheapest way to get the foam board is for the designer to buy an A1 sheet for £2.75 and cut this into eight A4-size pieces. He will have seven pieces and one spare.

KEY POINTS

- Paper and boards are widely available and one of the most inexpensive materials.
- The cost per mm^2 decreases as the sheet size increases.

CHAPTER 9
Timber

Chapter 9 Timber

KEY POINT

When designing, it is important to know and consider the physical and working properties of different timbers to make sure you select the best material.

You learnt about the different types of timber in Chapter 5. This chapter includes more in-depth knowledge on timber.

9.1 Physical and working properties

LEARNING OUTCOMES

By the end of this section you should know about and understand the physical and working properties of a range of different timbers and manufactured boards, including:
→ how easy different types of timber are to work with
→ how well they fulfil the required functions of products in different contexts.

KEY TERMS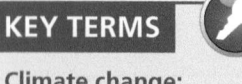

Climate change: A change in global climate apparent from the mid- to late twentieth century onwards.

Organic: Derived from living matter.

Renewable: A natural resource that is not depleted by use.

For centuries, our ancestors used wood to put a roof over their heads, walls under the roof and furniture inside the walls. It is nature's strongest, most readily available, **renewable**, warm and welcoming building material. Timber is the term given to wood that has been converted into useable forms such as boards, planks, sections. It is an **organic**, sustainable, natural and renewable material. It is strong, lightweight, adaptable and competitive with other materials in terms of cost. It actually contributes to reducing **climate change**, because it requires less energy to manufacture than any other building material.

You have already learnt in Chapter 5 timber can be classified into three 'families':
- Softwoods (from coniferous trees)
- Hardwoods (from deciduous trees)
- Manufactured boards (manufactured sheets of timber using both soft and hardwoods).

The following tables outline the physical and working properties of a range of common softwoods, hardwoods and manufactured boards, along with some examples of their typical uses.

Common softwoods

Table 9.1 **Physical and working properties of common softwoods**

Name	Physical and working properties	Typical uses
Redwood (Scots pine)	Straight grain, knotty, easy to work, finishes wells, durable; widely available and relatively cheap.	Suitable for all inside work. Woodturning. Most-used softwood in UK.
Western red cedar	Resists insect attack, weather and dry rot due to natural preservative oils. Lightweight, soft, knot-free, straight-grained, very durable. Attractive surface.	Outside joinery. Building cladding. Bathroom and kitchen furniture. Panelling walls.
Parana pine	Hard, straight grain; usually knot-free; fairly strong, works easily; fairly durable.	Internal building work such as staircases and built-in furniture.
Whitewood (spruce)	Easy to work, fairly strong, resistant to splitting. Can contain pockets of resin.	General construction work. Crates.

Common hardwoods

Table 9.2 Physical and working properties of common hardwoods

Name	Physical and working properties	Typical uses
Beech	Very tough, hard, straight and close-grained. Withstands wear and shocks, polishes well, liable to warp.	Most-used hardwood in UK. Toys, furniture, wooden tools. Good for steam bending.
Ash	Wide-grained, tough, very flexible. Finishes well.	Baseball bats, flooring, tool handles.
Elm	Tough, flexible, durable, water-resistant, liable to warp. Can be difficult to work due to its cross-grain.	Garden furniture (if treated), woodturning, furniture.
Oak	Heavy, hard, tough. Open grain, finishes well, good outdoors. Due to containing tannic acid it will corrode steel screws, leaving a blue stain.	Garden furniture, doors, floors, high-end furniture.
Mahogany	Easy to work, wide boards available, polishes quite well. Has interlocking grain, which makes it difficult to work.	Furniture, shop fittings, boat-building, doors, pool cues.
Teak	Hard, durable. Natural oils resist moisture, fire, acids, alkalis. Straight grain, works well, very expensive.	Laboratory benches, ships' decks, high-end furniture.
Balsa	A very soft and lightweight wood with a coarse and open grain.	Modelling structures such as buildings and bridges. Model boats and airplanes

Common manufactured boards

Table 9.3 Physical and working properties of common manufactured boards

Name	Physical and working properties	Typical uses
Plywood	Constructed of layers of veneer glued at 90 degrees to each other which makes it very strong and tough. Available for Structural, Exterior, Interior and Marine work. High strength to weight ratio. Relatively stable under changes in temperature and moisture. Cuts relatively easily but can splinter.	Used for strong structural panelling board used in building construction. Furniture making.
Flexible plywood	Designed to make curved parts. Made by gluing three layers of veneer with the grain running the same way. The middle veneer is extra thin. Called "Flexi- ply" or "bendy plywood" Easier to cut than regular plywood but can still splinter easily.	Curved furniture, bespoke shopfittings.
Marine ply	Moisture resistant form of plywood.	Boatbuilding, decking.
MDF	Easily machined and painted or stained. Smooth, even surface. Available in water and fire resistant forms.	Used mainly for furniture and interior panelling due to its easy machining qualities. Often veneered or painted.
Moisture resistant MDF	A version of MDF that is resistant to moisture. It is often coloured green	Used in areas where moisture is prone, such as Kitchens and bathrooms
Flame retardant MDF	A version of MDF that is resistant to fire. It is usually coloured pink or blue	Used in the building industry for buildings where there is an above average risk of fire
Blockboard	Similar to plywood but the central layer is made from strips of timber. Good resistance to warping and relatively easy to cut and finish. Edges are difficult to clean up so softwood edging strips are often used.	Used where heavier stronger structures are needed. Common for shelving and worktops.
Chipboard	Also known as Particle board. Made from wood chips and shavings glued/pressed together. Usually veneered or covered in plastic laminate. Easy to cut but not particularly strong. Lightweight and weak. Cuts easily but edges need covering. Prone to expansion when exposed to moisture where a moisture resistant version can be needed.	Used for kitchen worktops, usually veneered or covered with a plastic laminated. Shelving and general DIY work.

STRETCH AND CHALLENGE

Choose five common products made from timber. Use the internet to identify the timbers used to make each of these products. Think about the following:
- What does the product do?
- How is the product used?
- Where is it used?

9.2 Sources and origins

LEARNING OUTCOMES

By the end of this section you should know about and understand:
→ the sources and origins of a range of timbers and manufactured boards
→ the processes used to convert them into workable forms
→ the ecological and ethical issues associated with their conversion
→ the different ways in which timber can be recycled.

When a tree has been cut down, its branches are trimmed off and its bark removed before it is cut roughly into boards, planks or veneer. This process is known as **conversion**. There are two general methods of doing this: plain sawing and quarter sawing. A combination of the two is often used to reduce waste.

KEY TERM

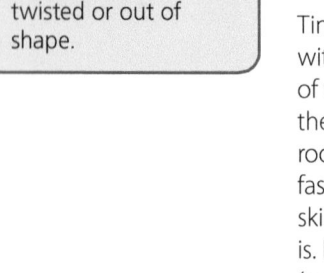

Warp: Become twisted or out of shape.

Plain sawing is the cheapest method but the outer planks tend to **warp**. Quarter sawing is more expensive as it requires more time and labour and produces more waste. Quarter sawn planks are more stable, however.

Timber contains a lot of moisture once it has been cut, which makes it very difficult to work with. For this reason it is usually dried out before use, a process called seasoning. The planks of wood are stacked on top of each other in such a way that air can circulate between them and reduce the amount of moisture in them. This is usually done outdoors under roofing to shield the timber from sunshine and rain. Alternatively, the timber can be dried faster under controlled conditions using a kiln. Natural seasoning is cheap and needs little skilled attention. It can take several years, however, depending on how thick the section is. Kiln seasoning takes only a few days or weeks and also kills any insect eggs in the wood (woodworm). However, kilns are expensive to build and run.

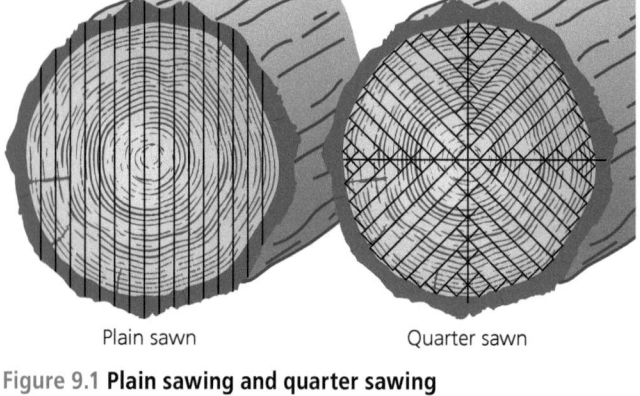

Plain sawn   Quarter sawn

Figure 9.1 **Plain sawing and quarter sawing**

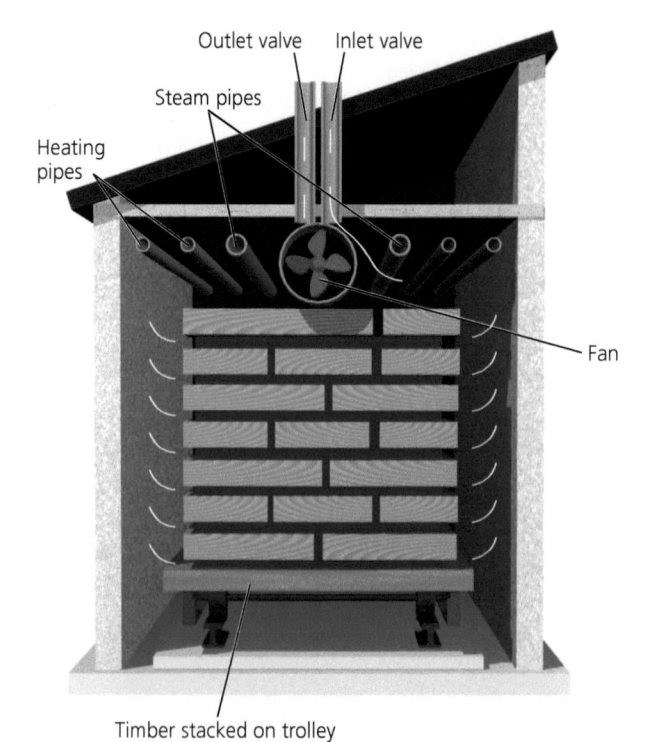

Outlet valve   Inlet valve

Steam pipes

Heating pipes

Fan

Timber stacked on trolley

Figure 9.2 **Kiln seasoning timber**

Timber sources

Timber is grown and processed all over the world. Below are the sources of the common softwoods and hardwoods.

Table 9.4 **Sources of common softwoods and hardwoods.**

Softwood	Sources
Redwood (Scots pine)	Northern Europe, Scandinavia, Russia, Scotland
Western red cedar	Canada, USA
Parana pine	South America (mainly Brazil)
Whitewood (spruce)	Northern Europe, Canada, USA
Hardwood	Sources
Beech	Europe (including UK)
Ash	Europe (including UK)
Elm	Europe (including UK)
Oak	Europe (including UK), Russia, Poland
Mahogany	West Africa, e.g. Nigeria, Ghana
Teak	Burma, India, Thailand

The ecological impact of using timber

Humans have been cutting down trees for thousands of years to clear land for farming and building and to use the wood for fuel. This process is called forestry. For thousands of years this has not been a problem, as the forests have continued to grow to replace the trees cut down. Therefore, timber could be described as a sustainable material. Over the past hundred years or so, however, this has not been the case, and the world's forests are steadily shrinking. This process is called **deforestation** and it has some important consequences, such as increasing soil erosion. Soil erosion causes barren land, flooding and landslides as well as destroying forest habitats. Forestry is sustainable as long as forests are allowed to replace themselves, or are replanted after felling.

The Forest Stewardship Council (FSC) was established in 1993 to promote responsible management of the world's forests. It does this by setting standards on forest products, along with certifying and labelling them as eco-friendly. If you use timber that is FSC-certified it means that it met the requirements of the FSC council and was gathered in an environmentally responsible and socially beneficial manner.

Manufactured boards

Manufactured boards were introduced to help reduce the amount of natural timber being used, and they have several advantages over solid natural timber:
- They are readily available in larger sheets of sizes up to 2440 × 1220mm from DIY stores and timber merchants.
- They are more stable than natural timber and less likely to twist and warp.
- They tend to be less expensive than natural timber.

Figure 9.3 **The Forest Stewardship Council (FSC) logo**

Veneers

Several manufactured boards make use of veneers. A veneer is a thin shaving of wood that is either cut from a log by rotary peeling or thinly sliced from a long block. Veneers are often applied to manufactured boards to improve their appearance and give them a smoother finish.

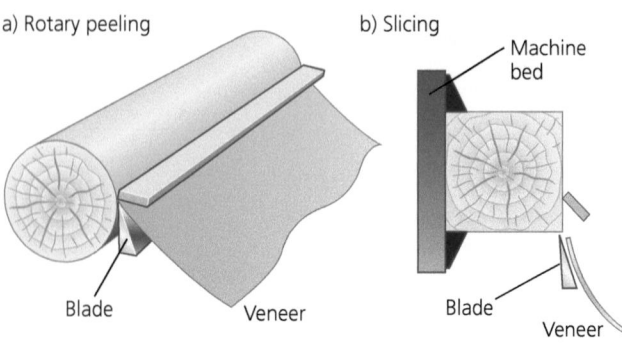

a) Rotary peeling

Blade   Veneer

b) Slicing

Machine bed

Blade

Veneer

Figure 9.4 Cutting wood veneers

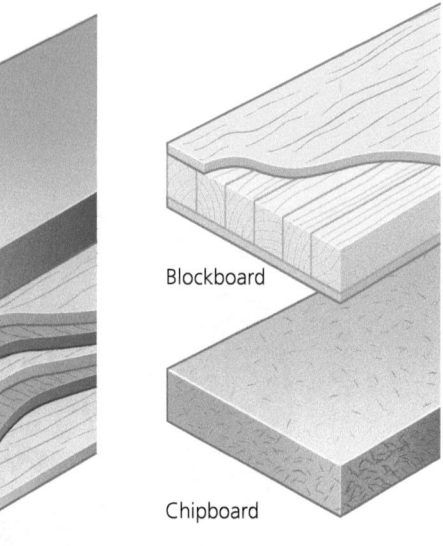

MDF

Plywood

Blockboard

Chipboard

Figure 9.5 Commonly used manufactured boards

KEY TERMS

Grain: The texture seen in wood.

Upcycling: Recycling into a higher-quality product.

Plywood

Plywood is made of three or more veneers of wood glued or 'laminated' together. Each layer of wood, or ply, is glued at right angles to the next in order to increase the strength of the finished piece. There is always an odd number of layers or plies so that the **grain** runs the same way on both outside pieces. This means the board is unlikely to bend, twist or warp. Plywood can be faced with a veneer of decorative hardwood to improve its appearance. Plywood is usually graded for interior or exterior use, depending on the glue used. Marine plywood is used in boatbuilding; regular plywood is used for doors and drawer bottoms; a rougher grade of plywood is used in the building industry.

Medium density fibreboard

Medium density fibreboard (MDF) is made by compressing tiny wood particles/fibres and adhesive together. It is generally denser than plywood and, comes in standard thicknesses from 2mm to 25mm. It is often painted or veneered, and is used to make indoor furniture such as wardrobes, bookcases and cupboards. Waterproof and fire-resistant versions are available for building work.

Blockboard

Blockboard is made by gluing strips of softwood of up to 25mm side by side. These strips are then sandwiched between two veneers of hardwood. Blockboard is used to make things like doors, shelves and tables.

Chipboard

Chipboard is made by gluing wood chips together. It is a fairly soft manufactured board and can crumble when worked with or exposed to wet conditions. There are different grades available: normal, medium and high density. High density chipboard is often used for kitchen worktops, where it is laminated with melamine (a polymer). As with MDF there are moisture- and flame-resistant versions available.

Recycling timber

There are a number of ways that timber products and the waste created from producing them can be used again.

- Reclaimed timber is the term used when old buildings are torn down and the timber used in their construction is put to other uses, such as flooring or to make items of furniture. Reclaimed railway sleepers are a favourite of landscape gardeners.
- **Upcycling** is the process of reusing materials in such a way as to create a product of higher value, such as furniture made from old pallets. It can also describe the process of 'freshening up' a product by modifying it.

- Downcycling is the opposite of upcycling; the timber is converted into a material of lesser quality, such as strips for blockboard or chips for chipboard.
- Biomass is a fuel developed from organic materials such as scrap timber and is burned to create electricity.

Eco-materials

Bamboo is a grass that grows quickly. An oak tree can take 120 years to grow to maturity before it is felled, while bamboo can be harvested in just three years. It is a flexible and lightweight material that is very strong. It is used predominantly as a building material in South East Asia but it can be cut and laminated into sheets and planks and has become a popular material for designers. It is considered an **eco-material**.

Figure 9.6 **Bamboo is considered an eco-material.**

ACTIVITY

Make your own manufactured board:
1. Mix some wood chippings with glue until you have a stodgy paste. Spread this between two small sheets of card and compress using a vice. Leave to dry overnight. Once dry, cut into square chunks so you can see the structure of the 'chipboard' you have made.
2. Do the same experiment with wood dust and glue to form your own version of MDF. Cut scrap wood into identical sections and glue together to form 'blockboard'. Add an outside membrane to hold it together and to give it a flat surface.

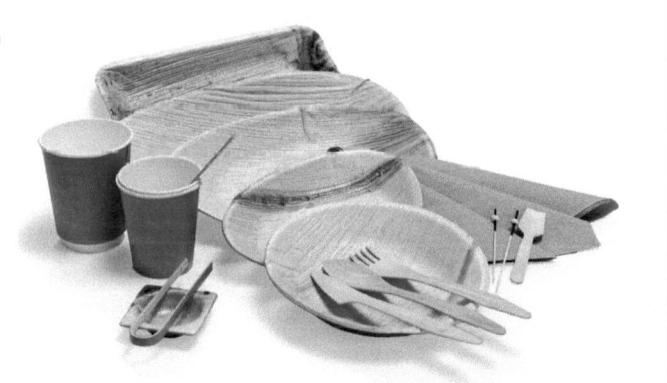

Figure 9.7 **A range of bamboo products**

9.3 Commonly available forms

LEARNING OUTCOMES

By the end of this section you should know about and understand:
→ the commonly available stock forms of timber and manufactured boards
→ a range of standard components used to assemble products made with timber and manufactured board.

Commercial forms and sizes of timber

After conversion and **seasoning**, timber is ready to be reduced again into smaller sections of common shapes and sizes that can be bought from timber merchants and DIY stores. This is done using a range of band saws, circular saws and planing machines. Timber is sold either rough sawn or planed. Planed timber can be either planed both sides (PBS) or planed all round (PAR).

The size of planed timber is described as the nominal (rough sawn) size but will actually be around 3mm smaller. PAR timber costs more than rough sawn timber and would only be used where accuracy and a smooth finish were required.

KEY POINTS

- Timber is only a sustainable material if you know where in the world it comes from and whether its source is maintained and looked after.
- Understanding the structure of manufactured boards will help you determine their suitability for a range of products.

KEY TERMS

Eco-material: An environmentally friendly material.

Seasoning: Adjusting the moisture content of timber to make it more suitable for use.

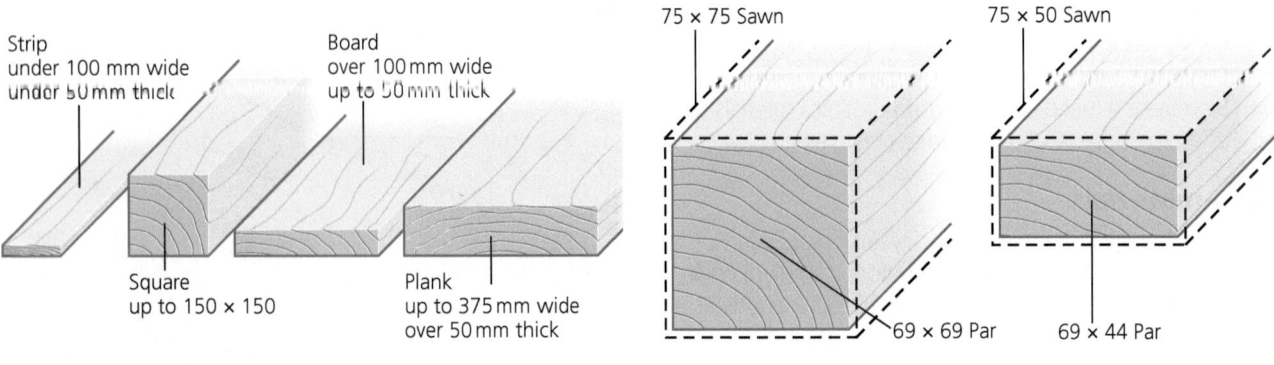

Figure 9.8 **Standard timber sections**

Figure 9.9 **Typical planed timber sizes**

KEY TERMS

Moulding: A pre-cut section of timber often used for skirting boards.

Timber yard: A place where you can buy timber.

Timber can be cut into a variety of shapes and sizes that are often described as planks, boards, strips and squares. With machinery on site, however, **timber yards** can often cut timber to any size/section required. Timber is usually available in lengths of 2.4m and 1.8m from DIY stores and longer lengths of 3.6m and 4.8m from timber yards.

- **Plank** is the term given to the largest sections, usually around 50mm thick and measuring anywhere from 225 to 375mm wide.
- **Boards** are smaller than planks, usually no thicker than 40mm and around 100mm wide or more.
- **Strips** are smaller than boards, less than 50mm thick and no wider than 100mm.
- **Squares** are smaller still and have the same height and width.
- **Dowelling** is a solid cylindrical rod of wood often called dowel rod. Dowels are used in a variety of products, for example as axles in toys, as supports for hanging items like keyrings and as structural reinforcement in furniture (dowel joints).
- **Timber mouldings** are available in a wide variety of shapes and sizes and can be used for decorative or constructional purposes. They are made by machining sections of timber with specially shaped cutters.

Commercial forms and sizes of manufactured boards

Manufactured boards are available in large sheets measuring 2440 × 1220mm. Half sheets are also available at 1220 × 1220mm. Each board is available in a variety of set thicknesses depending on the type of board, for example MDF and plywood are available in thicknesses from 3 to 25mm thick.

MDF can also be used for **mouldings** and is available in different shapes and sizes that can be veneered or painted and used as decorative architrave and skirting.

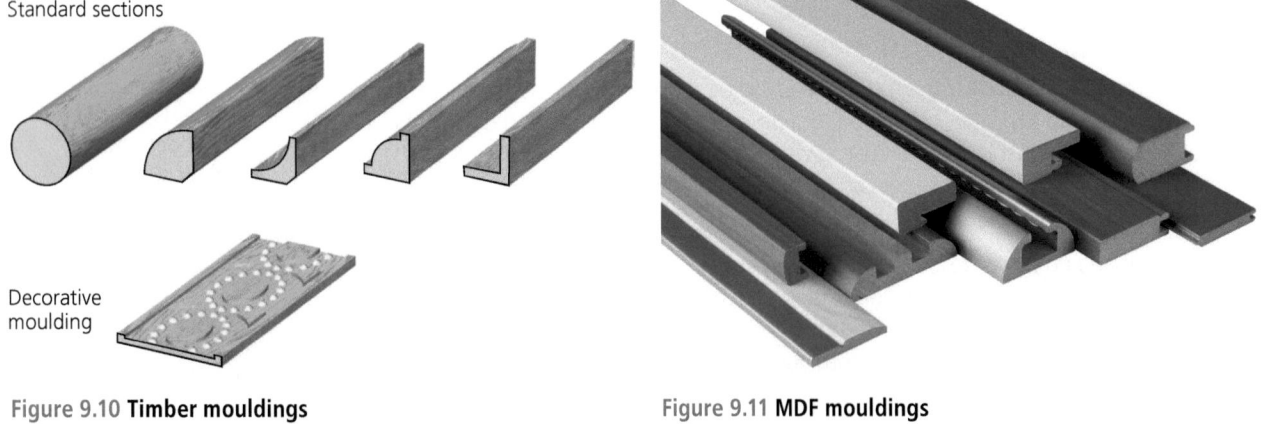

Figure 9.10 **Timber mouldings**

Figure 9.11 **MDF mouldings**

Standard components

There will be times when you will construct and assemble a product and will need to use a woodscrew, handle, hinge or bracket. These are all examples of pre-manufactured standard components. There are many thousands of different components but this section will look at examples that could be important during your NEA (non-examined assessment).

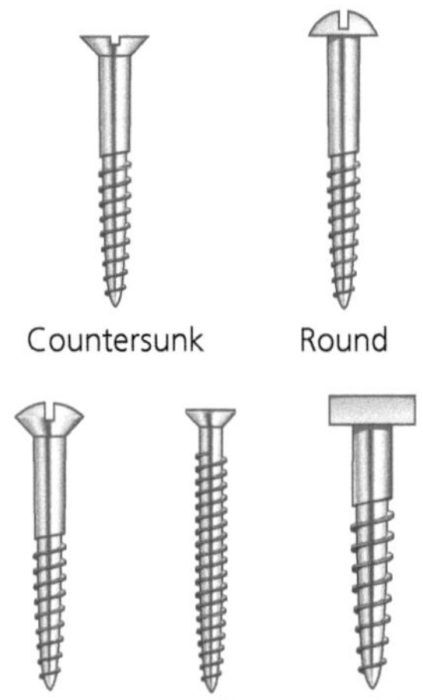

Woodscrews

Woodscrews provide a strong, neat method of fixing wood to wood, and metal and plastic to wood, and also for fixing hinges, catches and locks. There are many different types of screw, which are specified by length, gauge, material and type of head (e.g. 30mm, No. 5, brass, countersunk head). Screws are mainly made from steel (cheap and strong) or brass (better-looking and corrosion-resistant). Steel screws can be made more attractive and corrosion-resistant by galvanising, chrome-plating and black-lacquering (called black-japanning).

Screws have various types of head and screwdriver slots:

- **Countersunk head:** A countersunk head is used when you want the head of the screw to be level with the surface, for example when fitting hinges.
- **Round head:** A round head is used to fasten thin sheet materials such as metal or plastic to wood, such as brackets.
- **Raised head:** A raised head is used for decorative purposes, for example when fitting door furniture such as handles.
- **Twinfast:** A twinfast is used specifically on chipboard as it has two threads, which provide greater holding power.
- **Coach:** A coach is used where more holding power is required. It is tightened with a spanner. The metalwork vices in your workshop will be fastened to the benches using coach screws.

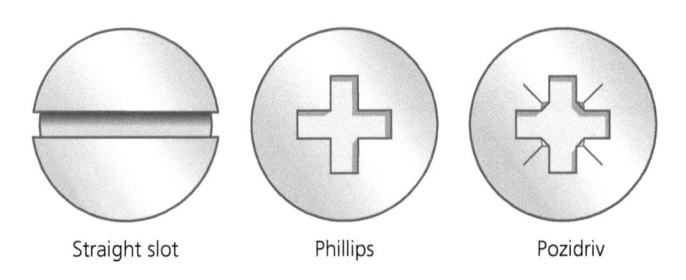

Figure 9.12 **Common types of woodscrew**

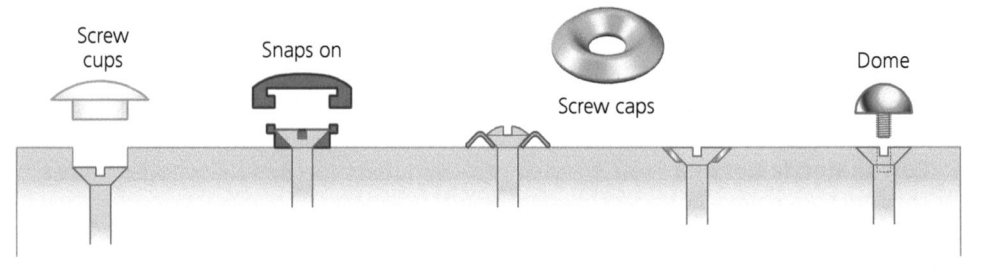

Figure 9.13 **Types of screwdriver slots**

Caps and cups

There are various types of caps and cups that can be used to cover or enhance the look of screwheads.

Figure 9.14 **Caps and cups**

Nails

Nailing is the quickest way to join pieces of wood. Most nails are made from steel but there are also galvanised types as well as brass ones.

- **Round wire nails:** These are used for general joinery work and are available in lengths from 12 to 150mm.
- **Oval wire nails:** These are used for interior joinery. They range in length from 12 to 150 mm. Because they have virtually no head, they can be hidden easily and filed.
- **Panel pins:** These are used with small-scale work and for pinning thin sheet material. They range in length from 12 to 50mm.
- **Masonry nails:** These are particularly useful if you want to fasten into brickwork or mortar.

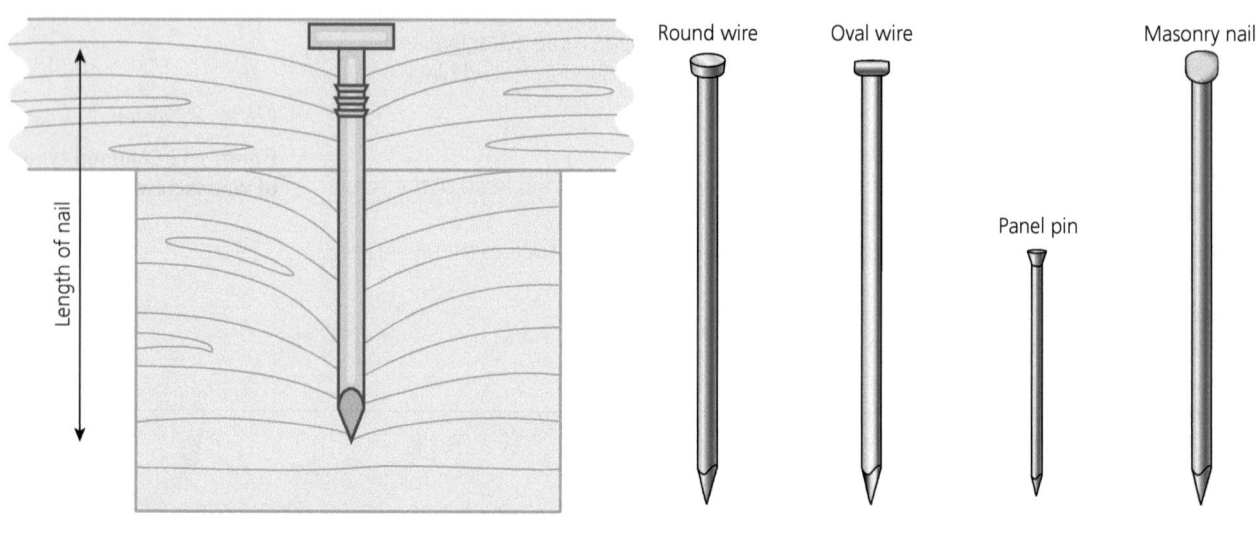

Figure 9.15 **Two pieces of wood fastened with a nail**

Figure 9.16 **Common types of nail**

Other fastenings

- **Clout nails:** These are used to fasten roofing felt to shed roofs. They are galvanised to prevent rusting.
- **Staples:** These are used to make packing crates and in upholstery work. Square types can be fired from a gun, whereas round staples are hammered in.
- **Cut tacks:** These are used in upholstery to fasten fabric to wooden frames. Because of their appearance they are usually hidden in use.
- **Hardboard pins:** These are used to fasten hardboard to frames such as the back of wardrobes and bookcases. Because of their diamond-shaped head they are hidden once hammered in.
- **Corrugated fasteners:** These are used to make quick corner joints in wooden frames. They are hammered across each piece of wood.

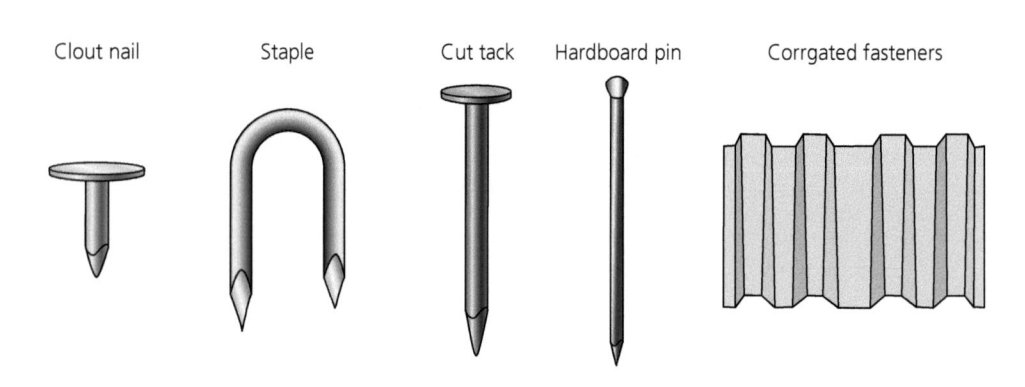

Figure 9.17 **Common types of fastening**

Hinges

A hinge is a moveable joint that swings as it opens and closes, like you would find on a door or window. There are several types suitable for use when making timber products:

- **Butt hinges:** One of the most commonly used hinges. They are used for doors, windows, and large and small boxes. They are commonly made from steel or brass and need recessing into the wood.
- **Piano hinge:** Piano hinges come in different lengths and are used where you need a lot of support along the edges of a long product. As the name suggests, they are used on the lids of pianos.
- **Butterfly hinge:** Used on lightweight doors. They are easy to fit and look more attractive than butt hinges.
- **Flush hinge:** Do not need a recess to be cut, but are not as strong as butt hinges.
- **Barrel hinge:** Screwed into the edge of the wood. They are easy to fit and allow doors to be removed easily.

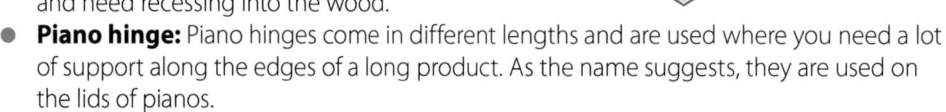

Figure 9.18 **Common types of hinge**

Stays

Stays are used on doors to control how they open, stay open and close. Stays can be used when doors open downwards in order to stop them dropping completely, and they can also be used to hold up doors that open upwards.

Drawer and cabinet components

Handles

When you make a product with a small door or drawer you will need a handle to help you pull it open. DIY stores stock hundreds of different types, available in a range of materials, finishes and styles.

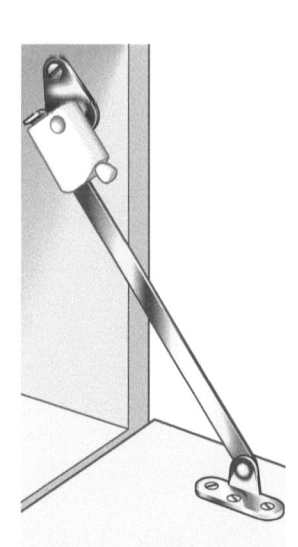

Figure 9.19 **Brass stay with adjustment on a door that opens down**

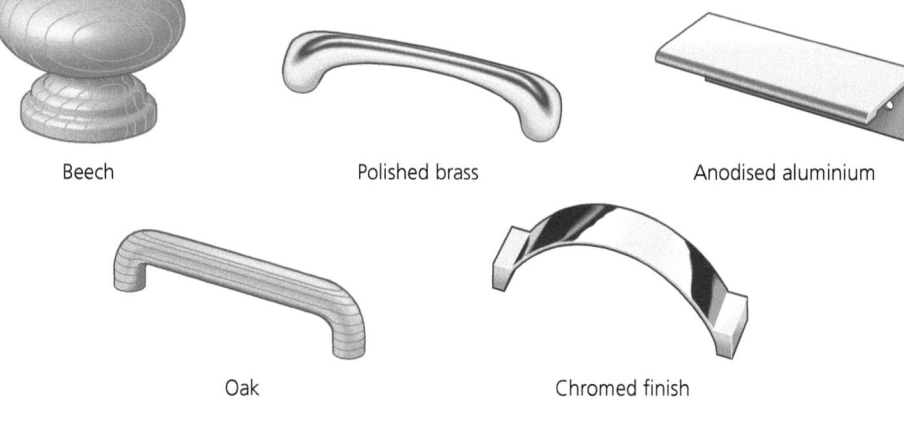

Beech   Polished brass   Anodised aluminium

Oak   Chromed finish

Figure 9.20 **Drawer/cabinet handles**

Runners

Drawers in products require runners to help them open and close. Runners range from simple nylon pieces fixed to the side that run in a groove, to complicated metal runners that support the whole drawer as it is pulled out (extension runners).

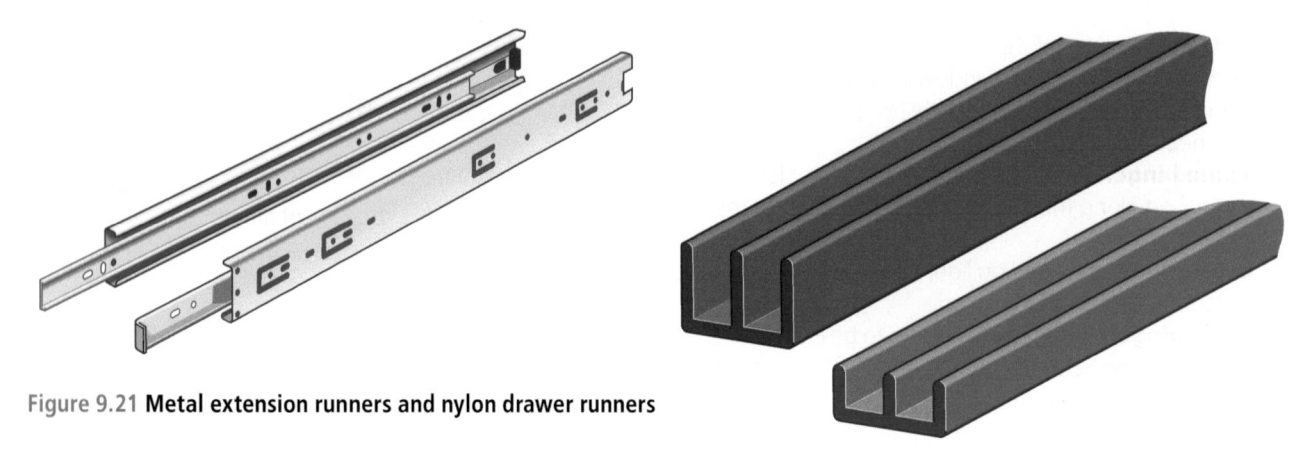

Figure 9.21 **Metal extension runners and nylon drawer runners**

Figure 9.22 **Plastic runners**

Door runners are required when sliding glass or plastic doors are used. These are extruded plastic 'channelling' that is fixed to the top and bottom of the cabinet.

Locks

There are several different types of lock available that you could incorporate into your products:

- **Hasp and staple** – are used with padlocks to fasten shed doors and gates.
- **Cam locks** – have a tab that turns and catches in a slot; they are used in display cabinets.
- **Magnetic locks** – are hidden locks that can require a magnetic key to open; they are often used to 'child proof' cupboards.

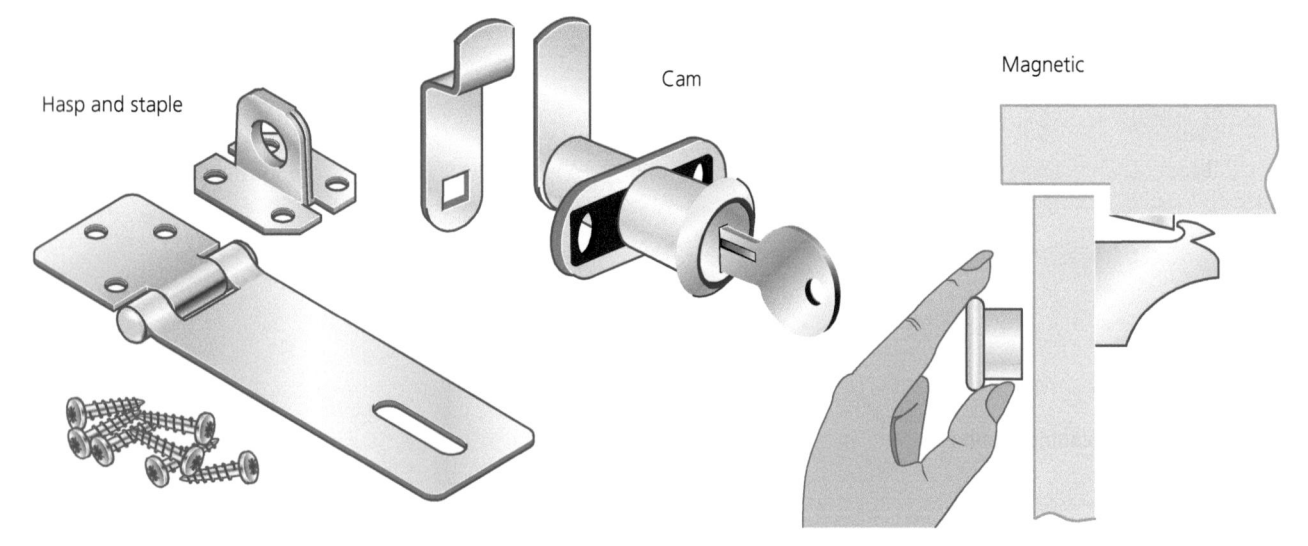

Figure 9.23 **Different locks**

Catches

Catches are used to keep lids closed on boxes or doors shut on cabinets. They are available in a range of materials depending on the type of product they will be used on.

- **Magnetic catches** – are very common and are available in a range of materials. Polymer ones can be found on kitchen cabinets, while high end cabinet work would use ones made from brass.
- **Spring catches** – tend to be used on more functional items as they are not as attractive.
- **Ball catches** – can be recessed into doors and are very neat. They are available in a range of sizes.
- **Toggle catches** – can be used to fasten the top and bottom of a product together.

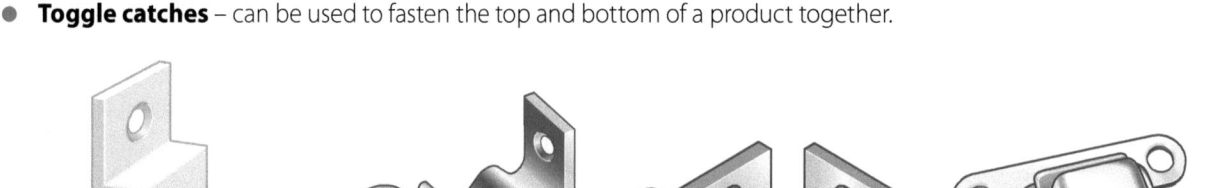

Magnetic catch   Spring catch   Ball catch   Toggle catch

Figure 9.24 **Common types of catch**

KEY POINT

An awareness of the different components available for use on timber products will help inform your design decisions when you are developing ideas.

ACTIVITY

1. Prepare a stock list of the wood materials in your workshop. Present this in the form of a table and include details such as: name of timber; type of wood (hardwood or softwood); forms and sizes; type and thickness of manufactured boards; shapes of timber mouldings.
2. Visit your local DIY store and look at the timber for sale there.

- How much difference is there between the prices of sawn timber and planed timber?
- Why do you think there is this difference?
- How many different shapes of timber mouldings does the store sell?

9.4 Manipulating and joining

LEARNING OUTCOMES

By the end of this section you should know about and understand the use of specialist techniques to shape, fabricate, construct and assemble high quality prototypes from timber and manufactured board, including:
➜ wastage   ➜ addition   ➜ deforming.

Wastage

Wastage involves cutting materials to the shape required and removing any excess material. It includes processes such as sawing, drilling, filing, sanding and turning.

Marking-out tools

Before cutting, bending, shaping and joining materials you will need to mark out your design. Marking out accurately is important to ensure different parts fit together correctly. A range of tools can be used when marking out.

Marking knife

This is used to mark lines on a piece of timber. It is thinner than a pencil mark and can sometimes help guide you when sawing. Marks are made across the grain and a sharp pencil can sometimes be used inside the line to make it show up more clearly.

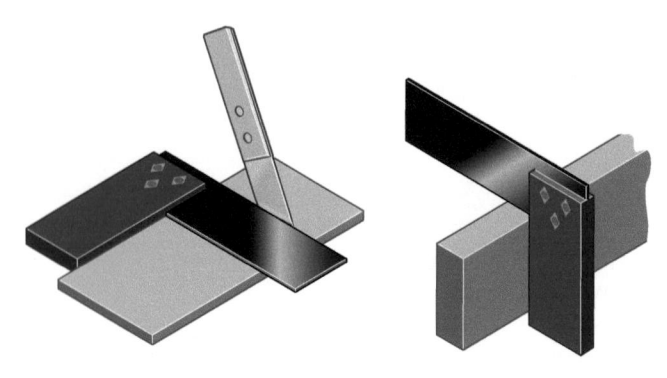

Figure 9.25 Try square

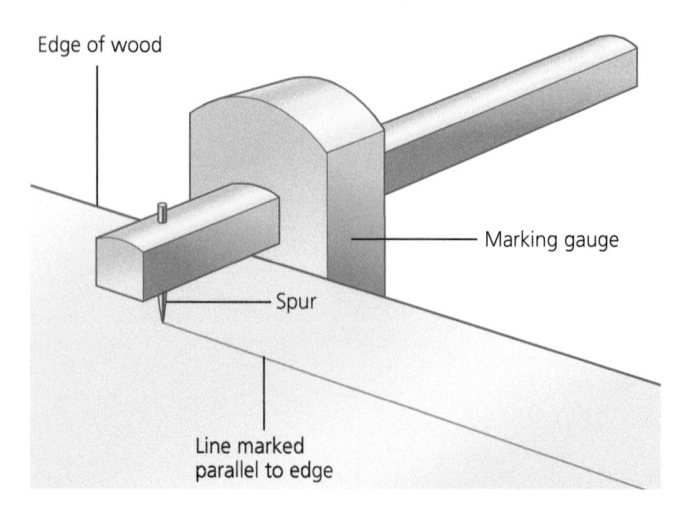

Edge of wood

Marking gauge

Spur

Line marked parallel to edge

Figure 9.26 Marking gauge

Try square

A try square is used to mark a line at 90 degrees to an edge. It can also be used to test how square an edge is by placing it on the corner.

Marking gauge

This is used on timber to draw a line along the grain, parallel to an edge, using a sharp point called a spur.

Cutting gauge

This is used on wood to draw a line across the grain, parallel to an end. It is different to a marking gauge in that it uses a small blade to mark across the wood, like a pizza cutter.

Mortise gauge

This is used in the same way as a marking gauge, but it has two spurs that can be adjusted to mark out the mortise or tenon on a wood joint.

Mitre square

This is used on timber to mark out or measure angles of 45 or 135 degrees.

Sliding bevel

This can be set to any angle, using a protractor, and then tightened in position.

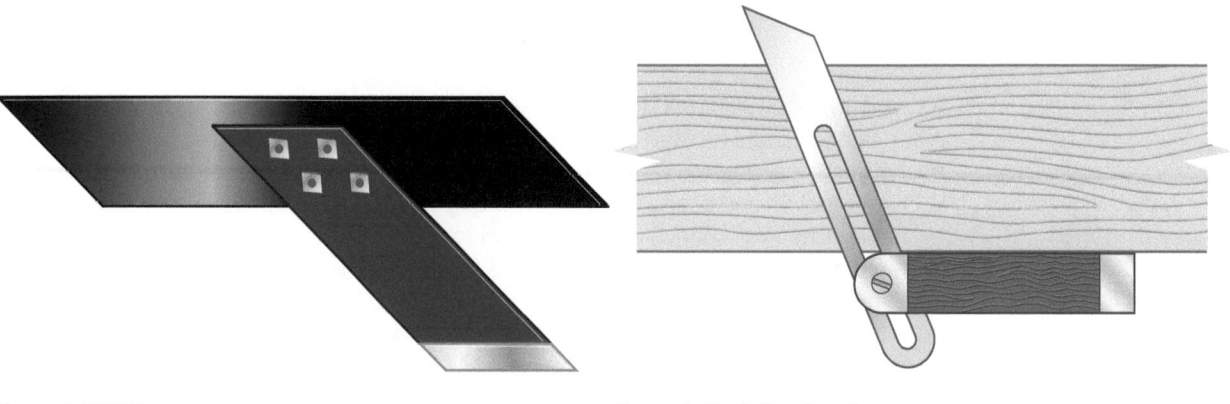

Figure 9.27 **Mitre square**

Figure 9.28 **Sliding bevel**

Saws

A 'handsaw' is the general name for a saw used to cut large pieces of timber by hand.

- A **cross-cut saw** is used to cut across the grain.
- A **ripsaw** is used to cut along the grain.
- **Tenon saws** are general purpose saws that are very good at making straight cuts in timber.
- **Coping saws** are used to cut curves in thin pieces of timber.

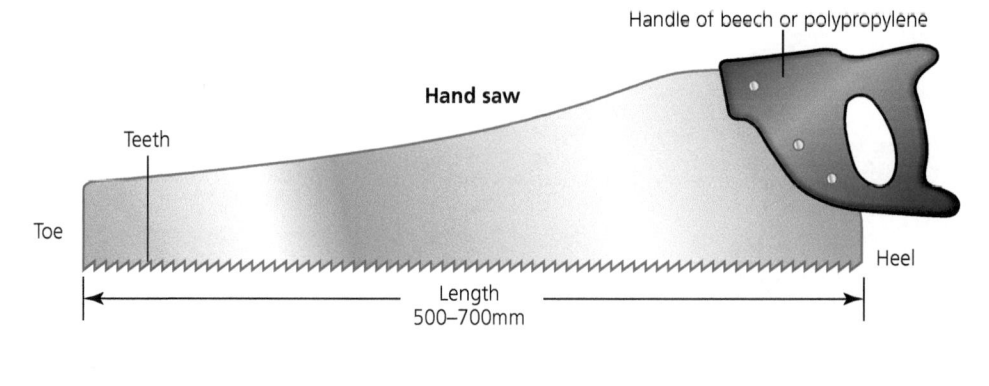

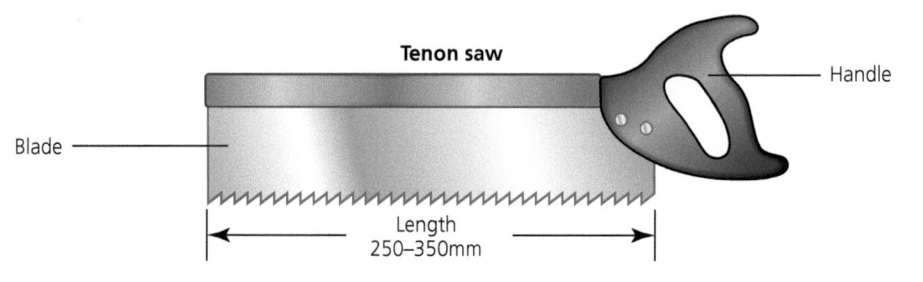

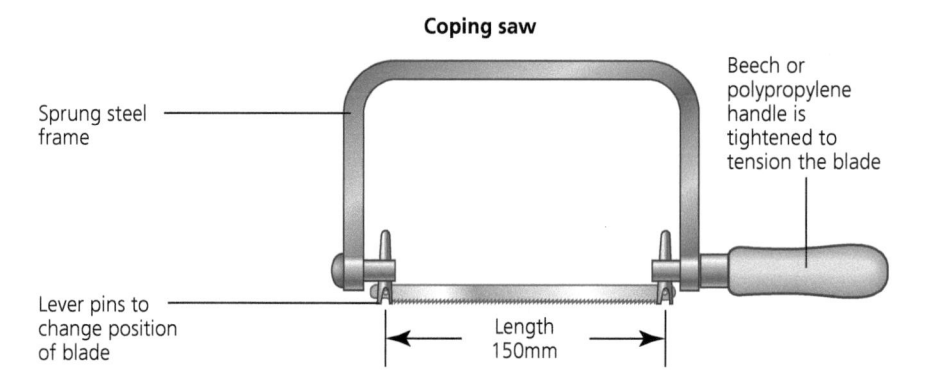

Figure 9.29 **Handsaw, tenon saw and coping saw**

A woodworker's vice is used to secure the workpiece when using tenon and coping saws. A bench hook could be used to cut smaller pieces of timber.

- A **jigsaw** is an electric saw and is useful for cutting shapes from manufactured board. Fine, medium and coarse blades are available for different purposes. The work is usually held across a workbench to make it easier to cut.
- A **fret saw** is a small electrical saw that is fixed to a bench and used for cutting curves in thin material. The blade used is thin and moves quickly up and down to cut through the work, so it is important that the piece is held firmly.
- A **band saw** is a large, heavy duty electrical saw with a band-shaped metal blade for cutting irregular shapes in timber and manufactured board. The width of the blade determines how tight a curve can be cut. Portable versions are available, which are fixed to benches. Larger models are freestanding and fixed to the floor.
- A **circular saw** is another large heavy duty electrical saw, and it uses a circular blade for cutting timber and manufactured board. Smaller handheld models are used where portability is needed. Larger freestanding saws are often used on building jobs, and even larger versions called table saws are used in factories and workshops.

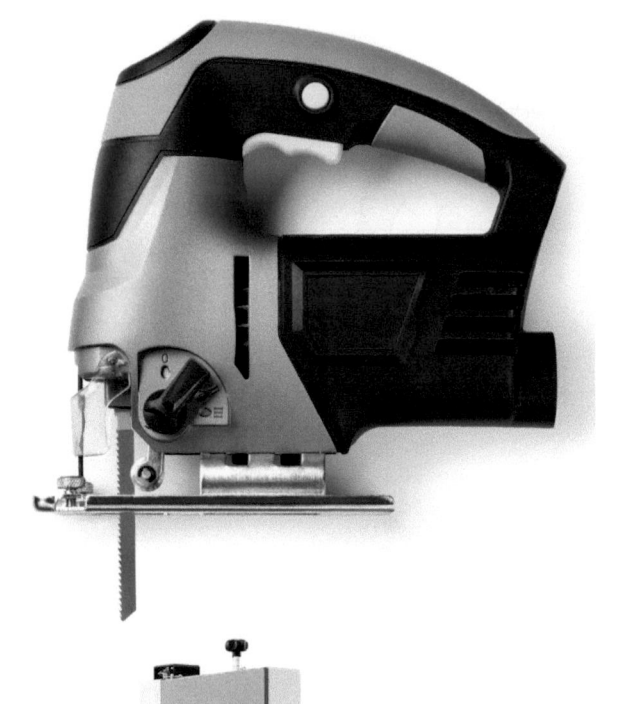

Figure 9.30 **Jigsaw, fret saw, band saw and circular saw**

Drills and drill bits

There are many different types of drills and drill bits used when drilling holes in timber and manufactured board. The drills and drill bits can be held either in a hand drill, a portable drill, a drilling machine, or a centre lathe.

- **Twist bits** are the most common type of drill bit and are used for drilling holes in timber, metal and plastic.
- **Countersink bits** provide the countersunk shape for screws so they do not protrude from the surface.
- **Forstner bits** provide smooth-sided, flat-bottomed holes.
- **Flat bits** provide fast and accurate drilling in solid pieces of timber.
- **Hole saws** have interchangeable cutting blades that enable you to drill holes of 20–75mm diameter.
- **Expansive bits** can be adjusted to drill shallow holes in wood of 12–150mm diameter.

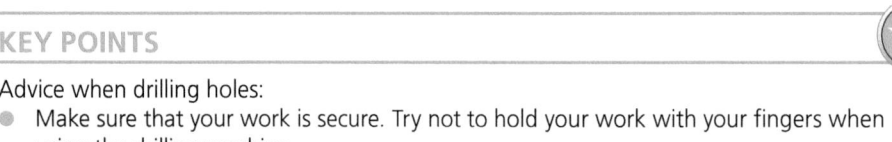

Portable electric drill

Chuck

Wheel brace or hand drill

Cordless rechargeable drill

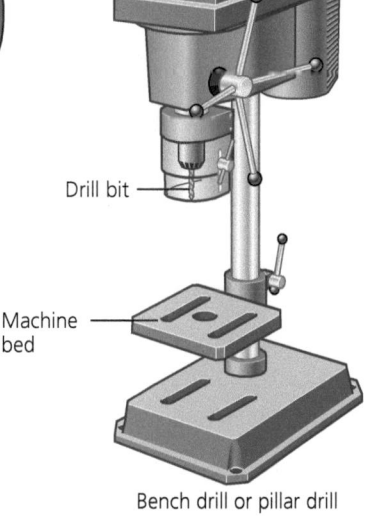

Drill bit

Machine bed

Bench drill or pillar drill

Figure 9.31 Hand drills and electric drilling machines

KEY POINTS

Advice when drilling holes:
- Make sure that your work is secure. Try not to hold your work with your fingers when using the drilling machine.
- Make sure that the drill bit is tight in the chuck, remove the key and ensure that the safety guard is in place.
- If you are drilling straight through material, place scrap wood underneath to drill into.
- When drilling large-diameter holes, drill a 'pilot' hole first.

Filing

Files are made from hardened and tempered high-carbon steel. They can be used to shape and remove/waste timber and some manufactured boards. Files are classified by length, shape and cut.

There are various grades of files: rough and bastard cuts for coarse work; second cut for general use; and smooth and dead smooth for very fine work before polishing. The shape of each file is designed to be used to produce a similar shape on the material it is cutting. There are flat, round, curved, square and triangular shaped files. A hand file has a flat edge without any teeth that allows it to file accurately in a 90-degree corner.

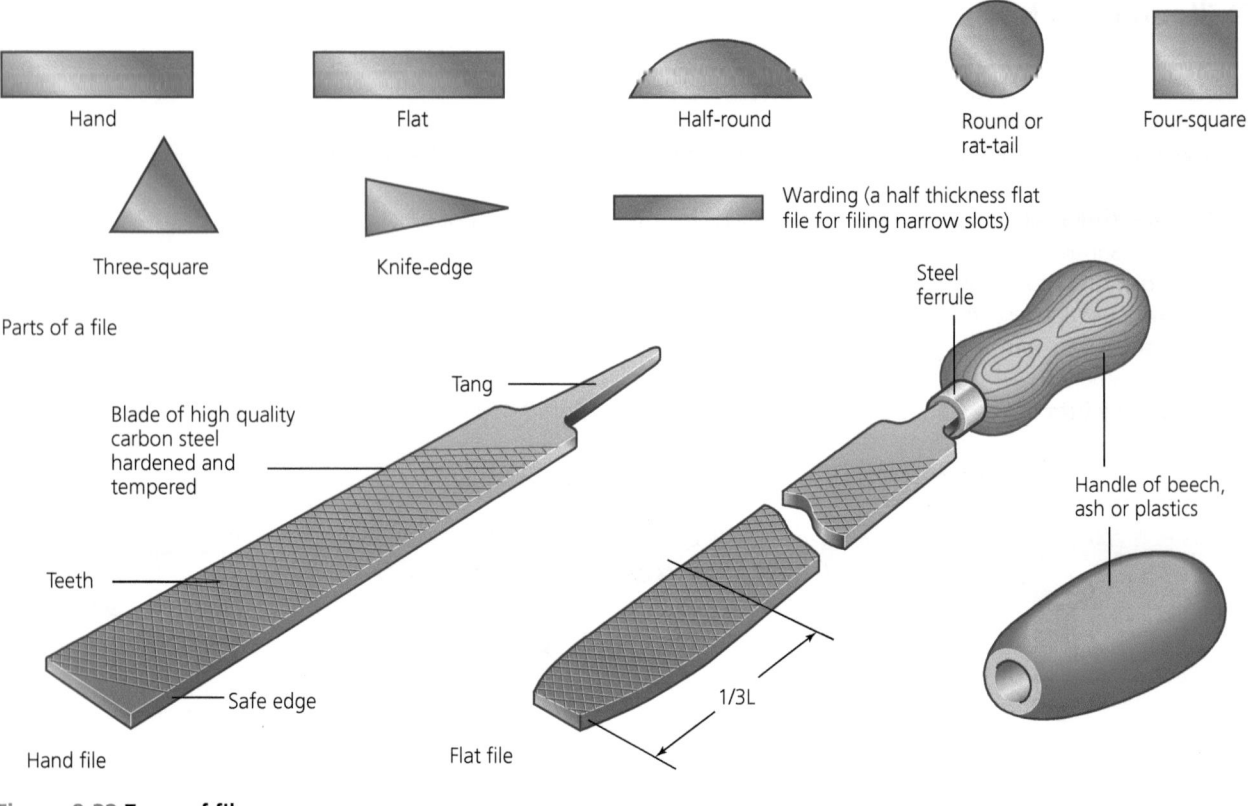

Figure 9.32 **Types of files**

> **KEY POINT** ★
>
> Advice when filing:
> - Cross filing uses the full length of the file and is used to remove waste quickly but does not leave a smooth surface.
> - Draw filing is used to produce a smooth surface after cross filing. Only part of the file is used so very little material is removed.

Needle files are small files used for intricate work and are available in a wide variety of shapes for different purposes.

Surfacing forming tools (surform tools)

Surform tools have a cutting action similar to a cheese grater and can remove wood very quickly. The most common shaped blades are flat, curved and round.

Chiselling

Chisels are used to remove small amounts of wood. Chisels are mainly used when cutting a joint in wood. Three basic types of chisel are used: the firmer chisel is a general all-purpose chisel that has flat sides and is used with a mallet. The bevel-edge chisel has bevelled or sloping edges that allow it to be used in corners, for example, when cutting out a dovetail joint. The mortise chisel has a thicker blade to lever out waste without the wood breaking. It is designed to be hit with a mallet.

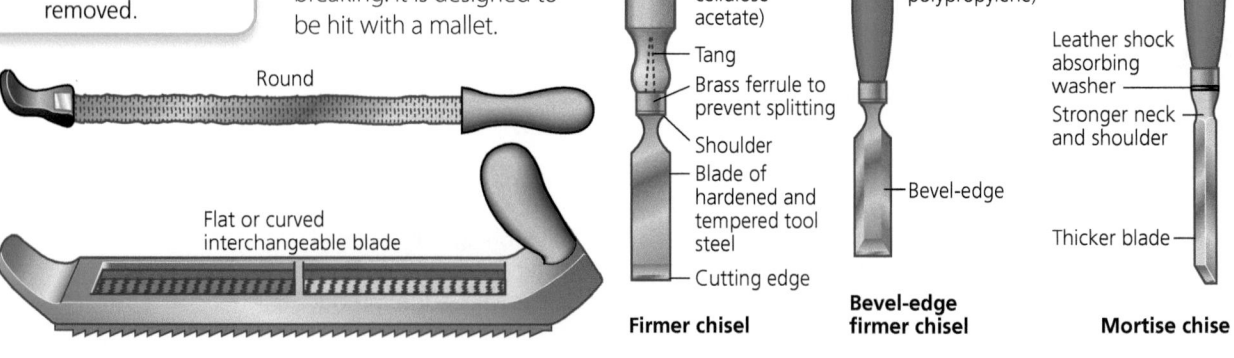

Figure 9.33 **Surform tools**

Figure 9.34 **Types of chisel**

KEY POINT

Advice when chiselling:
- Chisels can be hit with a mallet or manipulated by hand. Paring describes chisels that are used with hand pressure only.
- Safety is essential, and the most important rule is that you must keep both hands behind the cutting edge of the chisel at all times.
- Horizontal paring is when the wood secured in a vice. You would use horizontal paring to cut out a halving or housing joint.
- Vertical paring from above requires considerable pressure and a G cramp is needed to secure the wood. You may need to use a mallet to provide enough force to cut through the fibres of the wood.

Planing

Planing is carried out to remove excess wood and to produce a smooth surface finish. The cutting action of a plane is similar to that of a chisel held in a frame at a specific angle. The two most common planes are Jack planes and Smoothing planes.
- Smoothing planes are about 250 mm long; they have a blade that is ground and sharpened for fine finishing and for planing end grain.
- Jack planes are longer and heavier and are used for the quick removal of waste wood to make surfaces flat and to achieve the required size.

There are other 'special' planes that you may see in a workshop.
- Bullnose planes are used to plain into corners
- Rebate planes cut out rebates or grooves along the edges of timber pieces.
- Router planes are used to cut out groves across the grain of wood.
- Spoke shavers can used to round the edges on curved pieces of timber

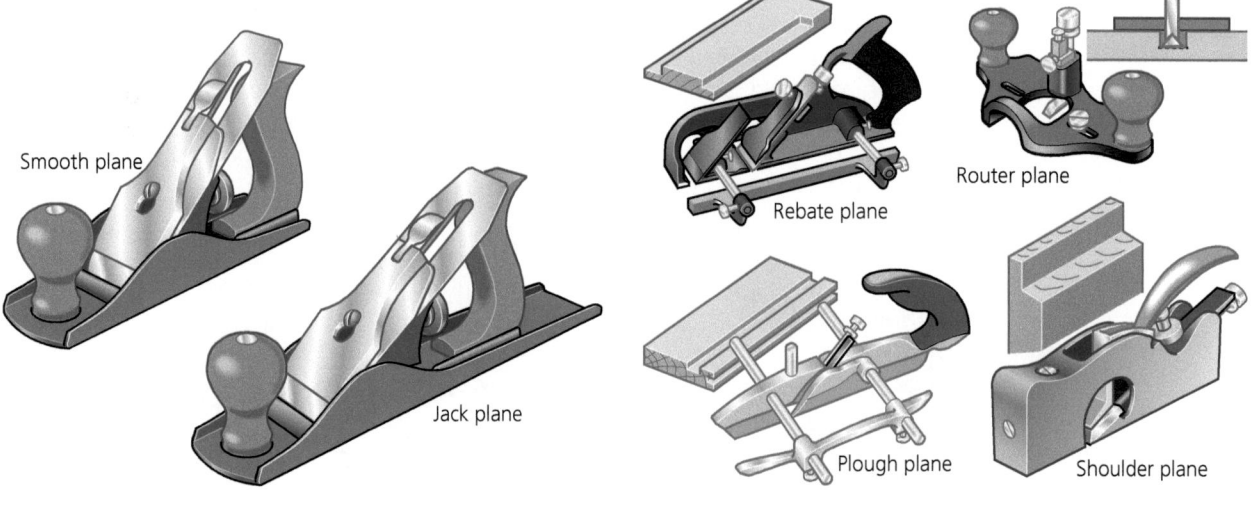

Figure 9.35 **Types of plane**

Figure 9.36 **Special planes**

KEY POINT

Advice when planing:
- Always plane along the grain, otherwise you will tear the surface of the wood.
- When planing the edge of a piece of timber it is important to secure the wood in a vice.
- When planing long thin strips of wood use a bench stop, as the bench itself will support the strip and stop it from bending.
- When planing end grain there is a high chance that the work will split when the plane goes across the edge. If the work is long enough plane from the edge to the middle. Otherwise consider cramping or gluing a small piece of wood to the end that will take the damage.

Hand-held portable planing machines called electric planers can be used when you need to remove small amounts of wood away from a workshop. When fitting a door for example.

Planer thicknessers are larger electric planers fixed to the floor and are used to plane pieces of timber to specific thicknesses. The work is pulled through the machine as a spinning tool blade rotates and removes a specified amount.

Figure 9.37 **A hand held electric planer and a planer thicknesser**

Sanding

Abrasive paper is used to prepare timber for a finish (see earlier section) this sanding is done by hand. Machinery can also be used to help speed up this process. Orbital sanders are portable, hand held sanding machines. Abrasive paper is attached to the pad of the machine by clips, the pad then vibrates in small circular motions. Palm sanders are small sanders that fit into the palm of your hand. The abrasive paper is attached to the pad using Velcro. The pad is triangular shaped so you can sand into corners with them.

 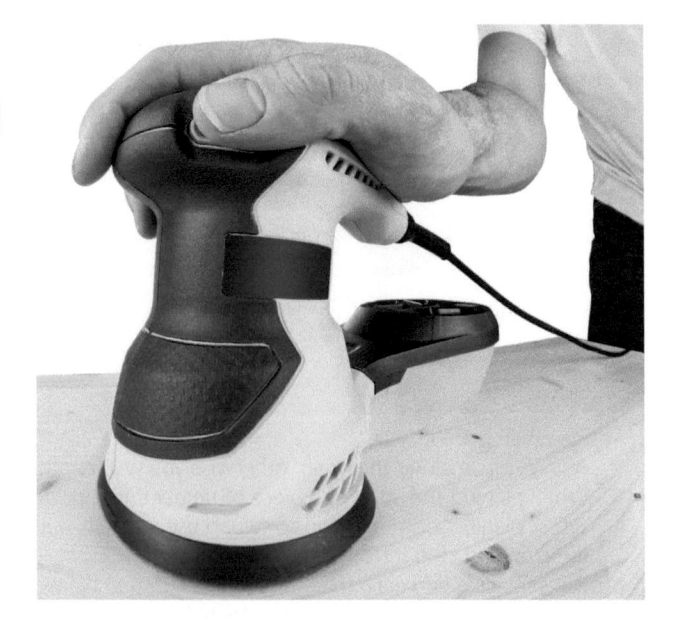

Figure 9.38 **an orbital sander and a palm sander**

Belt sanders have a motor that drives two drums that turn a continuous band of abrasive paper. They can be hand held or fixed. Some smaller units are fixed to bench tops whilst others have built in extraction and are fixed to the floor. As with orbital sanders, belt sanders are fitted with an extraction system that can gather dust in a cloth bag or be fitted to a vacuum system.

Disc sanders are bench mounted machines that have circle of abrasive paper fixed to a spinning disc. Due to the rotation of the disc they can only be used on one side, the other side is guarded. The work piece is placed on the machine's table and moved towards the spinning disc. Never attempt to sand work unless it is held firmly on the table.

Both Disc and Belt sanders are quite aggressive and can remove significant amounts of material so they tend to be used at the start of a project. You would then move on to hand sanding or orbital sanding to finish.

Figure 9.39 A disc sander and a hand held belt sander

Wood turning

The wood lathe is a machine that rotates a piece of timber and special tools are then used to cut into the wood as it turns to create a 'turned' shape. There are two types of turning that can be carried out on a wood lathe:
- between-centres turning
- faceplate turning.

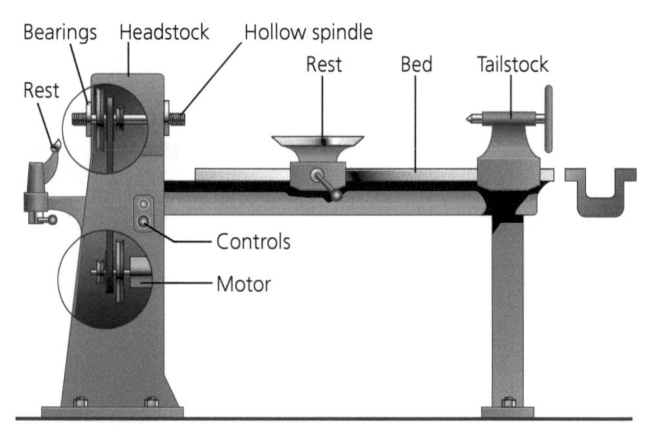

Figure 9.40 Wood-turning lathe

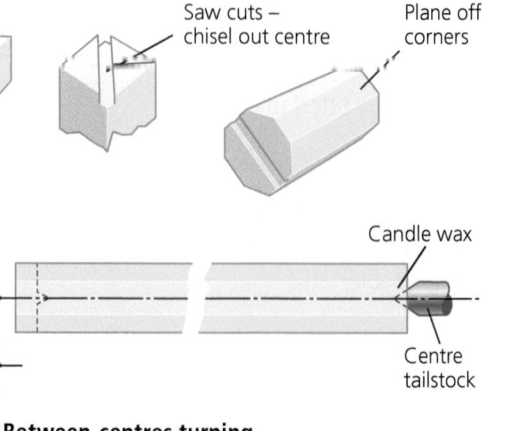

Saw cuts –
chisel out centre

Plane off
corners

Fork centre

Candle wax

Centre
tailstock

Figure 9.41 Between-centres turning

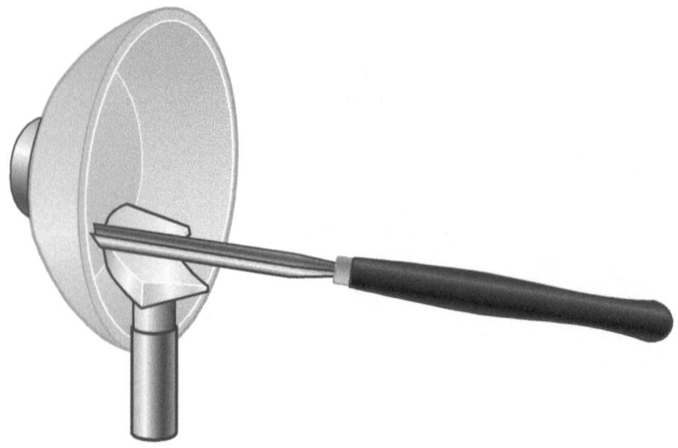

Figure 9.42 Turning a bowl using a faceplate

Between-centres turning

Between-centres turning is used to make products such as rolling pins and chair legs. You begin with a square section of timber which has had the corners planed off and a saw cut across the end (this makes it easier to start). A fork centre (specially shaped to grip the piece) is held in a hollow spindle on the headstock of the lathe. This is connected directly to the motor and turns the piece of timber. A dead centre (shaped to allow the piece to spin freely) is held in the tailstock at the opposite end. The tailstock is moved down to the workpiece, which is held firmly between both the headstock and tailstock. As the timber turns, tools are placed on the rest and moved towards it to cut and carve shapes into it.

Faceplate turning

Faceplate turning is used to make products such as bowls. The timber used for faceplate turning is bigger and flatter than that used for between-centres turning. The workpiece is screwed to a faceplate, which is connected to the outer side of the headstock. Specially shaped tool rests are used to allow the user to get their cutting tools inside the work.

Lathe tools are called chisels but they differ from the ones described earlier in this chapter. They feature long, round, curved handles that offer a better grip and sufficient leverage to enable the turner to control the cutting edge accurately. There are several different types of turning tool:

- **Gouges**: These usually have specially shaped cutting edges for performing particular cuts, such as bowl gouges, with concave, curved cutting edges to form the smooth, curved surface of a bowl.
- **Scrapers**: These are often flat or slightly curved chisels for removing wood from flat or cylindrical shapes, or for roughing out a shape.
- **Parting tools**: These are thin, vee-tipped tools for cutting off workpieces.
- **Spoon cutters** have a spoon-shaped cutting edge and are also often used for shaping bowls.
- Other tools you may encounter are skew chisels, fluted gouges, spindle gouges, and nose chisels.

Routing

A power router is an extremely useful and versatile tool that is used for cutting grooves in timber and manufactured board. A high-speed cutter is plunged into the material and then moved around. It can be used with guides for straight lines and with templates for more complex shapes. Different shaped cutters can be used to cut different shaped grooves or to produce decorative edges. A table router is a stationary version of a router. It is mounted underneath the work surface with the cutting tool protruding upwards. Pieces of work can be pushed against the cutter. A fence is often used to guide the movement on straight pieces.

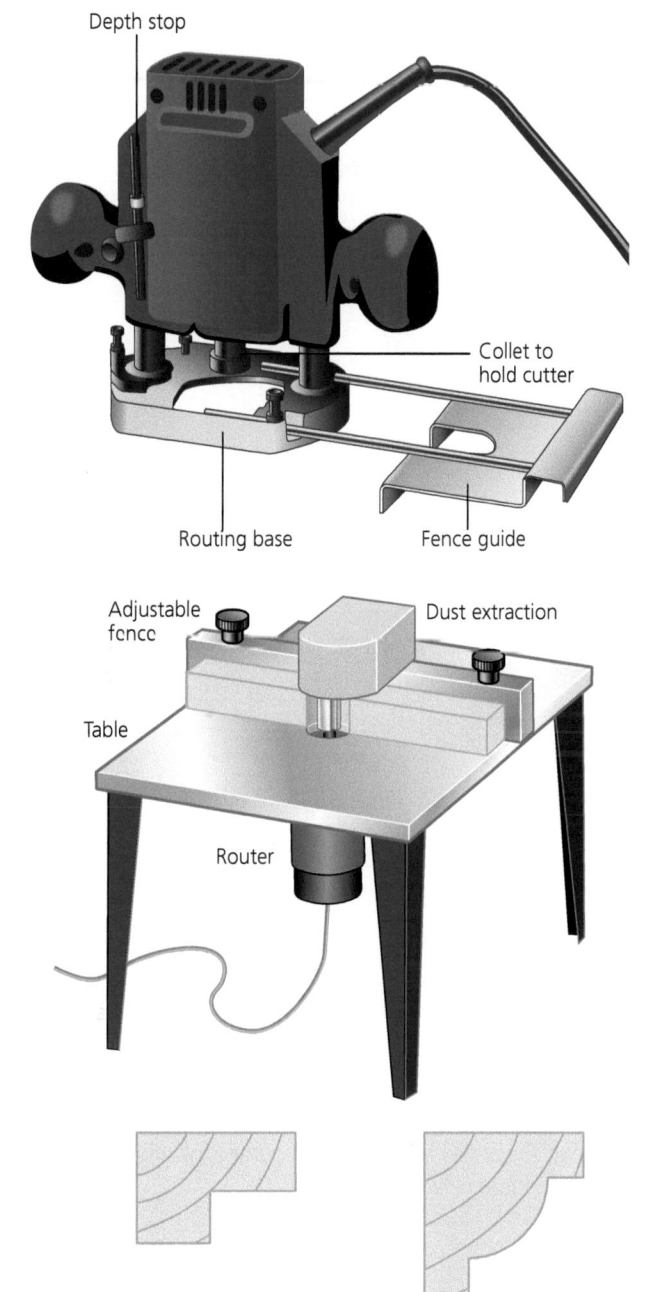

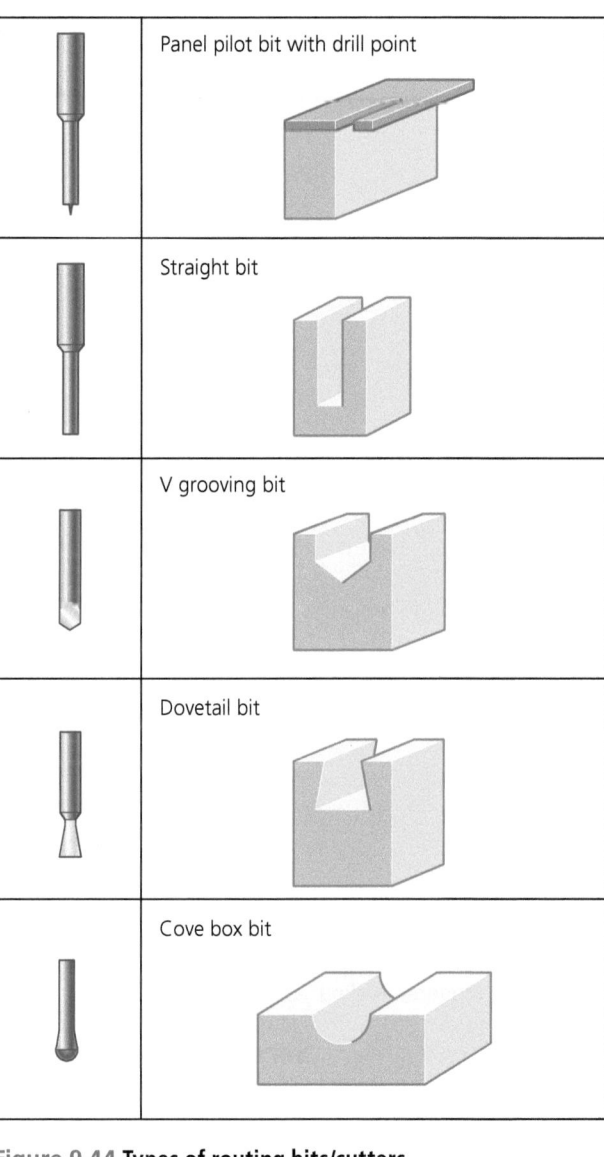

Figure 9.44 **Types of routing bits/cutters**

Rebate cut with straight bit

Profile cut with
Roman Ogee piece

Figure 9.43 **A handheld router and a table router**

Addition

Addition processes such as adhesion and joining are used to join materials together.

Nailing

When you nail a timber frame together, you should stagger the positions of the nails. This will help to avoid splitting the timber along the grain.

Hammering two nails into materials at opposite angles will make the joint stronger and more difficult to pull apart. This is called a 'dovetailing' nailing.

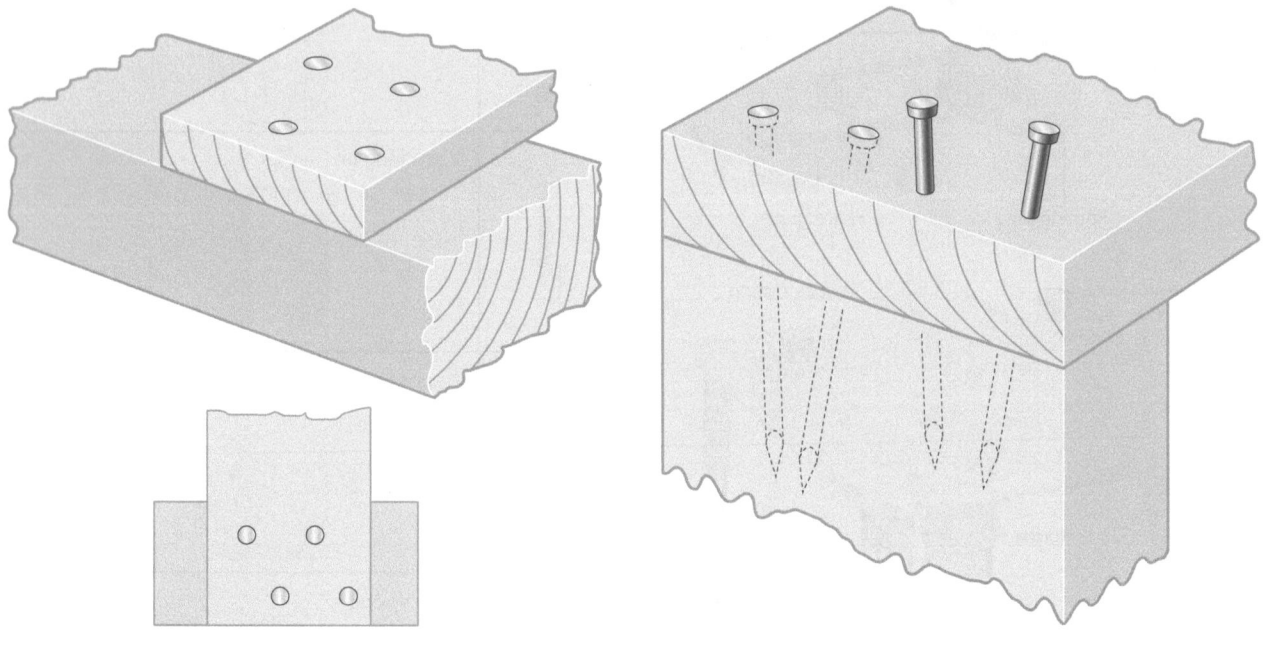

Figure 9.45 **Staggered nailing**

Figure 9.46 **Dovetail nailing**

Screwing

When screwing together two pieces of timber a pilot hole (which is smaller than the core of the screw) and a clearance hole (which is slightly bigger than the shank of the screw) are screwed. When using countersunk screws, the screw head sits in a countersink which is level with the surface of the wood, as shown in Figure 6.48.

1 Drill a pilot hole.
2 Drill a clearance hole.
3 Drill the countersunk for the head to 'sit' in.

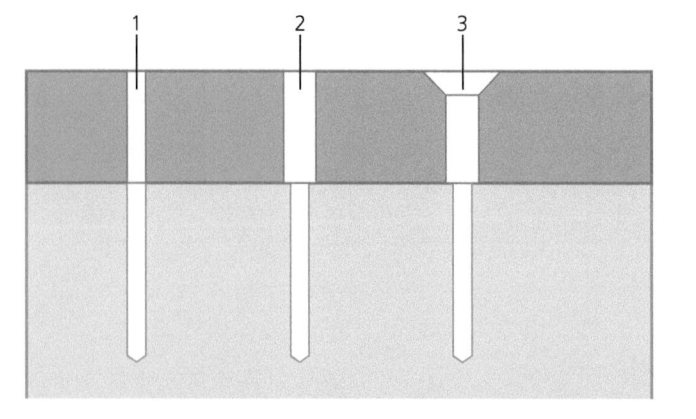

Figure 9.47 **Preparing two pieces of timber to be screwed together with countersunk screws**

Wood joints

In addition to using physical components to join work together, timber can also be manipulated to create specific joints to maintain a structure.

Joint	Illustration	Description	Example uses
Butt		A simple joint in which two pieces of timber are joined by placing their ends together and nailing or gluing them. This joint can be very weak and is often strengthened with corner pieces or a fastener.	Basic boxes or cabinets, building frames.
Dowel		A series of corresponding holes are drilled in the joint surface of two pieces of timber. Short dowels are then inserted with glue and the joint is clamped together until dry. Accuracy when drilling holes is important to make sure the two pieces of wood join up.	Chair and table legs, tabletops, cabinets and panels.
Comb or finger		Made by cutting rectangular cuts in two pieces of timber, which interlock at 90° a angle (like your fingers) when glued.	Tables and chairs, floorboards, roof and door construction.
Dovetail		Pins are cut in one piece of timber and interlock with a series of tails cut in another piece of timber. These are glued together. The joint is very resistant to being pulled apart.	Joining the sides of a drawer to the front, jewellery boxes, cabinets.
Half-lap		Two pieces of pieces of timber together by overlapping them. Material is removed from each piece of timber at the point of intersection. This joint is often reinforced with dowels or fastenings such as nails.	Framing

Joint	Illustration	Description	Example uses
Mitre		A joint made by cutting the two pieces of timber to be joined at an angle of 45°, so that a 90° corner is formed. Mitre joints are quite weak so keys can be used to reinforce them.	Picture frames, pipes, moulding,
Housing	Housing joints / Stopped housing	A joint made by cutting a channel in a piece of timber and inserting and securing another piece of timber into the channel. It can be used with manufactured boards such as MDF and particle board.	Fitting shelves and partitions into bookcases and cabinets
Mortise and tenon		A joint comprising of a mortise hole and a tenon tongue. The tenon is cut in one piece of timber (known as the rail) to fit exactly into the mortise hole in the other piece of timber. This is a strong joint.	Table and chair legs
Bridle		Similar to a mortise and tenon joint but with a longer tenon (as long as the depth of the timber) and a mortise that is cut to the whole depth of the timber so that the two pieces of timber lock together tightly. It is slightly stronger than the mortise and tenon, but less attractive as the end grain of the tenon is visible.	Legs or stiles to rails, frames
Corner halving		A joint in which channels that interlock are cut into the corner of two pieces of timber. They are often glued and nailed or screwed to reinforce them.	Frames

Joint	Illustration	Description	Example uses
Cross-halving		Constructed in the same way as the corner halving joint, but the joint is made in the middle of the timber to allow any internal parts of the frame to 'cross' each other inside the frame.	Strengthening rails for tables and chairs, trellis, box compartment dividers
Biscuit		Small slots are cut into the edges of the boards using a biscuit joiner. An oval-shaped compressed wood fibre biscuit is then glued into the slot and the two boards are clamped together This joint can be used with manufactured boards such as MDF and particle board.	Table tops

Joints used in the construction of timber products can be broadly classified into three groups: carcase or box construction (joints that could be used at the corners of a product), frame construction (used to make doors and windows) and stool construction (used when making small stools or tables and joining rails to legs).

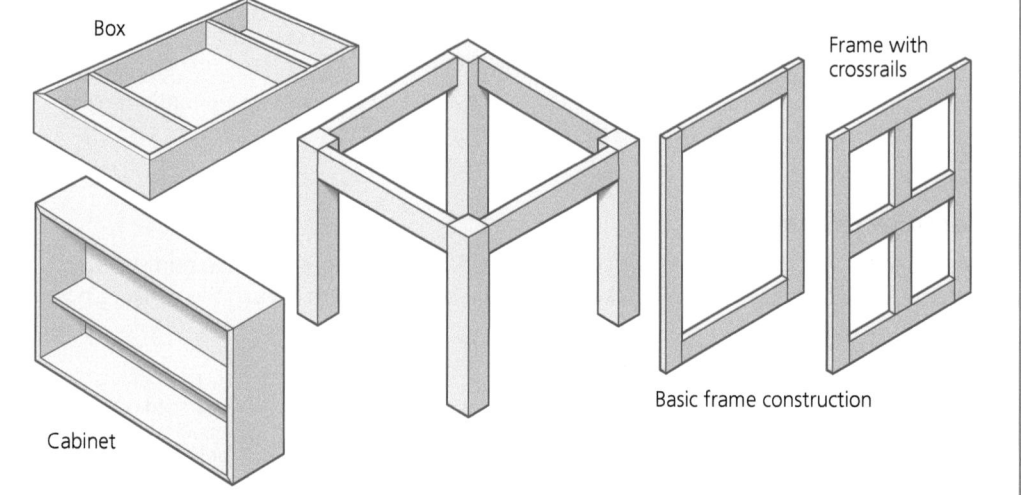

Box

Cabinet

Frame with crossrails

Basic frame construction

Figure 9.48 **Box, stool and frame construction**

Deforming and reforming

Deforming and reforming is the process of changing the shape of timber. It includes processes such as steaming and pressing

Bending

A very basic method of bending timber into curves is called **kerfing**. Putting a series of evenly spaced saw cuts on the back of the timber allows the piece to be bent. This method is really suitable only if just one side of the curve is to be visible, such as for the curved body

STRETCH AND CHALLENGE

Working with a partner, cut twelve pieces of timber 60 x 60 x 20mm. Select six different joints from the table above and make three each using these pieces of timber. Discuss the processes used and the advantages and disadvantages of each joint. Photograph the joints you have made and create a resource sheet for use lower down in the school.

Create a skills board with examples of a selection of different joints suitable for box, frame and stool construction.

KEY TERM

Kerfing: Making saw cuts so that wood can bend.

KEY TERMS

Former: A block made to hold material in the shape required.

Laminating: Building up a shape in thin layers.

Steam bending: Softening the fibres of wood with steam to allow it to bend.

of a guitar. The principle has also been applied to manufactured boards, with versions of flexible (bendy) MDF and plywood available for larger projects such as a curved reception desk or shop counter.

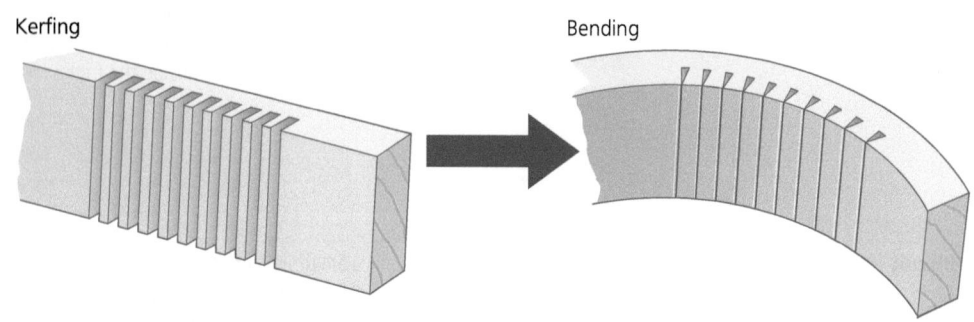

Figure 9.49 **Kerfing wood**

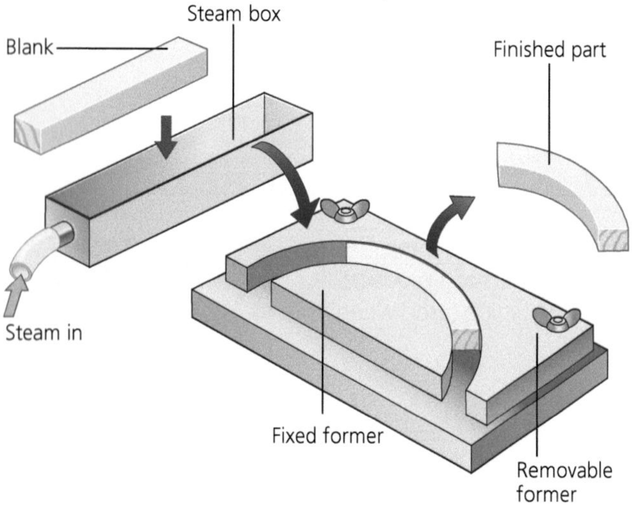

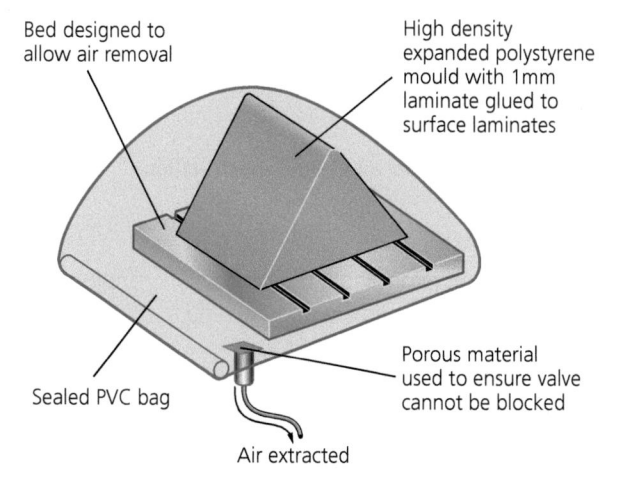

Figure 9.50 **Steam bending timber**

Steam bending

Another method of deforming timber that has been used for many years is **steam bending**. This involves placing the timber piece in a sealed box that is kept filled with steam. As the timber absorbs the hot moisture it becomes softer and easier to bend.

The steamed timber is bent into the required shape around a special **former** and clamped firmly in place so that it keeps its shape while it dries out. The shaped wood needs to dry out thoroughly before it is unclamped.

Laminating

Another method of producing curved shapes in wood is **laminating**. Thin layers of wood (called veneers) are glued together and held around a former until the glue sets. Formers can be held together with G or sash clamps and flexible steel bands.

Vacuum pressing

One-sided formers can be used with a vacuum bag, which is connected to a pump that removes the air from inside. This allows air pressure on the outside to hold the veneers in place as the glue dries.

Figure 9.51 **Laminating wooden shapes using a former and a vacuum bag**

9.5 Structural integrity

LEARNING OUTCOMES

By the end of this section you should know about and understand how and why timber/manufactured board products need to be reinforced to withstand forces and stresses.

Also, the processes that can be used to ensure the structural integrity of a product, including:
→ how triangulation is used to reinforce timber products and structures
→ how knock-down fittings can be used to reinforce products made from manufactured board.

KEY TERMS

Cellulose: An organic compound, structurally important in plant life.

Compressive strength: The resistance of a material from breaking compression/squashing.

Tensile strength: The resistance of a material to breaking under tension/stretching.

Trees are living structures and they grow by producing hollow tube-like cells composed mainly of **cellulose**. Softwoods are composed almost entirely of tube-like cells, whereas hardwoods have a more complex structure. Some contain fibrous material, which adds mechanical strength and makes the wood harder.

From a structural point of view all timber can be thought of as bundles of parallel tubes, like a bundle of drinking straws. These tubes are made from the same material but have varying wall thicknesses. The **tensile strength** of wood in general is quite high. Weight for weight, some timbers have a tensile strength greater than mild steel.

The **compressive strength** of timber is much lower. Wood is also very weak across the grain in terms of tension and compression. Hardwoods tend to be stiffer than softwoods because of their fibre content. You must remember, however, that the terms 'hardwood' and 'softwood' relate to the trees they come from and not to their mechanical properties.

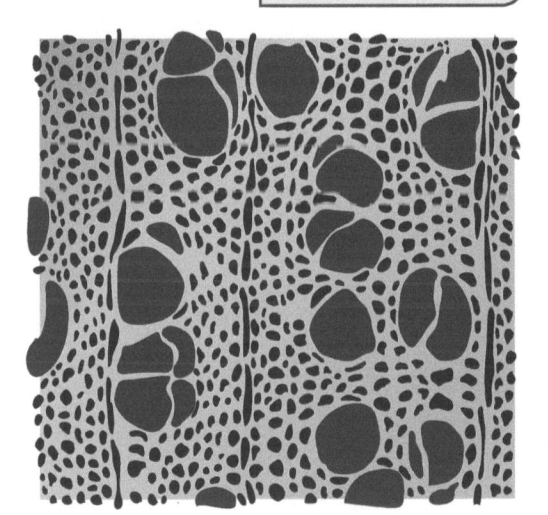

Figure 9.52 **A magnified section of wood**

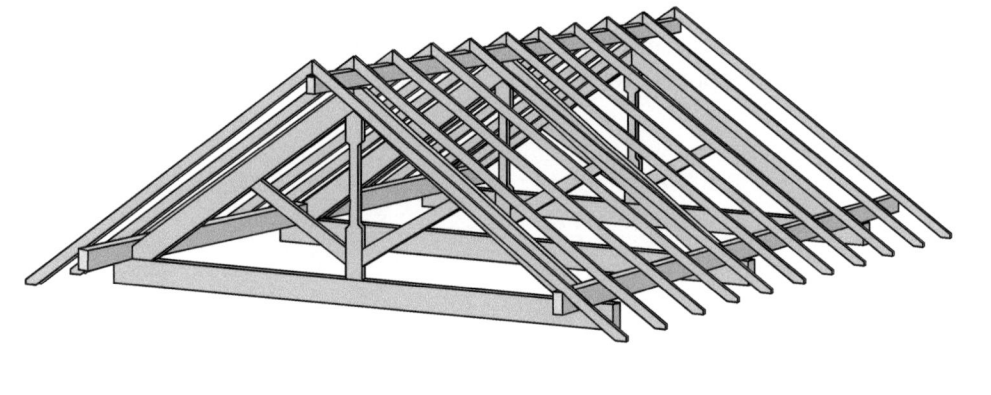

Triangulation

Maintaining structural integrity in timber products depends on how well individual pieces are joined together, and as such wood joints are looked at specifically in a later section (see Section 6.7, page 149). The basic principle of triangulation, however, can be applied to various timber products such as gates and roof trusses. Triangulation involves using triangular shapes to give stability to structures.

Timber frame products can also be reinforced by using the principle of triangulation or by using premade components such as T-plates, flat corner irons and corner braces.

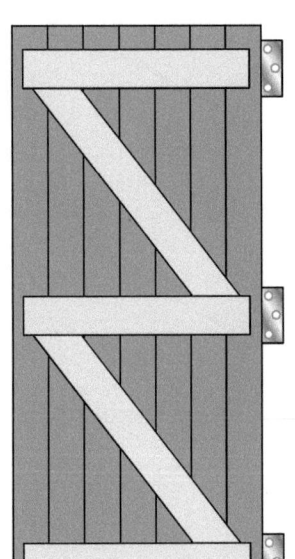

Figure 9.53 **Triangulation in roof trusses and the back of a shed door**

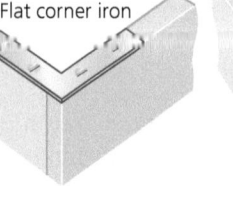
T-plate

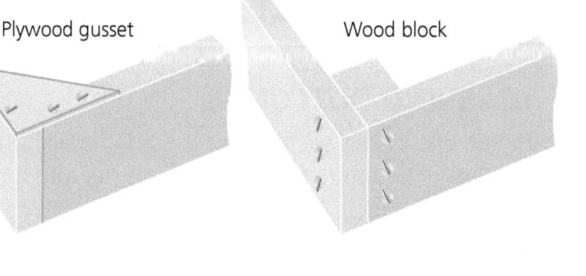
Flat corner iron | Plywood gusset | Wood block

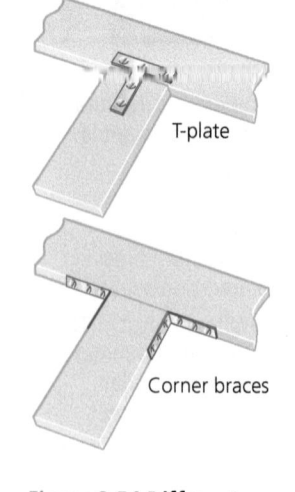
Corner braces

Figure 9.54 Different ways to reinforce timber frame structures

Knock-down fittings

Knock-down (KD) fittings are a modern way of joining materials quickly and easily without the use of clamps and glue. They allow products to be dismantled and assembled whenever necessary and reinforce products' structural integrity. KD fittings therefore allow for products such as wardrobes and kitchen units to be sold as flat packs and assembled at home.

KD fittings provide simple construction techniques for materials such as laminated MDF and chipboard. There many different types of KD fitting available and several include the use of dowel or metal pins to locate part of the joint. Most can be fitted with basic tools such as a hammer, screwdriver and drill.

Two block fitting

Two block fittings are made from plastic. A bolt passes through the first fitting into the thread of the second. As the bolt is tightened it draws the two fittings together. The pins help keep the fitting straight. This gives a very strong joint and it can be dismantled using a screwdriver.

Rigid joint

These are normally moulded in plastic, which makes them strong. Screws pass through the four holes, which hold the sides at each corner firmly together.

Connector bolt

These fasteners have a 6mm thread system and are used to join wardrobes and cabinets together. As the bolt head will be visible, they have a brass or anodised finish. They are used with a screw that is pushed through from the opposite side. The fastener is then tightened with an Allen key.

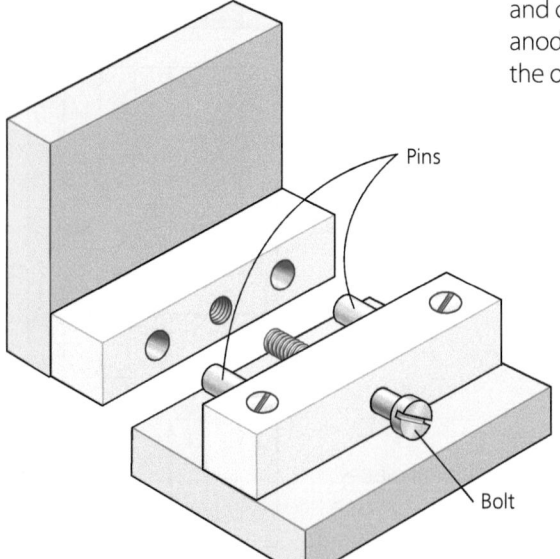

Figure 9.55 A two block fitting

Figure 9.56 A rigid joint

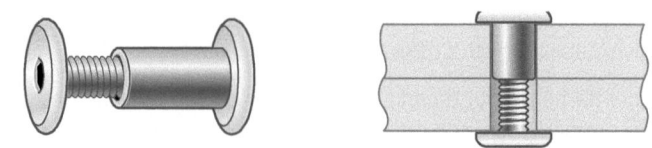

Figure 9.57 A connector bolt

Cross dowels

These are strong enough to be either permanent or temporary joints. The cylinder is inserted into the first side of a cabinet in a pre-drilled hole. The screw is then pushed through the hole in the second side until it meets the cylinder. It can then be tightened with an Allen key until both sides of the cabinet pull together.

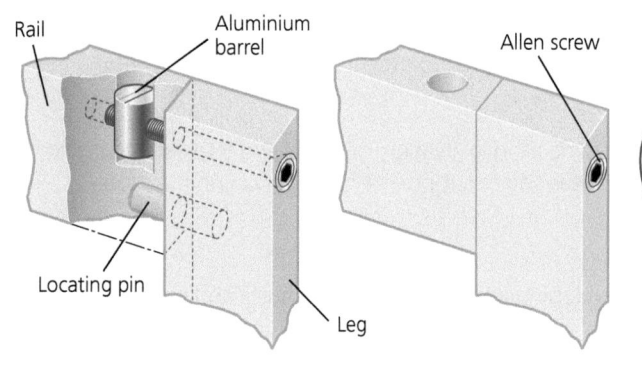

Figure 9.58 A cross dowel

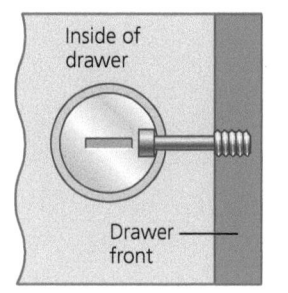

Figure 9.59 A cam lock

Cam locks

The disc fits into a recess in the first side of the cabinet. It is rotated by inserting a screwdriver into the slot in its side. The shaft is screwed into the second side of the cabinet. The collar of the shaft is passed through the hole in the second slot in the disk. When the disk is rotated the shaft is locked in position. This keeps both sides of the cabinet locked together.

Bench top joiners

Bench top joiners are used to temporarily join large flat pieces of wood. A flat hole is drilled in each piece with a flat channel to allow the bolt to pass through. The joiner is then inserted and tightened up using a spanner.

Table plate

The 'table plate' fitting is used to assemble rails to legs in flat pack tables.

Figure 9.60 A bench top joiner

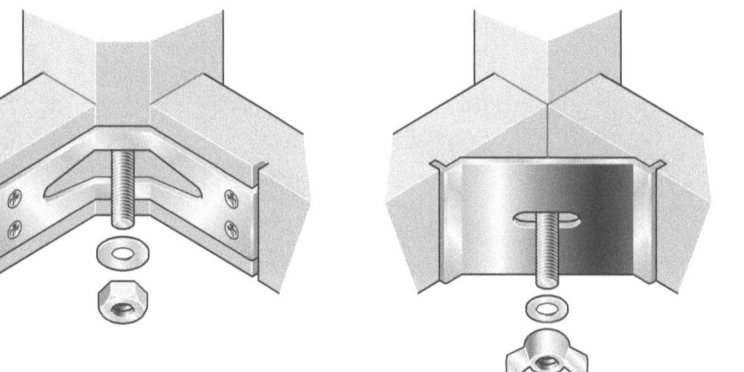

Figure 9.61 A table plate

KEY POINT

Think carefully about how your product will be assembled and disassembled. Will it remain structurally sound when it's put together?

ACTIVITY

1 Take a bundle of art straws and glue them together length ways to explore the structural strength of wood. Experiment with them by hanging weights on the end of the string (tension), cutting short straws and standing on a bundle of them (compression) and bending them.
2 Create a skills board with examples of different knock-down fittings.
3 Create a series of structures by hot gluing small pieces of wood sections together – experiment with different ways to reinforce the joints. Which method is the most successful.

9.6 Making iterative models

LEARNING OUTCOMES

By the end of this section you should know about and understand:
→ the processes and techniques that can be used to produce early models made from timber and manufactured board
→ how the use of timber and manufactured board can support iterative modelling.

Timber and manufactured board can be used in a variety of useful and interesting ways to help you explore your early design ideas and to support iterative designing in school.

Sheet materials

Sheet materials are excellent for developing initial ideas. They are quick and easy to work and are not too expensive.

MDF

MDF is available in a range of different thicknesses. Thinner sheets of 2 to 10mm can be manipulated fairly quickly in a workshop to produce early models. MDF can be cut by hand with a coping saw or by machine using a fretsaw or band saw. A special grade of MDF suitable for laser cutting is needed for more accurate/intricate work.

Figure 9.62 **A model made using a thin sheet of MDF, dowel and square section**

Plywood

Thinner sheets of plywood can also be used for quick models but the finish will not be as good as with MDF, since plywood tends to splinter. Like MDF, a special grade of 'Laser Ply' is needed for work on the laser cutter. Laser Ply is bonded with specially modified adhesives that produce less smoke when cutting, which is necessary as smoke reduces the efficiency of the laser.

Pre-cut sections

Small lengths of square timber and **dowelling** can be useful for modelling frame-type structures. These can be brought into or cut to size in the workshop. They can be cut quickly by hand using a junior hacksaw and a bench hook.

Balsa

Balsa wood has been used for modelling for a long time. It is the main wood used for model aeroplanes as it is so lightweight and easy to manipulate. Balsa modelling kits contain a variety of sheets and sections that can be cut with a craft knife and small hand saw.

Pre-formed timber pieces

Matchstick modelling has been around since the eighteenth century, when navy prisoners used to carve small 'matches' from pieces of wood to build small scale models. Today the matches are pre-made (without a combustible head) and can be bought as part of a replica model kit or in bags of various quantities. As with pre-cut sections matches can be used to quickly explore frame structures, but their short length can be limiting. They can be cut with a special safety cutter. Lollipop sticks can also be bought in bulk and used in much the same way, and these are available in a variety of colours and can also be used to quickly explore rough ideas.

Figure 9.63 **Balsa modelling packs come in a variety of shapes and sizes.**

Fixing

The quickest and easiest way to fix timber and manufactured board together when modelling is by using a hot glue gun, as the glue cools and dries in seconds. Care must be taken when handling a hot glue gun, though, and so a 'cool' glue gun is sometimes a safer option. PVA wood glue can also be used but its drying time is usually overnight, which may not be fast enough. Specialist solvent glues such as UHU can be used for fixing but these must be used in a well ventilated area because of the fumes they produce. Non-solvent-based glues are also available, but their drying time is longer.

KEY POINT

Timber and manufactured board can be manipulated quickly and easily and can be used for a wide variety of iterative modelling. Be careful not to let pre-formed pieces such as matchsticks and lollipop sticks limit your design possibilities.

ACTIVITY

Select an iconic chair design and build a scale replica of it using the materials in this section.

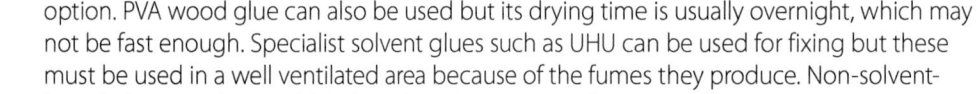

KEY TERMS

Dowelling: Round section wood.

CAD/CAM: Computer-aided design/manufacture.

9.7 Finishes

LEARNING OUTCOMES

By the end of this section you should know about and understand the processes used for finishing and adding surface treatments to materials and products for specific purposes, including:
→ function
→ environmental factors
→ aesthetics.

There are two reasons to add a finish to timber products:
- **Aesthetics:** You want to improve how the product looks. This could be to enhance the natural look of the timber or to change its colour.
- **Function:** You want to protect the product from damage by increasing its durability and protecting it from environmental factors such as moisture and sunlight.

Preparing timber for finishing

Sanding should be kept to a minimum (especially on hardwoods), so use a plane to achieve the best possible finish before sanding.
- Use abrasive paper wrapped around a cork block to avoid damage to the wood.
- Sand with moderate pressure and always in the direction of the grain.
- Take care not to round corners or rub hollows in flat surfaces.
- Sand with coarse paper at first, working down through the grades and finishing with a fine paper.
- Use a fine brush in the direction of the grain to remove the dust after sanding.

Figure 9.64 Using abrasive paper on timber

Abrasives

There are three types of abrasive that you can use on timber:
- **Glass paper** – consists of ground glass, graded into sizes and glued on to backing paper. Sheet size is 280 × 230mm.
- **Garnet paper** – consists of hard stone, crushed, graded and glued on to backing paper. It is more expensive than glass paper.
- **Wire wool** – used as a very fine abrasive in the final stages of polishing.

Grades of abrasive

Abrasives are graded according to the size of the particles within them. These sizes are referred to as grits.
- **Extrafine:** This grade of sandpaper is used between coats of paint or varnish. Grits of 240, 320 and 400 are termed very fine, while extra- or superfine sheets with grits of up to 600 are available for polishing jobs.
- **Fine:** Fine abrasive papers have a grit range of 120 to 220. For most workshops, fine will suffice for final sanding before the work is finished.
- **Medium:** Some final shaping can be done with medium, which has a grit range of 60 to 100. General sanding work is often best done with medium-grade sandpaper.
- **Coarse:** Rough shaping is done with coarse paper, as is the removal of previous finishes. The grits are typically in the 40 to 50 range.
- **Extra coarse:** This stuff is really rough, used for removing paint and varnish that you think might never come off. The sanding of old floors also sometimes requires extra coarse sandpaper. Only to be used on the toughest jobs.

Varnishes and lacquers

Varnishes and lacquers are available as oil-, water- and solvent-based. Varnishes are normally clear or transparent and are offered in **satin**, **matt** or **gloss** finishes. Spray cans are suitable for small projects and there are special types available for coating metals.

Oil

Teak and comparable timbers are naturally oily. Applying teak or linseed oil provides an improved appearance within the grain of the timber. This protects the wood very well for external use. Vegetable oil can also be applied to wood that comes in contact with foods, for instance a spatula or salad server.

French polish

French polish is a traditional finish, which is accomplished by mixing shellac in methylated spirits. French polish is applied by brush and cloth and the finish is built up in several layers, which accomplishes a very deep finish. Wax is then applied over the French polish to improve the shine.

Wood stains

Wood stains are used to improve the colour of wood and they also help to show up the grain. Wood stains are applied using a cloth. They are available in numerous colours but work well only if the stain is actually darker than the colour of the wood. Wood stain, if used on its own, only colours the wood and so does not protect it against moisture. A coating of wax or varnish over the wood stain is required to make it weatherproof. Stains are available in water- or solvent-based forms.

Sanding sealer

Sanding sealer is normally a solvent-based product like varnish, which is used to seal the wood to moisture. Sanding sealer is a quick-drying liquid that seals the surface and raises the wood fibres so they can be sanded with fine glasspaper. Sanding sealer is good for a first coat before applying varnish or wax.

Paint

Timber items can be painted with water- or oil-based enamel paints, chalkboard paint, spray paint and acrylic paint. Different types of paints produce different effects on wood.

- **Enamel paints** are permanent paints that create a hard, glossy surface that is durable and hard to chip. Enamel paints work well on wood furniture and other hard surfaces, such as glass or tile. Enamel paint colours do not blend well.
- **Spray paint** refers as much to the method of applying the paint as to the paint itself. Paint sprayed from an aerosol can is typically used on wooden furniture for application on areas that brushes can't reach easily. It is a very fast method of applying paint.
- **Acrylic paint** is easy to use on wood and creates deep, saturated colours. Often, acrylic paint does not need a second coat when applied to wood.

> **KEY TERMS**
>
> **Gloss:** A shiny, reflective surface finish.
>
> **Matt:** A dull, non-reflective surface finish.
>
> **Satin:** An in-between gloss and matt finish.

Figure 9.65 **A selection of finishes for timber**

KEY POINT

Think carefully about the environment your product will be used in and the external factors it will come into contact with when you select an appropriate finish.

1 Cut three different pieces of timber into long strips.

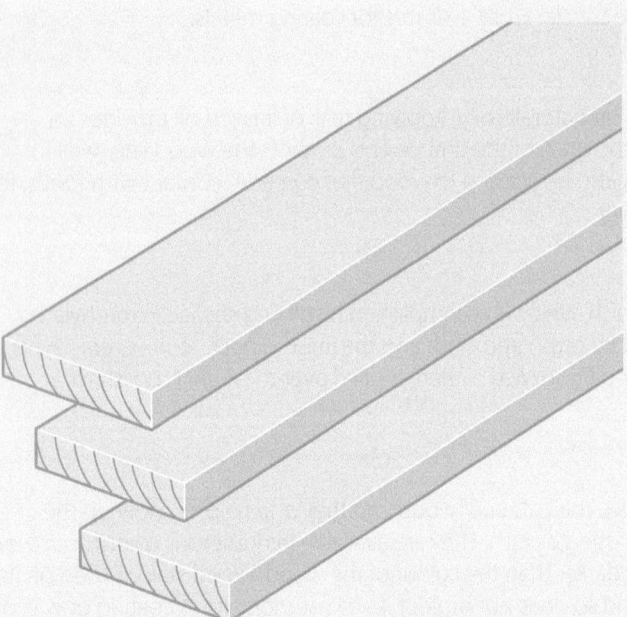

2 Place pieces of masking tape to divide the strips into areas.

3 Apply a different finish to each of the exposed areas and record the results – how does it affect the look and feel of the timber? How water-repellent is each finish?

9.8 Using digital design tools

LEARNING OUTCOMES

By the end of this section you should know about and understand:
→ the use of 2D and 3D digital technology and tools to present, model, design and manufacture solutions.

There has recently been a trend of huge growth in timber furniture production, which results in shorter design and production techniques, and most furniture is becoming increasingly complex. As a result, consumers are spending increasing amounts on exclusive furniture. It is a big challenge to design such complex furniture and to find how to do it the quickest way possible. CAD/CAM that is aimed specifically at designing and making products out of timber can help with this.

Computer-aided design

Various software applications are used widely in design and technology to create and explore ideas, from simple 2D sketching programs that help designers quickly visualise ideas, to advanced 3D modelling software that allows them to create fully realised products. Add-ons for popular CAD programs enable designers to work on furniture and other timber products. Timber materials can be assigned to parts in the design and a wide variety of **standard fixings** can be used to assemble the product in the computer. Computer Aided Engineering (CAE) can be used to calculate stresses and strains in the design of components for timber structures, such as roof trusses.

KEY TERM

Standard fixings: Items such as nuts, bolts and knock-down fittings.

Figure 9.66 **CAD being used for structure design**

Custom laminate

Laminate flooring is manufactured board (usually MDF) with a printed wood effect and plastic wear layer on the top. The printed paper layer can be digitally printed with any design, ranging from wood textures to tile effects.

Bendable wood

Thin sheet material can be cut on a laser cutter to create a bendable piece of wood. The effect is achieved by making thin cuts on both sides of a single sheet of wood. The start of each row of cuts shifts by half the length of the previous row. The result creates enough torsion to allow the wood to bend like a hinge.

CNC machines

A Computer Numerical Control (CNC) router is a computer-controlled cutting machine related to the handheld router used for cutting hard materials. Complex shapes can be cut from sheets of manufactured board, and certain carved effects can be obtained by moving the cutting tool over the work. More complex multi-axis machines can effectively carve any shape in timber. Other CNC machines are described in the next section.

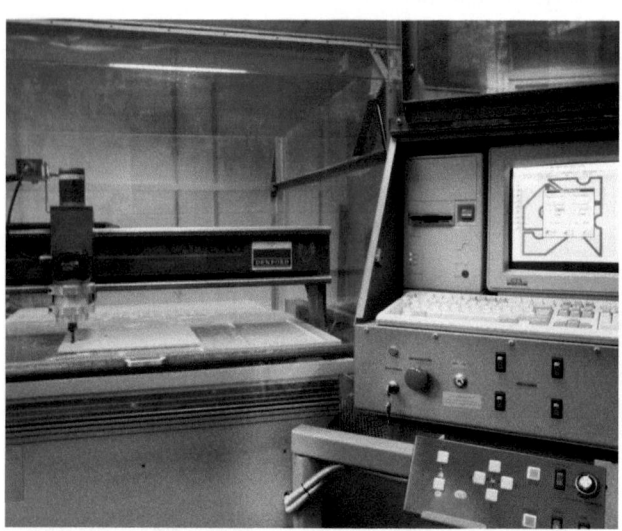

Figure 9.67 **A CNC router**

Figure 9.68 **3D printed 'wood'**

KEY TERM

Fused deposition modelling (FDM): Building up a 3D shape by laying down material in layers.

3D printing

A recent development is the use of fine wooden particles for 3D printing. A granular powder made from wood chips can be fused together by a laser using a process called laser sintering. These same particles can be added to the polymer filament used in **fused deposition modelling (FDM)** printers. The resulting products can be sanded and painted like wood.

KEY POINT

CAD/CAM is being used more and more in the design and manufacture of timber products. Think about how you can use computers to explore design possibilities and to enhance the manufacturing of the products you design.

STRETCH AND CHALLENGE

Describe how CAD and computer-aided manufacture (CAM) work together to produce a quality product.

9.9 Manufacturing methods and scales of production

LEARNING OUTCOMES

By the end of this section you should know about and understand:
→ the methods used for manufacturing timber products at different scales of production
→ manufacturing processes used for larger scales of production.

Methods for manufacturing timber products at different scales of production

Manufacturing products with timber and manufactured board can be done at varying scales of production. The hand tools and machines described in earlier sections would be used extensively for one-off products. For small batches, jigs and templates would be used more in order to ensure quality and accuracy. As the scale of production rises towards larger batches and mass production, the level of automation and the use of specialist machinery would increase accordingly.

Lean manufacturing with timber and manufactured board

With other resistant materials such as plastic and metal, the off-cuts and waste can often be recycled. Timber waste can also be recycled, into manufactured boards such as blockboard and chipboard.

Several furniture-makers make use of lean manufacturing to minimise the waste they produce and make savings. One system involves cataloguing every piece of off-cut timber/board and tracking these via computer. The off-cuts are barcoded and stored on shelves. As work progresses the computer system identifies whether a suitable piece of timber/board is already available in the 'spares bin', thus avoiding the need to cut into fresh pieces.

Manufacturing processes used for larger scales of production

Jigs and fixtures

Some machine operations are performed easily. The machine is set and the work cut. Some operations, however, require the tool to be guided and/or the work to be held in a specific way. A device that guides the tool is called a jig and a device that holds the work is called a fixture.

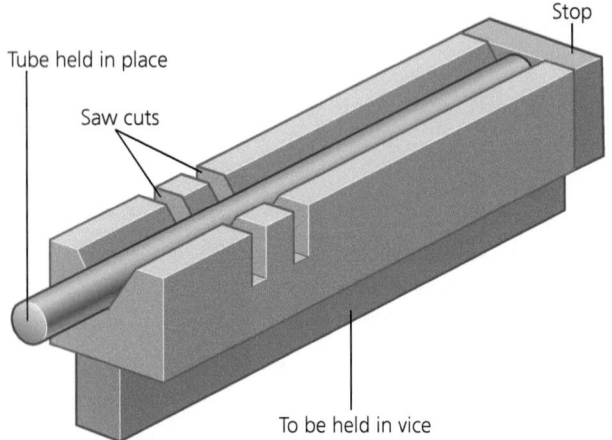

Figure 9.69 **Sawing jig**

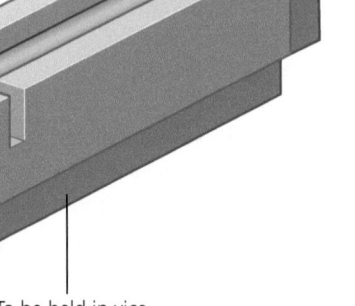

Figure 9.70 **Drilling fixture**

Figure 9.71 **An acrylic template used to mark out a guitar shape**

Templates and patterns

Templates are used to mark out awkward or irregular shapes repeatedly. Any thin, fairly rigid material is suitable for a template, such as card. If the template is used constantly then a more resistant material should be used, such as acrylic or thin metal sheet. A pattern is something fixed directly on to the material to guide cutting. For example, if you were cutting an ornate decorative pattern you could print the design on to paper and glue that to the work.

Sawing machines

Specialist sawing machinery may be needed as the number of pieces required increases. Such machinery can vary from larger table saws to machines made to make specific types of cuts repeatedly in sections of timber and manufactured board, such as beam saws and vertical panel saws.

Steam bending machines

When producing batches of components that are steam bent, specific machines can be used to perform the tasks repeatedly and to a consistent standard. Very often these will be custom-built for the component and will use hydraulic pressure to bend the wood around a former and hold it until it has dried.

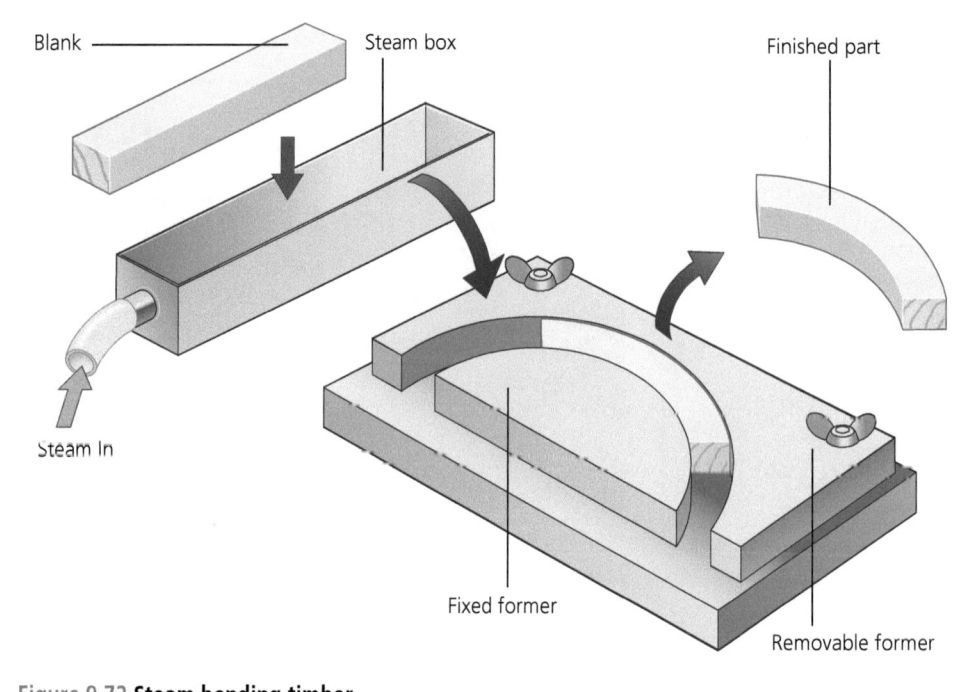

Figure 9.72 Steam bending timber

Laminating presses

Much like steam bending machines, custom-built laminating presses can be used to press and hold laminated shapes together while their adhesive dries. For batch production, different formers can be swapped out depending on the job at hand.

Roller coaters

When laminating timber and manufactured board, a great deal of adhesive is required and each piece must be coated evenly for them to laminate together efficiently. Roller coaters are machines that apply a thin, even coating of adhesive on flat pieces. They can also be used to apply decorative laminate.

CNC machines

As with other materials, timber and manufactured board can be manipulated with specific CNC machines. As the volume of pieces needed increases, so does the complexity of the machine.

- **Industrial CNC routers** make up roughly 80 per cent of the CNC routers in circulation. Many companies own several of these machines to produce the bulk of their products, such as furniture, doors, signs, etc. The machines are usually very large and very expensive. They are built for long operating hours, however, and include such features as fast cutting speeds, automatic tool changers and advanced dust collection systems.
- **Mid-range CNC routers** are usually smaller and lighter than industrial models and do not come fitted with a dust extraction system as standard (one would need to be purchased separately).

Figure 9.73 Diagram of a roller coater

- **Multi-axis CNC routers** can rotate their cutting head to create fully 3D parts (regular machines cut out in 2D). They are very expensive and require specialist software to run.
- **Specialist CNC routers** can be used to carve intricate designs in wood, which can be used for decorative purposes.
- **CNC wood lathes** can be used to create identical pieces, such as those used in stair balusters.
- **CNC drilling machines** are used in the furniture industry to drill holes for knock-down fittings.

STRETCH AND CHALLENGE

Think about the jigs you could create to help increase your accuracy when repeating processes in quantity production, such as batch production.

STRETCH AND CHALLENGE

Create a reference table of the processes in this section. Use the following headings:
- Process
- Scale of production
- Example products.

Figure 9.74 **A CNC machine for use with timber and manufactured boards**

9.10 Cost and availability

LEARNING OUTCOMES

By the end of this section you should know about and understand:
→ how the cost and availability of specific timbers can affect decisions when designing
→ how to calculate the quantities, sizes and costs of materials required in a design.

How the cost and availability of specific timbers can affect design decisions

During the design and development of a new product many factors will influence your choice of materials, and this section looks at how the cost and availability of timbers and manufactured boards can affect your material choices.

Pricing

KEY TERM

Commodity – a raw material that can be bought and sold such as coffee or gold.

Like other **commodities**, the price of timber is variable. Timber prices can also vary depending on your location. What may be a lengthy (and expensive) import process for a tropical hardwood by one person may be considered a local (and cheap) utility wood by another. Less than 10 per cent of the timber used in Britain is home-grown or domestic timber.

Domestic timber refers to timber that is harvested locally and usually the least expensive. Transportation costs play a large factor in the ultimate price of timber. Scarcity, ease of access and drying costs, however, also play a role in timber prices, resulting in a range of prices for domestic timber. The overall price range for domestic timber tends to be less expensive than for imported timber.

The value of timber is based on the value of the wood delivered to a processing facility, minus the harvesting costs and transportation costs.

Figure 9.75 **Harvesting timber in the UK**

Harvesting costs

These are the costs incurred when cutting the trees down, and they include the cost of personnel and vehicles. Timber size and species also affect the costs of harvesting. Large trees are less expensive to harvest than small trees. Hardwood trees require more time to prepare for conversion than softwoods and are, therefore, more expensive to harvest.

Weather conditions can also influence harvesting costs. Wet weather probably increases harvesting costs more than any other single factor. Wet weather may reduce the number of trees that can be taken out of a forest by skidders (the vehicles used), which can also damage the soil more easily in wet weather.

Figure 9.76 **Importing timber from abroad**

Transportation costs

The cost of moving workers and equipment to and from various timber forests can vary greatly and will affect harvesting costs. As the amount of timber moved from a forest increases, however, the cost per unit of production goes down. Transportation from the forest to the timber mill (where the trees are cut into planks) is another significant cost.

Availability

Timber that is hard to find tends to be expensive, but this is not always the case. Sometimes a timber species can go out of fashion – what was a popular wood 20 years ago might no longer be found for sale. This is particularly true for **imported** timber, as exporters will not waste time and money importing timber that will not sell. Also, certain types of timber may simply become scarce (difficult to find).

Depending on the size of the tree the timber may be available only in certain sizes or forms. Some timbers are primarily sold as a veneer, while others may be sold primarily as flooring planks, or turning blanks, or plywood, and so on. Any limitations in the available forms of the wood will be noted.

KEY TERM

Import – to bring goods in from another country

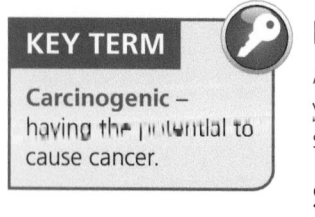

KEY TERM

Carcinogenic – having the potential to cause cancer.

Environmental concerns

As explained in Section 6.2, there are several environmental concerns that may influence your choice of timber or manufactured board, such as whether the timber is from a sustainable source. You may also wish to reuse or upcycle timber from another product.

Safety concerns

All dust is a safety hazard and must be controlled when working with timber and manufactured boards. The dust from several species of timber will irritate your skin and eyes if you come into contact with it. Also, the dust from some species of timber has been shown to be **carcinogenic**. Nearly all parts of the yew tree are considered toxic and poisonous to humans, so great care should be exercised if working with this wood. This will certainly affect your choice of timber when designing and making products.

Figure 9.77 **Working with wood can be harmful, for example if it gives off a lot of dust**

Calculating the quantities, sizes and costs of materials required in a design

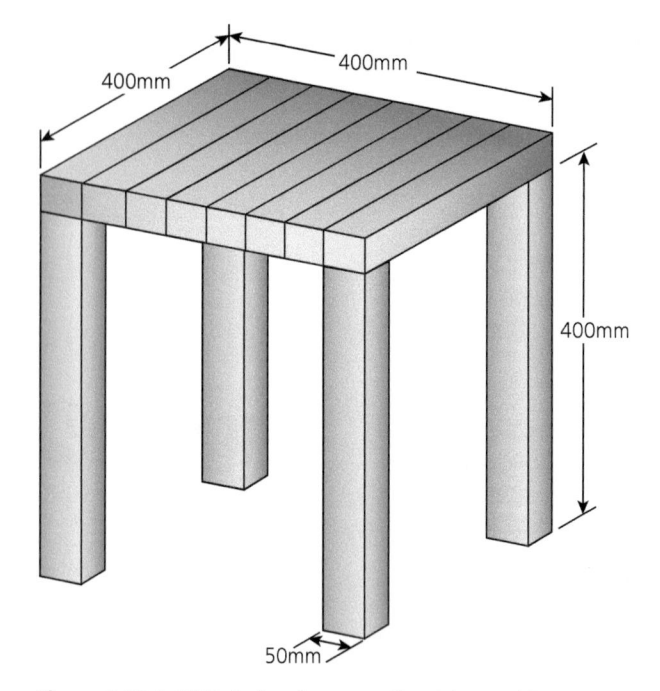

Figure 9.78 **A CAD design for a small outdoor table**

When making any timber-based product, you will need to calculate how much wood you need, what standard sizes are available, and therefore how much the material will cost.

This illustration shows the design for a simple outdoor table made from teak (suitable for outdoors) .

Calculating sizes of materials

Once you know how much timber you need, the next step would be find out what size sections of timber are available from your supplier. The illustration below shows an example of a typical supplier's price list for teak.

The first thing you might spot is that they only sell timber by the metre. To produce the table (allowing for waste) we have calculated that 5.28 metres of teak are needed, so we would need to round up to 6 metres.

The design for the table also shows that planed teak that is 50mm square is needed. The supplier doesn't sell planed teak in the exact size required, so you would have to decide whether to alter the size – down to 45 x 45mm – or to buy 75mm sawn and then cut/plane it down to 50mm, which could be expensive/wasteful.

Calculating costs of materials

You would likely want to compare the cost of both options to help you decide which timber to buy, and therefore being able to calculate the costs of materials is important.

Teak prices – sawn and planed *All prices are per running metre*										
Teak – Sawn										
Width **Thickness**	25mm	38mm	50mm	75mm	100mm	125mm	150mm	175mm	200mm	225mm	Sq. m.
12mm	£6.79	£9.06	£10.06	£15.09	£20.12						
18mm	£7.64	£10.19	£11.32	£16.98	£22.64	£28.30	£33.96				
25mm	£8.49	£11.32	£12.58	£18.87	£25.15	£31.44	£37.73	£44.02	£50.31	£56.60	£247.48
32mm		£14.15	£15.02	£23.58	£31.44	£39.30	£47.16	£55.03	£62.89	£70.75	£309.35
38mm		£16.24	£18.05	£27.07	£36.09	£45.11	£54.24	£63.16	£72.18	£81.20	£355.08
50mm			£24.06	£36.09	£48.12	£60.15	£72.18	£84.21	£96.24	£108.27	£473.44
63mm				£39.37	£52.50	£65.62	£78.74	£91.87	£104.99	£118.12	£516.48
75mm				£47.25	£63.00	£78.74	£94.49	£110.24	£125.99	£141.34	£619.78
100mm					£73.64	£92.05	£110.46	£118.87	£147.28	£165.69	£724.51
Teak – Planed										
Width **Thickness**	20mm	32mm	45mm	70mm	95mm	120mm	145mm	170mm	195mm	220mm	Sq. m.
6mm	£7.99	£10.26	£11.26	£16.29	£21.32						
12mm	£8.84	£11.39	£12.52	£18.18	£23.84	£29.50	£35.16				
18mm	£9.69	£12.52	£13.78	£20.07	£26.35	£32.64	£38.93	£45.22	£51.51	£57.80	£256.48
25mm		£15.35	£16.92	£24.78	£32.64	£40.50	£48.36	£56.23	£64.09	£71.95	£318.35
32mm		£17.44	£19.25	£28.27	£37.29	£46.31	£55.34	£64.36	£73.38	£82.40	£364.08
45mm			£25.26	£37.29	£49.32	£61.35	£73.38	£85.41	£97.44	£109.47	£482.44
60mm				£40.57	££53.70	£66.82	£79.94	£93.07	£106.19	£119.32	£525.48
70mm				£48.45	£64.20	£79.94	£95.69	£111.44	£127.19	£142.94	£628.78
95mm					£74.84	£93.25	£111.66	£130.07	£148.48	£166.89	£733.51

Figure 9.79 Typical price list for teak

To calculate the costs in this example you will need to multiply the price per metre of teak by the number of metres required:

6 metres of sawn teak 75 × 63mm thick is £39.37 per metre: 6 × 39.37 = £236.22

6 metres of planed teak 45 × 45mm thick is £25.26 per metre: 6 × 25.26 = £151.50

A small change at the design stage could save you money – over £80 in this case if you decided to use the 45 × 45mm teak.

Sometimes the cost of your chosen material may be too expensive, and therefore you may want to compare it with the cost of other suitable timbers. Paying over £150 for a small table, for example, may seem too expensive and therefore another suitable timber such as Cedar could be considered.

Returning to the price list in Figure 9.81 we can see that the same company sells planed cedar at 50 × 50mm (the required size) for only £1.89 per metre. The cost of the 6m of cedar needed to make the table would be:

6 × 1.89 = £11.34

This is much cheaper than producing the table in teak. But how long would it last for? Would it look as nice? Would it withstand bumps and knocks as well? How often would it need to be treated? Don't consider cost as an isolated requirement when designing products – be sure to balance your research and select the best all round material for your product that will meet the needs of your stakeholders.

ACTIVITY

1. Try to design and make a useful product that can be made only from off-cuts found in the workshop.
2. Locate the price guide for a local timber merchants and calculate the cost of the table in this section if it were made from:
   a) a low cost wood
   b) an imported luxury wood.

CHAPTER 10
Metals

You learnt about the different types of metals in Section 5.1. This chapter includes more in-depth information on metals.

10.1 Physical and working properties

LEARNING OUTCOMES

By the end of this section you should know about and understand the physical and working properties of specific metals, including:
→ how easy they are to work with
→ how well they fulfil the required function of products in different contexts.

Metal properties

There are many different types of metal and they have a wide range of properties and characteristics, as detailed below. The general characteristics of metals, however, are that they:
- have a high melting point
- have high **tensile strength**
- have a lustrous shiny finish
- are **malleable**
- are **ductile**
- are good conductors.

This chapter covers the main categories and types of metal available to designers and that you may have experience of or use in school. As you have already learnt in Chapter 5, there are two main categories or types of metal:
- **Ferrous** – metals that contain iron.
- **Non-ferrous** – metals that do not contain iron.

Ferrous metals have a high iron content. The word 'ferrous' is from the Latin word *ferrum*, meaning 'containing iron'. Ferrous metals corrode quickly and easily unless they are treated with a suitable surface coating such as paint, oil or wax. Surface rust on a metal is a clear indicator that it is ferrous. Another way ferrous metals can be identified is that they are attracted to magnets.

Non-ferrous metals do not contain any iron at all. As a result, they do not rust as easily and are not attracted to magnets.

As well as the two main categories above, there is another group of metals known as **alloys**. Alloys are metals mixed with other metals and other substances to improve certain properties.

All ferrous metals are technically alloys because, although they are mostly made of iron, they contain small amounts of other metals or substances that give them their individual properties and characteristics.

KEY TERMS

Alloy: A metal made by combining two or more metals to give greater strength or resistance to corrosion.

Tensile strength: The resistance of a material to breaking under tension/stretching.

Malleable: Able to be hammered or pressed into shape without breaking or cracking.

Ductile: Able to be bent or deformed without losing toughness.

Ferrous metal: A metal that contains iron.

Non-ferrous metal: A metal that does not contain iron.

Ferrous metals

Carbon steel

Carbon steel is iron mixed with a small amount (between 0.05 and 2.0 per cent) of carbon. There are three different grades of carbon steel, depending on the amount of carbon content.

Table 10.1 **The properties, working characteristics and uses of ferrous metals.**

Name	Properties/working characteristics	Composition	Uses
Low carbon steel or mild steel	• Inexpensive compared to other metals • Tough • East to cut, drill and weld	0.05% to 0.3% carbon and 99.7 to 99.95% iron	General building and engineering (nuts, bolts, girders), car body panels, gates
Medium carbon steel	• Similar properties to low carbon steel • Slightly harder and less ductile	0.30 to 0.60% carbon and 99.4 to 99.7% iron	Large vehicle parts such as axles, gears and crankshafts
High carbon steel (or 'carbon steel')	• Much harder and stronger than mild steel • More **brittle**	0.6 to 1.4% carbon and 98.6 to 99.4% iron (plus small quantities of other metals such as silicone and copper)	High strength cables, springs, saw blades, drills, etc.
Tool steel (die steel)	• Extremely hard • Resistant to heat	98.5 to 99.3% iron and between 0.7 to 1.5% carbon (plus small quantities of other metals such as tnugsten and vanadium)	Cutting tools, machine parts
Stainless steel	• Resistant to wear and corrosion	Iron, nickel and a minimum of 10.6% chromium	Cutlery, surgical instruments, kitchen utensils, specialist vehicle parts
Cast iron	• High **compressive strength** • Extremely brittle • Extremely resistant to corrosion and oxidisation	2 to 6% carbon and 94 to 98% iron	Car engine blocks, manhole covers, kitchen saucepans
Wrought iron	• Strong • High resistance to corrosion • Attractive 'patina' as it ages	Almost 100% iron mixed with tiny slivers of iron silicate (also known as 'slag')	Ornamental gates, fences, garden benches

Non-ferrous metals

Non-ferrous metals do not contain any iron. This means that they do not corrode as fast or as easily as ferrous metals, and they are not attracted to magnets.

There are over fifty different non-ferrous metals. The periodic table has 67 metal elements that are non-ferrous so it would be impossible to cover them all in this book. This chapter looks at the main non-ferrous metals available to designers and that you may have experience of or use in school.

KEY TERMS

Brittle: Likely to snap, crack or break when bent or hit with an impact.

Compressive strength: The resistance of a material to breaking as a result of compression/squashing.

KEY POINT

The metal lead is not used in pencils. Pencil 'lead' is actually a mixture of clay and graphite.

Table 10.2 **The properties, working characteristics and uses of non-ferrous metals**

Name	Properties/working characteristics	Uses
Aluminium	• Lightweight • Attractive natural finish • Malleable • Good conductor of heat and electricity • Can be polished to a mirror finish	Drinks cans, foil, automotive parts, cooking utensils, window frames, ladders, roof joists, body shells of cars and aircraft
Copper	• Soft and extremely ductile • Malleable • Excellent conductor of heat and electricity • Good resistance to corrosion	Electrical cables, water pipes, cooking pots, jewellery, statues, roof coverings
Lead	• Soft and malleable • Dense (heavy) • Good resistance to corrosion • Extremely resistant to acid	Car batteries, fishing and diving weights, weather protective flashing around house roofs
Zinc	• Weak (low compressive and tensile strength) • Brittle • Poor conductor of electricity and heat • Extremely resistant to corrosion • Retains its shiny silvery finish	As a coating on products made from ferrous metals such as steel buckets, watering cans, automotive parts, bolts and screws, etc.
Tin	• Ductile • Malleable • Easy to break • Very resistant to corrosion • Retains its shiny appearance when exposed to moisture	As a coating on the surface of other metals to prevent them corroding, such as food cans or 'tin' cans

Alloys

An alloy is a metal that has been mixed or combined with other substances to make it stronger, harder, lighter, or better in some other way.

Steel is technically an alloy because it is made from iron mixed with very small quantities of other materials such as carbon (see the information on Mild, carbon and stainless steels, on page 79). Most other alloys are non-ferrous but many still contain small traces of iron.

There are so many different alloys that it would be impossible to cover all of them in this book. This chapter looks at the main alloys available to designers and that you may have experience of or use in school.

Table 10.3 **The content, properties, working characteristics and uses of alloys**

Name	Approximate contents	Properties/working characteristics	Uses
Brass	• 65% copper • 35% zinc	• Durable • Good corrosion-resistance	Musical instruments, door knockers, letterboxes, handles, boat fittings
Pewter	• 85–99% tin • 1–15% copper, lead and antimony (a hard, brittle metal with a bright silvery finish)	• Tough • Malleable • Polishes to bright finish	Drinking tankards, jewellery, picture frames, ornaments Often used as a cheaper alternative to silver
Duralumin	• 94% aluminium • 4% copper • 1% magnesium • very small quantities (les than 1%) of manganese and silicon	• Lightweight but much stronger than aluminium • extremely strong • corrosion resistant • difficult and expensive to extract	Aircraft frames, car chassis, speedboats, door handles, hand rails, etc.
Solder	• 70% lead • 30% tin Note: Modern solder does not contain lead and is approximately 99% tin and 1% copper	• Malleable • Stronger than lead and tin • Low melting point • Good electrical conductivity	Electrical connections on printed circuit boards and pipework joints in plumbing
Alnico	• 50% Iron • 15–26% nickel • 8–12% aluminium • 5–24% cobalt	• Extremely magnetic • Good electrical conductivity • Good corrosion-resistance	Magnets in loudspeakers, guitar pickups
Bronze	• 88% copper • 12% tin • Small amounts of other metals such as aluminium, zinc, lead and silicon	• Malleable • Ductile • Good heat and electrical conductivity • Attractive golden-brown colour	Sculptures, statues, musical instruments, trophies, medals
Gunmetal	• 80–90% copper • 3–10% zinc • 2–11% tin	• Tough • Durable • Very low friction (slippery surface)	Valves, hydraulic equipment, bearings, bushes, gear wheels and fittings
Sterling silver	• 92.5% silver • 7.5% copper	• Shiny and attractive • Ductile • Extremely soft • Tarnishes over time if not cleaned	High-quality cutlery, medical instruments, jewellery, musical instruments

Shape-memory alloys

Shape-memory alloys (SMAs) are metal alloys that return to their original shape when heated. The most common SMAs are made up of nickel and titanium. These are known as nitinol alloys. Nitinol is used by the military and in many specialist medical products. (For more information on SMAs see Chapter 2.)

STRETCH AND CHALLENGE

Complete the table below by filling in the missing information.

Name	Type	Appearance	Properties	Uses
Mild steel			Tough, easy to cut, drill and weld	
			Soft, ductile, excellent heat and electricity conductor	Electrical cables, water pipes, jewellery
		Dull blue/grey		Roof flashings, fishing weights
	Alloy		Strong and lightweight	
Solder				Electrical connections on printed circuit boards

10.2 Sources and origins

LEARNING OUTCOMES

By the end of this section you should know about and understand why it is important to understand the sources and origins of metals, including:
- → the processes used to extract metals into useable form
- → the ecological, social and ethical issues associated with processing metals
- → the lifecycle of metals
- → recycling, reuse and disposal of metals.

Sources and origins of metal

Seven metals (known as the metals of antiquity) are believed to have been discovered and used as far back as 6000BC:

- Gold
- Silver
- Tin
- Mercury
- Copper
- Lead
- Iron

Early metal working and shaping was accomplished by hammering the soft metals into shape. Later, melting and casting of metals with low melting points was carried out. Alloys were discovered in the Bronze Age (2300–700BC), when man found that mixing metals (initially copper and tin, making bronze) made a metal much stronger.

Much later, other metals such as zinc, platinum and bismuth were discovered during the medieval period (fifth to fifteenth centuries), but the vast majority of metals have been discovered only since the beginning of the industrial age (eighteenth century). Technological advances in metal **smelting** and processing led to the creation of new alloys such as steel, which was strong and easy to work but also cheap to manufacture. Steel was produced in huge quantities and used to build railways, buildings, bridges and many other structures. Around the end of the nineteenth century, other processes allowed the widespread use of aluminium, which had previously been extremely expensive to produce. While various other metals have been discovered as recently as the 1940s, steel is still the most commonly used metal today.

Extraction and conversion of metals

Metals are found in **ores**. These are naturally occurring rocks in the Earth's crust that contain metal or metal compounds such as gold and **metal oxides**. Metal oxides are metals that have been oxidised by (reacted with) air or water, such as iron oxide and aluminium oxide. Ores are dug out of the ground by mining, but in order to be turned into a metal form that can be used they must be separated from whatever they are mixed with. This process is known as **extraction**.

Some metals can be found on their own as metals, and do not require any extraction. These are called native metals. Gold is an example of a metal that can sometimes be found in native form.

There are different methods of extraction used for different metals. The method used is determined by the metal's reactivity. The higher the reactivity, the harder it is to extract the metal.

There are three main methods for extracting metal:
- Chemical reaction
- Carbon or carbon monoxide reduction
- **Electrolysis**.

KEY TERMS

Electrolysis: Extraction of metals by melting and passing electric currents through it.

Extraction: Separating metal ores from oxides and other contaminants.

Smelting: Melting down metals into molten liquid.

Table 10.4 **Methods of extracting metals**

Ease of extraction	Metal	Extraction method used
Hard to extract ↕ Easy to extract	Potassium Sodium Magnesium Aluminium	Electrolysis
	Zinc Iron Tin Lead	Carbon reduction
	Copper Silver Gold	Chemical reaction

Extraction by chemical reaction

Some metal ores can be extracted by heating them in very hot air. When they are heated, thermal gas–solid reactions occur such as **oxidation**, reduction and sulphation. In a process called **roasting**, the sulphides in the metals are converted to oxides, and sulphur is released as sulphur dioxide gas. For example, copper is found in an ore called copper sulphide. When this is heated in oxygen, the sulphur dioxide escapes as a gas, leaving behind the pure copper.

This roasting process was originally done by burning wood on top of ore. When the ore reached a certain temperature, the sulphur in the ores would burn and become the source of fuel instead of the wood. The roasting process releases large amounts of acidic, metallic and other toxic compounds that can have extremely harmful effects on the environment, such as **acid rain**.

Extraction by carbon or carbon monoxide reduction

Carbon reduction is mainly used to extract iron but can also be used for other metals such as zinc, tin, lead and copper. Carbon reduction uses a blast furnace heated by a source of carbon (usually coal or a cheap form of coal called coke). The coke is used to heat the blast furnace to temperatures above 1500°C. When the furnace reaches this temperature the carbon (coke) reacts with oxygen in the hot air and produces carbon dioxide. The carbon dioxide then reacts with more of the hot carbon to produce carbon monoxide gas, which in turn reacts with the iron oxides in the metal, reducing them to iron.

The molten iron liquid pools and is collected in the bottom of the furnace, from where it is drained off into moulds.

Limestone is also added to the furnace with the iron. The limestone gets broken down by the heat of the furnace into calcium oxide, which also reacts with the iron ore and removes impurities in it. These impurities form a molten slag that floats on top of the molten iron, and this is also drained off. Slag was once considered a useless waste product but is now used in road base material, asphalt, concrete and even glass.

> **KEY TERMS**
>
> **Acid rain:** Sulphuric acid mixed with rain that is harmful to humans and the environment.
>
> **Oxidation:** Discolouring, tarnishing and/ or rusting of metal through reaction with air or water.
>
> **Roasting:** Heating metals to release toxic compounds.

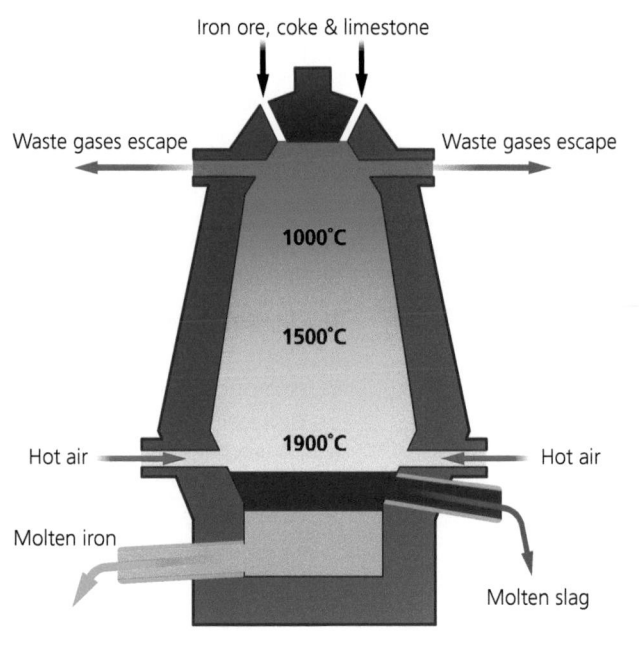

Figure 10.1 **A blast furnace**

Extraction by electrolysis

Electrolysis uses an electrical current that is passed through a liquid that conducts electricity. Two electrodes are placed in the liquid to create the current. The positive electrode is called the anode and the negative is called the cathode. Ions in the molten liquid are attracted to one of the electrodes. The positive electrode attracts negatively charged ions, which are non-metals. The negative electrode attracts positively charged ions, which are metals. By using electrolysis in molten metals, the non-metal elements are separated and extracted from the metal.

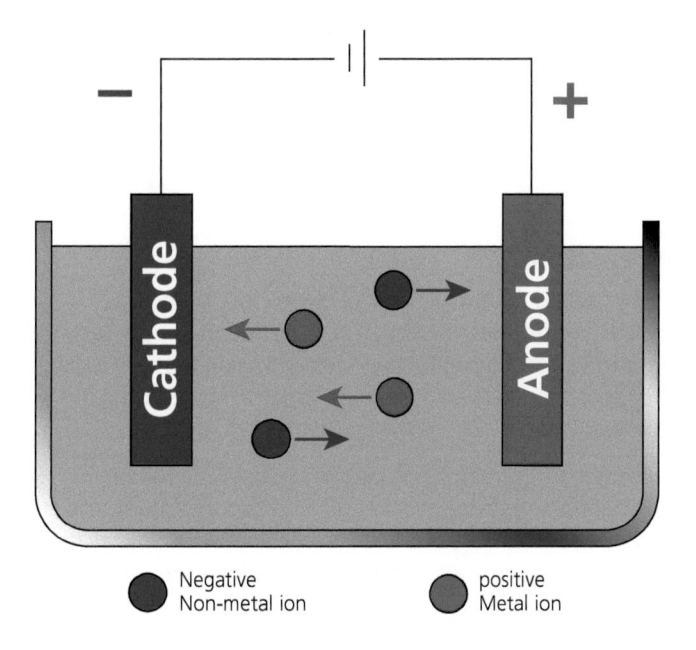

Figure 10.2 **Electrolysis**

Ecological, social and ethical issues

Environmental damage from mining

The process of extracting metal ores from the ground and processing them into useable metals has some harmful environmental effects.

When metal ores are taken from the ground by mining and quarrying, this damages the landscape and results in lots of waste material. Digging up large areas of natural countryside ruins the look of the landscape, leaving these areas scarred or disfigured. This also destroys the natural habitats of wildlife such as birds and small animals.

Once mining has finished some re-landscaping can be done to restore the look of the area, but it may take many years before the 'natural' look of the landscape and wildlife returns.

The mining and quarrying of metals can also have negative social effects:
- Dust created from the mining and processing can spread to surrounding residential areas.
- Noise from the machinery and lorries can be disturbing to local residents as mining and processing can take place at all hours.
- Disused quarries can fill with water, which can be hazardous.
- Mine workings can collapse, causing subsidence under houses and buildings that can be difficult and expensive to fix.

Environmental damage from extraction

The extraction of sulphur dioxide from metals such as copper can have devastating environmental effects. Sulphur dioxide gas is released into the atmosphere during the extraction process. When it rains, this then dissolves in rain water and reacts with oxygen in the air, forming sulphuric acid and acid rain. Acid rain has many harmful effects on the environment, including:

- damaging forest ecosystems, including trees, plants, animals, fish, soil and water supplies
- causing humans to experience breathing problems and reducing oxygen in the heart
- damaging buildings and statues through corrosion.

Lifecycle and recycling of metals

Almost all metals can be recycled and, unlike many other materials, metal can be recycled over and over again without a loss in quality or strength. In the UK over 11 million tonnes of metal are recycled every year and metal accounts for 8 per cent of all recycled materials.

Recycling metal instead of mining and extracting virgin steel drastically reduces CO_2 emissions, air pollution, water use and water pollution.

The first step in the metal recycling process is the collection of the metals. Scrap metal is collected in many different ways. Domestic metals from tins and household appliances can be collected with normal refuse. Companies and individuals can take larger quantities of metal to recycling centres or scrap metal merchants, who will pay a set amount per tonne. Vehicles that are no longer roadworthy are dismantled and scrap metal produced by industry is collected by specialist companies.

Whichever way the metal is collected, it is then sent to a processing plant where it is separated into its different types. Once separated the different types of metal are heated up and melted back down into liquid form using smelters. Smelters are furnaces that can melt large amounts of metal at a very high temperature. Various different smelters are used because of the different melting points of each type of metal.

Once the metals have melted completely, they are cast into small bars called ingots and are then ready to be re used.

ACTIVITY

Investigate your local recycling centre by visiting it or looking at its website. Find out how much metal is recycled there every year.

Figure 10.3 **Metal being smelted**

STRETCH AND CHALLENGE

Draw a lifecycle diagram for an aluminium drinks can.

10.3 Commonly available forms

LEARNING OUTCOMES

By the end of this section you should know about and understand why it is important to know the different available forms of specific materials, including:
→ weights and sizes
→ stock forms.

Sizes and stock forms of metals

Metals are available in many different stock forms depending on the type of metal.

Sheet metal

Most types of metal are available in sheet form. Sheet sizes range from around 500 × 500mm up to around 2.5 × 1.25 metres. Many suppliers will supply sheets cut to a specific size. Common sheet thicknesses are from 0.025mm up to 12.7mm.

Sheet steel and other commonly used metals, such as aluminium, are also available with different shaped perforations that create a mesh-like pattern.

KEY TERM

Profile: The shape of a metal bar.

Stock profiles

Commonly used metals such as steel, aluminium, copper and brass are available in a range of different **profiles** as shown in the diagram below.

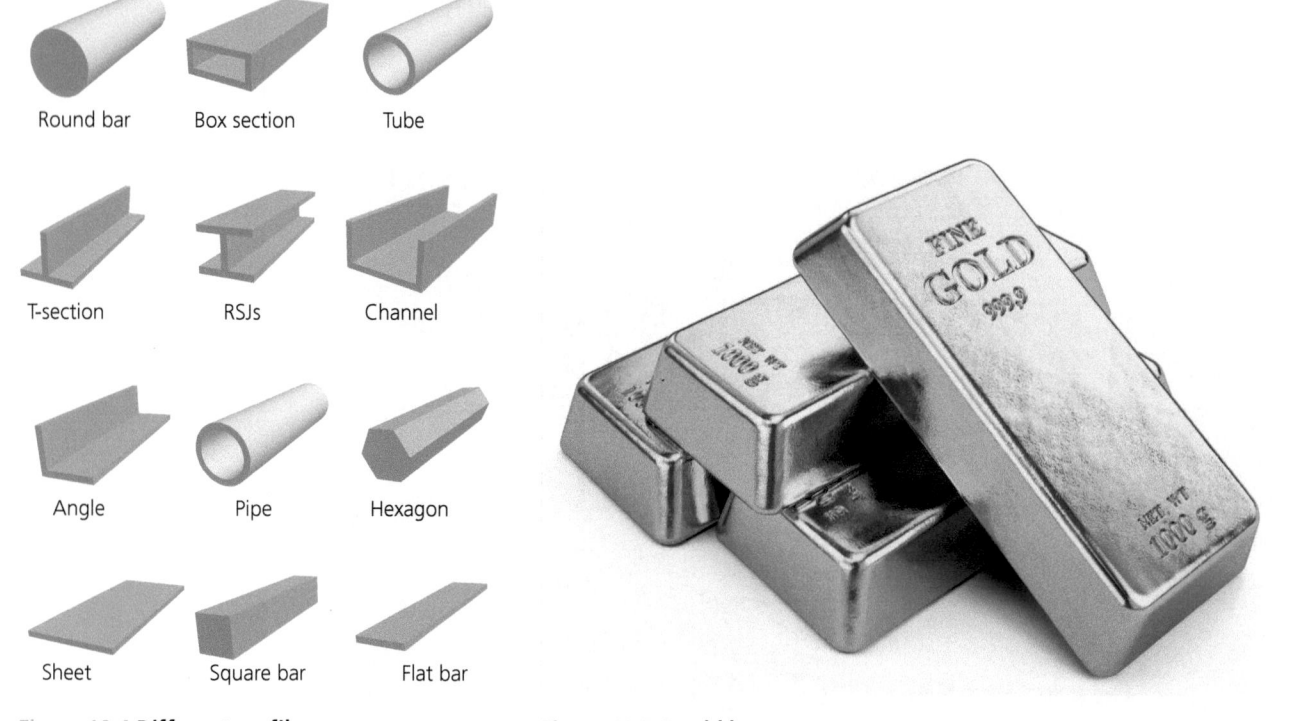

Round bar	Box section	Tube
T-section	RSJs	Channel
Angle	Pipe	Hexagon
Sheet	Square bar	Flat bar

Figure 10.4 **Different profiles**

Figure 10.5 **A gold bar**

Precious metals

Precious metals such as gold and silver can be bought in bars ranging from 1 gram up to 12.5 kilograms in weight. They can also be bought in very thin sheets called 'leaf' and as thin wire in various diameters.

10.4 Manipulating and joining

LEARNING OUTCOMES

By the end of this section you should know about and understand how materials can be manipulated and joined in different ways in a workshop environment when making final prototypes, including:

→ a range of specialist techniques used to shape, fabricate, construct and assemble high-quality prototypes using metal in a workshop.

Wastage: cutting and shaping

Wastage involves cutting materials to the shape required and removing any excess material. It includes processes such as sawing, drilling, shearing and turning.

There is a wide range of hand and power tools available for cutting and shaping metal.

Marking out

Before a metal can be cut it must be marked out. Traditional methods for marking out metal were to use a scriber and **engineer's blue**. Engineer's blue or 'marking blue' is a dye or stain used to colour the surface of the metal in a dark blue colour. The scriber is then used a clean, extremely sharp line by scratching off the blue dye to show the surface of the metal beneath. Modern metalworkers still use scribers but the use of thin permanent markers is also now extremely common. When marking out holes that will later be drilled, the centre of the hole is marked using a centre punch, which puts a slight indent into the metal that 'holds' the drill in the correct spot and stops it 'wandering' around on the metal.

> **KEY TERM**
>
> **Engineer's blue:** A blue dye used for marking out on metals.

Figure 10.6 **Marking out tools**

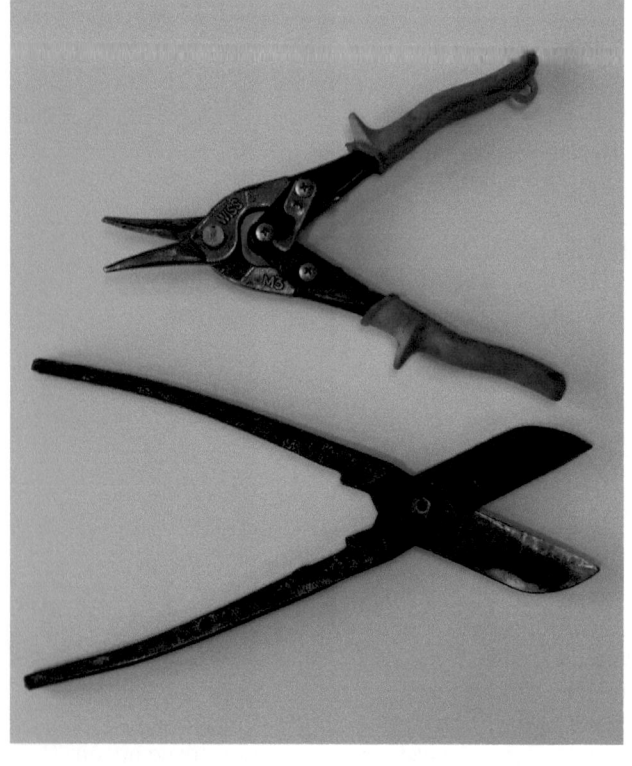

Figure 10.7 **Traditional tin snips and aviation tin snips**

Hand tools

Metal shears or tin snips

Metal shears or tin snips are available in most schools. There are straight or curved types for cutting straight lines or curved shapes out of thin sheet metal up to around 2mm thick. Traditional tin snips can be difficult to use and require a lot of pressure to operate. Aviation tin snips work in the same way but are easier to use and give a more accurate cut.

Bench shears

Bench shears are larger shears mounted on to a bench. They work in the same way as tin snips but have much larger and thicker blades. The long handle that operates the blades gives much more leverage and bench shears can be used for cutting straight lines in thicker sheet metals and flat steel bar up to around 6mm.

Throatless shears

Throatless shears are similar to bench shears but they have no throat (the blades separate completely along their entire length). The lack of a throat allows the metal to be moved around freely and the shears to produce curved or straight cuts. Bench shears and throatless shears are widely available in schools.

Nibbler

A nibbler works in a similar way to tin snips but distorts the metal less. Nibblers remove a thin section of metal between 3 and 6mm wide, called a **swarf**, which forms a tight spiral as it cuts and is wasted. Hand-operated nibblers are relatively inexpensive to buy and widely available. Powered nibblers are also available, which use compressed air, and nibbler tool attachments are available that fit on to electric drills.

Figure 10.8 **Throatless shears**

Figure 10.9 **A nibbler**

Hacksaw

Hacksaws have a C-shaped adjustable frame that can accommodate blades of different sizes. The frame has a screw-type mechanism that holds the blade under tension. Hacksaw blades have very fine teeth and can cut metal bars and rods. C-shaped hacksaws are not suitable for sheet metal but panel hacksaws are available that have different blade holders, allowing them to cut sheet metal. Junior hacksaws are smaller versions of the hacksaw, with smaller sprung frames and thinner blades. They are used for finer cuts in thinner metals.

Figure 10.10 **A hacksaw**

Files

Files are small hand tools that are used to remove small amounts of material from metal, usually after it has been cut by another tool. Files are also used to smooth the edges of metal and remove unsightly cut marks from the edges in a similar way that abrasive paper is used on wood. There are different file types available to suit different shapes, such as flat files (rectangular), square, triangular, round or half-round, and rat tail. One or more surfaces of the file is cut with sharp, parallel teeth. The depth and spacing of the teeth give the file its roughness. The deeper and further apart the teeth are, the rougher the file will be. Most metalworking files have a pointed end called a **tang** that a handle can be fitted to, although many files now have plastic handles instead of tangs. Files are widely used in most schools for filing metal along with other materials such as wood and edges of acrylic.

Needle files are exactly the same as normal files but much smaller and finer. They are used for intricate cutting and shaping where a larger file will not fit.

> ### KEY TERMS
>
> **Swarf:** Waste material when cutting out metals.
>
> **Tang:** The pointed end of a metal file, to which the handle is attached.

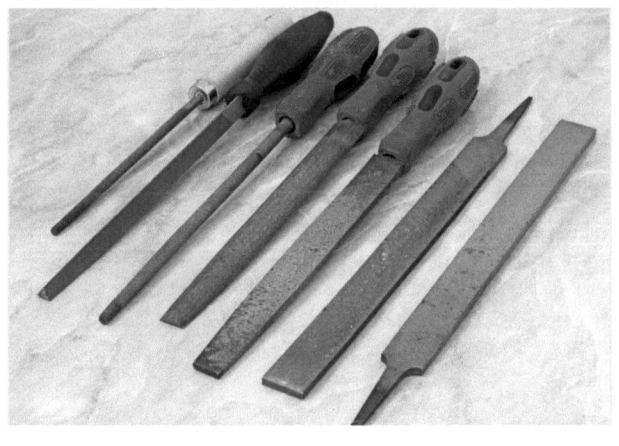

Figure 10.11 **Different files**

Figure 10.12 **An electric angle grinder**

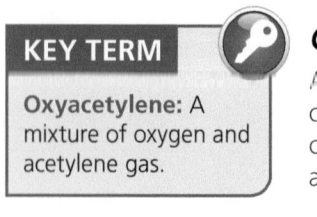
Guillotine

A guillotine is a much larger piece of equipment that is used in industry. It works by first clamping the material down, then using a fixed blade to shear the sheet metal. Guillotines can cut sheet steel and flat bar up to 10mm thick but can only make straight cuts all the way across the whole of the metal. They are usually foot-operated or hydraulic.

Power tools

There is a number of power tools that are purely powered versions of hand tools, such as mechanical hacksaws and air-powered nibblers. There are also metal-cutting blades that can be fitted to jigsaws, chopsaws and bandsaws.

Angle grinder

Angle grinders are versatile tools that can be powered by electricity, compressed air or even petrol engines. Angle grinders have a spinning disc to which different tools for cutting, grinding or sanding can be attached. Different-sized grinders are available that can take cutting discs of different sizes, from 100mm in diameter up to 300mm. Even the smallest grinders are capable of cutting sheet steel and metal bars up to 20mm in thickness.

Figure 10.13 An oxyacetylene torch

Oxyacetylene torch

Oxyacetylene torches are used for oxy-fuel cutting (also known as oxyacetylene cutting or oxy cutting). An oxyacetylene torch mixes acetylene gases and oxygen together, which are then ignited. The pure oxygen when mixed with the acetylene increases the flame temperature to over 3500°C. The torch heats the metal until it is cherry red, then oxygen is blasted on to the heated area by pressing the oxygen-blast trigger. The oxygen reacts with the metal, forming iron oxide and producing heat. This heat continues the cutting process. As the metal burns, it turns to liquid iron oxide and drips from the cutting area.

Plasma cutters

Plasma cutters are heavy duty tools used to cut through steel and other thick metals. Plasma cutters are a relatively new metalworking tool that was first used in industry but is now widely available to the individual hobbyist. There is a wide range of different sizes and power ratings available and many schools now have small, low-power cutters. Plasma cutters work by sending an inert gas at an extremely high speed from the nozzle tip down to the metal surface. The gas has an electrical arc running through it that converts the gas into plasma. The plasma melts the metal in a fraction of a second, effectively cutting it. Because of the intense bright light and heat generated, suitable safety equipment must be worn. Leather gloves and apron along with green lens goggles and a face mask protect the user from eye damage and risk of burns.

Figure 10.14 A plasma cutter in use

Drilling

Drilling metal can be done using a pillar drill or a hand drill. A wide range of drill bits specifically for drilling holes in different metals is available. Standard metal drill bits are suitable for drilling in soft metals like copper and aluminium. Harder metals such as stainless steel require chrome vanadium, cobalt or titanium carbide bits. Standard metal drill bits are available in diameters from 1 to 13mm. When drilling metal the speed of the drill should be slower than for materials such as wood or plastic. As a general rule, the harder the metal, the slower the drill should be.

It is good practice when drilling larger holes to start with a small hole and use progressively larger sizes until the required size is reached. Special **stepper drills** are available that drill larger and larger holes without the need to constantly change drill bits. Before drilling a hole in metal, the centre of the hole should be punched to prevent the drill bit wandering over the surface of the metal. A hand vice should always be used when drilling metal as it has a tendency to snag the drill as it comes out of the opposite side of the metal.

Turning

The metal **lathe** is used for 'turning' solid pieces of metal into useable objects by spinning and applying cutting tools to the surface. The metal 'blank' is held in a chuck with three or four clamps, depending on the shape of the blank. The cutting tools are mounted on to a tool-post, which can be moved along the workpiece and in and out towards the centre of rotation by winding small handles that move the tool in very small, precise increments. Lathes are used to create circular metal objects such as candle sticks, crankshafts, wheel spacers, handles and many other machine parts. Lathes can also cut threads on metal objects to high levels of accuracy and much quicker than traditional methods. Lathes are used in engineering to make and modify specialist machine parts to very fine levels of accuracy.

> **KEY TERMS**
>
> **Lathe:** A machine used for making cylindrical-shaped objects by spinning.
>
> **Stepper drill:** A drill bit that drills progressively larger and larger holes.

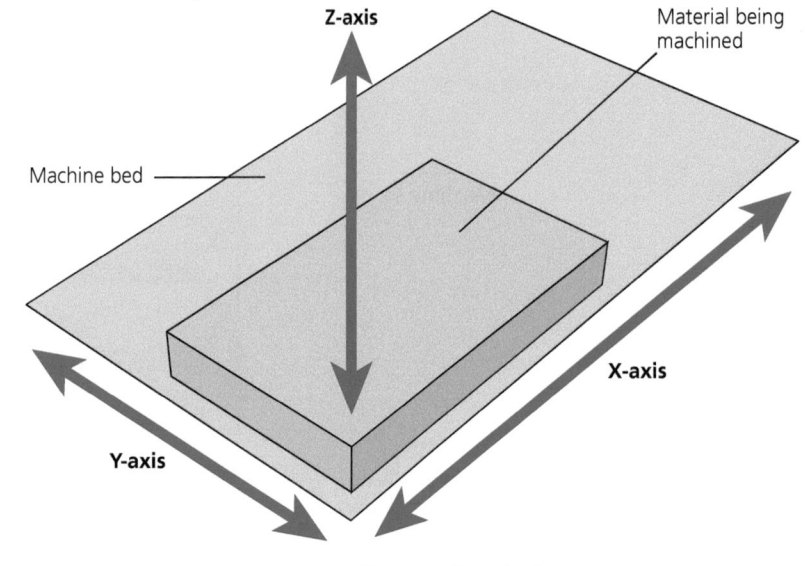

Figure 10.15 **A metalworking lathe**

Figure 10.16 **A milling machine bed**

Milling

Milling machines use a rotary cutting tool, which is used in engineering to shape and create metal machine parts and other metal items. The metal part is clamped on to a flat bed that can be moved in two different perpendicular axes by turning handles similar to those on a lathe. The milling tool is lowered to the level of the part, which is then moved along or across the cutter along the two axes using the turning handles. As the milling cutter enters the metal, the cutting edges (flutes or teeth) of the tool shave off material by performing many separate, small cuts. This is accomplished by a cutter with many teeth, spinning at high speed. The speed of the cutter will depend on the type of metal being milled. The speed at which the material is passed along the cutter (also known as the feed rate) will also vary.

Addition: joining metals

There are many ways of joining metals together. Some are permanent and some are called temporary methods. Temporary methods allow the metal to be 'un-joined' at a later date. Permanent methods are not easily undone without cutting, damaging or even destroying the metal parts.

Temporary methods

Nuts and bolts

One of the most common ways of temporarily joining metal together is by using nuts and bolts. Before a bolt can be inserted, a suitable-sized hole must be drilled in both pieces of metal. The bolt is then passed through both holes and the nut is threaded on the end of the bolt and tightened. As the nut gets tighter it forces the two pieces of metal together. To help spread the force over a slightly wider area and get a better join, washers are also used either side of the two metal pieces.

There is a wide range of different nuts, bolts and washers available. Different combinations of length, diameter, head shape, material, finish and thread pitch can be used for different applications.

Table 10.7 **Applications for different bolts and screws**

Bolt type	Image	Applications
Hex bolts		Used in machinery and construction. Can be fully or part threaded. Tightened with spanner or socket.
Machine screws		Used in machinery, construction and household products and appliances. Fully threaded. Tightened with a flat or crosshead screwdriver.
Thread cutting machine screws		Used in construction, bodywork, steel frame buildings. Point is 'self-tapping' so can be bolted into a non-threaded hole.

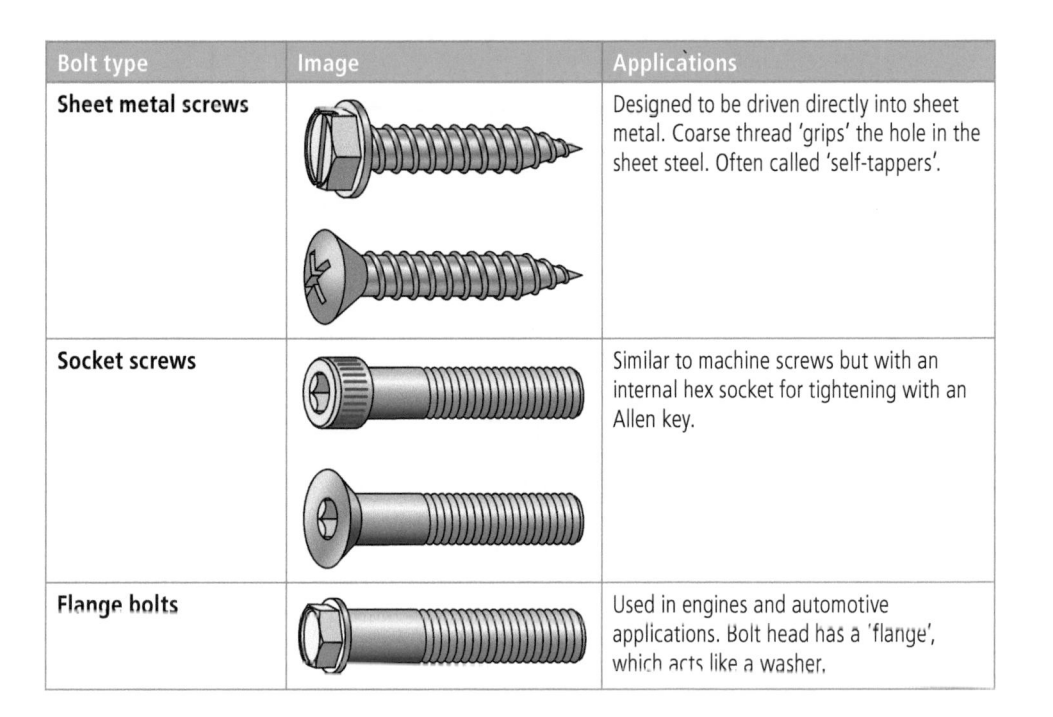

Bolt type	Image	Applications
Sheet metal screws		Designed to be driven directly into sheet metal. Coarse thread 'grips' the hole in the sheet steel. Often called 'self-tappers'.
Socket screws		Similar to machine screws but with an internal hex socket for tightening with an Allen key.
Flange bolts		Used in engines and automotive applications. Bolt head has a 'flange', which acts like a washer.

Permanent methods

Pop rivets

Pop rivets are cylindrical steel tubes that can be inserted in metals to join them together. Like for bolts, a suitable-sized hole must be drilled in both pieces of metal before the pop rivet can be inserted.

Pop rivets have two separate parts: the pin and the sleeve. To use pop rivets you also need a pop rivet gun. Once the pop rivet has been inserted into the two pieces of metal the gun pulls the pin through the sleeve, distorting and spreading the end of the tube so that it cannot slip through the hole and holding the two pieces of metal together.

Figure 10.17 **A pop rivet and rivet gun**

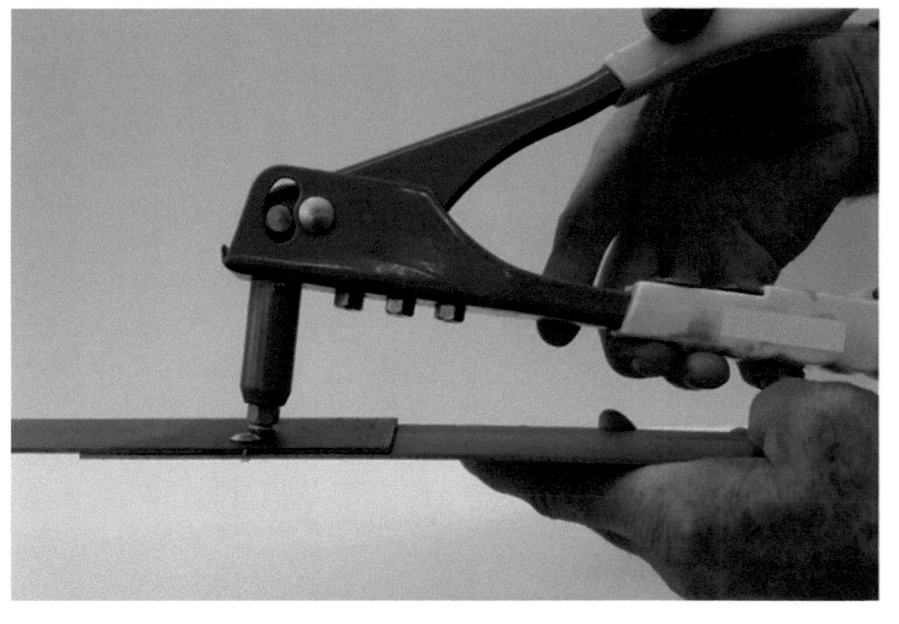

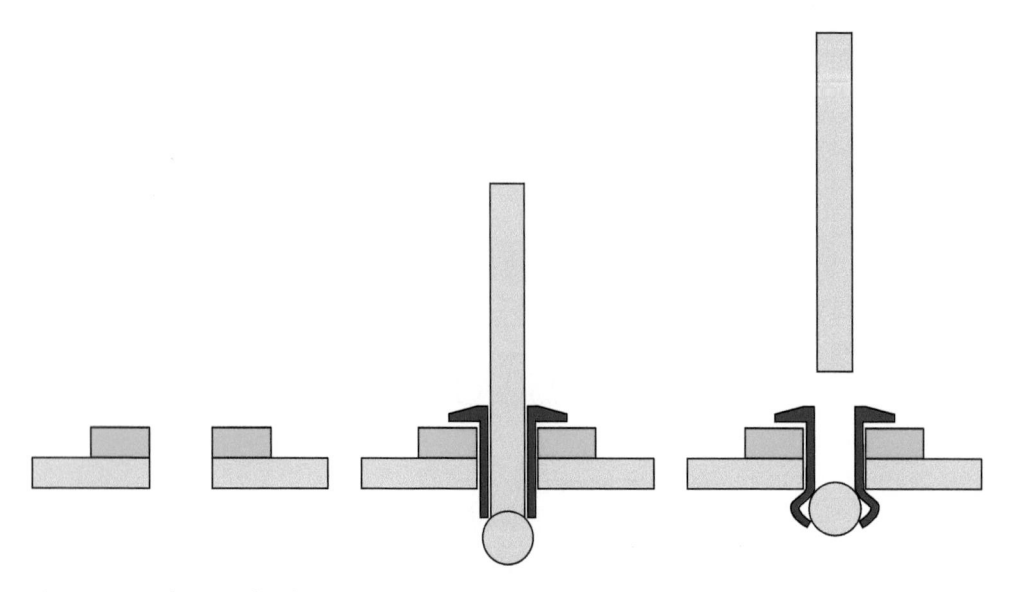

Figure 10.18 **The pop riveting process**

1   A hole is drilled or punched in both sheets of metal.
2   The pop rivet is inserted through a hole in both sheets.
3   The gun pulls the pin through the tube, pushing the walls out. When the ball can go no further it breaks off and the pin falls out.

Gluing (bonding)

Metals can be joined by strong adhesives called epoxy resins. Epoxy resins have two parts: the resin itself and the hardener or activator, which when mixed together in correct proportions react chemically to form an extremely strong and tough adhesive. There is a wide range of different epoxy resin products available. The epoxy and hardener are supplied in two separate tubes or syringes with different-sized nozzles. By squeezing out equal lengths of each the correct mixture proportion is dispensed, which the user then mixes and applies.

Figure 10.19 **A syringe epoxy resin dispenser**

Soldering

Soldering is a process that involves melting metal (called solder) on to and into the point where two metals meet in order to create a joint. There are three forms of soldering, which require increasingly high temperatures and in turn provide an increasingly strong joint strength:

- Soft soldering, which uses a tin–lead or –copper alloy solder.
- Silver soldering, which uses an alloy containing silver.
- Brazing, which uses a brass alloy for the filler.

Soft soldering uses a solder that is an alloy of two metals (mainly tin, with a smaller proportion of copper or lead), which has a low melting point (between 180 and 200°C). The two pieces of metal to be joined are placed together and the area is heated. Solder is then fed into the area, which melts and flows around the joint, filling any small gaps and 'enveloping' the two metals. The heat is then removed and as the joint cools the solder hardens and fixes the joint together.

Soft soldering is used in electronics for the joining of electronic components, electrical wires and cables, as it provides a solid connection without overheating sensitive electronic components. Electrical soldering heats the joint using a soldering iron, which has a fine tip that can concentrate heat in a small area. The solder comes in the form of a thin 'string' that is fed into the joint as required. The solder also contains flux, which is a chemical that removes contaminates and helps the solder 'flow' into the joint.

Figure 10.20 **A soldering iron joining electronic components** Figure 10.21 **A silver soldering torch in use**

Silver soldering is most commonly used to join semi-precious and precious metals such as platinum, gold, silver, copper and brass. Silver solders are alloys of mostly silver with different mixtures of other metals and have melting points of between 600 and 750°C. Silver soldering uses a soldering torch powered by gas such as butane, which provides a small high-temperature flame that can be applied to the desired area. The solder is placed on to the joint along with the flux before it is heated. As the joint heats up the solder and flux flow into and around the joint and the heat is then removed.

KEY TERM

Soldering: Melting solder around two or more metals to create a joint.

Brazing is a form of soldering that uses very high temperatures. Brazing can join different metals such as bronze, steel, aluminium, wrought iron and copper that have different melting points. Brazed joints create an airtight and watertight bond that can be easily plated over to create a seamless appearance. Brazing requires temperatures of over 1000°C; to achieve this an oxyacetylene torch is used. When brazing, the **flux** is applied to the joint in the form of a paste before it is heated. The joint is then heated and the braze, which can be in the form of a rod, wire or disc joint, is then fed into and along the join, where it melts and flows around the metal before being allowed to cool.

Welding

Welding is the most effective way of permanently joining two metals together. Welding melts the two metals being joined and fuses them together, effectively making them into one single piece. There are different types and ways of welding but the two main types used are:

- gas or torch welding
- arc welding.

Figure 10.22 Brazing

🔑 **KEY TERMS**

Brazing: Soldering at very high temperatures.

Flux: A chemical that removes contaminates when soldering.

Welding: A fusion of metals caused by intense heat.

Gas or torch welding uses an oxyacetylene torch to melt the two metals and the filler material, fusing them together. The process is similar to brazing but the temperature used in welding is higher and, unlike brazing, it is the two metals being joined that are actually melted rather than just the filler rod. Gas or oxyacetylene welding is often called braze welding because of the similarity in techniques. Gas welding is not widely used today and has largely been replaced by arc welding techniques, such as mig or tig welding (see below). Gas welding is used mainly for maintenance work as no electricity is required and for welding delicate parts and joints in soft metals like aluminium.

Arc welding uses electricity to melt the metals and the welding filler by creating an electrical arc. An electrical arc is like a tiny lightning bolt that passes between two metals, generating intense heat that melts the two metals and the filler material, fusing them together.

STRETCH AND CHALLENGE

A box made from sheet steel is shown below.

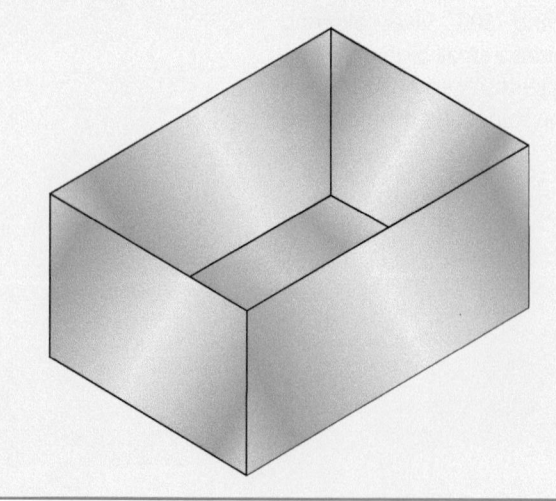

Complete the table below to show the processes and tools used to make the box in the workshop.

Process	Tools/equipment
Cutting the sheet steel to shape	
Smoothing the edges	
Bending the steel into shape	
Fixing the sides together	

Folding

Many metal objects are made by folding sheet metal into shape. Metal can be folded by simply bending it over in a bench vice and hammering the bend flat to get a neat crisp fold. This is suitable for small objects but larger metal sheets require a folder.

There are many different types of metal folder available but they all work in more or less the same way. The metal is placed on to the bed of the machine and clamped or held in place along the crease line. The bed is then hinged at the crease line, forcing the sheet metal upwards and forming a neat bend or fold up to angles of 120 degrees.

Special folders are available that clamp and fold the metal on more than one side to form box shapes. These are sometimes known as box-and-pan brakes. After bending the sides are fixed together by soldering, welding or riveting to form a solid box.

Figure 10.23 **A folding machine folding metal**

Pressing

Metal pressing (sometimes known as stamping) is a method of forming sheet metal into a three dimensional shape by pressing it between two shaped dies. The pieces of sheet metal called 'blanks' are pressed into the same shape as the surface of the dies. Metal pressing can be used to shape and cut metal in different ways such as cutting, embossing, bending, and flanging. It is widely used in industry to make a wide variety of metal products such as car body panels, brackets and metal enclosures.

Metal pressing uses large and powerful hydraulic presses to provide the force needed to press the metal into shape.

Casting

Casting is one of the oldest metal shaping processes and has been used to make metal objects for thousands of years. Casting metal requires the metal to be heated up to a molten liquid state. It is then poured into a mould where it cools and solidifies into a solid metal shape. The metal shape known as the 'casting' is then removed. Usually the casting will have various 'nibs' that need further finishing by filing, machining or grinding smooth. Casting is often used to create metal shapes that would be difficult or impossible to make using other methods.

Casting is still widely used in industry to create a wide range of products. There are many different and specialised methods used in industry but casting can also be done on very small scales in a school workshop using metals such as pewter and moulds made from sand or even MDF.

Sand Casting involves the use of a bucket or other container of damp sand. When a suitably shaped object is pressed into the sand it leaves a printed indent of its exact shape which the molten metal can be poured into. Once the metal has solidified it can be dug out and the sand simply brushed or washed off.

MDF moulds can be made by milling or chiselling out shapes in blocks of MDF. A pouring channel and an air channel must also be chiselled or milled out from the shape to the edge of the material. The blocks of MDF are then bolted or screwed tightly together and the molten metal can be carefully poured into the mould through the channel. Once solidified the blocks can be unscrewed or bolted and separated to release the casting.

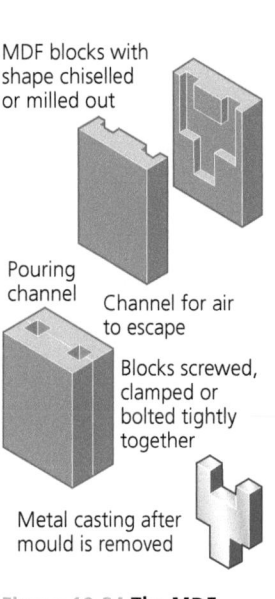

MDF blocks with shape chiselled or milled out

Pouring channel

Channel for air to escape

Blocks screwed, clamped or bolted tightly together

Metal casting after mould is removed

Figure 10.24 **The MDF mould process**

10.5 Structural integrity

LEARNING OUTCOMES

By the end of this section you should know about and understand how and why metals need to be reinforced to withstand forces and stresses, including:
→ how and why metals can be reinforced to withstand external forces and stresses
→ the processes that can be used to ensure the structural integrity of a product.

Reinforcing metal

Shaping metal

Metals can be shaped to improve their structural integrity. Section 5.4 looked at how folding or curving a piece of paper can give it more structural integrity. In the same way, sheet metal can be curved, folded or formed into shapes to give it increased strength and rigidity.

A normal length of flat steel bar will bend quite easily one way but not the other. A folded section is difficult to bend in either direction. This principle is used to increase the structural integrity of steel bars and beams used in construction.

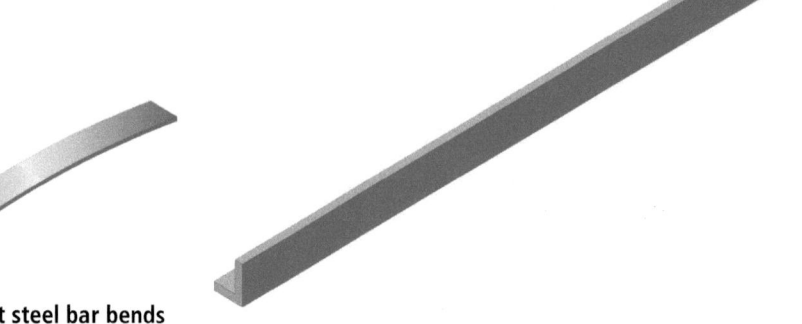

Figure 10.25 **A flat and bent steel bar. A flat steel bar bends easily this way around. Turning the steel on its edge makes it much more difficult to bend.**

Figure 10.26 **A folded steel bar. By folding the metal it becomes much more difficult to bend.**

T-section, I-section, U-channel and L-angle are common shapes given to metals to increase their structural integrity.

ACTIVITY

Take a normal flat metal ruler, and a metal safety ruler. Look at the section of each.

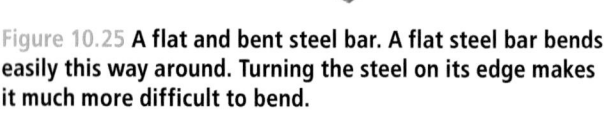

Try to flex the flat metal ruler (but don't completely bend it) – you'll notice that it's very easy!

Now try to bend the safety rule. The curves in the metal make it much harder to bend.

If you flattened the safety rule it would be as easy to bend as the other one.

Processes to ensure structural integrity

As described earlier in this chapter, the most common way of altering the structural integrity of metals is by mixing them with other metals or substances to create alloys.

Alloys can be created that combine the properties of one metal with another. For example, Duralumin (an alloy made up of aluminium and copper) has the same lightweight properties as aluminium but is much stronger due to the small amount of copper added.

There are other treatments and processes that can be applied to metals to make them harder, tougher or more ductile.

Hardening and tempering

Hardening and **tempering** are treatments that can be done to ferrous metals to increase their hardness and **toughness**. The two processes are usually done together. This is because the hardening process also makes the metal more brittle, so after hardening the metal is tempered. The tempering process slightly reduces the hardness of the metal but increases its toughness, making it less brittle and less likely to snap.

The hardening process involves gradually heating the metal until it is glowing red, which occurs at around 1050°C. Once it reaches this 'red hot' stage, it is then cooled very quickly by immersing it in cold water (called quenching). This gradual heating and rapid cooling is what hardens the metal.

The tempering process is done immediately after hardening, once the metal has been removed from the cold water and cooled to room temperature. The metal is heated gradually once again to a temperature of 175–350°C for around two hours. It is then allowed to cool gradually, which tempers the metal.

The temperature used for tempering depends on the required outcome of the finished metal. Higher temperatures will result in a softer metal with greater toughness. Lower temperatures will produce a harder but more brittle metal.

Case hardening

Case hardening is another method of hardening ferrous metals. It is easier than the hardening and tempering process and is usually done on metals with a low carbon content. Case hardening only affects the outer surface of the metal so its other properties are not altered.

Like the normal hardening process, case hardening gradually heats the metal until it is glowing red, which occurs at around 1050°C. Once it reaches this 'red hot' stage it is dipped into case hardening compound. Case hardening compound can be bought in powdered form.

The compound sticks to the hot surface of the metal and the metal cools down a little. It is then re-heated gradually to the previous temperature and quickly cooled by quenching in cold water.

The process can be repeated to further increase the outer hardness of the metal.

Annealing

Annealing is a process that is used to reduce a metal's hardness and increase its ductility, making it easier to work.

The annealing process is relatively simple and involves gradually heating the metal until it glows red, then removing the heat and allowing it to cool very slowly. Annealing is usually done on ferrous metals but can also be done on copper, silver and brass.

In industry, large gas-fired ovens or furnaces are used for the hardening, tempering and annealing processes. The metals are often left inside the ovens after they are shut off so that they cool very gradually in a controlled manner.

KEY POINTS
- Metals get harder the more they are worked.
- Annealing can be done at any point to make metals easier to work.

KEY TERMS

Annealing: Reducing the hardness of a metal and making it more ductile.

Case hardening: Hardening the outer surface of a metal.

Tempering: Using heat to make a metal less brittle.

Toughness: Ability to resist breaking, bending or snapping.

KEY TERM

Quenching: Rapidly cooling hot metal by immersing in cold water.

Large saltwater troughs are often used for **quenching** because they reduce bubbling and cool metal faster than just water.

Hardening, tempering, case hardening and annealing can all be carried out in a school workshop using a brazing hearth and oxyacetylene torch.

ACTIVITY

Watch your teacher complete the following demonstration:
1 Take a length of mild steel around 30cm long. Using the brazing hearth and oxyacetylene torch, harden one end of the rod or bar.
2 Once cooled, place the bar or rod in a vice. Take a hacksaw and cut a small piece off the untreated end. Now cut a piece off the treated end. How much harder was it to cut?
3 Try case hardening another piece of steel bar or rod and repeating the exercise. Which process makes the steel hardest?

STRETCH AND CHALLENGE

Create a step-by-step guide using diagrams and notes to show someone how to do one of the above processes.

10.6 Making iterative models

LEARNING OUTCOMES

By the end of this section you should know about and understand:
→ how processes and techniques are used to produce early models to support iterative designing.

Processes and techniques for iterative models

Metals can be used in a variety of useful and interesting ways to help you explore your early design ideas and support iterative designing in school.

Metal wire and thin metal rods are excellent for developing initial ideas. They are quick and easy to bend and fold and are relatively inexpensive.

Welding rods come in 900mm lengths and can be bent into different shapes by hand or with pliers to form frameworks or supports for model ideas. They can be twisted together to make more intricate shapes and designs. Thin metal rods can be very useful for creating pivots, joints and axles in working models of mechanisms or moving parts.

In the same way, strips of thin metal such as copper and aluminium can be bent by hand or with hand tools to form simple shapes that can be the basis for models of ideas. Thicker metal can be bent and formed using a vice and hammer or metal folder (see the next section on manipulating and joining).

Steel or aluminium mesh is a commonly used metal material for modelling. There are many different types and varieties of mesh available. Chicken wire is a very thin mesh made from thin wire. It can be easily cut, bent and twisted into different shapes by hand and is often used to form the basic shapes of 3D models. It can then be covered with papier mâché or other modelling materials if required. With practice, mesh can be made into extremely intricate shapes.

Thin sheet metals such as tin and aluminium foil are widely available and inexpensive. Foil can be used in modelling to cover Styrofoam or cardboard shapes or shapes made in other materials in order to give them a metallic look, to make them reflect light or to add strength. Foil is easy to cut and shape like paper but will hold its shape better once creased or folded. Foil can be scrunched up and moulded by hand to form all manner of shapes that can be extremely useful for modelling, including sculptures and items of clothing. Due to its electrical conductivity it is also extremely useful for modelling situations using small electrical components such as bulbs, LEDs, motors and so on.

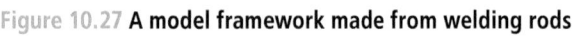

Figure 10.27 **A model framework made from welding rods**

Figure 10.28 **Chicken wire mesh models**

Figure 10.29 **A foil model**

ACTIVITY

Take some thin metal wire, chicken wire or pipe cleaners (they are metal wire inside the furry coating). By twisting and bending the wire, design and model a children's climbing frame.

ACTIVITY

Experiment with foil and try making some animal or human sculptures using just your hands.

10.7 Finishes

LEARNING OUTCOMES

By the end of this section you should know about and understand how metals can be finished for different purposes, including:
→ function
→ aesthetics.

Most metals require a finish of some kind to protect them from damage or to improve their appearance. The type of finish will depend on the type of metal and how the product is to be used. For example, a steel machine part may not require a finish that looks nice but will need some sort of protection from rusting. This may be as simple as a spray of oil. A steel bicycle frame, however, will need to have a nice appearance and be protected from corrosion, so a suitable surface coating needs to be applied.

Ferrous metals in particular are much more susceptible to corrosion and will begin to corrode almost immediately if left unfinished. Some metals may require no finishing of any kind. This is called a **self-finish**.

> **KEY TERM**
>
> **Self-finish:** The natural finish of a metal.

Painting

Painting is one of the easiest and quickest methods of finishing metal. Painting is commonly used on ferrous metals to prevent corrosion. Before any paint can be applied the metal must be clean and free of any rust. If surface rust has already occurred it can be removed using abrasive paper or a wire brush. It must then be cleaned to remove any oil, grease or other substance that may prevent the paint adhering to the surface.

A primer should first be applied to prevent rust reforming, help the paint stick to the metal surface better and add an extra layer of protection.

A wide range of different paints are available for metal:
- Water-based (latex or acrylic) paints
- Solvent-based (mineral spirit or thinner) paints
- Oil-based paints.

Figure 10.30 Metal paints

Traditionally, oil- or solvent-based paints were the only suitable paints for ferrous metal as water-based paint would not give suitable protection from corrosion. New advances in paint technology, however, have meant that water-based paints are now equal to oil- and solvent-based coatings.

Oil- and solvent-based paints contain high levels of solvents that release volatile organic compounds (VOCs) into the atmosphere. VOCs are carcinogenic and harmful to the environment. Because of the health risks and environmental hazards, paint manufacturers are developing low or non-VOC paints. These water-based paints are eco-friendly and not harmful to humans.

Metal paint can be applied by brush or roller or by spraying. A wide range is available from DIY outlets.

Specialist painters such as car bodywork workers use special 'two pack paints', which consist of a polymer resin that is mixed with a hardener called polyisocyanate resin.

Polymer resin is a clear liquid plastic. When it is mixed with the hardener, a chemical reaction takes place and it begins to harden or 'set' within a few minutes. The hardened and fully set resin mix creates a durable, glossy and waterproof coating. Two pack paints are much harder than solvent-based paints and more resistant to petrol, acid rain and sunlight. (For more information on polymers see Chapter 13.)

Many paints are available that claim to give rust protection on ferrous metals without the use of a separate primer. These types of paint have a zinc-based primer mixed with the paint that seals the rust to prevent it from spreading. In practice these types of paint are not as long-lasting or rust-resistant as using a separate primer and top coat, but they do offer a 'quick fix'.

Lacquering

Lacquer can be used as an alternative to paint where the natural colour of the metal is required to show through. Lacquer dries to a completely clear, colourless film that prevents oxidisation and tarnishing of the metal's surface. It is often used on metals such as brass, copper, bronze and silver that have lustrous, attractive finishes, rather than on ferrous metals.

Like paint, lacquer is available in tins for applying by brush and in cans for spray application. When applying lacquer it is vital that the surface of the metal is not tarnished and is free from any contaminates such as wax, grease and metal polish. If the surface is not prepared properly the clear lacquer can go a yellowy or milky colour after a short period of time.

Stove enamelling

Stove enamelling is similar to spray painting but the drying or 'curing' process is speeded up by using heat. The metal is prepared and sprayed in the same way but is then baked in an oven at temperatures from 150 to 200°C. The resulting finish is one of the finest paint finishes available for metal. It combines a high-quality finish with a tough, durable corrosion-resistant and long-lasting surface. A wide range of colours and effects is available and metallic colours can also be stove enamelled.

Stove enamelling is suitable for high temperature environments and applications through the use of special, heat-resistant paints. Because of this, stove enamelling is often used for metal products such as radiators, kitchenware and wood-burning stoves.

Figure 10.31 **A stove enamelled product**

Powder coating

Powder coating uses a dry powder made from a polyester, polyurethane, epoxy or acrylic polymer providing a glossy, shiny finish that is much more durable than paint or laquer (for more information on polymers, see Chapter 10).

Before metal can be powder coated it must be cleaned to remove oil, dust, grease and rust (most commonly by shot blasting or sand blasting the metal).

The metal to be coated is hung from a wire that has a small negative electrical charge applied to it. The powder is sprayed on to the metal using an electrostatic gun or corona gun. The gun gives the powder a positive electrical charge so that it is attracted to the negatively charged piece of metal. This means that there is very little overspray as the powder is drawn to the metal and clings to it.

> **KEY TERMS**
>
> **Powder coating:** A coating of electrically charged powder that is baked on to achieve a tough finish.
>
> **Stove enamelling:** A paint coating that is baked on to achieve a tough, durable finish.

Once the metal item is coated in a layer of powder it is heated or baked in a large oven at 200°C for approximately ten minutes. As it is heated the powder melts and then chemically reacts to form a strong chemical structure. The melted powder forms a smooth, even coating over the metal. Once cool the coating hardens and forms an extremely tough and resilient coating.

Powder coating is often used on household appliances such as washing machines, dishwashers and fridges, as well as bicycle and motorcycle frames, alloy wheels on cars, and structural girders.

Figure 10.32 Powder coating a car wheel

Enamelling

Enamelling is an ancient process that was traditionally used on metal jewellery. It is still used for coating metal jewellery today but since the nineteenth century it has also been used as an industrial coating on many other metal products, such as kitchenware, dishwashers, washing machines, baths and sinks.

The enamelling process involves powdered glass that is heated in an oven to between 750 and 850 °C. The glass powder then melts and spreads across the metal in a smooth, even layer. When the metal cools the glass hardens to a smooth, durable coating.

Most modern industrial enamel (vitreous enamel) is applied to steel but it can also be applied to gold, silver, copper, aluminium, stainless steel and cast iron.

Vitreous enamel is smooth, hard, chemically resistant, durable and scratch-resistant, has long-lasting colour fastness, is easy to clean, and cannot burn. A wide range of colours is available and different colours can be mixed to create new shades, similar to paint. Transparent, opaque and translucent finishes are available.

A disadvantage of enamel is that it can crack or shatter when the metal is stressed or bent.

Figure 10.33 Enamelled jewellery

Plating

Metal **plating** is a process that lays a thin surface coating of one metal on to another type of metal; it is done for the following reasons:

- Decoration
- Corrosion inhibition
- To improve solderability
- To harden
- To improve wearability
- To reduce friction
- To improve paint adhesion
- To alter electrical conductivity.

There are several different plating methods.

KEY TERM

Plating: Coating one type of metal with another to improve appearance or corrosion-resistance.

Electroplating

Electroplating is widely used in industry for coating metal objects. Electroplating increases the life of metal and prevents corrosion. It is also used in making inexpensive jewellery, by coating cheaper metal such as copper with a thin outer layer of a more precious metal such as gold or silver.

Electroplating uses an electric current to make dissolved metal atoms form a metal coating on another metal.

The metal that is going to be coated or plated is given a negative electrical charge. The metal it will be coated with is given a positive electrical charge.

Both metals are then immersed in a salt solution of the metal that will be used to plate the object (e.g. silver plating uses silver nitrate solution). The metallic ions of the solution are positively charged and are attracted to the negatively charged metal object. Once they connect, they revert back to their metallic form on the surface of the metal, creating a thin coating.

Electroless plating

Electroless plating deposits metal on an object without the use of an electric current. It is often used to plate unusually shaped objects, which are difficult to plate evenly with electroplating as it deposits metal evenly along edges, inside holes and in cavities. The electroless plating process uses hydrogen, which reacts with the metal ions to produce a negative charge and deposit metal on the surface of the part. The most common electroless plating method is electroless nickel plating. Gold, silver and copper can also be applied by electroless plating.

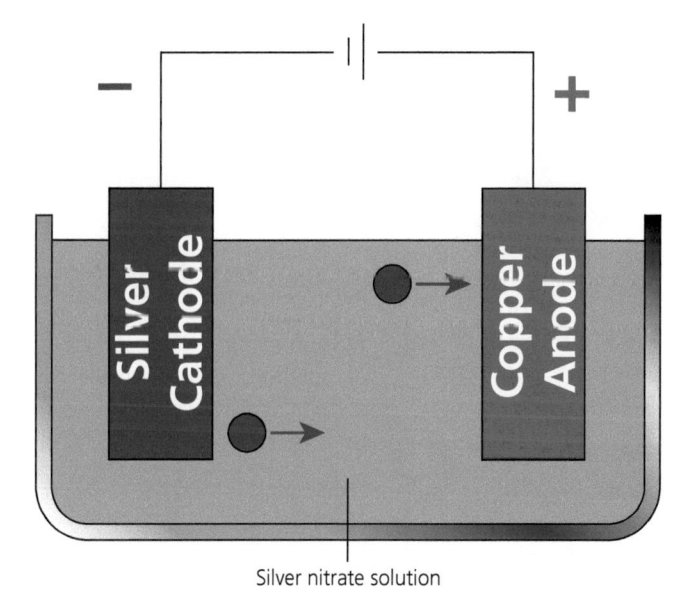

Silver nitrate solution

Figure 10.34 Electroplating

Table 10.6 Plating is done using many different metals

Metal plate	Metal plated over	Uses and applications
Gold plating	Copper Silver Nickel	Jewellery, electrical connectors, printed circuit boards
Silver plating	Tin Antimony Copper	Jewellery, cutlery, candlesticks, musical instruments
Chrome plating	Steel Nickel Copper	Car parts, bath taps, kitchen gas burners, wheel rims
Zinc plating	Steel	Screws, nails, bolts, roofing sheets, outdoor building hardware
Tin plating	Copper Nickel Iron	Food cans

Figure 10.35 **A galvanised metal surface, showing 'spangles'**

KEY TERM

Galvanising: Coating steel with zinc to stop it corroding.

Galvanising

Galvanising is a method of metal plating that applies a coating of zinc to ferrous metals to prevent the metal beneath corroding by preventing oxygen and moisture reaching it.

The most common method of galvanising is called hot dip galvanising. Zinc is heated to a temperature of 450°C so that it melts and becomes molten zinc. The ferrous metal is then lowered and immersed in the tank of molten zinc. A reaction occurs between the ferrous metal and the molten zinc, creating a series of metal alloy layers across the surface of the ferrous metal; this completely coats and penetrates the surface to create a firm bond. After five minutes the metal is removed from the molten zinc and allowed to cool. Galvanised metal has a crystallised pattern on the surface called 'spangles'.

As for most other types of metal finish, the metal being galvanised must be clean and completely free from grease, oil and corrosion for the chemical reaction to work properly. Before the galvanising process, metal parts are washed in a series of different solutions to remove all contaminates to ensure a good finish is achieved.

Galvanising is used on a wide range of products, such as roofing sheets, metal buckets, industrial fencing, gates and motorway Armco barriers.

Dip coating

Dip coatings on metals involve creating a thermo polymer coating on the surface of the metal. A variety of different thermo polymers such as polyethelene, nylon and PVC can be used.

The metal to be coated is heated to between 250 and 400°C and the thermo polymers in powdered form can be applied to the metal in a variety of different ways.

Fluidised bed coating

Fluidised bed coating is the most commonly used method of dip coating, and small fluidised bed dip coaters are available in many schools. The fluidised bed is created by blowing air into a tank containing the powder. The air makes the powder look and behave like a liquid. The metal item is heated to the correct temperature and completely immersed into this fluidised bed. The powder melts on contact with the metal and the item is then lifted out of the fluidised bed and cooled to leave a high-quality polymer coating.

Electrostatic spray

This method is similar to the powder-coating process. The thermo polymer powder is electrically charged and sprayed on to the earthed metal item, then placed in an industrial oven and heated until the powder melts. The item is then cooled to leave a high-quality polymer coating.

Flock spraying

Flock spraying is a simple process. The metal is heated to the correct temperature and the thermo polymer powder is sprayed on to the hot metal using compressed air. The powder melts on contact with the metal and is then left to cool to leave a high-quality polymer coating.

Extrusion

Extrusion is used for coating metal wire. The wire is passed through a tank of molten polymer, which coats the wire, and it is then cooled to leave a continuous polymer coating.

Polishing

Polishing and buffing are used for smoothing the surface of a metal to enhance the appearance, prevent contamination and corrosion, remove oxidation and create a reflective surface. Polishing is a more aggressive process, which removes deeper scratches and blemishes, whereas buffing is less harsh, and removes the finer scratches and gives a smoother, brighter finish.

When polishing metals, a rough abrasive (60 or 80 grit) will be used first. As the polishing process continues finer and finer abrasives are used, through to extremely fine grit abrasives such as 400 or more, leaving smaller and finer scratches that are not visible to the naked eye.

Once the metal has been polished through all the abrasives, it can be buffed. Buffing may be done by hand or using a polishing wheel, a high-speed polishing machine or another machine tool that can be used for buffing or polishing, like an electrical drill. There is a wide range of different polishing and buffing mops available along with a large selection of different buffing compounds. Different metals require different combinations of mops and compounds to achieve a good finish.

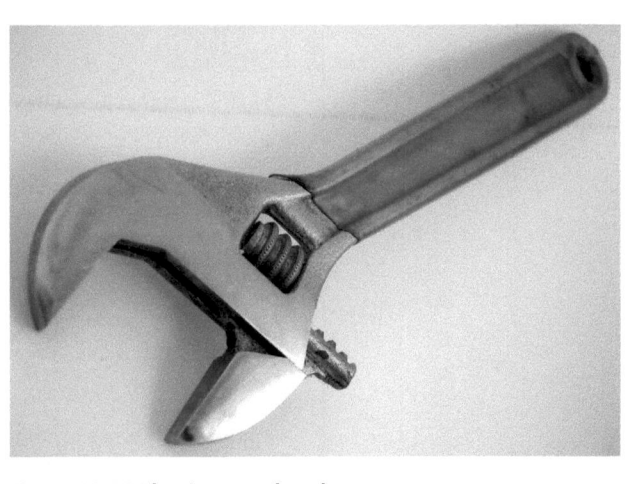

Figure 10.36 **Plastic-coated tools**

	Plastics			Silver, gold and thin plastics			Nickel and chrome plate			Copper, brass, aluminium, pot metal and other soft metals			Steel and iron			Stainless steel		
Buff type	A	B	C	A	B	C	A	B	C	A	B	C	A	B	C	A	B	C
Sisal										X			X			X		
Spiral sewn								X			X			X			X	
Loose												X			X			X
Canton flannel							X			X								
String	X	X	X															
Compound	A	B	C	A	B	C	A	B	C	A	B	C	A	B	C	A	B	C
Black										X			X			X		
Brown											X							
White								X					X	X				
Blue	X	X	X					X			X						X	
Green																	X	X
Red								X			X					X		

Figure 10.37 **A polishing compound colour chart**

STRETCH AND CHALLENGE

1 Use the internet to look at different types of jewellery and silverware. Compare the prices of plated products to the prices of pure solid items.
2 What are the benefits to the consumer of electroplating?

ACTIVITY

Ask your teacher for some 'scrap' pieces of steel. Clean the steel up using an abrasive pad. Apply different finishes such as paint, varnish and so on to each item but leave one piece of steel without a finish applied. Leave the scrap pieces of steel outside for a few days. Evaluate how effective the finishes have been in protecting the metal from corrosion compared to the metal without a finish applied.

ACTIVITY

Find examples of metal products at home, school or outside. Identify what type of finish has been applied, and explain why.

KEY POINTS

- Most finishes are applied to ferrous metals to prevent corrosion.
- Polishing compounds work by using mild abrasives to remove scratches from metal.

Product	Finish	Explain why the finish has been applied
Climbing frame in a park	Galvanising	Will not rust, looks attractive, will not wear off.
Fence made from steel	Painting	Can be done 'in-situ'
Bicycle frame		
Radiator		
Silver jewellery		

2 Choose two other examples of metal products in the home, at school or outside. Identify what finish has been applied, and explain why.

10.8 Using digital design tools

LEARNING OUTCOMES

By the end of this section you should know about and understand how industry professionals use digital design tools when exploring and developing design ideas, including:

→ the use of 2D and 3D digital technology and tools to present, model, design and manufacture solutions.

Developing ideas and manufacturing products

KEY TERMS

CAM: Computer-aided manufacture.

CNC: Computer numeric control.

Origin: The starting point for a CNC or CAM machine, from which all coordinates are taken.

A range of digital tools can be used for developing ideas and manufacturing products in metal. Computer-aided design (**CAD**) is a digital design tool that can be used by designers to produce working drawings using computer softwre, rather than by hand. It also allows designers to create both 2D and 3D visualisations of concept. Computer-aided manufacture (**CAM**) can be used in various forms with metals. Many of the CAM machines are computer-controlled versions of the existing metal-shaping tools as described in Section 7 of this chapter:

● **CNC** milling machines
● CNC lathes
● CNC plasma cutters.

These types of CAM equipment carry out the same tasks and work in a very similar way to the hand-operated tools, but the movement of the cutting tool is controlled by a computer. Specialist programs are used to programme the machinery and are then loaded on to the machine.

CAM machines – for example lathes and tools for cutting sheet materials, such as plasma cutters can cut to extremely high accuracies of up to one-thousandth of a millimetre. They are able to produce detailed products of consistent size much quicker than a human operative.

STRETCH AND CHALLENGE

Create a series of designs for some chess or game pieces that could be made out of aluminium on a CNC lathe.

Laser cutters

Laser engraving and laser cutting machines are CAM machines that can be used for cutting thin sheet metals although not all laser cutters are capable of cutting metals. Laser cutters can accurately mark and engrave most polymers and some manufactured timbers such as plywood and MDF. Larger and more powerful types are available that can cut sheet metals, ceramics, glass and even stone.

Laser cutters are suitable for cutting thin sheet materials only and the depth of cut cannot accurately be programmed. Therefore they are of use only for engraving the surface of a material or cutting all the way through it.

Figure 10.38 **A laser cutter cutting material**

10.9 Manufacturing methods and scales of production

LEARNING OUTCOMES

By the end of this section you should know about and understand how processes vary when manufacturing products to different scales of production, including:
→ the methods used for manufacturing at different scales of production.

Scales of production

Deciding on a suitable **scale of production** and manufacturing method for your metal parts depends on a number of factors:
- **Form:** What shape are your parts? This will be limited by the restrictions of the manufacturing processes available.
- **Budget:** Part cost and tooling cost. Some methods have expensive set-up costs; others are quite cheap, but there's usually a trade-off. Most high volume manufacturing processes are expensive to tool but offer cheap parts – the opposite is also true: low volume processes are cheap to set up but parts are expensive.
- **Time:** Tooling takes time; manufacturing takes time. More expensive tooling usually means longer set-up time.
- **Material:** Material choice is determined mostly by form, function and cost.

One-off production

One-off production of metal products is expensive and time-consuming. One-off production requires skilled workers who are experienced with metalworking and can use a range of techniques.

If a metal product is produced as a one-off the material will first need to be cut to size. If the product is made from a number of different separate pieces the materials for these will need to be cut using appropriate tools. If certain parts are made from sheet metal and other parts are made from rod, bar or channel, these will require different cutting techniques and equipment. Shaping, bending or folding metal is difficult to do without specialist tools but one-off products often need to be made using basic tools as it is not cost effective to invest in expensive jigs and so on for just one product.

KEY TERMS

One-off production: Making only one or a small number of products. Usually used for specialist products.

Scale of production: The number of products being produced in one go.

Joining metals to create one-off products would most likely be carried out using a mig welder. Specialist firms can do one-off welding jobs but these can be expensive.

Although many buildings that use steel frames and cladding are one-offs, the materials used are standard mass-produced items such as steel beams, girders, standard-sized sheet metals and universal fixings such as bolts and rivets.

Some common metal products that are made as one-offs are bespoke items such as furniture, metal gates, custom vehicles and parts, and architectural fittings including fire escapes and staircases.

Batch production

Batch production involves making a product stage by stage in batches. This can be done by the same person or by having different people carrying out each stage. Batch produced metal products may have a person who is responsible for cutting the pieces of metal to size. Then another person shapes and drills the metal parts, another assembles the product, and another applies a finish. Batch production is often used in the manufacture of similar metal products that may have a common section but several variations, such as garden gates or hand rails.

KEY TERMS

Batch production: Making a set number of products that can have some variations.

Mass production: Large-scale production of an item or product.

Mass production

Mass production is the most commonly used form of production for metal products. Production often involves the assembly of a number of smaller sub-assemblies of metal components, which when assembled are fed on to the main production line. Metal parts such as fasteners, brackets, hinges and other hardware may be bought from other companies to speed up the manufacturing process. Many of the assembly and production processes use automation instead of humans. CNC machines can also be used to make certain parts but these usually require a human operative. Automated production lines have machines that can cut, fold and press metal into shape. Robots are used to position, weld and rivet metal parts together as well as to apply finishes. The vast majority of metal products we use on a day-to-day basis, such as cutlery, cars, pens, chairs, railings, etc., have been mass produced.

Figure 10.39 An automated production line and robot welder

STRETCH AND CHALLENGE

Research the different assembly and sub-assembly lines used in a car or motorcycle production line. Draw a diagram showing how the different sub-assemblies feed into the main production line.

10.10 Cost and availability

LEARNING OUTCOMES

By the end of this section you should know about and understand how cost and availability of specific metals can affect decisions when designing, including:
→ the significance of the cost of specific metals
→ how to calculate the quantities, costs and sizes of metals required in a design
→ consideration of suitable tolerances and minimisation of waste.

The significance of cost

During the design and development of a new product there are many factors that will influence your choice of materials. This section looks at how the cost and availability of different metals can affect your material choices.

As with any other materials, the more that you buy, the cheaper it becomes. This is called bulk buying or **bulk discounting**. A large manufacturing company that requires lots of sheet steel and buys hundreds of sheets per year will pay less per sheet than a smaller firm or individual who may buy only one or two.

The prices of individual metals can go up and down by significant amounts depending on the current supply and demand and the predicted future supply and demand. The prices of different metals are a little like the stock exchange and can sometimes change on an hourly basis. There are many websites that show up-to-date metal prices.

Because of the many different types of metals, prices and quantities, metals are separated into five different market groups:
- Base metals – such as copper, tin.
- Steel and ferro-alloys.
- Minor metals – such as indium, tungsten, molybdenum.
- Platinum group metals (PGMs) – such as platinum and osmium.
- Precious metals – such as gold and silver.

Although the market prices of metals affect the cost for manufacturers, they usually buy specific grades, forms and quantities of metal and in certain stock forms (see Section 7.3). The prices of metals on the markets will affect the prices of these stock forms, but other factors that go into manufacturing of the metal product are not affected by its price. Therefore a large price increase in the market may result in only a small increase to the manufactured product. For example, even if the price of copper goes up significantly, the price of a copper pipe may increase only a little, as the labour, energy and transport costs to make the pipe remain the same.

Reducing costs

When the availability of a certain metal becomes extremely limited and/or when the prices make using it for a design uneconomic, there are alternatives:
- Using a similar metal or alloy that costs less.
- Plating a cheaper metal with more expensive metals.

Precious metals often fluctuate in price. When gold prices are high, you will tend to see numerous TV adverts from companies offering to buy unwanted and broken gold jewellery. When a metal is being used for a product because of a particular quality it may have, there are frequently alternatives that may be very similar but much cheaper and/or more available. A good example of this is pewter, which has a similar colour to silver and can be polished to a high shine to look almost the same. In the case of industrial metals such as steel, a lower grade may be chosen to save money where this will not affect the safety or functionality of the product.

> **KEY TERM**
>
> **Bulk discount:** Reduced price of items for buying lots at a time.

189

Figure 10.40 **A silver-plated item**

Plating is commonly used in jewellery-making to reduce costs. By electroplating a less expensive metal such as copper, jewellery can be given a thin coating of gold and made to look and feel like the real thing. In the same way, mild steel – which is much cheaper than stainless steel – can be plated in metals such as zinc to stop it rusting.

Plated metals are a cheaper alternative to the real thing, but the plating does not last indefinitely and will wear off after a period of time to reveal the base metal. If, however, the lifecycle of a product requires it to last only a limited period of time, then plating can be a viable alternative for designers when cost or availability prevents the use of their chosen metal.

How to calculate the quantities, cost and sizes of metals required

Calculating the quantities of materials needed to make a product and the cost of these materials is important when designing and making. If a design costs more or almost as much to produce as it is likely to sell for it is not 'cost effective' to make.

Example

A metal shelf bracket is made from three pieces of flat steel bar as shown below.

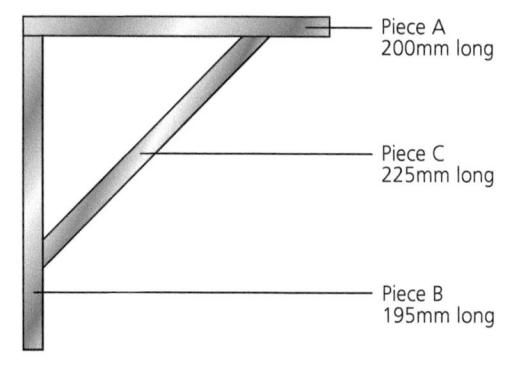

Piece A
200mm long

Piece C
225mm long

Piece B
195mm long

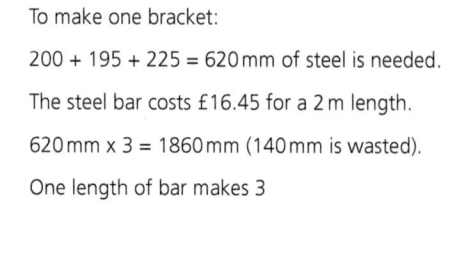

To make one bracket:

200 + 195 + 225 = 620 mm of steel is needed.

The steel bar costs £16.45 for a 2 m length.

620 mm x 3 = 1860 mm (140 mm is wasted).

One length of bar makes 3

ACTIVITY

A component made from 5mm steel plate is shown below. The components will be cut from a sheet of steel plate 1000 × 2000mm by a CNC water jet cutter.

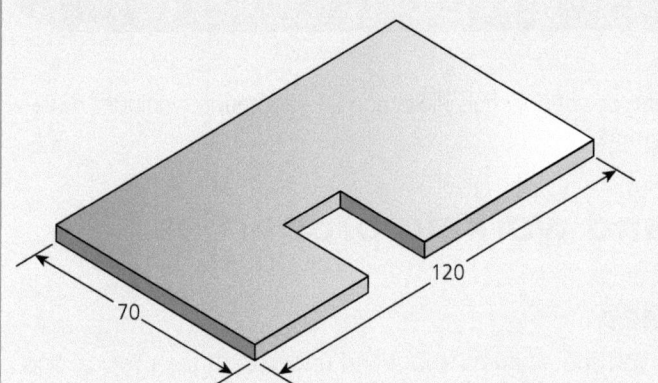

120

70

● Calculate the best way to minimise waste by getting as many as possible out of one sheet.
● If the 1000 × 2000mm sheet costs £200, calculate the cost per component.
● The water jet cutter costs £28 an hour to run. It takes 90 seconds to cut out one component. By how much does this increase the manufacturing cost per component?

Consideration of suitable tolerances and minimisation of waste

Designers should ideally look at ways to minimise waste materials when using metals, but if this is not possible the scrapping and recycling of the waste can recover some of the material costs.

The vast majority of metals used in industry are recycled. It is far cheaper to recycle an existing metal than to mine, extract and process new. Specialist metal recycling companies purchase and collect scrap metal from businesses and manufacturing plants to recycle. Like the market prices of metals, the value of scrap metals including steel, aluminium and copper rise and fall accordingly.

You learnt about the different types of polymers in Section 5.1. This chapter includes more in-depth information on polymers.

11.1 Physical and working properties

LEARNING OUTCOMES

By the end of this section you should know about and understand the physical and working properties of a range of polymers, including:
→ how easy they are to work with
→ how well they fulfil the required functions of products in different contexts.

KEY TERMS

Monomer: A molecule that can be bonded to other identical molecules to form a polymer.

Polymer: A substance that has a molecular structure built up from a large number of similar units (monomers) bonded together.

Synthetic: A manufactured substance that imitates a natural product.

Plastics and polymers

Manufactured plastics and natural materials such as rubber or cellulose are composed of very large molecules called **polymers**. Polymers are constructed from relatively small molecular fragments known as **monomers** that are joined together.

Rubber and cellulose are examples of natural polymers that have been known about and used since ancient times. **Synthetic** polymers, including the large group known as plastics, grew in importance in the early twentieth century. Chemists' ability to engineer them to achieve particular properties (strength, stiffness, density, heat resistance, electrical conductivity) has greatly expanded the many roles they play in the modern industrial economy.

Why choose polymers?

Synthetic polymers are generally easy to process and they can be cost effective. This means that high-quality products can be manufactured at relatively low cost. Polymers that are now available have a wide range of properties, some of which are very specialised.

Generally, polymers are:
- lightweight
- waterproof
- tough
- electrical and/or thermal insulators
- resistant to atmospheric degradation (they neither rust like metals nor rot like timber).

In addition, by varying the additives they can be made:
- transparent or opaque
- rigid or flexible.

Polymers can also be engineered to have specialised properties, such as water solubility, electrical conductance, high temperature resistance and biodegradability, and even to be piezoelectric (to have the ability to generate electricity when bent or twisted). The sections that follow describe how common polymers are classed.

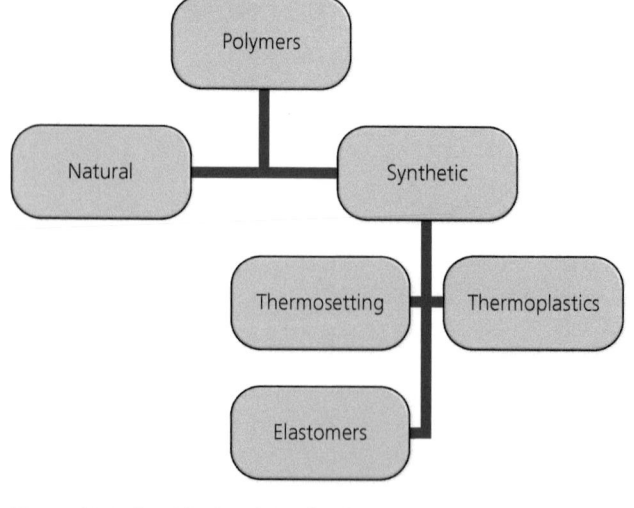

Figure 11.1 **The relationship of polymers with plastics**

Common thermopolymers

Table 11.1 **Common thermopolymers**

Recycling Code	Common name	Properties/working characteristics	Uses
♲ 1 PETE	Polyethylene terephthalate (PET)	Clear, tough and shatter-resistant, PET has good moisture and gas barrier properties	Soft drink bottles, mineral water bottles, fruit juice containers and cooking oil bottles
♲ 2 HDPE	High-density polythene (HDPE)	Range of colours, hard, stiff, good chemical resistance, high impact	Milk crates, bottles, pipes, buckets, bowls
♲ 3 PVC	Polyvinyl chloride (PVC)	Stiff, hard, tough, good chemical and weather resistance, uPVC (un-plasticized PVC) has strong resistance to chemicals and sunlight	Pipes, guttering, roofing sheets, window frames
♲ 4 LDPE	Low-density polythene (LDPE)	Range of colours, tough, flexible, good electrical insulator and chemical resistance	Washing-up liquid, detergent and squeezy bottles, bin liners, carrier bags
♲ 5 PP	Polypropylene (PP)	Hard and lightweight, good chemical resistance, can be sterilised, good impact, easily welded together, resistance to work fatigue	Medical equipment, syringes, crates, string, rope, chair shells, containers with integral (built-in) hinges, kitchenware
♲ 6 PS	Polystyrene (PS)	Range of colours, stiff, hard, lightweight, safe with food, good water resistance	Disposable plates, cups, fridge linings, model kits, food containers
♲ 6 PS	Expanded polystyrene (EPS)	Lightweight, absorbs shock, good sound and heat insulator	Sound and heat insulation, protective packaging
♲ 7 OTHER	Nylon	Hard, tough, resilient to wear, self-lubricating, resistant to chemicals and high temperatures	Gear wheels, bearings, curtain-rail fittings, clothing, combs, power-tool cases, hinges
♲ 7 OTHER	Acrylic	Stiff, hard, clear, durable outdoors, easily machined and polished, good range of colours, excellent impact resistance (glass substitute), does scratch easily	Illuminated signs, aircraft canopies, car rear-light clusters, baths, Perspex sheet
♲ 7 OTHER	Thermoplastic elastomers (TPE)	A combination of thermoplastics and elastomers. Flexible and tough. After stretching and bending they will return to close to their original shape	Watch straps, scuba diving masks, remote control buttons
♲ 9 ABS	Acrylonitrile Butadiene Styrene (ABS)	Tough, high impact strength, lightweight, scratch-resistant, chemical resistance, excellent appearance and finish	Kitchenware, safety helmets, car parts, telephones, food mixers, toys

Common thermosetting polymers

Table 11.2 **Common thermosetting polymers**

Common name	Properties/working characteristics	Uses
Urea-formaldehyde	Stiff, hard, brittle, heat resistance, good electrical insulator, range of colours	White electrical fittings, domestic appliance parts, wood glue
Melamine-formaldehyde	Stiff, hard, strong, range of colours, scratch and stain resistance, odourless	Tableware, decorative laminates for work surfaces, electrical insulation
Phenol-formaldehyde	Stiff, hard, strong, heat resistance	Dark electrical fittings, saucepan and kettle handles
Epoxy resin	Good chemical and wear resistance, heat resistance to 250°C, electrical insulator	Adhesives such as Araldite® used to bond different materials such as wood, metal and porcelain
Polyester resin	Becomes tough when laminated with glass fibre, hard and strong but brittle without reinforcement	GRP boats, chair shells, car bodies

Common elastomers

Table 11.3 **Common elastomers**

Elastomer	Properties/working characteristics	Uses
Silicone	Excellent heat and oil resistance	Flexible baking trays, bathroom sealant
Neoprene	Weather resistance, flame retardant	Wetsuits, knee and elbow pads
Butadiene rubber	Resistant to abrasion and cracking	Tyres, golf ball cores
Fluoroelastomer	Durable, chemical resistance	Apple Watch Sport straps

KEY POINT

When designing, it is important to know and consider the physical and working properties of different materials to make sure you select the best material. For example, it wouldn't be good to use acrylic for a product that needed to be flexible.

STRETCH AND CHALLENGE

Use the internet to gather a range of images of polymer products and try to identify the polymers used. Think about the following:
● What does the product do?
● How is the product used?
● Where is it used?

11.2 Sources and origins

LEARNING OUTCOMES

By the end of this section you should know about and understand:
➜ the sources and origins of a range of polymers
➜ the processes used to extract and convert polymers into a useable form
➜ the ecological, social and ethical issues associated with processing polymers
➜ the lifecycle of specific polymers
➜ recycling, reuse and disposal of specific polymers.

Sources

Natural polymers

Before there were plastics and synthetic polymers, natural polymers existed to help make life possible. For example, animals' horns and hooves could be moulded into a variety of shapes, and this material was used to make waterproof containers. Natural rubber latex is made from the resin from rubber trees, and was used to make primitive rubber bands and glues. Shellac is a resin secreted by the lac bug in India and Thailand. The dried flakes were mixed with ethanol to produce the first varnish used to protect wooden products.

Synthetic polymers

The first partially synthetic polymer was called Vulcanite – a thermosetting polymer made by chemically modifying natural latex rubber by adding sulphur (a process called vulcanisation). It was first used in the late nineteenth century to make products such as combs and buttons.

Figure 11.2 **A wooden box finished with shellac**

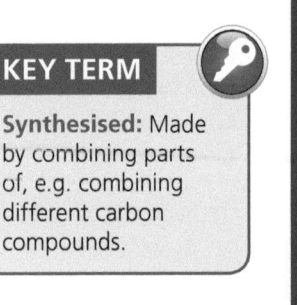

KEY TERM

Synthesised: Made by combining parts of, e.g. combining different carbon compounds.

Bakelite was the first entirely synthetic polymer and was developed in the early twentieth century. It was lightweight and durable and could be moulded into almost any shape, so its use quickly expanded as manufacturers realised its potential. Bakelite was used to make jewellery boxes, kitchenware, lamps, clocks, radios, telephones and many more products.

Extraction and conversion

Most modern polymers are totally synthetic. The chemicals from which they are made are **synthesised** from carbon compounds that are largely obtained from crude oil. The crude oil is first fractionally distilled and then some of the products are processed further to produce the chemical. Only a small amount (about 4 per cent) of total crude oil production goes into making polymers.

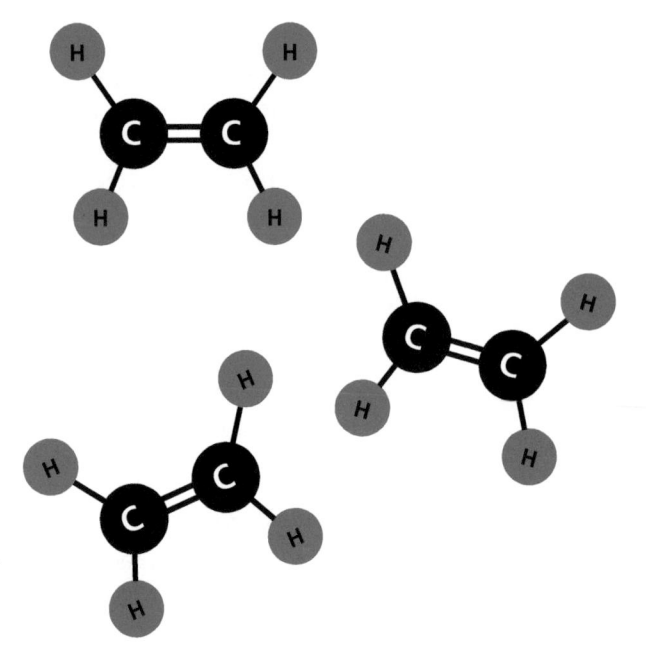

Figure 11.3 **A Bakelite radio**

One of the chemicals obtained from crude oil is ethene. This is used to make the plastic polythene, and if you look at how polythene is made you can learn about the general structure of plastics and begin to understand their properties.

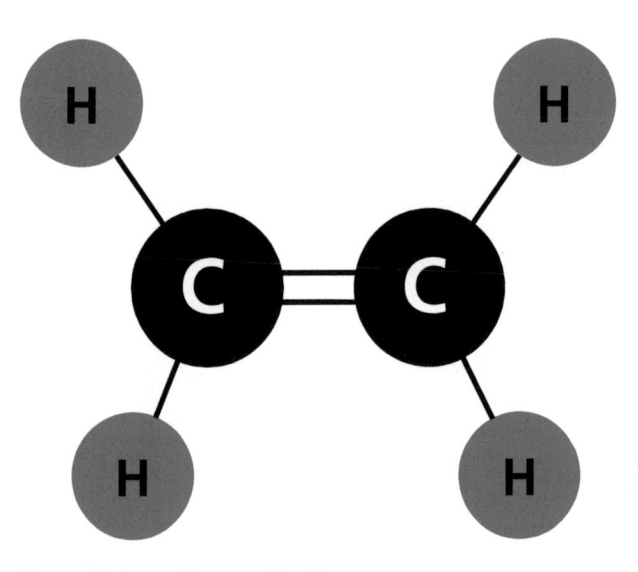

Figure 11.4 **An ethene molecule**

Figure 11.5 **Ethene molecules floating as a gas**

KEY TERMS 🔑

Additives: Substances added to polymers to improve their mechanical properties.

Catalysts: Chemicals that cause a reaction to happen.

Polymerisation: The process of joining small molecules to form polymers (long chains).

An ethene molecule (see Figure 11.6) is made of four hydrogen atoms and two carbon atoms – the lines represent the bonds between them. Ethene gas is made up of millions of these molecules floating around.

Polythene is made by getting these ethene molecules to join together to form long chain molecules. This is done using other chemicals called **catalysts**. During the process, thousands of ethene molecules join together to form polythene molecules.

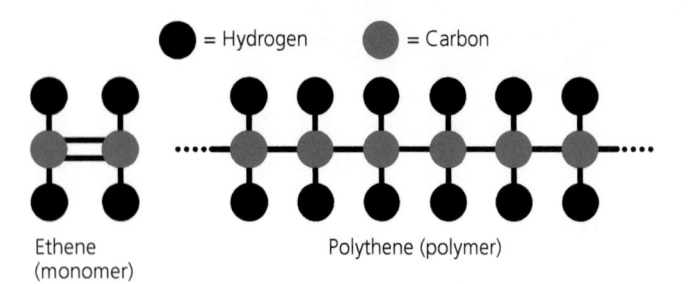

● = Hydrogen   ● = Carbon

Ethene (monomer)

Polythene (polymer)

Figure 11.6 **Ethene molecules joined together to form polythene**

Polythene molecules do not need to be made to join together; they attract each other and become tangled and twisted together, which forms the solid, high-density polythene.

Small molecules like ethene that can do this are called monomers (mono = single). Once joined together they become a polymer (poly = many). The process of joining them together is called **polymerisation**.

Additives

The mechanical properties of polymers can be improved by the use of **additives**:

● **Plasticisers** are substances added to polymers to improve their flow properties for moulding. PVC is used for tough drainpipes, yet by adding a suitable plasticiser it can be made flexible enough to make a garden hose.
● **Pigments** add colour.
● **Stabilisers** help prevent damage from ultraviolet light, which can cause several polymers to become brittle.
● **Fillers** are powdered solids that bulk up polymers and reduce costs, but they can also improve strength by reducing brittleness and increasing resistance to impact.
● **Catalysts** can be used to speed up the synthesis of polymers. They include epoxy resin, which hardens in one hour compared to the usual 12.
● **Antioxidants** prevent oxidation.

Ecological, social and ethical issues associated with processing

Only a small amount (4 per cent) of crude oil production goes into making polymers. The vast majority goes towards providing energy for heating, transport and electricity (86 per cent). World production of polymers has been rising steadily, however, and is now in the region of 300 million tonnes a year, and it will continue to rise approximately 4 per cent each year. During the manufacture of polymers significant quantities of toxic chemicals are produced. Producing a 500ml PET bottle generates more than a hundred times the toxic emissions to air and water than making the same bottle out of glass.

Safety concerns

Producing plastics can be hazardous to workers. Serious accidents have included explosions, chemical fires, chemical spills, and clouds of toxic vapour. These kinds of occurrences have caused deaths, injuries, evacuations and major property damage.

Environmental and health effects

Many chemical additives that give plastic products desirable performance properties also have negative environmental and human health effects. Phthalates are added to plastics to make them softer (plasticisers) and several of them have been classed as carcinogens – substances capable of causing cancer.

The most obvious form of pollution associated with polymers is the waste products that are either dumped or sent to landfill. Between 10 and 20 million tonnes of plastic ends up in the ocean every year. This plastic debris damages marine ecosystems. Animals such as seabirds, whales and dolphins can ingest or become entangled in plastic matter, and floating plastic items – such as discarded nets, docks and boats – can transport microbes, algae, invertebrates and fish into non-native regions, affecting the local ecosystems.

It is estimated that every adult male in the UK generates around one tonne of waste each year. Around 9 per cent of this is polymer-based packaging waste such as milk bottles and food packaging. This waste could actually be recycled and made into new polymer-based products.

Figure 11.7 **The majority of polymer products end up in landfill**

Lifecycle

Polymers are a very stable material and tend to stay in the environment for a long time after they are discarded, especially if they are shielded from direct sunlight by being buried in landfills. They decompose very slowly, especially as most polymers contain antioxidants to resist attack by chemicals. Different kinds of polymer degrade at different times, but the average time for a simple PET bottle to degrade is at least 450 years. Some might even take a thousand years. While they slowly degrade they can release significant quantities of toxic chemicals such as trichloroethane and methylene chloride.

Recycling, reuse and disposal
Recycling

Recycling is an important factor in conserving natural resources and greatly contributes towards improving the environment. Recycling recovers materials used in the home or industry for further uses. It reduces the amount of energy and natural resources (such as water, petroleum and natural) needed to create new plastic. It also reduces the amount of material dumped in landfill sites.

Many polymers, such as PET, can be recycled without any loss of quality. In fact, the plastic from PET drinks bottles can be recycled into clothing products such as fleeces. Several products, especially those that come into contact with food, must use new or virgin plastic for health and safety reasons.

Other products are allowed to use a certain percentage of recycled polymer mixed in with virgin material. Recycling a product into a lower quality product is called **downcycling**. The HDPE used in milk bottles is an excellent polymer for downcycling and has been used to create products such as playground equipment and picnic benches.

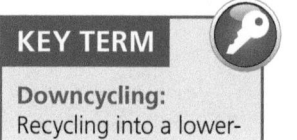

KEY TERM

Downcycling: Recycling into a lower-quality product.

Polymer recycling codes

If you look carefully at a plastic product you will see a symbol and a number. These identify the type of polymer used to make the product. For recycling purposes it is essential to know which plastic is which.

PETE	HDPE	PVC	LDPE	PP	PS	OTHER
polyethylene terephthalate	high-density polyethylene	polyvinyl chloride	low-density polyethylene	polypropylene	polystyrene	other plastics, including acrylic, polycarbonate, polyactic fibers, nylon, fiberglass
soft drink bottles, mineral water, fruite juice container, cooking oil	milk jugs, cleaning agents, laundry detergents, bleaching agents, shampoo bottles, washing and shower soaps	trays for sweets, fruit, plastic packing (bubble foil) and food foils to wrap the foodstuff	crushed bottles, shopping bags, highly-resistant sacks and most of the wrappings	furniture, consumers, luggage, toys as well as bumpers, lining and external borders of the cars	toys, hard packing, refrigerator trays, cosmetic bags, costume jewellery, CD cases, vending cups	

Figure 11.8 **Polymer recycling codes**

Figure 11.9 **Pencil cases made from reused PET drinks bottles**

Reuse

Despite it being better than creating new plastic, recycling still uses a lot of energy. The waste polymer needs to be sorted, cleaned and chipped before being melted down into useable granules. Reusing a container or packaging is therefore even better than recycling. For example, once you have consumed a bottle of water, refill it from a tap and reuse it. Reuse plastic jars by washing them out and using them for storage. Use a reuseable wrapper for packed lunches rather than continual use of cling film. Sell or give away old toys instead of throwing them away. Consider adapting an existing plastic product into something more useful.

Reduce

As a consumer, think about the products you buy and try to choose products with minimal packaging. In 2014 over 7.6 billion single-use plastic bags were given to customers by major supermarkets in England. That's something like 140 bags per person, or the equivalent of about 61,000 tonnes in total. From October 2015 large shops in the UK have had to charge 5p for every single-use plastic carrier bag, which has encouraged shoppers to buy thicker, more reuseable, 'bags for life'.

As a designer, think about the products you design and how you can reduce the amount of polymer material being used. For example, the new Nestle Eco-Shape bottle uses 30 per cent less plastic and a thinner label than the previous version.

Figure 11.11 **An over-packaged banana**

Figure 11.10 **This PET bottle has been designed to use 30 per cent less polymer.**

The slogan 'Reduce, reuse and recycle' appears on the waste hierarchy, which is a set of principles for the efficient use of resources. The largest part of the pyramid (the base) is the worst place to be, as that is what we do now – we dispose of most of our waste. The best place to be is at the top, where we completely eliminate the need for the waste. In practice, this is the difference between buying a regular banana from a local greengrocer or a shrink-wrapped version on a tray from a major supermarket.

Disposal

Environmentally friendly polymers

Environmentally friendly polymers contain polymers that are **biobased** (from a renewable resource) or **biodegradable**.

Polylactic acid (PLA) is a biobased polymer derived from cornstarch that breaks down into harmless chemicals when composted. It has been used to make disposable items such as cups, cutlery and food containers.

Biodegradable bags are made from biodegradable plastics that contain additives that cause them to decay more rapidly in the presence of light and oxygen (moisture and heat also help). Unlike bioplastics, biodegradable plastics are made of normal (petrochemical) plastics and don't always break down into harmless substances: sometimes they leave behind a toxic residue and that makes them generally (but not always) unsuitable for composting.

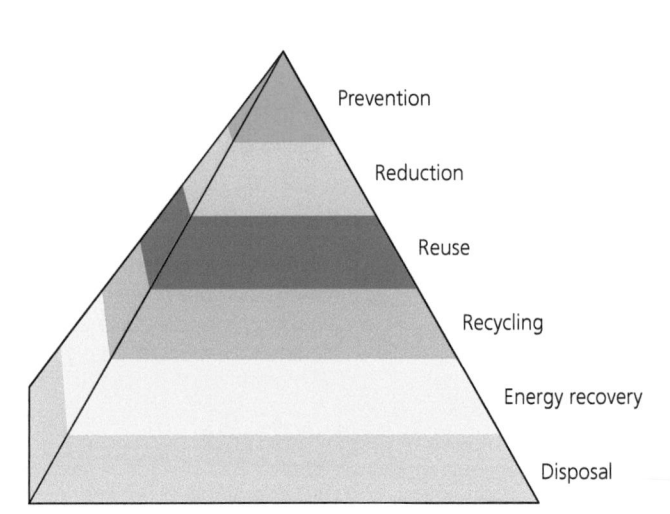

Prevention

Reduction

Reuse

Recycling

Energy recovery

Disposal

Figure 11.12 **The three 'Rs' in the waste hierarchy**

KEY TERMS

Biobased: A product made from a renewable resource.

Biodegradable: The ability of a substance or object to break down naturally in the environment through the action of micro-organisms, thereby avoiding pollution.

Figure 11.13 **PLA food packaging**

Figure 11.14 **A biodegradable carrier bag**

Recycling and the law

There are many laws and regulations that require consumers and manufacturers to consider the environment when a product comes to the end of its life.

Some examples of ways in which these laws encourage manufacturers and consumers to consider the environment when buying or disposing of products include:

- Electrical items such as refrigerators and washing machines now have a label on them that describes their energy efficiency rating. This helps consumers make an informed choice about whether the product they choose is better for the environment.
- Consumers are encouraged to take their old devices to collection points, from which manufacturers arrange collection. As a result designers have had to make products easier to dismantle, reuse and recycle.
- New vehicles do not use toxic materials and have all polymer parts labelled to help with recycling. Vehicle manufacturers publish information about how to dismantle their vehicles.

KEY POINT

Polymers have been manufactured since the early twentieth century. They are versatile materials whose properties can be altered by various chemical additives. Despite their amazing properties they can cause serious health issues and ecological problems for the planet. It's important that you always design with sustainability and safety in mind.

ACTIVITY

Gather a range of polymer products and identify the polymer used by finding the polymer recycling code.

ACTIVITY

Design and make a product that reuses a polymer product that would normally be disposed of. For example, a scoop from an old milk bottle, a phone case from woven crisp packets or a stationary organiser from a drinks bottle.

11.3 Commonly available forms

LEARNING OUTCOMES

By the end of this section you should know and understand:
→ that polymers are available in a range of stock forms
→ that there is a wide range of standardised components made from polymers.

Stock forms

Stock forms of polymer are basically the forms that it can be bought in and that can be stored 'in stock', ready for use for moulding, cutting, bending, etc.

- **Sheets** are often used for signage, roofing and panels. They can also be used for line bending.
- **Granules** are used for injection moulding as they melt quicker when fed into the moulding machines.
- **Rods** are generally used for CNC machining.
- **Extruded sections** are similar to rods, where a continuous profile is maintained throughout the section. A wide range of shapes is available, such as T, H and C sections. Several companies specialise in custom plastic extrusions.
- **Tubes** can be made in any shape – round, square and triangular are examples you may come across in school.
- **Foamed** plastics are available in sheet or roll form, such as Plastazote®, or block form for expanded polystyrene.
- **Powdered** polymers are often used to provide a coating on metal surfaces.
- **Reels** of plastic wire are used in certain 3D printers.

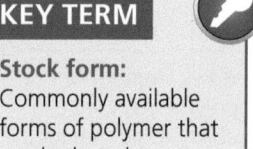

KEY TERM

Stock form:
Commonly available forms of polymer that can be bought.

Figure 11.15 **Some common forms of polymer**

KEY TERM

Standardised component: An individual part or component, manufactured in thousands or millions, to the same specification.

KEY POINT

A key strength of polymers is their ability to be moulded and formed into any shape. An awareness of the stock forms, however, will help your iterative modelling.

Standard components

There are several **standardised components** commonly manufactured from polymers:

- **Nuts and bolts** – usually made from nylon.
- **Washers** – used to secure fittings.
- **Wall plugs** – used to provide anchors for screws in walls.
- **End caps** – used to hide the heads of screws or to close pipes/tubing.
- **Plastic gear wheels** – used in various toys and mechanical devices.

Figure 11.16 **An assortment of polymer fixings**

STRETCH AND CHALLENGE

Prepare a stock list of the polymer forms available in your school workshop. Present this as a table and give details of the following: name of the polymer, type (thermo or thermosetting), form (sheet, rod, square, etc.), size, total length.

STRETCH AND CHALLENGE

Produce an information/example board with samples of the different polymer forms available to students in your school workshop.

11.4 Manipulating and joining

LEARNING OUTCOMES

By the end of this section you should know about and understand:
→ a range of specialist techniques used to shape, fabricate, construct and assemble high quality prototypes using polymers in a workshop.

Wastage

Wastage is the process of cutting away material to leave a desired shape. It is called this because the material removed is often thrown away.

Marking out

Before a design can be cut it must be marked out. There are a number of ways to mark out on a piece of polymer. A chinagraph pencil is often used as it can be wiped off easily, but the line can sometimes be too thick to follow effectively. Sometimes a non-permanent marker can be used, but the lines can be smudged easily. A permanent marker can be used on quick early models where progress is more important than a high quality finish.

KEY TERM

Wastage: Cutting away material to leave a desired shape.

Figure 11.17 Cutting acrylic on a scroll saw

Figure 11.18 Cutting acrylic on a laser cutter

Cutting

Scissors can be used on thin foam sheets and 0.5mm HIPS. Craft knives can be used to cut thicker sheets of HIPS by first scoring along the line then flexing the material until it snaps along the line.

Coping saws can be used to cut curves in thin plastic sheet, and junior hacksaws can be used to cut sections. Scroll saws can also be used for curved shapes, but care must be taken to ensure that the plastic 'dust' does not heat up and re-seal the cut. Leaving the protective film on or applying sticky tape to the area that is to be cut can help prevent this happening. Your teacher may use a band saw for cutting larger, more precise straight lines and shallow curves.

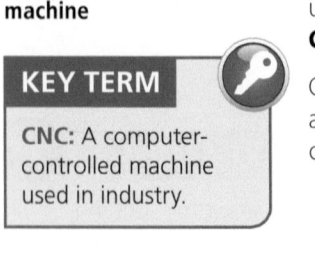

Figure 11.19 A CNC milling machine

KEY TERM

CNC: A computer-controlled machine used in industry.

A laser cutter can be used to cut any 2D shape out of acrylic sheet. Its only limitation is the size of acrylic sheet the machine can take. It can also engrave designs into acrylic for decoration. Several laser cutters have a 'deep engrave' setting that can be used to create shallow channels.

Milling and turning

Milling machines can be used to cut slots and grooves in blocks of suitable polymer materials. The workpiece is fixed to a table that can move backwards, forwards, up and down under a fixed cutter. A variety of different cutters can be used, depending on the type of job. **CNC** milling machines can be used to cut more complex shapes.

Centre lathes can be used to make round components. The workpiece is held securely and rotates while a cutting tool removes material. CNC lathes can be used to cut more complex shapes.

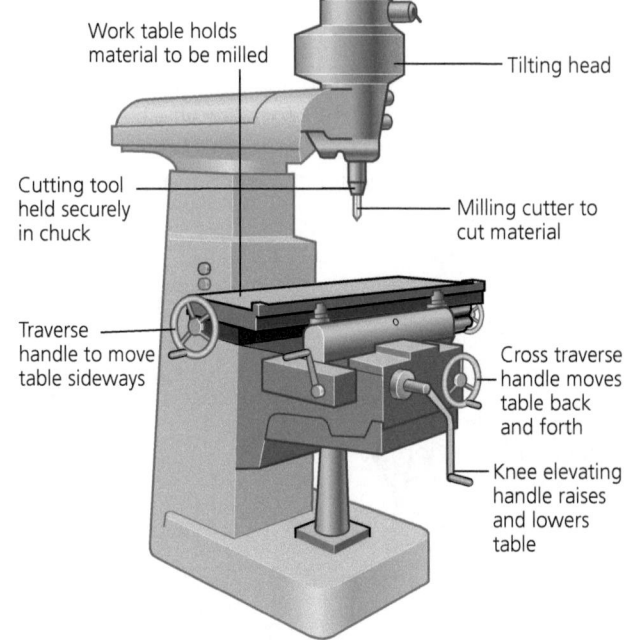

Motor inside

Milling cutter to cut material

Traverse handle moves table sideways

Work table holds material to be milled

Handle raises and lowers table

Horizontal milling machine

Work table holds material to be milled

Tilting head

Cutting tool held securely in chuck

Milling cutter to cut material

Traverse handle to move table sideways

Cross traverse handle moves table back and forth

Knee elevating handle raises and lowers table

Vertical milling machine

Figure 11.20 Milling machines

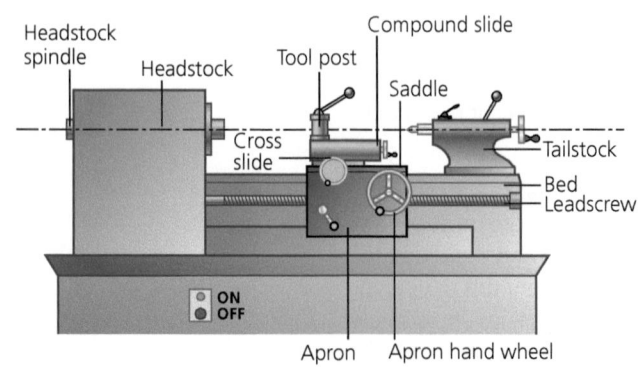

Headstock spindle

Headstock

Tool post

Compound slide

Saddle

Cross slide

Tailstock

Bed

Leadscrew

ON
OFF

Apron

Apron hand wheel

Figure 11.21 Centre lathe

Figure 11.22 A CNC lathe for training use

Multi-axis machining

Milling machines move on three axes – up/down, left/right and forward/backward. A multi-axis machine can move in four or more ways in order to manufacture complex parts out of polymers. Imagine a lathe that could stop partway through while a milling machine came in to cut some slots.

Drilling

While drilling through polymers can be done using any power drill, it's important to have the correct drill bits so you don't chip or crack the material (see Figure 9.31, page 127). Drill bits for polymers are available in a variety of sizes and are different from bits used for wood or metal as the tip has a shallower cutting angle and helps the bit carve its way through the polymer. When drilling plastic, the larger the hole, the slower drill speed you should use because high speeds can melt the plastic. Also, always reduce the drill speed as the drill bit exits the material. Before drilling, it's best to clamp down the plastic securely to a solid surface and back up the piece you are drilling with a spare piece of plywood underneath. This way the drill bit enters the plywood when it exits the plastic and avoids chipping the surface on the bottom.

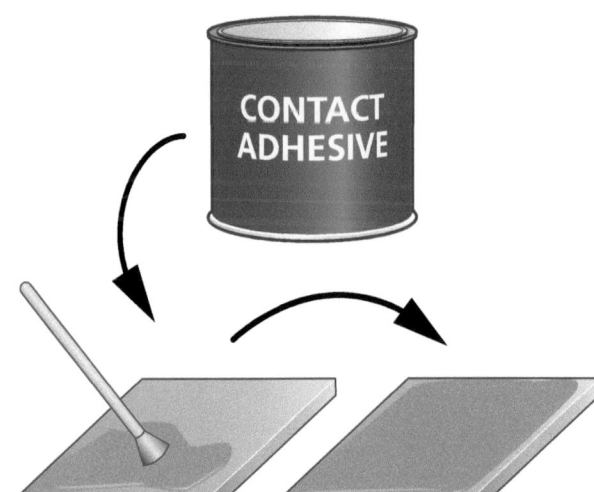

Figure 11.23 A multi-axis machine

Addition

As with all materials there are several permanent and temporary ways of joining polymers together.

Adhesion

Contact adhesive such as Evo-Stik can be used to join thin polymer sheets to other materials. Epoxy resin can be used to glue some polymers together but the surfaces need to be roughened first. Polystyrene cement is used to join modelling kits – such as AirfixTM kits – together, but can be quite messy. Tensol is a solvent-based adhesive used to join acrylic and sets quickly. DCM is a thinner version of Tensol that can also be used on acrylic and HIPS. Depending on its strength, double-sided sticky tape can be used as a semi-permanent method of joining polymers together.

Contact adhesive is applied to both surfaces, allowed to dry and then pressed together

Figure 11.25 Using Tensol to join acrylic sheets

Figure 11.24 Using contact adhesive to join acrylic sheets

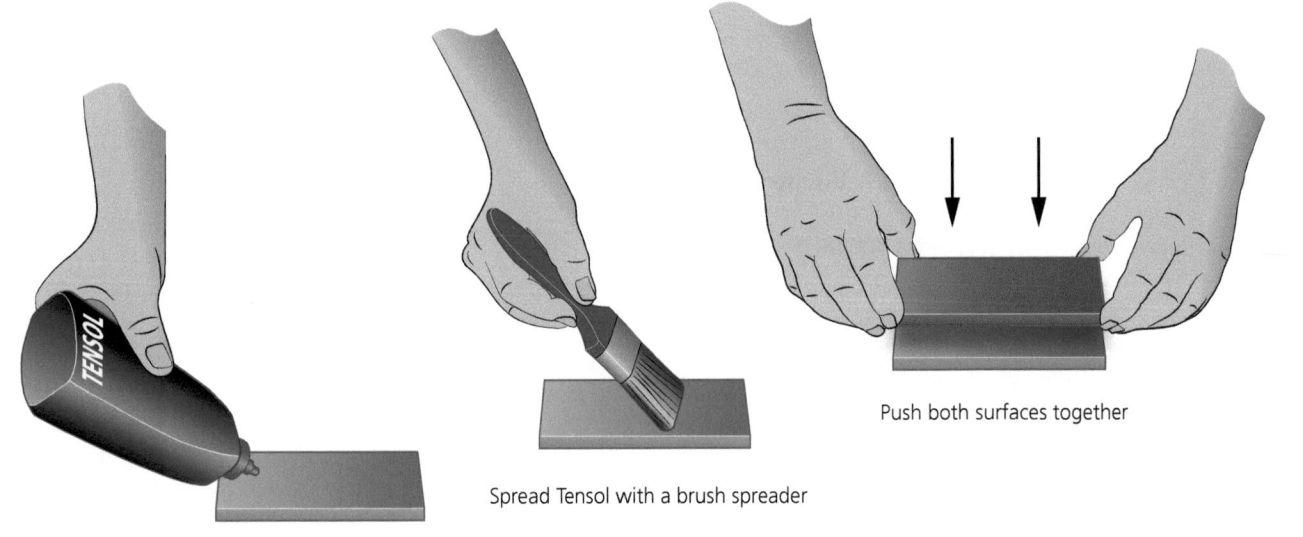

Pour Tensol cement on to the plastic

Spread Tensol with a brush spreader

Push both surfaces together

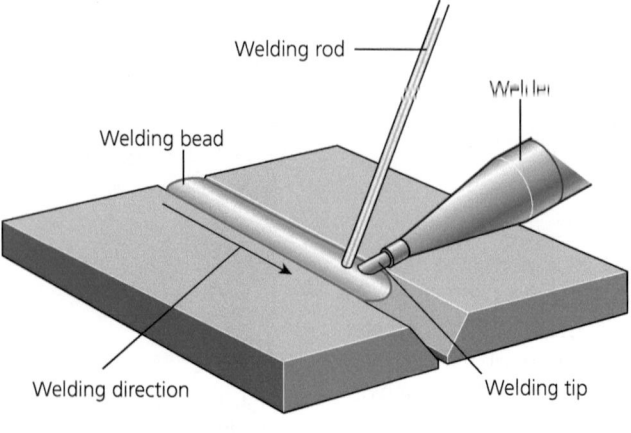
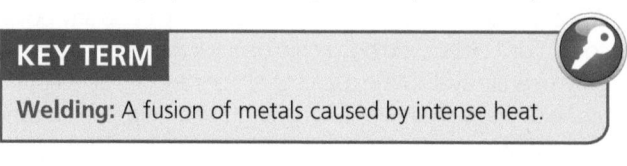

Welding rod

Welder

Welding bead

Welding direction

Welding tip

Figure 11.26 Heat welding plastic

Heat welding

Heat **welding** can be carried out on specific polymers using a hot air welding tool. Different accessory nozzles can be used for different polymers and applications. A tacking nozzle is first used to hold the materials in place before moving on to an extrusion tool that heats and feeds a thin wire of polymer into the join for a permanent join.

> **KEY TERM**
>
> **Welding:** A fusion of metals caused by intense heat.

Ultrasonic welding is similar to heat welding and uses a high frequency electromagnetic wave to soften the polymers for joining. It is most commonly used on clamshell packaging and polymer film sealing.

Polymers can also be joined using a wide range of mechanical fixings. These include machine screws and bolts, self-tapping screws, rivets and various spring fasteners and clips.

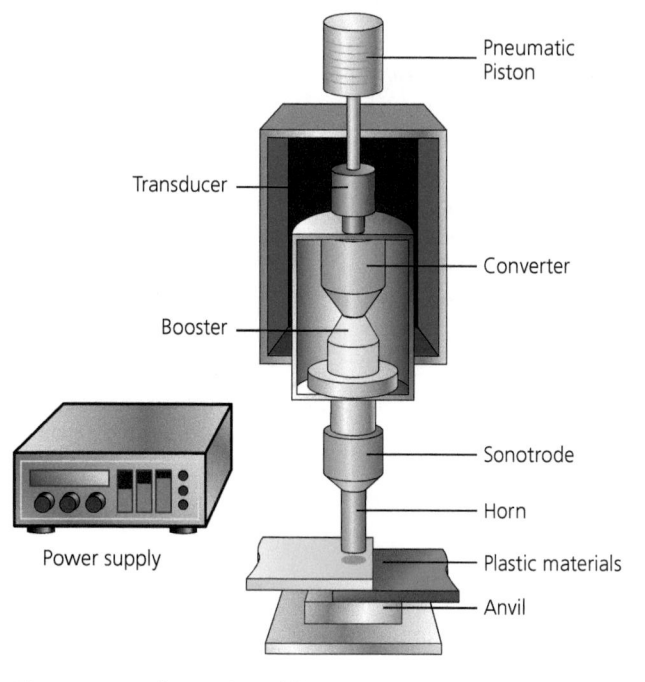

Pneumatic Piston

Transducer

Converter

Booster

Sonotrode

Horn

Power supply

Plastic materials

Anvil

Figure 11.27 Ultrasonic welding

Deforming and reforming

Depending on the scope of your workshop, you will have access to a range of machinery that can be used to deform and reform polymers.

Line bending

Line bending is a process used to create simple bends in polymer sheet such as acrylic and HIPS. The sheet material is heated along a line using a machine called a strip heater. Once the material has softened the heater can be removed and the sheet bent to the desired angle using a former or bending jig. It must then be held in place until the polymer has cooled and become hard again.

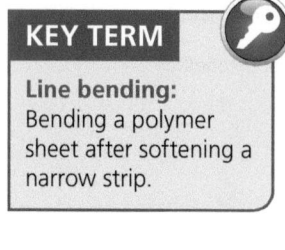

> **KEY TERM**
>
> **Line bending:** Bending a polymer sheet after softening a narrow strip.

Drape forming

Drape forming is used if a large curve or bend is required. The polymer sheet is usually heated in an oven to ensure the whole piece is heated evenly. Once softened, it is draped over a former and a piece of cloth is pulled tightly across it to hold it in shape until the polymer has cooled.

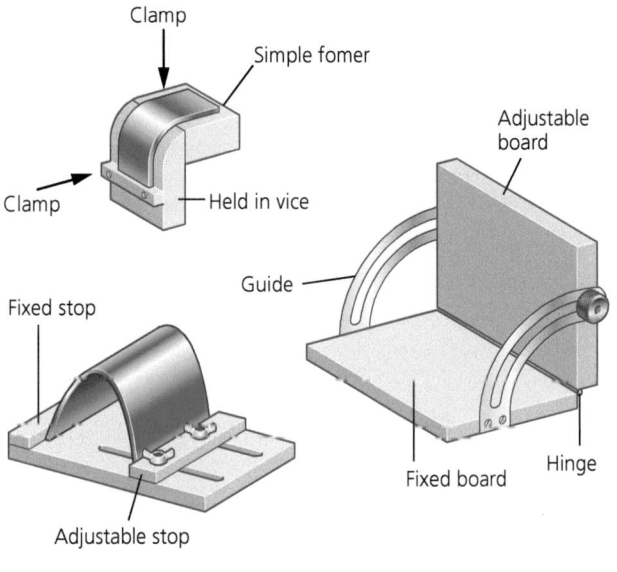

Figure 11.28 **Line bending**

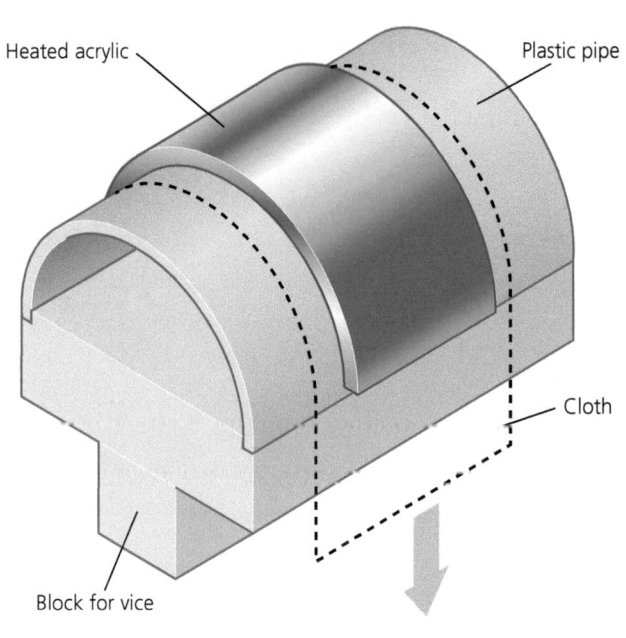

Figure 11.29 **Drape forming**

Press moulding

Press moulding is used to produce more complex shapes such as trays and dishes. It can be used on thin polymer sheet materials such as acrylic and HIPS and also on foamed polymers such as Plastazote®. Press moulding uses two-part formers called a yoke (the upper piece) and a plug (the bottom piece). The polymer sheet needs to be heated in an oven to ensure it is heated evenly. It is then positioned over the plug and pushed down over it using the yoke. These are then clamped together until the material has cooled. It is important that the plug has angled sides and rounded edges so that the finished piece can be easily removed from it. Once the piece has cooled you can remove the excess plastic.

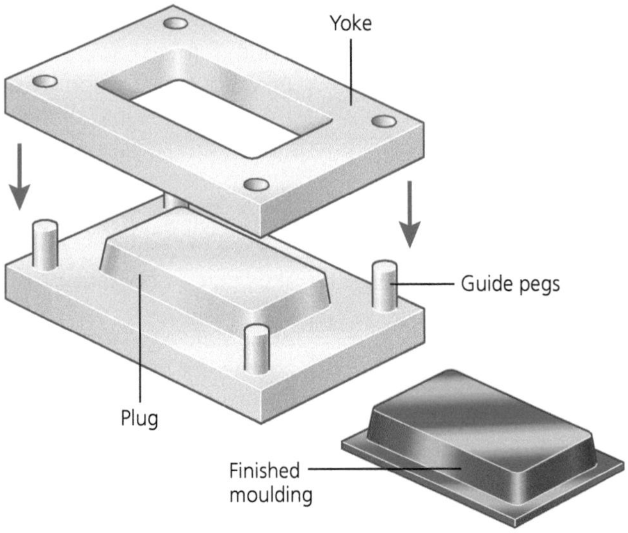

Vacuum forming

Vacuum forming is quite similar to press moulding but it uses a special machine. The plastic sheet is heated evenly until soft, and then air pressure is used to shape it over a **mould**. The heating element is similar to the grill on a cooker and is usually moveable to allow access to the machine. A sheet of suitable polymer (usually HIPS but sometimes acrylic) is fixed across the top of the machine by clamping. This must form an airtight seal. Below the plastic sheet in the air chamber is the mould. When the sheet is hot and soft the heater is moved out of the way, the mould is raised and the air between the mould and the plastic sheet is evacuated by an air pump.

Figure 11.30 **Press moulding**

KEY TERMS

Mould: Made to the shape required to be used many times.
Press moulding: Forming a hollow shape from a softened polymer sheet.
Vacuum forming: Producing thin hollow items over a shaped mould.

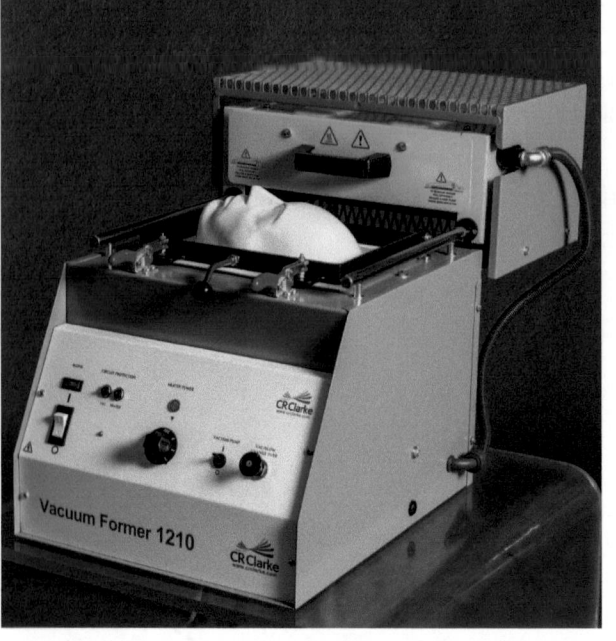

Figure 11.31 **Vacuum-forming machine**

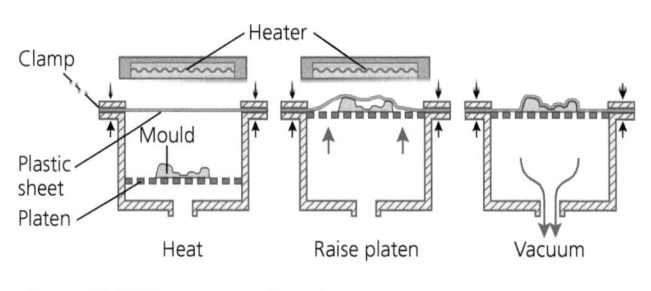

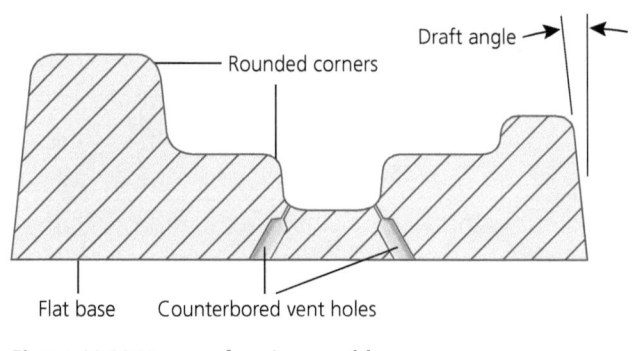

Figure 11.32 **The vacuum-forming process**

Figure 11.33 **Vacuum-forming mould**

KEY POINT

Despite lack of access to industrial polymer production methods such as injection and blow moulding, polymers can be used successfully in a school workshop using a broad range of techniques. Always think about what you can do with a polymer as you design and develop your ideas.

The air pressure on the outside of the sheet then presses the plastic into close contact with the mould. The shape of the mould must be carefully designed to allow the plastic sheet to be easily removed and the mould reused. The sides must be slightly tapered to produce a draft angle and corners should be radiused (rounded off). Sometimes vent holes are required to allow the plastic to form into small recesses.

Vacuum-formed components always have a shell of the same thickness throughout. This method of construction is extensively used in the packaging industry where it is used to form clear plastic into shapes that allow a product to be viewed. Much larger machines are used to form acrylic baths.

STRETCH AND CHALLENGE

The figure below shows details of the base of a lamp. State a suitable polymer for the base of the adjustable lamp and give two properties or characteristics that make the material suitable for this use.

Describe in detail how the base would be manufactured in a batch of 250. Include details of any jigs and/or formers used. Use annotated diagrams to support your answer.

11.5 Structural integrity

LEARNING OUTCOMES

By the end of this section you should know about and understand:
→ how polymers can be reinforced to withstand external forces and stresses
→ the processes that can be used to ensure the structural integrity of a product.

Thermosetting polymers

Thermosetting polymers produce a hard but brittle material. This can be made stronger and tougher by using other materials to reinforce it.

Glass-reinforced plastic

Glass-reinforced plastic (GRP) is the reinforcement of polyester resin using strands of glass fibre. This gives the material a much higher tensile and compressive strength. It provides a light, hard-wearing surface with a thin section, which has excellent resistance to corrosion. The glass fibre used is available as a woven mat or loose strands.

Figure 11.34 A fibreglass product

Thermopolymers

Thermopolymers tend to be moulded as thin hollow shapes. They are used for a wide range of products that are exposed to various forces that test their flexibility and toughness.

If you look closely at a simple product such as a disposable cup you will see that it has been designed to include 'lips' and 'ridges', which make the shape more rigid and less likely to twist, flex and ultimately crack.

Injection-moulded products can be strengthened by increasing their wall thickness. This would lead to higher production costs, however, so they are carefully designed to include features such as stiffening ribs that help **reinforce** and strengthen the product and increase rigidity.

KEY TERM

Reinforce: To strengthen or support an object or substance with additional material.

KEY POINT

Just because polymers can be moulded into any shape doesn't mean that they are suitable in all situations. An awareness of the stresses and strains that a part will come under will influence the design of a part and the addition of extra material to help strengthen and reinforce it.

Figure 11.36 **An injection-moulded piece with stiffening ribs**

Several thermopolymers, such as low-density polyethylene (LDPE) and polypropylene (PP), are attacked by ultraviolet light and will become brittle over time – this process is known as **UV degradation**. Pigments and dyes can also be affected, and you will have seen many outdoor polymer products that have faded over time.

To prevent this manufacturers add UV **stabilisers** that absorb the UV radiation and dissipate it as low-level heat. These chemicals are similar to those used in sun cream.

Figure 11.35 **A disposable cup with ridges and a rolled lip**

 KEY TERMS

Stabiliser: An additive added to polymers to help them withstand UV degradation.

UV degradation: The weakening of polymers when exposed to the ultraviolet light in sunlight.

ACTIVITY

Vacuum-form a basic shell design, or find a disposable plastic cup. Use offcuts of high-impact polystyrene (HIPS) to add stiffening ribs to reinforce the shape and stop it flexing.

ACTIVITY

Take a disposable plastic cup and use scissors to remove the lip at the top – how does it affect the structural integrity of the cup?

11.6 Making iterative models

LEARNING OUTCOMES

By the end of this section you should know about and understand:
→ the processes and techniques used to produce early models to support iterative designing.

Polymers can be used in a variety of useful and interesting ways to help you explore your early design ideas and to support iterative designing in school.

Sheet materials are excellent for developing initial ideas. They are quick and easy to work and are not too expensive.

Acrylic sheets

These are probably the most common form of polymer to be found in schools. Acrylic sheet can be cut by hand or machine but this can require a great deal of skill and the edges will take considerable time to clean and tidy up. This process is not particularly useful for quick **iterative modelling**.

Using a suitable CAD program such as 2D Design and a laser cutter will result in more precise and useful models that are a step on from and more resilient than card. These models can be used to explore linkages and other moving parts. Layers of acrylic can also be built up to form a more complex model for testing.

Offcuts of acrylic from the laser cutter can often be used to quickly represent specific aspects on a variety of models.

> **KEY TERM**
>
> **Iterative modelling:** Repeated modelling to develop an idea.

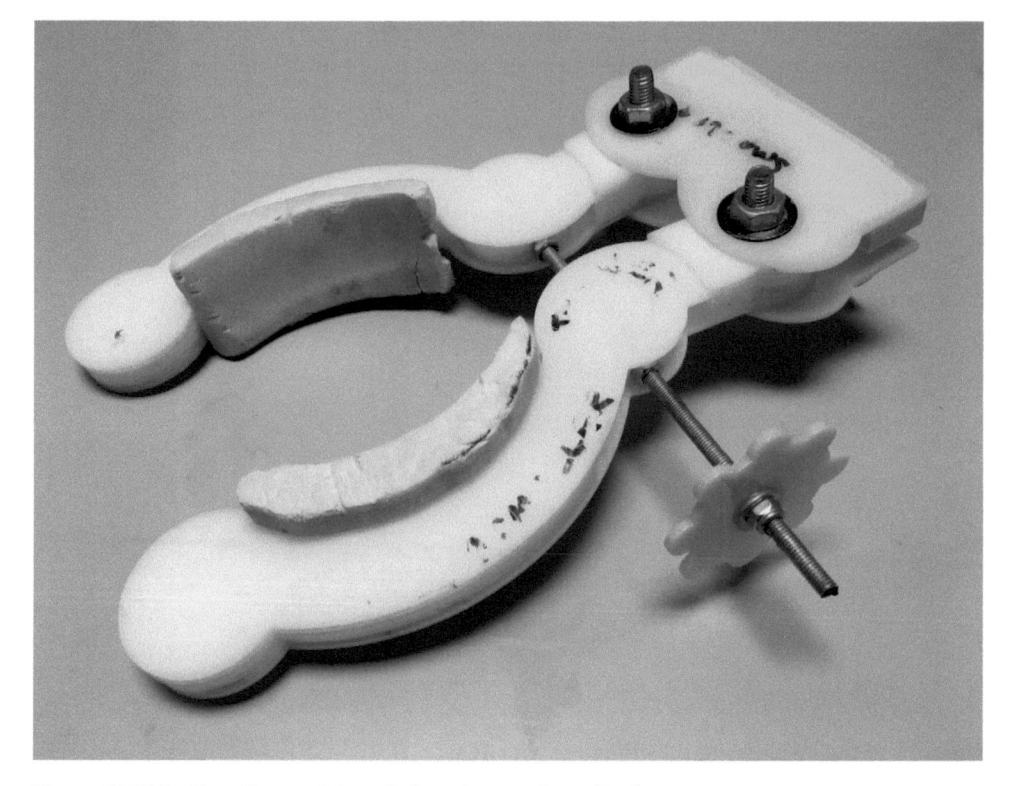

Figure 11.37 An iterative model made from layers of acrylic sheet

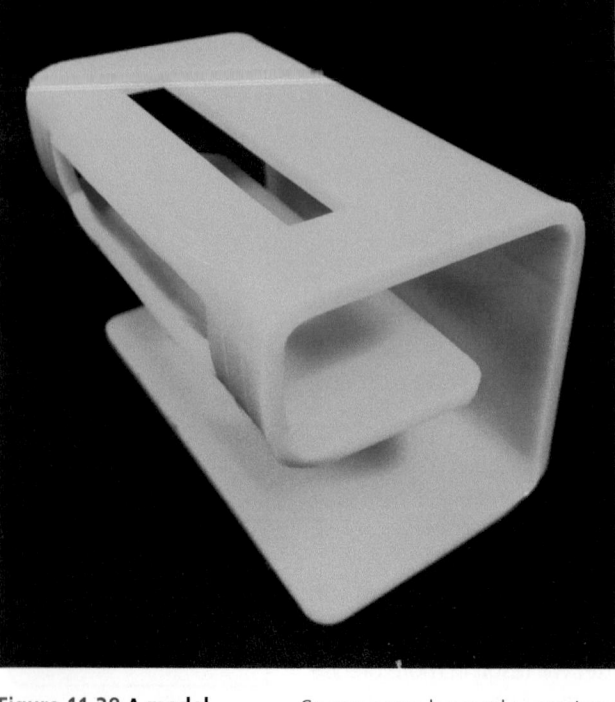

A bending strip heater could be used to create simple angled pieces, or if more complex shapes are required the acrylic shape can be heated in an oven and formed around a simple wooden mould. For even more rapid modelling, small pieces of acrylic can be manipulated quickly by hand while wearing welder's gloves for protection.

Acrylic sheet can be quickly glued together with **dichloromethane** (DCM) or another suitable polymer glue, but you must ensure that surfaces are completely flush for the glue to hold. A hot melt glue gun can also be used but the results do not look as clean and precise.

Drilling can be used to create holes but you must be careful to ensure that suitable 'plastic bits' are used (see Section 11.4).

Styrene sheets

HIPS sheets are commonly used for vacuum forming. The sheets can be manipulated quickly, however, which also makes it an ideal material for certain types of iterative modelling.

Figure 11.38 A model made using laser-cut acrylic and a line bender

KEY TERM

Dichloromethane: A solvent used to join polymer pieces together.

Styrene can be cut by scoring and then bending until it snaps along the line. This makes it ideal for rapid box construction-type models. Simple curved shapes can also be achieved with this 'score and snap' method. Note that styrene must not be cut on a laser cutter.

As with acrylic sheet, styrene can be bent on a line bender. Depending on its thickness this will be faster than for acrylic so extra care must be taken.

DCM is again the best method for gluing styrene parts together. Use a syringe applicator or thin brush to run the liquid along the joint, then hold the pieces in for a few seconds.

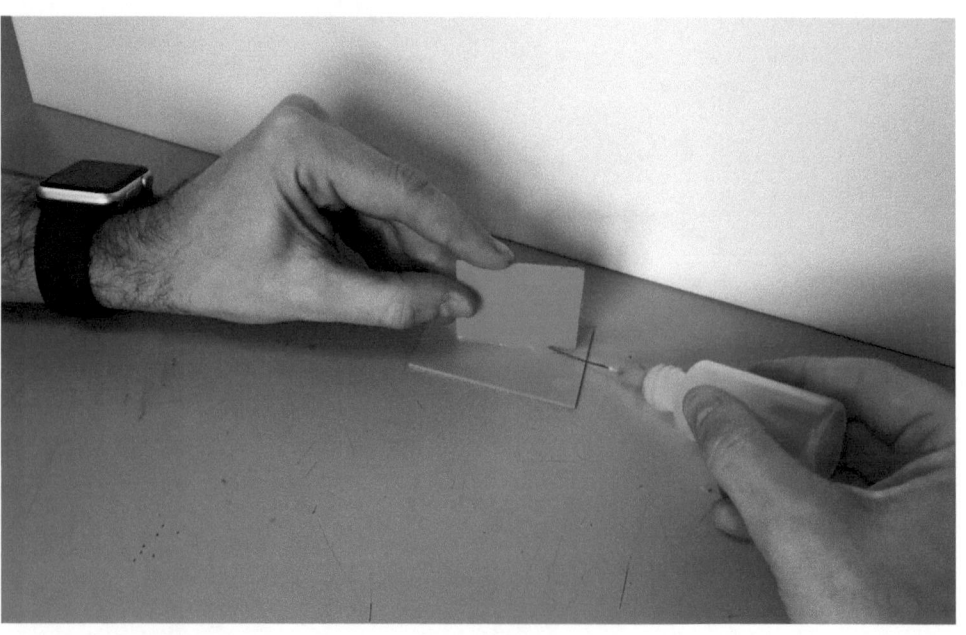

Figure 11.39 Gluing pieces of HIPS together with DCM

Drills can be used for small holes in styrene; if the sheet is thin enough a hole can also be used. For larger holes a reaming tool can be used to enlarge a hole that has already been drilled.

Foam sheets

Foam sheets are available in a variety of sizes and thicknesses, from 2mm-thick A4 craft sheets to thicker Plastazote® sheets like the type used in camping mats. Thin foam can be used to represent more flexible materials on a model, such as textile elements, padding or grip detailing.

These sheets can be cut quickly and easily using scissors or a craft knife and can be glued using craft foam. A cool melt gun can be used on certain pieces depending on the size and shape. Some craft foam sheets come with a self-adhesive backing and this can be an excellent way of adding grip details to certain models. You can make your own self-adhesive pieces by applying double-sided tape to the back of sheets before cutting them.

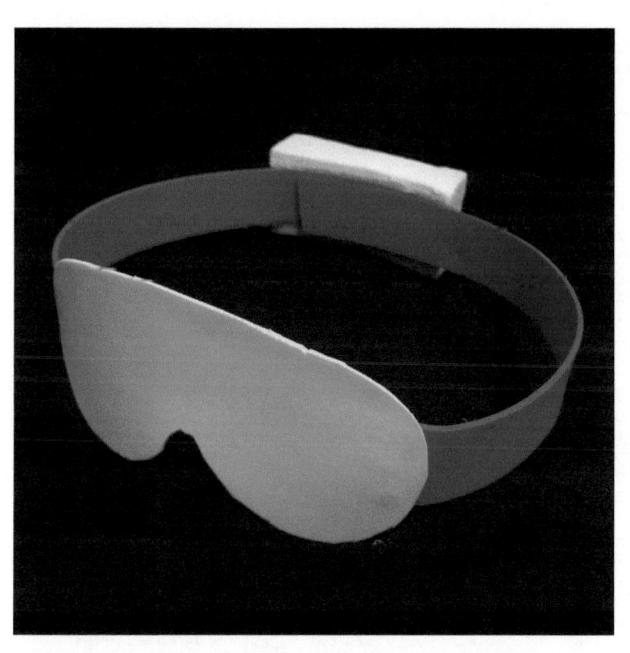

Figure 11.40 **A simple model made with foam sheet**

Figure 11.41 **A foam board model**

Sticky-backed vinyl

Sticky-backed vinyl can be cut by hand using scissors or a craft knife to create a variety of interesting shapes that could be used for decorative or informative details on your models. A broad range of colours is available as well as mirrored, metallic and printable types that could be used to create your own sticker designs. More precise and intricate shapes can be designed on a suitable CAD program and then cut on a CAM knife cutting machine.

Foam board

Foam board is a thin sheet of expanded foam sandwiched between two layers of cardboard. It is usually approximately 5mm thick, available in white or black and in sheet sizes from A4 up to A1.

It is best cut with a sharp craft knife and safety ruler in straight lines, but curves can be achieved with a little more care and practice. It can be glued with an all-purpose adhesive such as UHU, or with a glue gun if time is an issue.

Foam board is often used for quick box-like constructions such as initial architectural models, and interesting effects can be achieved if you laminate it with printouts of textures before cutting it.

Block model: An informal model that captures the form of an idea – tends to have no moving parts.

Polystyrene foam

Polystyrene foam can be used for simple **block models**. These are very useful for developing the general feel and appearance of an idea. It is often used by designers to explore and represent shapes that could be complicated and difficult to draw. Often referred to simply as 'blue foam' due its colour, it is an excellent material to use for models that require 'hands on' analysis when exploring ergonomic suitability.

Blue foam is relatively cheap and easy to work with and tends to come in large, thick sheets. Initial blocks of material can be cut either with a band saw or by hand using a panel saw. These blocks can then be shaped further by machine or hand saws and sandpaper. Details can be added using files, craft knives and marker pens.

Figure 11.42 A blue foam block model

Figure 11.43 A model made from Lego

Figure 11.44 An example of a model made with a 3D printer

Acrylic shapes

Acrylic rod is available in a range of colours and diameters and can be quickly cut with a junior hacksaw to the desired length. Acrylic tubing can also be useful when representing specific design elements. Modelling suppliers can provide a range of different acrylic shapes that might be useful when modelling, such as domes, angles, channels and acrylic fixings such as hinges, nuts and bolts.

Modelling kits

Model-making kits were originally produced as toys. Lego and Lego Technic are examples of structural kits and these can be used to quickly explore and develop structures. They are also useful for developing solutions to mechanical problems. Lego Technic has a wide range of gear wheels and mechanical components, which can be used to explore and test out mechanical solutions.

3D printing

The use of suitable 3D modelling software and a 3D printer can also be beneficial when iterative modelling. Designs can be changed quickly and easily and then output to a suitable printer for testing. Production time can vary, though, depending on the size and complexity of the design.

KEY POINT

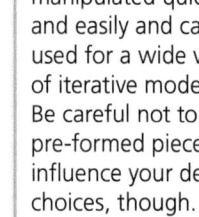

Polymers can be manipulated quickly and easily and can be used for a wide variety of iterative modelling. Be careful not to let pre-formed pieces influence your design choices, though.

ACTIVITY

Create a model skills board by making a series of small models using each of the methods in this section.

STRETCH AND CHALLENGE

Design and make an instruction sheet for creating models using each method for use lower down the school.

Coloured card

- Coloured card is supplied in a range of colours, shades and thicknesses.
- Card can even be bought with different colours on the front and back surfaces.
- It is suitable for colourful models, especially the early stages of making a series of models.
- A good example of card as a modelling material is in the manufacture of a prototype board game.

The equipment needed:

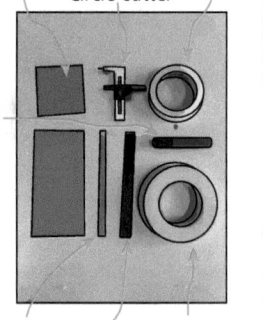

2 x red card
Sellotape
Circle cutter
Pen
Thin strip of red card
Double-sided tape
Thin strip of black card

What you need to do:

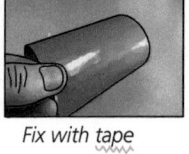 Roll the card into a tube

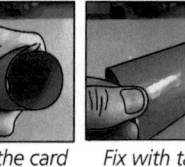 Fix with tape

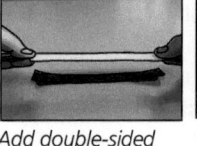 Add double-sided tape to the black card

 Trim off the excess with scissors

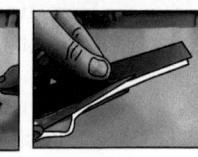 Peel off the double-sided tape

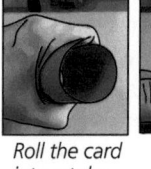 Fix the black card to the base of the tube – do the same with the red strip and stick around the opposite end

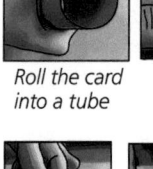 Cut a circle using the circle cutter

Cut from the edge to the centre to make a shallow cone – fix with 4 spots of hot glue

 Add some details with the pen – and you have a postbox!

Finished with time to spare? Try adding some details (like a handle) using the permanent pen.

Figure 11.45 **Example instruction sheet for creating a model**

11.7 Finishes

LEARNING OUTCOMES

By the end of this section you should know about and understand:
→ the processes used for finishing plastics for specific purposes.

Industry finishes

Polymers are usually described as **self-finishing**, which means they require no further finishing. Injection-moulded parts, for example, are formed in a mould that is either textured or highly polished – this finish is imparted to each product produced by the mould. Most plastics are manufactured with a very good surface finish.

Figure 11.46 **A co-injection-moulded toothbrush**

KEY TERMS

Co-injection moulding: An injection-moulding process that uses two different polymers.

Flash: Excess material formed between the joint of a mould.

Gate: The entry point for molten plastic to flow into a mould.

Self-finishing: A material that requires no further coatings or finishing processes.

Certain products (such as toothbrushes) are **co-injection moulded**, which produces a two-colour product combining a hard polymer with a softer elastomer for improved grip.

Once a product has been injection moulded, the part may be subjected to one or more finishing processes:

- **Degating** – injection moulding involves injecting a pressurised flow of molten polymer through a channel system of runners and gates into a mould cavity. The **gate** and the runner typically remain attached to the part when it's ejected from the mould. Degating is the process of removing this excess material.
- **Deflashing** – during moulding, excess material called **flash** may leak out between the mould cavity halves. Deflashing is the removal of this excess material, often using a knife or other cutting utensil.
- **Cleaning** – polymer parts often require some form of cleaning after moulding. Mould release agents may leave a residue, and grease and dirt can be picked up from the machine. Cleaning is accomplished by spraying or dipping parts in a mild detergent solution then rinsing and drying.

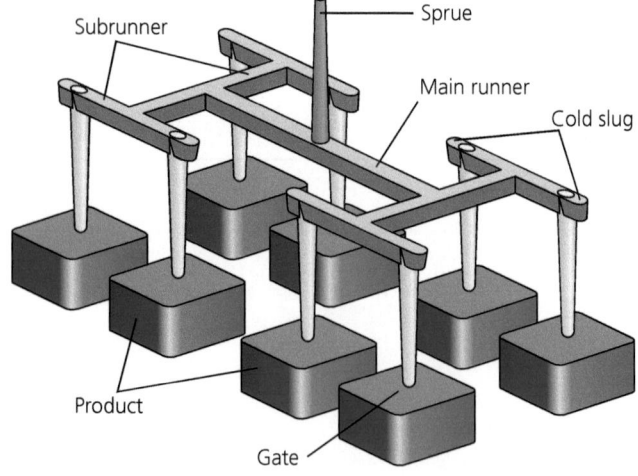

Figure 11.47 **The parts of an injection-moulded product**

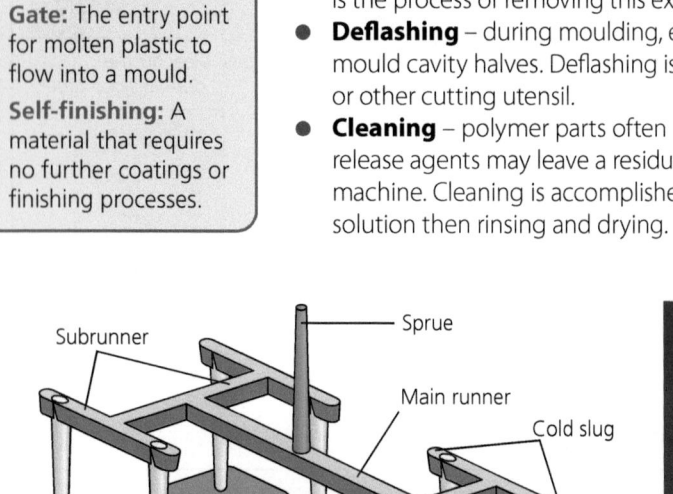

Figure 11.48 **Flash on a plastic box**

- **Decorating** – some polymer parts may require applied decoration after moulding. This may include plating, vacuum metallising, printing and painting.

Figure 11.49 A painted action figure

School-based finishes

In the school workshop you may be required to clean up and polish the edges of acrylic. This can be done by first draw-filing the edges, then moving up through progressively higher grades of wet and dry paper. A final polish with a buffer or by hand with a suitable polishing compound will be the final stage.

When vacuum forming, once you have trimmed the part you may need to clean up the edges and this can be done by scraping a steel ruler along the edges.

You might also have access to a sublimation printer at school. This is a printer that uses heat to transfer dye on to sheets of plastic and could print a specific design or texture.

> **KEY POINT**
>
> Due to the fact that they are moulded, polymers can be given any texture for functional or aesthetic reasons. Modern printing technology can also be used to further enhance their aesthetic appeal.

ACTIVITY

Take a permanent marker and add a design to a sheet of HIPS. Try vacuum forming with it to see how the design deforms around various moulds.

ACTIVITY

Make a keyring from acrylic. Cut a shape out of acrylic sheet and try to get the edges to a mirror finish.

11.8 Using digital design tools

LEARNING OUTCOMES

By the end of this section you should know about and understand:
→ the use of 2D and 3D digital technology and tools to present, model, design and manufacture solutions.

Industry professionals use a range of digital tools when exploring and developing design ideas that will be manufactured using polymers.

Rapid prototyping

KEY TERMS

CAM: Computer-aided manufacture.

Rapid prototyping: The process of making a 3D shape from a digital file.

Rapid prototyping is a process of making three-dimensional solid objects from a digital file. The creation of a 3D printed object is achieved using additive processes. In an additive process an object is created by laying down successive layers of material until the object is created. Each of these layers can be seen as a thinly sliced horizontal cross-section of the eventual object.

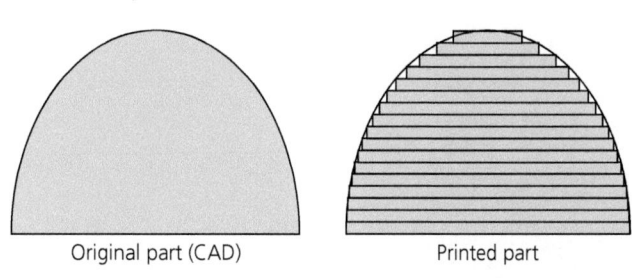

Original part (CAD)   Printed part

Figure 11.50 **Rapid prototyping builds objects up in layers**

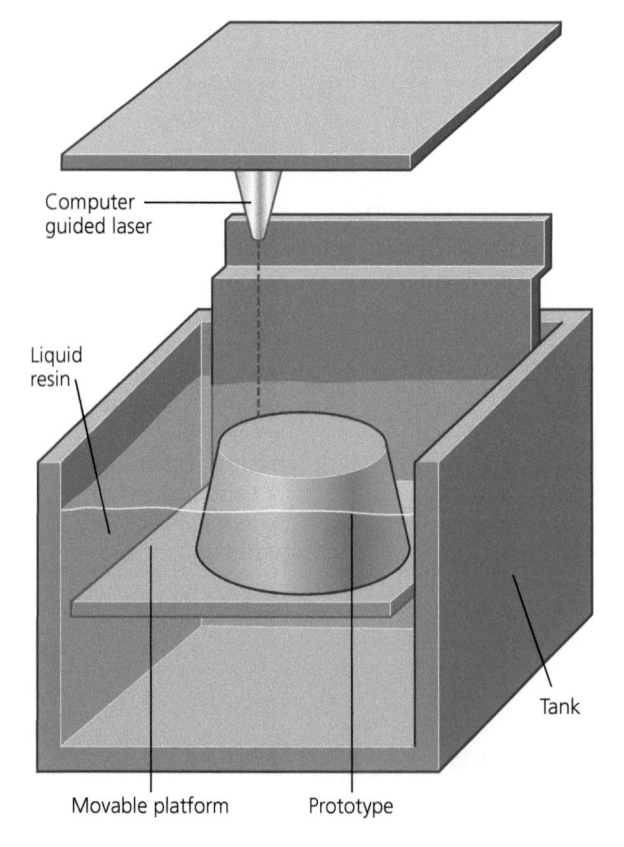

Computer guided laser

Liquid resin

Movable platform   Prototype

Digital manufacture

Computer-aided manufacture (**CAM**) can be used in various forms with polymers.

Stereo lithography

The component is built up in layers on the platform. The liquid is a photopolymer. When it is exposed to ultraviolet light from the laser beam, it cures or solidifies. The platform moves downwards, and the sweeper passes over the newly formed layer, breaking the surface tension and ensuring that a flat surface is produced for the next layer. Subsequent layers are laid down, and bind together. The part is then removed from the vat.

Laser sintering

This is a totally different manufacturing process, in which powder is spread over a platform by a roller. The laser then sinters selected areas, which makes the powdered polymer melt and then harden.

Tank

Figure 11.51 **Stereo lithography**

Fused deposition modelling

The build material for fused deposition modelling (FDM) is a polymer filament; this is passed through a heating element, melted and extruded. Each slice of the model is drawn from a continuous length of the molten filament. Typical build materials are acrylonitrile-butadiene-styrene (ABS) and polylactic acid (PLA). These are now commonly used in most schools.

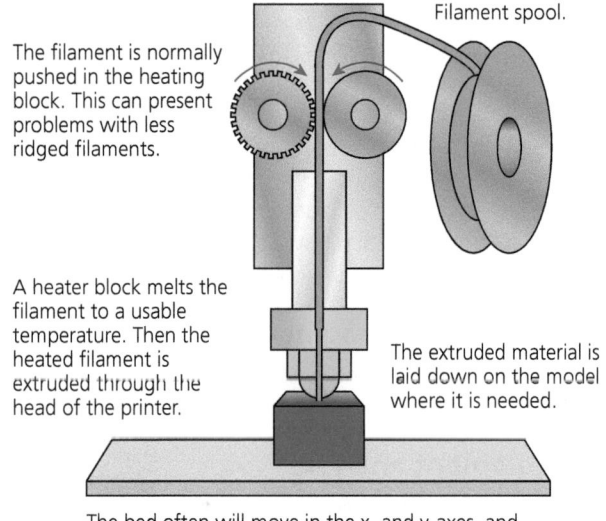

The filament is normally pushed in the heating block. This can present problems with less ridged filaments.

Filament spool.

A heater block melts the filament to a usable temperature. Then the heated filament is extruded through the head of the printer.

The extruded material is laid down on the model where it is needed.

The bed often will move in the x- and y-axes, and the extruder will move in the z-axis.

Figure 11.52 Fused deposition modelling (FDM)

3D printing

3D printing is the term used for all the above processes, but it actually describes machines that use a powder-based printing system. A thin layer of build material, typically plaster- or starch-based, is gradually glued together by a print head that can also colour the material using regular printer ink. The final product is removed from the powder, cleaned and the outer surface strengthened by dipping in Superglue.

Figure 11.53 A 3D printer

Figure 11.54 3D printed models

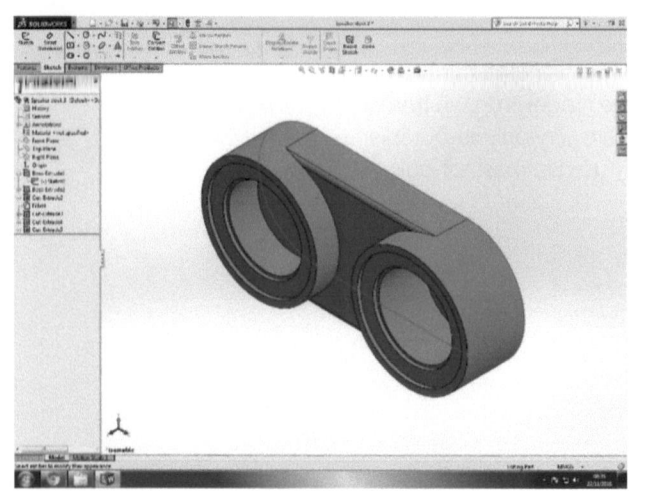

Figure 11.55 **On-screen modelling**

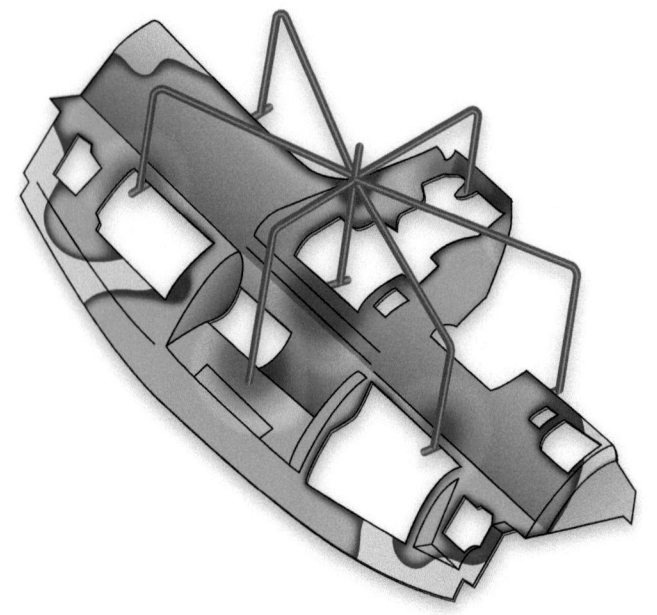

Figure 11.56 **Injection-moulding simulation**

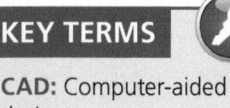

KEY TERMS

CAD: Computer-aided design.

CAE: Computer-aided engineering.

FEA: Finite element analysis.

Digital technologies: CAD, CAE and FAE

Computer-aided design

Computer-aided design (**CAD**) software applications are used throughout design and technology to create and explore ideas, from simple 2D sketching programs that help you quickly visualise ideas, to advanced 3D modelling software that allows you to create fully realised products. Using computers to help you design allows you to make quick alterations to components. It allows you to incorporate commonly used components from a built-in or online library. You can collaborate on designs with other professionals from around the world and render product designs in a variety of different colours and materials.

Computer-aided engineering

Computer-aided engineering (**CAE**) is the broad use of computer software to aid design engineers in analysis tasks.

Finite element analysis

Finite element analysis (**FEA**) allows you to model products in a virtual environment and test for weaknesses. This could include stressing your design to find weaknesses. With polymer products this is particularly useful for finding areas that would need reinforcing with stiffening ribs or for establishing areas where wall thickness would have to be increased to insulate against heat from an internal motor.

Virtual moulding

Simulation software can be used to see how a liquid polymer would flow into and around a mould cavity, quickly identifying potential design faults within a product that may cause the mould to fail. Polymer injection pressure can be measured to determine whether machinery could actually cope. Cooling time can also be calculated. Parts can be virtually arranged to ensure only a single 'shot' of injection moulding is required. Changes can be made to the model and the simulations quickly re-run to establish their effect.

Draft analysis

For polymer products to successfully release from a mould cavity, a draft angle is required. Computer simulation can also calculate the optimum draft angle for the part to ensure rapid production time.

KEY POINT

No other material works as well with CAD as polymers. The ability to generate a 3D model and output it to a 3D printer allows for rapid product development that mirrors industrial practice like no other technique.

STRETCH AND CHALLENGE

Create a series of designs for an organic shaped stool using layers of corrugated card as your modelling material.

11.9 Manufacturing methods and scales of production

LEARNING OUTCOMES

By the end of this section you should know about and understand:
→ the methods used for manufacturing at different scales of production
→ manufacturing processes used for larger scales of production
→ methods of ensuring accuracy and efficiency when manufacturing at larger scales.

Scales of production

Deciding on a suitable scale of production and manufacturing method for your polymer parts depends on a number of factors:

1 **Form:** What shape are your parts? This will be limited by the restrictions of the manufacturing processes available.
2 **Budget:** Part cost + tooling cost. Some methods have high set-up costs. Some are quite cheap but there's usually a trade-off. Most high-volume manufacturing processes are expensive to tool but offer cheap parts – the opposite is also true: low-volume processes are cheap to set up but parts are expensive.
3 **Time:** Tooling takes time, manufacturing takes time. More expensive tooling usually means longer set-up time. Injection-moulding tools can take 10–16 weeks to manufacture.
4 **Material:** Material choice is determined mostly by form, function and cost; it'll also depend on what manufacturing technique you choose, for example if you want a melamine formaldehyde (MF) bowl, you won't be able to injection mould it.

Fabricating: Making parts by gluing, turning, carving or welding.

GRP: Glass-reinforced plastic.

One-off production: Making only one or a small number of products. Usually used for specialist products.

One-off/low-volume polymer production methods

One-off production involves making only one or a small number of products.

GRP layup

The lay-up technique for glass reinforced plastics (**GRPs**) involves a comparatively simple profile mould of metal, wood or plaster and the following processes:

1 Liquid polyester resin, mixed with a catalyst (or hardener), is applied to the mould to form a pre-gelled coat.
2 Glass fibre in mat or woven fabric form is laid on the first gelcoat and a liquid polyester resin/catalyst mix is sprayed on until the fibre layer is saturated.
3 When the resin mix has hardened, the moulding is removed from the mould. Curing (setting) can take place in the cold or can be speeded up by heating.

GRP moulding

Two other techniques used with GRP are the rubber-bag and matched-die moulding methods, in which pressure is applied to the top surface of the moulding during processing. Various compositions of polyester resin/catalyst/glass fibre are used to produce mouldings in both these pressurised processes. By heating, comparatively fast hardening of the resin is possible.

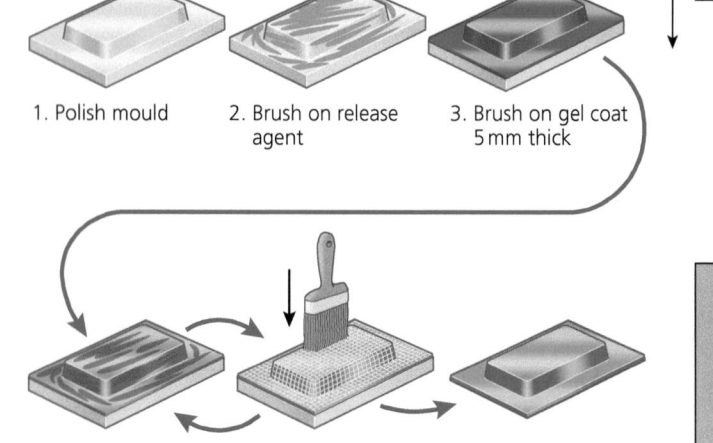

1. Polish mould
2. Brush on release agent
3. Brush on gel coat 5 mm thick
4. Brush on lay-up resin
5. "Stipple" glass fibre mat into resin
6. Trim moulding to final shape

Figure 11.57 GRP layup

Figure 11.58 Matched-die moulding of GRP

Resin
Catalyst
Mixing head
Vent
Fibreglass mat
Mould (closed before injection)

Fabricating

In **fabricating**, parts are made from either gluing, turning, carving or welding existing materials.

CNC machining

CNC machines are an example of subtractive manufacturing, which means material is cut away from a block. The shape of the parts possible is limited as internal structures are difficult to create.

Figure 11.59 Turning nylon rod on a centre lathe

3D printing

3D printing is actually cheaper than CNC but can take longer. Printing is from the bottom so support structures may need to be incorporated. These will also need removing once the machine has finished printing. See Section 8.8 for more information on 3D printing.

KEY TERM

Batch production: Making a set number of identical products.

Batch/medium volume polymer production methods

Batch production involves making a set number of identical products.

Vacuum forming

Sheets of plastic are warmed and sucked on to a mould. There is no need for pressure or high temperatures so moulds can be made from cheap materials. The shape, however, is determined by a single-sided mould and can be quite limited. See Section 11.4 for more details on the vacuum-forming process.

Casting

Liquid plastic in the form of a resin (and a hardener) is poured into a mould and then solidifies. Moulds are cheap and can be reused a number of times. Parts tend to be quite small but can be solid/thick.

Rotational moulding

The rotational moulding process consists of rotating a heated mould containing plastic paste or powder. As the plastic melts and the mould rotates, the plastic coats the surface of the mould cavity with an even layer of plastic. The mould is then cooled before opening. Typical objects made this way include farm tanks, barrels, septic tanks and large hollow toys.

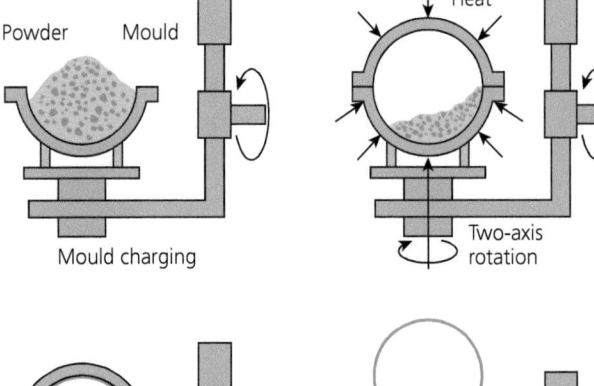

Figure 11.60 Rotational moulding

Figure 11.61 Outdoor toys made using rotational moulding

KEY TERM

Mass/high volume production: Producing very large numbers of products.

Mass/high volume polymer production methods

Mass/high volume production involves producing very large numbers of products.

Compression moulding

Compression moulding is the process most often used for shaping thermosetting plastics. The plastic 'moulding powder' is heated and compressed into shape. Typical objects include children's tableware, electric plugs, sockets and light switches.

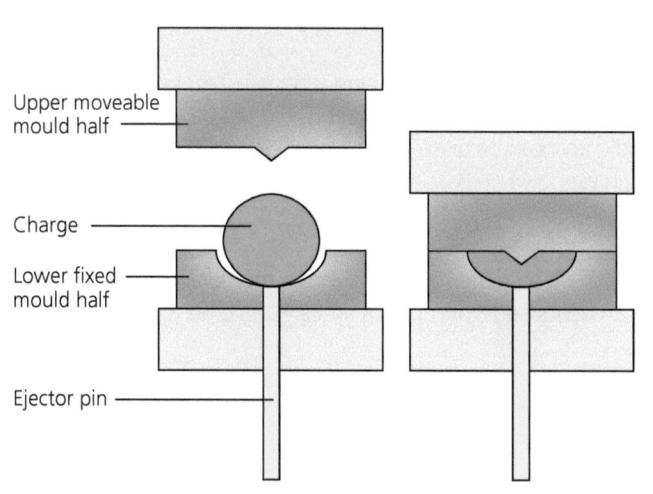

Figure 11.62 **Compression moulding**

Injection moulding

Injection moulding is the process whereby the molten plastic is injected into a mould via an injection screw or ram, cooled, and the object is then ejected. Typical objects produced this way include model kits, audio/video cassettes, bottle crates, buckets, car bumpers, dashboards, gear wheels, phone cases, plastic chairs and washing-up bowls.

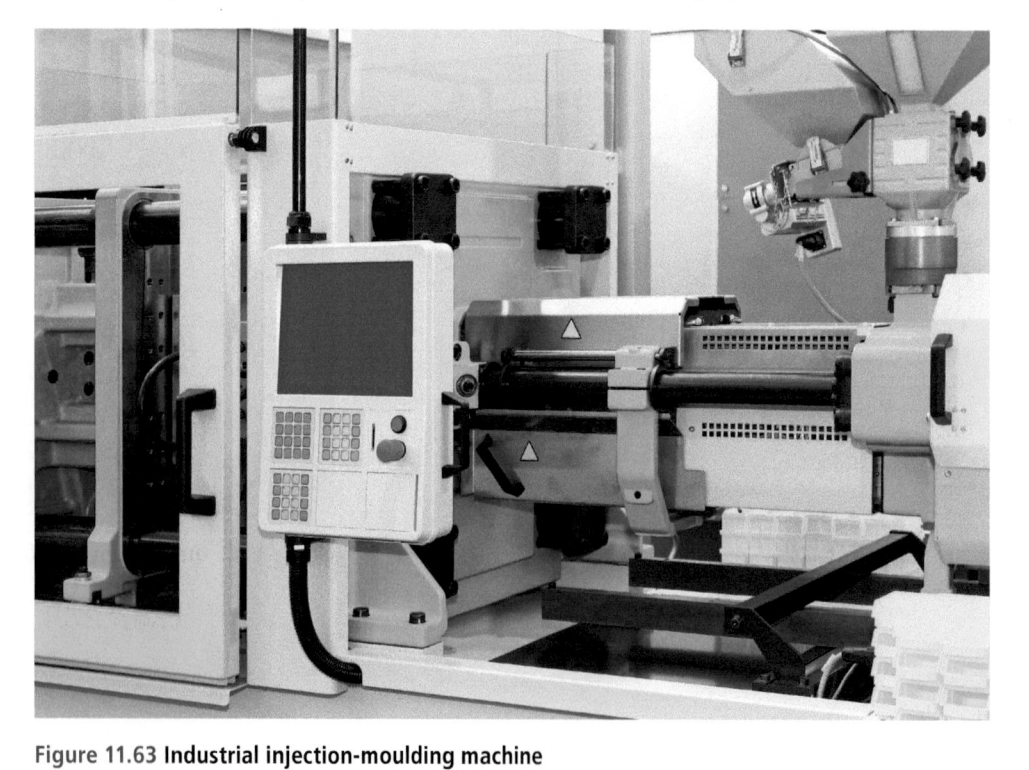

Figure 11.63 **Industrial injection-moulding machine**

Extrusion blow moulding

Extrusion blow moulding is the process whereby a short tube of melted plastic is extruded and trapped in a mould, then air is blown in so that the plastic takes the shape of the mould cavity. Typical objects include bottles made from poly(chloro-ethene), PVC, poly(propene), PP or high density poly(ethane), HDPE, petrol tanks, drums.

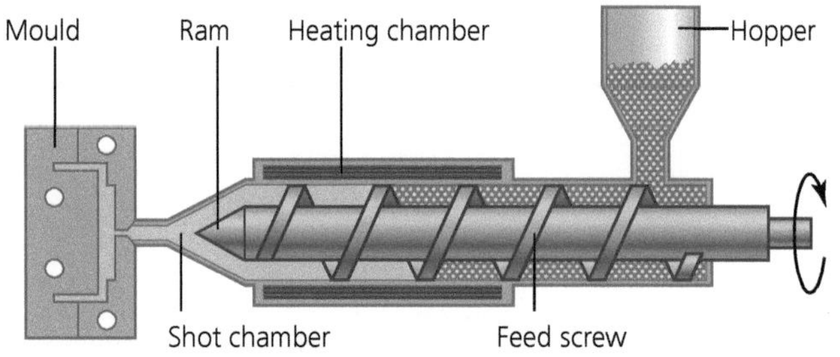

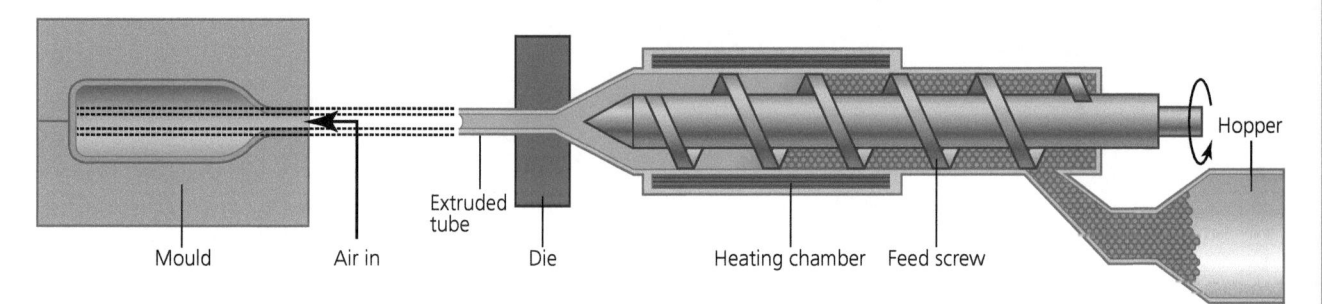

Figure 11.64 **An injection-moulding machine**

Figure 11.65 **Diagram of a combined extrusion/blow-moulding machine**

Thermoforming

Thermoforming, or sheet moulding, is a process used for vacuum forming for mass production, whereby an extruded sheet of plastic is heated and then shaped by pressure and/or by vacuum. Typical objects include chocolate box trays, refrigerator linings, packaging trays, vending cups, groups of yoghurt pots, baths and acrylic sinks.

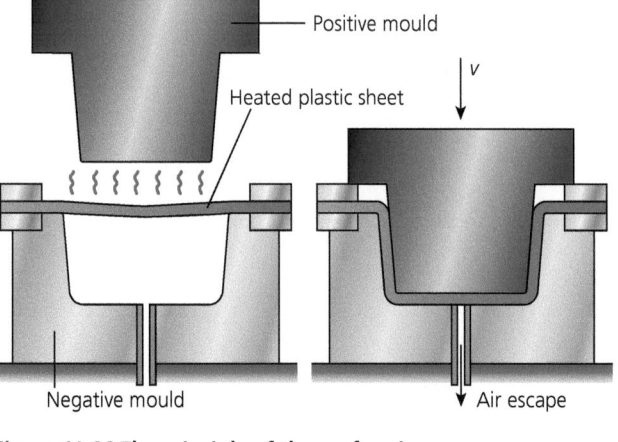

Figure 11.66 **The principle of thermoforming**

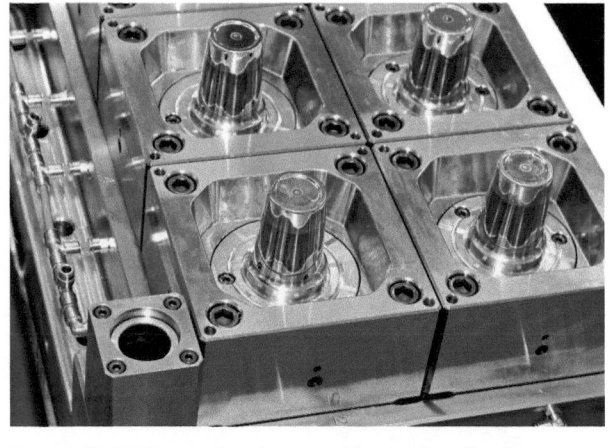

Figure 11.67 **Thermoforming mould used for plastic cups**

Extrusion

Continuous extrusion is the process whereby molten plastic is pushed continuously through a shaped hole (profile or die) before being cooled. Typical objects include curtain track, drain pipes, garden hose, guttering, rods, rulers, sheets, tubes and pipes, unplasticised poly(chloro-ethene), uPVC window frames. The shape of the die determines whether a solid rod, hollow pipe or plastic sheet is made.

Sheath extrusion is used for making cables. Electric wires are fed through an extrusion line together with the plastic they are going to be coated with. Typical objects include electric and optical cables.

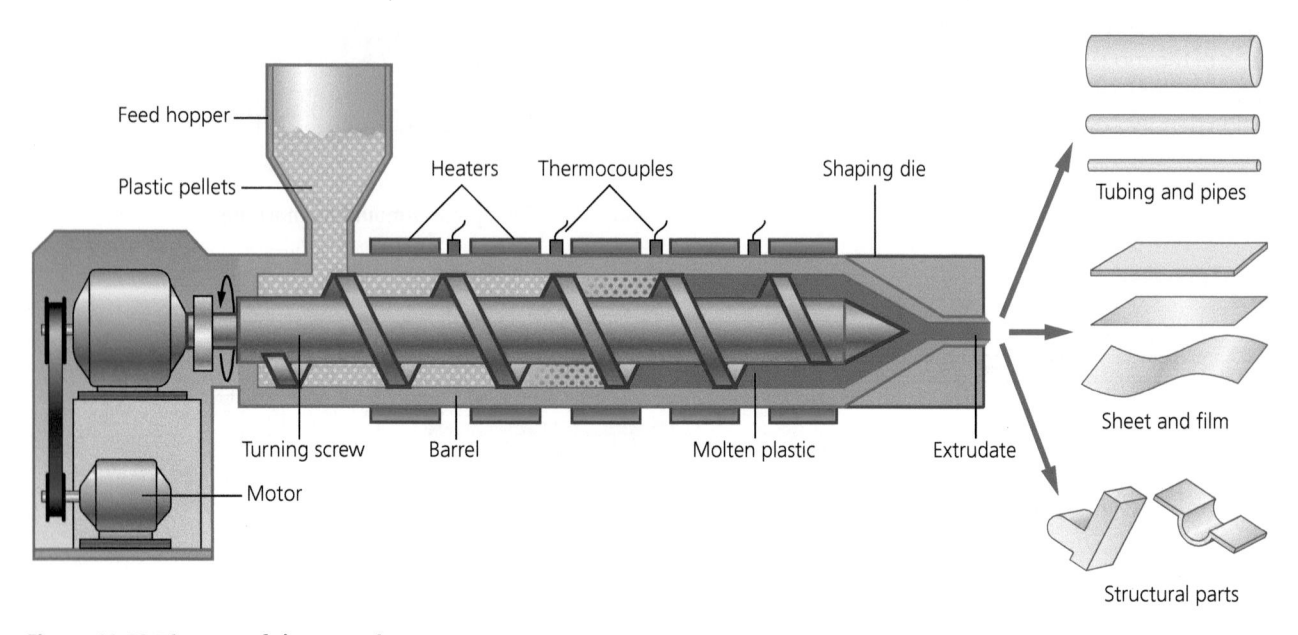

Figure 11.68 Diagram of the extrusion process

Calendering

Suitable thermoplastic compositions are passed through heated metal rollers with progressively smaller gaps to produce continuous film and precision thin sheet. The method is used to produce poly(chloro-ethene), PVC flexible film in widths of up to 4m, thin PVC and poly(phenylethene), and PS rigid foils for use in thermoforming processes or making sheet material. Embossing techniques can also be incorporated into the rolling process.

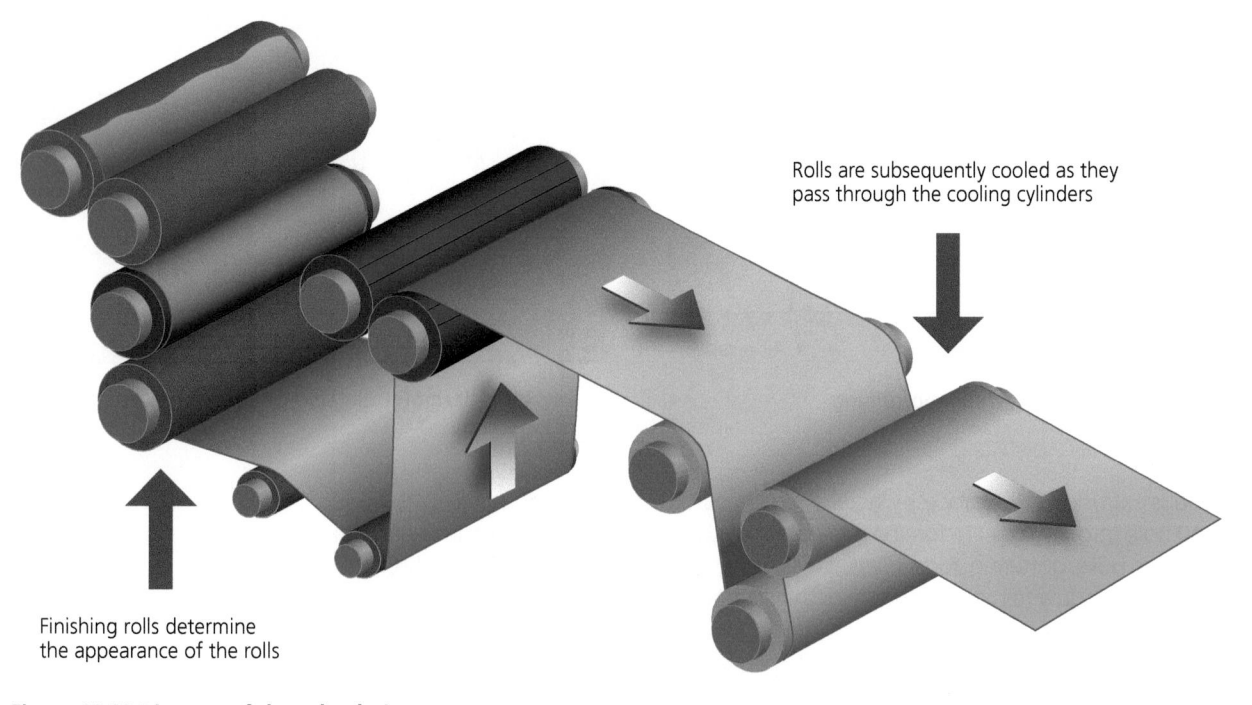

Figure 11.69 Diagram of the calendering process

Blown film extrusion

Tubular sheet or sheath extrusion is the process whereby extruded plastic is expanded into a sheath and then wound on to reels. Typical objects include packaging films, 'plastic' bags, greenhouse covers.

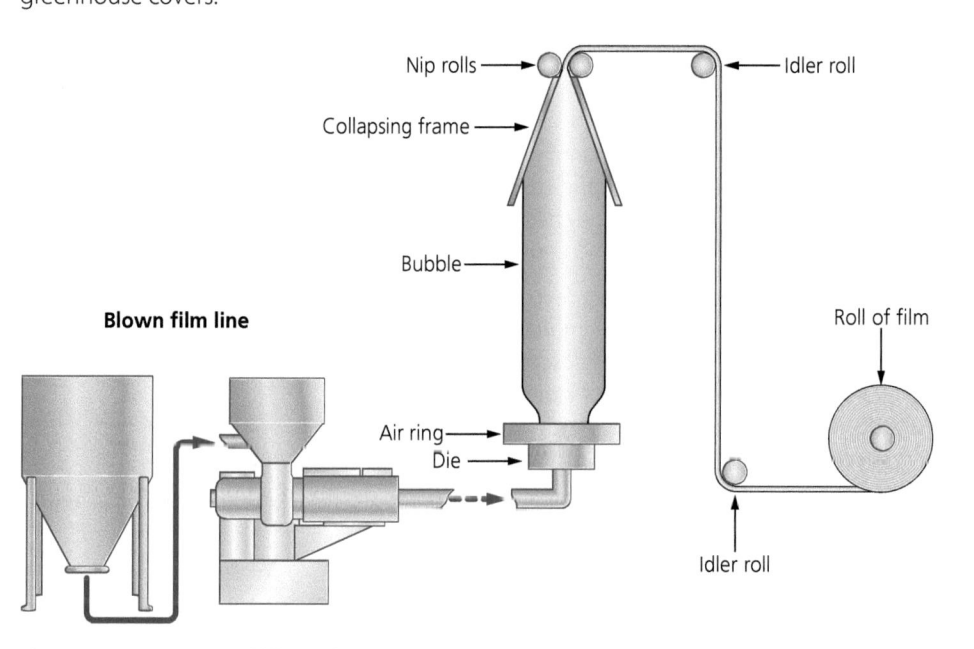

Figure 11.70 **Diagram of blown film extrusion**

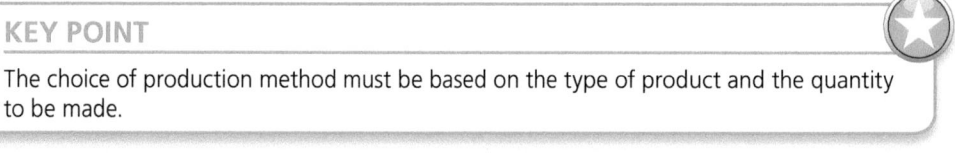

KEY POINT

The choice of production method must be based on the type of product and the quantity to be made.

STRETCH AND CHALLENGE

Create a reference table of the processes in this section. Use the following headings:
- Material category
- Process
- Scale of production
- Example products.

11.10 Cost and availability

LEARNING OUTCOMES

By the end of this section you should know about and understand:
➜ how the cost and availability of specific polymers and components can affect decisions when designing.

The significance of cost

During the design and development of a new product there are many factors that will influence your choice of materials. This section looks at how the cost and availability of polymers can affect your material choices.

Raw materials

Many of the polymers described in this chapter are the ones you are most likely to come across as a designer, and they all cost roughly the same. Manufacturing them requires crude oil so the rising price of oil affects the price of new or virgin polymer pellets.

Costs can be controlled slightly by using recycled plastic, which is around 25 per cent cheaper. Many manufacturers are trying to increase the amount of recycled polymer in the products they produce.

Four-fifths of all polymer products manufactured end up in landfill, so the economic success or failure of using polymers relies on the cost of the raw materials (petroleum and natural gas) used to make virgin polymers and the cost of recycling versus disposal. The cost of recycling can vary from city to city and country to country.

New and exciting polymers are being developed all the time, and high-performance polymers cost more than regular polymers. These include:
- Polyacetals – used in medical devices and mechanical gears.
- Kevlar – a fibre used in bulletproof vests.
- Polyetheretherketone (PEEK) – a very hard-wearing and long-lasting polymer used in the aerospace industry; one of the most expensive commercial polymers.

Your choice of polymer will affect your design development and the **viability** of your design.

KEY TERMS

Stakeholder: A person with an interest in the success of a product.
Viability: The ability to work successfully.

Responding to stakeholder needs

A key strength of using a polymer in your product is the ability to fine tune its properties in direct response to what your **stakeholders** require. You have seen that additives can be introduced to polymers to produce a specific result. With regard to stakeholders, the following may occur:
- Plasticisers may be requested by the developers/manufacturers to ensure the polymer flows quickly and efficiently into a mould to keep cycle times to a minimum.
- Increasing the amount of recycled polymer may be required by managers who want to reduce costs.
- Pigments may be required by users, to give them more choice of colours in the final product.

Manufacturing

The most common methods of polymer production are injection moulding, blow moulding and thermoforming. The process you choose depends on a number of factors, including the part size and complexity, the material selected, how many you want, how much you want each part to cost, and how soon you want it.
- **Injection moulding** is complex and requires a specific mould to be made, and for that reason it can cost more. The volume of parts produced is high, though, so costs can be spread over the total number of parts made.
- **Blow moulding** is simpler than injection moulding and is generally less expensive.
- **Thermoforming** is a single-sided process and simpler than blow moulding. It is used for lower-volume production and offers lower tooling costs.

Figure 11.71 Industrial injection moulds cost thousands of pounds

Example

If you wanted to produce 1000 washers per year, you would use a single cavity mould, meaning it would make one washer per machine cycle. In that case, the mould would probably cost £1000–£2000. On the other hand, if you are going to need 100,000 Xbox controllers every month, then you would need a 12-cavity hardened 'family' mould, which would make four fronts, four backs and four sets of buttons every cycle. This would cost you around £70,000 to £80,000.

Calculating quantities, costs and sizes of materials

Working with the commonly available forms of polymers seen in section 8.3 will require you to make a variety of calculations.

Sheet material such as acrylic can be used for linebending and it may be necessary to calculate the overall sizes needed for a piece. The example below is a menu holder for a café.

What size piece of clear acrylic would be needed? The answer is **200mm x 750mm**.

Explanation

To calculate the overall length of the single piece of acrylic needed you would add each measurement: 100 + 50 + 300 + 300 = 750mm (the extra 300mm is need to create the bend at the back to hold the menu in place)

Therefore, the overall dimensions of the single piece of acrylic needed would be **200mm x 750mm**.

Using a lasercutter often requires you to 'nest' your items together in the smallest space possible to minimise waste. For example, imagine you have a design for a pen made from layering two pieces of lasercut acrylic together like the example below.

The overall size of each piece is 50mm x 100mm.

The laser cutter can take sheets of material up to 600mm x 300mm. How many pens could you make from a single sheet?

The answer would be **18 pens**.

Explanation

The width of each piece is 50mm so you can calculate that you would fit a row of twelve by dividing the length of material in the laser cutter (600mm) by the width of each pen piece (50mm) **600 divided by 50 is 12**.

The length of each piece is 100mm so you can calculate that you would fit columns of three by dividing the width of the material (300mm) by the length of each pen piece (100mm) **300 divided by 100 is 3**.

Multiplying the rows by the columns will give you: **3 x 12 = 36 pieces.**

And finally each pen is made up of two halves so **36 divided by 2 is 18.**

If a sheet of acrylic (600mm x 300mm) costs £4.95, each pen would cost **27.5p** to make. (£4.95 divided by 18)

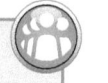

ACTIVITY

Based on the measurements of the menu holder above how many pieces would you get from a sheet of acrylic measuring 1000mm x 500mm?

How many sheets would you need to make 50 menu holders?

If a sheet of 3mm clear acrylic 1000mm x 500mm costs £9.85. How much would each menu holder cost to make?

Could you calculate how much wastage you have left at the end?

Calculating the cost of plastic products

There are several online calculators that can help customers establish a rough estimate for several polymer production processes, and especially for injection moulding.

The cost typically depends on the cost of the mould required, which depends on the size of the product, the wall thickness and how many products you want to come out of each injection.

For example, a single product roughly 10 × 5 × 1cm with a weight of 10g would cost approximately £2000 just for the mould. It would cost another £100 for the polypropylene pellets and roughly £400 for the labour/energy to give you approximately 2500 pieces. Each piece would cost £1 to make, and it would take around six days to produce 2500 pieces (if the machine was producing one item a minute through a regular working day).

Table 11.4 **How the cost of making goes down as the number of parts made increases**

Number of parts made	1000	2000	5000	10,000
Fixed cost (£)	2000	2000	2000	2000
Variable cost (£) @ 50p per part	500	1000	2500	5000
Total cost (£)	2500	3000	4500	7000
Cost per part (£)	2.50	1.50	0.90	0.70

The fixed cost is the cost of making the tools and fitting them into the machine. The variable cost is the cost of actually making the parts, including the raw material, labour and energy costs.

Controlling the cost at the design stage

Injection-moulded parts can be carefully designed using CAD and changes can be made to features, such as wall thickness and the number of components formed in a single injection. These can all contribute to establishing the viability of a polymer part by lowering costs wherever possible.

Figure 11.72 **Making changes at the design stage**

Calculating shrinkage

All polymers that are moulded by heat will shrink as the polymer cools after moulding. Working to these **tolerances** is important in order to maintain accuracy and quality in polymer products. Shrinkage is not always easy to predict and there are several factors that can affect the shrink rate, including the type of polymer used, the wall thickness, the mould temperature and the pressure of the polymer in the mould.

Tolerances and minimising waste

Lean manufacturing principles can be applied to polymer manufacturing to reduce the amount of waste at all stages of design, development and manufacture. For example:

- **Designing parts correctly:** This will avoid quality issues further down the line that would take time to correct.
- **Minimising material:** Judging the correct wall thickness affects the overall cost of a product and this can be explored at the design stage using modelling software.
- **Reducing the cycle time:** Investing time in advanced mould design that is able to cool and eject parts faster can result in greater product runs.
- **Reusing waste material:** Excess material from processing can be fed back into the moulding machines.

Figure 11.73 Waste from injection moulding can be recycled

KEY POINT

Think carefully about your stakeholders when developing your ideas. A key strength of using a polymer is the ability to adapt it for a wide variety of conditions.

KEY TERMS

Lean manufacturing: A systematic method for the elimination of waste within a manufacturing system.

Tolerance: An allowable amount of variation of a specified quantity, especially in the dimensions of a part.

STRETCH AND CHALLENGE

Use the internet to find a range of products that have been made using recycled polymers. Compare their uses and costs against products made from virgin polymers.

STRETCH AND CHALLENGE

Look back at a product you have designed and made. Use the lean manufacturing method to identify areas where you could have saved waste.

CHAPTER 12
Fibres and fabrics

You learnt about the different types of fibres and fabrics in Section 5.1. This chapter includes more in-depth information on fibres and fabrics.

12.1 Physical and working properties

KEY TERMS

Absorbency: The ability of a fibre to suck up moisture. This can be beneficial in products like towels. An absorbent fibre, however, is also more prone to staining.

Crease-resistance: A fibre's ability to recover after being wrinkled. In general, natural cellulose fibres, like cotton, have low crease-resistance and natural protein and synthetic fibres have high crease-resistance.

Fabric specification: A list of guidelines used by designers to ensure the correct fabric is used for a product. The list will include the key fabric properties required, as well as guidance on colour, texture and pattern.

Function: What a product will do and how it will work.

Handle: The level of comfort against the skin. Silk has an excellent handle, as it has a soft and smooth texture. Coarse wool does not have a good handle, as its texture can be rough and irritating.

LEARNING OUTCOMES

By the end of this section you should know about and understand the physical and working properties of a range of natural and synthetic fibres, including:
→ how easy they are to work with
→ how well they fulfil the required functions of products in different contexts.

Physical and working properties of fibres

Each natural and synthetic fibre comes with a range of properties – such as **absorbency**, stain-resistance, **crease-resistance** and insulation – that can be put to good use in the design of new products.

A fibre's properties will directly affect the end use of a product. A designer will work to a **fabric specification** and consider the physical and working properties of fibres and fabrics carefully, before selecting an appropriate material to use. This is important in making sure the finished product will **function** correctly.

The chart below shows what a bath towel needs to do, and what fabric properties are therefore required.

Table 12.1 **The fabric properties required for bath towels**

What does it need to do?	Properties required
Comfortable to handle	Good **handle**
Soak up moisture to aid drying	Absorbent
Keep us warm	Insulating
Be easy to care for	Crease-resistant
Withstand regular use	Durable

Figure 12.1 **The fibres in bath towels are selected for their working properties.**

Chapter 12 Fibres and fabrics

Natural fibres

Natural fibres are produced for use in an extensive range of textile applications. They are chosen specifically for their **physical and working properties**. In general, all natural fibres are absorbant.

Table 12.2 **The specific physical and working properties of a range of natural fibres**

Fibre	Properties/working characteristics	Uses
Cotton	Strong Cool to wear Creases easily Easy to handle and sew Will not stretch Burns easily	Medical dressings T-shirts Socks Denim jeans Cosmetic pads Nappies Bed sheets Upholstery Canvas Car tyre cords Fishing nets
Linen (flax)	Strong Cool to wear Natural lustre (shine) Creases easily Easy to handle and sew Will not stretch Burns easily	Bed sheets Table coverings Tents Skirts Suits Upholstery Canvas Wallpaper Bank notes
Wool	Strong Insulator Crease-resistant Shrinks easily Difficult to handle and sew Stretches easily Some flame-resistance	Coats Jumpers Sportswear Blankets Socks Insulation Sound-proofing Snooker tables Carpeting
Silk	Good handle Good insulator (cool in summer and warm in winter) High natural lustre (shine) Crease-resistant Difficult to handle and sew Low stretch (can be stretched out of shape) Burns slowly	Evening wear Ties Handkerchiefs Bed sheets Medical dressings Parachutes Sutures (stitches) Wall coverings

KEY POINT

All natural fibres share a lot of physical and working properties. For example, they are all absorbent. Make sure you know the few differences between the fibres when you revise this topic.

KEY TERM

Physical and working properties: The properties of a fibre, such as absorbency, strength or resistance to chemicals that affect the way the fibre is used.

Synthetic fibres

Synthetic fibres have a diverse range of properties. These polymer-based fibres can also be engineered to mimic the properties of naturally occurring fibres, making them extremely versatile.

Table 12.3 The physical and working properties of a range of synthetic fibres

Fibre	Properties/working characteristics	Uses
Polyester	Non-absorbent Strong Good handle Poor insulator Durable Crease-resistant Will not stretch Melts easily	Clothing Pillow filling Upholstery padding Bedding Carpeting Thread Ropes Boat sails
Nylon (polyamide)	Non-absorbent Abrasion-resistant Very strong Some elasticity Durable Resistant to chemicals and perspiration	Seat belts Tents Parachutes Rucksacks Shoelaces Toothbrush bristles Umbrellas Life jackets Tights Underwear Carpeting
Acrylic	**Water-resistant** Quick-drying Strong Good insulator Resistant to chemicals and perspiration	Fleece Ski jackets Blankets Rugs Outdoor furniture Knitwear Cleaning cloths
Viscose (rayon)	Absorbent Good insulator Creases easily Will not stretch Weak fibre, particularly when wet	Blouses Sportswear Shirts Blankets Curtains Table cloths Upholstery
Elastane	Non-absorbent Excellent elasticity Resistant to chemicals and perspiration Quick-drying	Swimwear Sportswear Denim jeans Leggings Tights

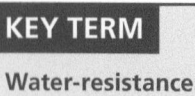

KEY TERM

Water-resistance: The ability of a fabric to resist water droplets for a period of time, though they will eventually soak into the fibre. A waterproof fibre resists water droplets indefinitely. In recent years, super-hydrophobic coatings have been developed that actually repel water.

Mixed/blended fibres

As we learnt in Chapter 5 sometimes fibres are mixed or **blended** together, to give a fibre that benefits from the properties of more than one fibre. This practice is very common and it can improve the function, aesthetic value and cost of the final product.

Figure 12.2 **Bales of raw fibre, prior to processing and blending**

Table 12.4 **Some common fibre blends and their physical and working properties**

Mix/blend	Benefits from fibre 1	Benefits from fibre 2
1 Polyester / 2 Cotton	Strength, durability, crease-resistance, stain-resistance, low cost	Handle, absorbency, cool to wear, strength, **drape**
1 Cotton / 2 Elastane	Handle, absorbency, cool to wear, strength, drape	Crease-resistance, stretch, snap-back, flexibility
1 Acrylic / 2 Wool	Quick-drying, strength, resistance to chemicals, low cost	Insulation, crease-resistance, absorbency, drape

STRETCH AND CHALLENGE

Complete the following activities for each of the products in the list below.
1   List the key fibre properties needed to make the product suitable for use.
2   Based on their properties, suggest a suitable blend of fibres for the product.
● Ski jacket
● Rucksack
● Deckchair
● Child's sleepsuit
● Garden parasol

KEY TERMS

Blending: Mixing fibres of different origins together in order to improve the properties of the finished yarn.

Drape: The way a fabric hangs under its own weight. Chiffon is an example of a material with excellent drape; felt is a material that does not drape well, due to the stiffness of its structure.

KEY TERMS

Natural cellulose fibres: Fibres that come from plant-based sources, for example cotton and linen.

Natural protein fibres: Fibres that come from animal-based sources. These include hair, fur or silk fibres.

Natural polymers: Chains of protein or cellulose molecules (monomers), such as keratin, glucose or fibroin. These chains are the basis of all natural fibres.

12.2 Sources and origins

LEARNING OUTCOMES

By the end of this section you should know about and understand:
→ the sources and origins of a range of fibres
→ the processes used to produce fibres, yarns and fabrics
→ the ecological, social and ethical issues associated with the production of fibres and fabrics
→ the lifecycle of natural and synthetic fibres
→ recycling, reuse and disposal of natural and synthetic fibres and fabrics.

Origins of natural fibres

Natural fibres can be classified into two categories:
● **natural protein**
● **natural cellulose**.

Figure 12.3 **Keratin protein chains from hair or fur fibres.**

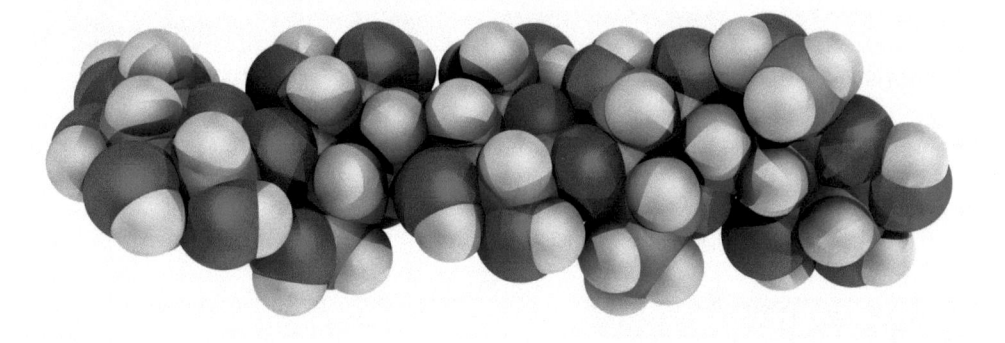

Figure 12.4 **Natural plant fibres contain cellulose.**

Natural protein fibres come from the animal kingdom in the form of hair, fur or excretions. Hair and fur fibres consist of keratin proteins, much like human hair. Silk fibre is excreted by silkworms and consists mostly of fibroin proteins, which give silk its unique properties.

Natural cellulose fibres come from plant-based sources and are made up of a natural polymer called cellulose. **Natural polymers** are chains of single molecules, and consist of specific sugars, such as glucose molecules. Cellulose fibres are found in the seeds, stems and leaves of a range of plants. Cellulose is also the primary substance in wood.

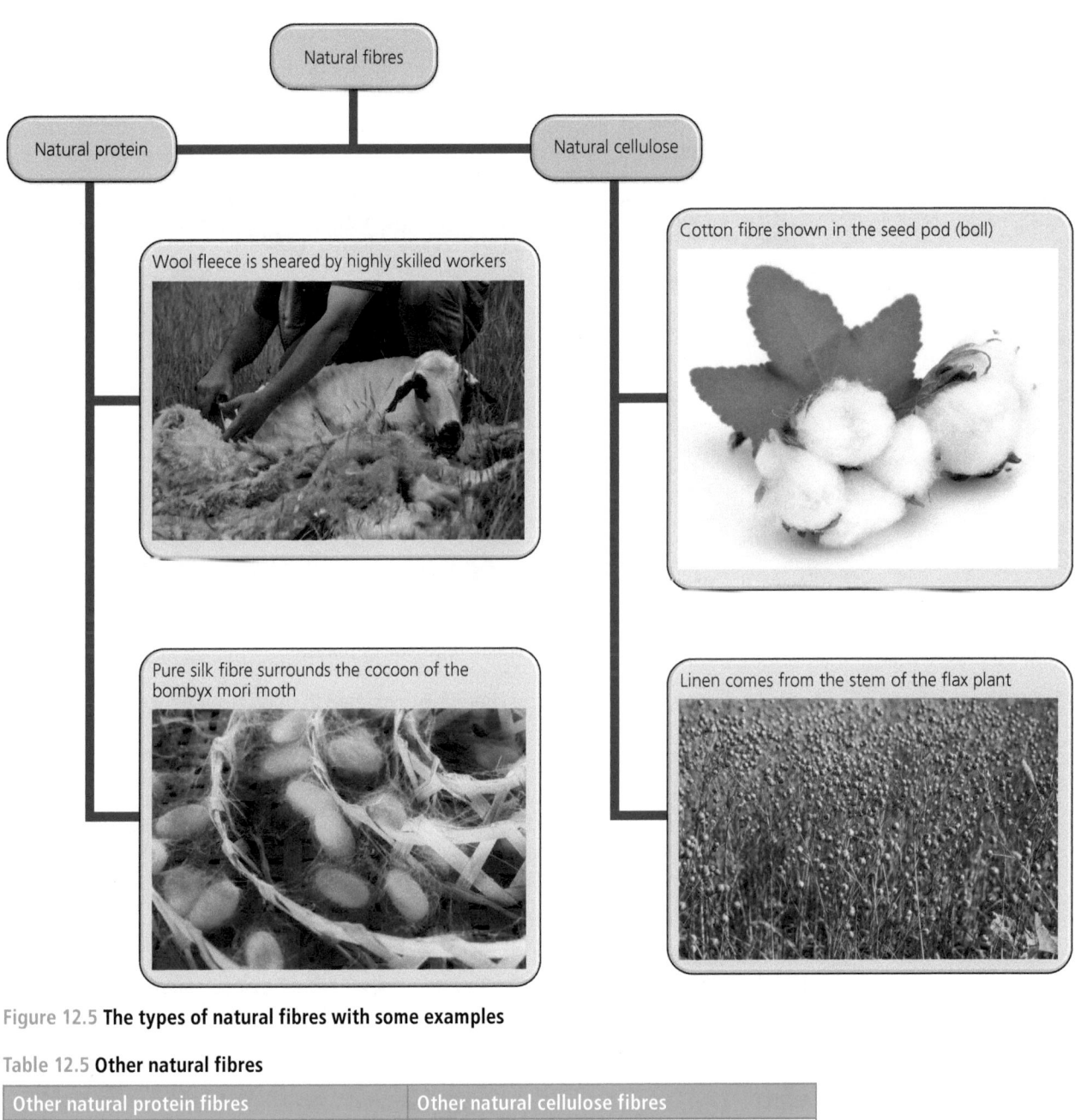

Figure 12.5 **The types of natural fibres with some examples**

Table 12.5 **Other natural fibres**

Other natural protein fibres	Other natural cellulose fibres
Alpaca	Ramie
Camel	Coir
Mohair	Hemp
Cashmere	Jute
Angora	Sisal

Origins of synthetic fibres

Synthetic fibres are also made up of polymer chains. These chains are formed by polymerising hydrocarbon monomers (single molecules that bond together) that have been extracted from petrochemicals such as coal and oil. Because they are sourced from **finite resources**, synthetic fibres are not **sustainable** and cause significant ecological damage.

Figure 12.6 **Synthetic polymer pellets are heated and extruded to form filament fibres.**

Figure 12.7 **Cellulose extracted from wood pulp is used to create regenerated fibres.**

Natural polymers (cellulose) can be chemically extracted from plant-based sources, and processed to form a different type of synthetic fibre, known as a **regenerated fibre**. These fibres are more sustainable and share some of the same properties as natural fibres.

The process of polymerisation can be adapted to produce fibres with specific properties – these are called **technical textiles**. An example of a technical textile is Kevlar®. In the production of this abrasion-resistant fibre, the long polymer chains are joined together in parallel rows, forming extremely strong bonds.

The table below lists some common synthetic and regenerated fibres.

Table 12.6 **Common synthetic and regenerated fibres**

Synthetic fibres from petrochemicals	Regenerated fibres from cellulose
Polyester	Viscose
Nylon	Acetate
Acrylic	
Elastane	

KEY TERMS

Finite resources: Non-renewable sources that cannot be replaced in a sufficient timeframe to allow further human consumption. Examples include crude oil and natural gas and coal, as these resources have taken millions of years to form.

Regenerated fibres: Fibres derived from natural cellulose sources, using synthetic chemical processes. These include viscose and bamboo fibre.

Sustainable: Renewable sources that can be replaced in a sufficient timeframe to allow continued human consumption. These resources include cotton, linen and wool. The processes used to cultivate and extract these fibres must be considered when deciding how sustainable a resource is.

Technical textiles: Textiles manufactured specifically for their performance properties instead of their aesthetic value. Examples of technical textiles include Kevlar and Stomatex®.

Extraction and conversion – natural fibres

Cotton

Brazil, Pakistan, Turkey and the USA are the world's leading producers of cotton crops. Cotton is intensively farmed across huge areas of land, using fertilisers, pesticides and large amounts of water.

Figure 12.8 **Thousands of hectares of land are used for cotton crops.**

Figure 12.9 **Machinery harvesting cotton bolls.**

The fibre is harvested from the seed pod (boll) of the cotton plant. A single mature cotton plant can produce around a hundred cotton bolls annually. This roughly equates to one bale (225kg). To put this into perspective, it takes around 1kg of cotton to make a single pair of jeans. Over 1 billion pairs of jeans are produced worldwide, each year.

The cotton harvest begins in July and continues until November. Cotton picking can be completed by hand or machine, depending on the scale of production on the farm.

Once picked, the cotton goes through the ginning process, which separates the fibre from the seed. Raw cotton bales are dried to remove any moisture content that could cause fibres to clump. The dry fibre is passed through a gin stand, where circular saws are used to remove the fibre from the seed. The cotton is baled up and shipped to a different factory for further processing.

Most cotton is used to produce yarns for the manufacture of fabric. For this, it must be cleaned and bleached to achieve consistency in the finished product. The cotton fibre goes through several stages of cleaning before it can be spun into yarn.

Linen

Canada, Russia and France are the world's leading producers of flax, which is where the linen fibre is found. These plants flourish in cooler climates, where they are harvested annually. The fibre is found in the stem of the flax plant, and this is also known as a bast fibre.

Figure 12.10 Flax plants are uprooted rather than cut to preserve their quality.

Figure 12.11 Raw linen fibre must be treated before use.

To harvest the linen, the stems are uprooted using machinery that pulls up the stem, then lies the plant on its side. It is important that the plants are not cut, as this removes the natural sap in the stem and affects the quality of the linen. Once uprooted, the stems are retted. The retting process involves leaving the stems out in the elements to rot away the tough, outer layer of the stem, and to dry them out. Once retting is complete the stems can be broken and the fibres can be removed from inside. These fibres then go through a series of processes to clean and prepare them for manufacture into yarn.

Figure 12.12 Wool comes from a wide variety of animals, including llamas and alpacas.

Wool

Wool production is centuries old and can be traced back to 4000bc. Australia, New Zealand and the USA are the world's largest producers of wool. The processing of wool fibre has remained unchanged through history, but the use of modern technology means we can now process over 2 million tonnes of wool annually.

The process begins in spring, when the majority of sheep are sheared. This highly skilled process involves the use of clippers to remove the whole fleece from the sheep. Workers attempt to do this quickly, to cause as little distress as possible to the animal.

Fleeces are separated and graded, as a single fleece will yield different types of wool. The outer wool on the fleece is thicker and coarser, which makes it more suitable for carpet or insulation. Meanwhile the wool on the inside, next to the skin of the sheep, is finer and softer. This wool is more suitable for clothing such as suits or jumpers.

Wool is naturally greasy. It contains a natural oil called lanolin, which must be removed before the wool fibre can be used. A process called scouring is used to remove the wool grease. This involves a series of baths that use warm water and detergents. The waste lanolin is often used in skincare products.

The wool is cleaned, dried and stored, ready for further processing.

Figure 12.13 **Raw wool fibre must be cleaned thoroughly before further processing.**

Silk

China is the world's leading producer of silk, and silk production there dates back to the Neolithic period. Silk is produced by the silkworm, which is actually the caterpillar of the *Bombyx mori* moth.

Figure 12.14 **The *Bombyx mori* moth**

Figure 12.15 **Boiled cocoons are unravelled to remove the silk fibre.**

Silkworms are bred in captivity and are fed a diet of mulberry leaves throughout their short lives. As it grows, the silkworm moults several times. After the fourth moult, the worm encases itself in a cocoon of silk fibre and begins to pupate. It is at this point that the cocoons are boiled, in order to kill the pupa inside. The process of boiling also helps to remove the seracin, a sticky gum that helps the silk fibres stick together when producing the cocoon. The cocoon must not be cut, in order to retain the length of the silk fibre.

Workers pull the silk fibre, which unravels off the cocoon, and wind it on to bobbins. The silk fibre is then cleaned and dried in preparation for further processing. The manufacture of silk is a labour-intensive process, making it one of the most expensive fibres available.

Recent innovations in the production of more sustainable natural fibres have led to the development of more unusual fibre sources. These include soured milk, fermented wine, spider silk, corn husks and lab-grown bacteria.

Extraction and conversion – synthetic fibres

Synthetic fibres are all produced using extrusion processes. Manufacturers begin with polymer solution or melted pellets of polyester or a similar synthetic fibre. The solution is placed in a tank above a spinneret. The spinneret is a metal extrusion plate that contains many tiny holes, and acts like the spinneret that a spider uses to produce its web.

The solution is forced through the holes in the spinneret using air pressure, or a syringe-like system, forming the solution into long individual fibres.

Melted polymer fibres require cool air to set them as they emerge from the spinneret. Regenerated fibres require a chemical solution to set the fibres as they exit the spinneret. The spinnerets are then placed in a chemical bath and the finished fibre is thoroughly cleaned off them before they are used again to ensure no trace of chemicals remains.

Once they are set and dried, the synthetic fibres are wound on to bobbins ready for further processing.

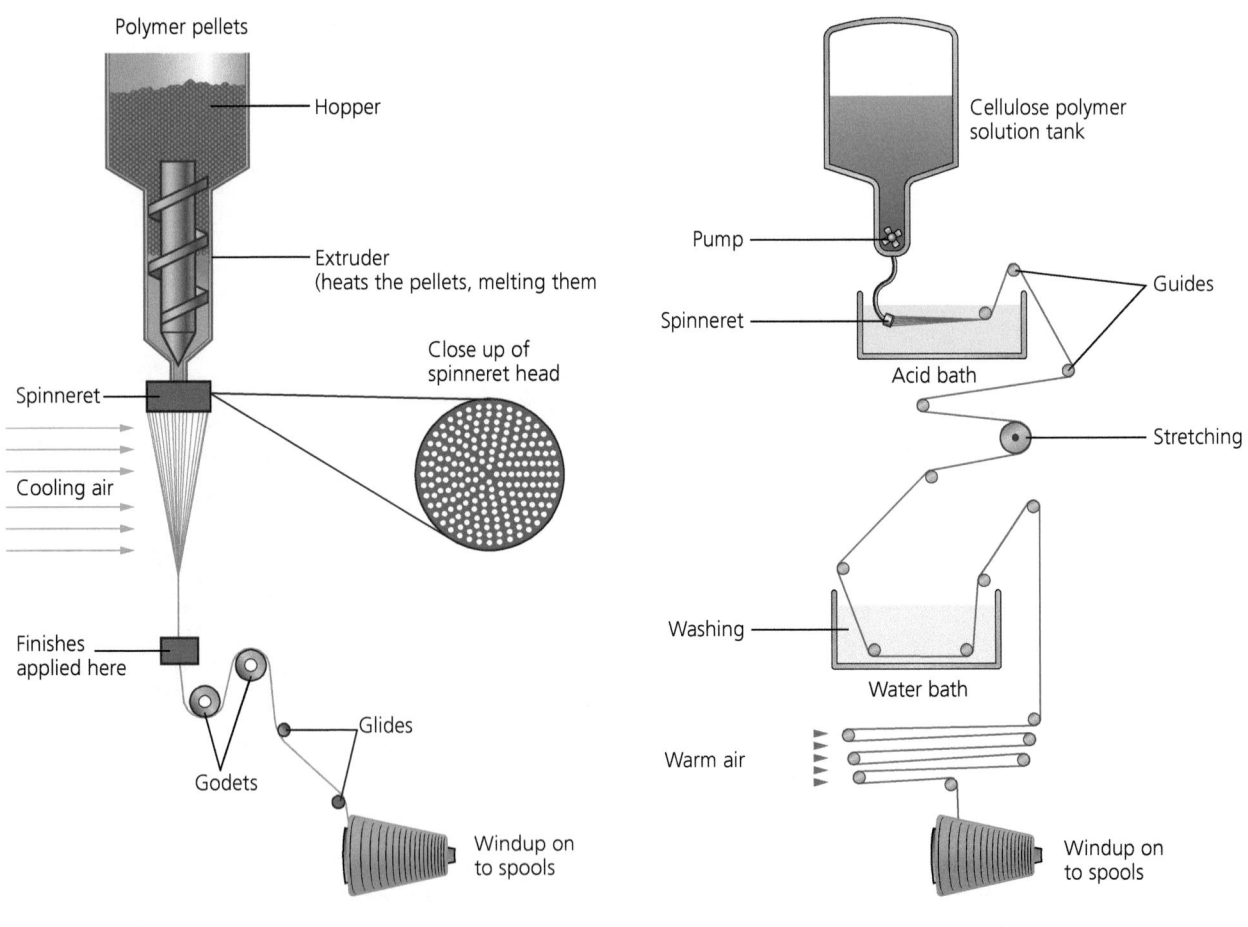

Figure 12.16 The melt spinning process

Figure 12.17 The wet spinning process

Staple and filament fibres

There are two types of fibre used in fabric and yarn production: staple fibres and filament fibres.

- **Staple fibres** are short fibres with a crimp (wavy texture). Staple fibres can be spun into yarns, felted or bonded. These fibres mesh together well, due to their crimped shape. All natural fibres, apart from silk, are staple fibres.
- **Filament fibres** are long, smooth fibres. Filament fibres are very strong, but do not mesh together well, due to their smooth surface. All synthetic fibres begin life as a filament, but they can be cut short and heated to give them the appearance of a staple fibre. Silk is the only natural filament fibre.

KEY TERMS

Filament fibre: A long, smooth, fibre of synthetic or natural protein origin.

Staple fibre: A short fibre with crimp (wavy texture); most natural fibres are staple fibres, although synthetic fibres can be manufactured to form the staple shape.

Under the microscope

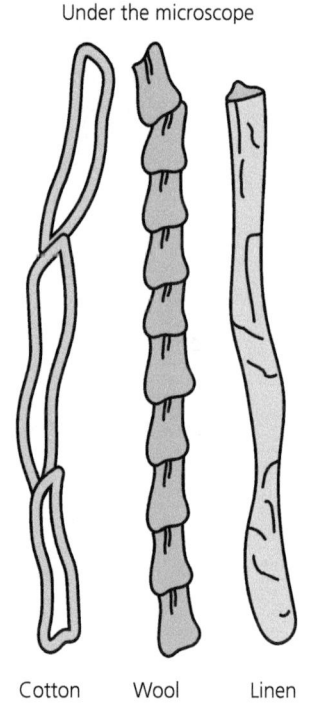

Cotton   Wool   Linen

Figure 12.18 Staple fibres

Long, smooth

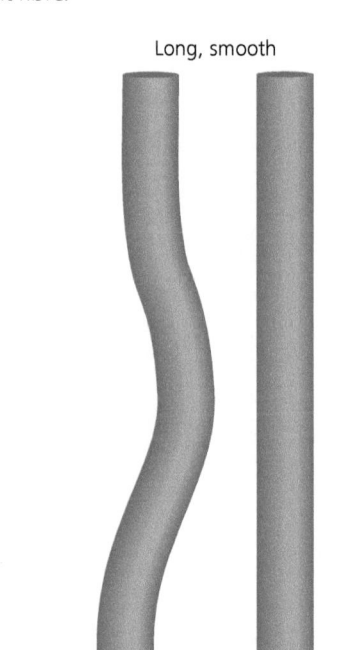

Silk   Polyester

Figure 12.19 Filament fibres

Microfibres

Microfibres are tiny filament fibres made from synthetic polymers such as polyester or nylon. They are approximately one-hundredth of the diameter of a human hair. Microfibres are very lightweight and versatile. They have a number of useful properties including excellent strength-to-weight ratio, good drape, breathability, water-resistance and durability. These fibres are used in a number of applications including clothing, cleaning cloths and insulation. They are of particular benefit when used for cleaning as the fibres attract dirt and dust, making them more efficient than regular cleaning cloths. They are also well suited to printing methods, as the size of the fibre does not distort images, giving good definition.

From fibre to yarn

Spun yarns are created by pulling and twisting a mass of staple fibres, causing them to mesh together. It is this twisting process that gives yarn strength.

The diagram below shows the manufacturing process for making spun yarns from natural or synthetic staple fibres.

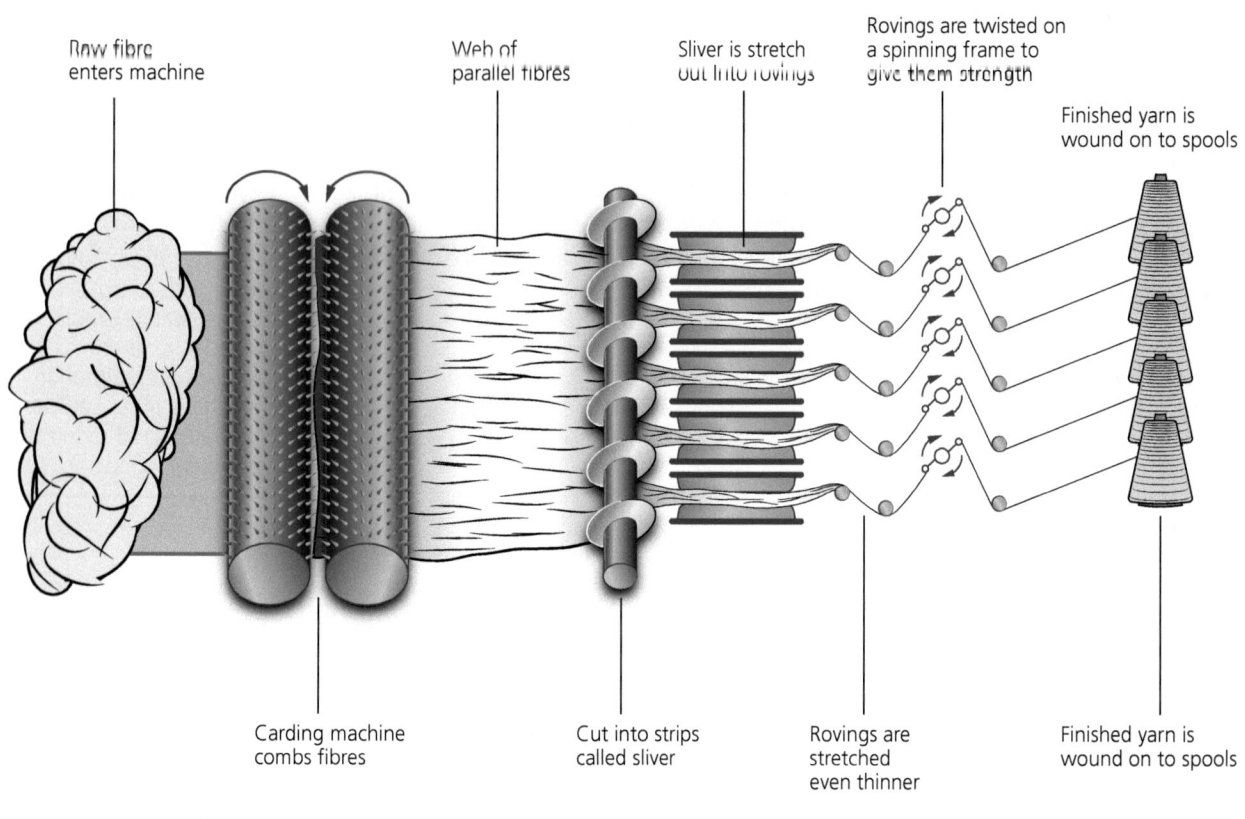

Raw fibre enters machine

Web of parallel fibres

Sliver is stretch out into rovings

Rovings are twisted on a spinning frame to give them strength

Finished yarn is wound on to spools

Carding machine combs fibres

Cut into strips called sliver

Rovings are stretched even thinner

Finished yarn is wound on to spools

Figure 12.20 **Process for manufacturing spun yarns**

ACTIVITY

You will need: a ball of cotton wool.

A ball of cotton wool is essentially a mass of cotton fibres.
1 Look closely at the surface of the cotton wool – you should see the ends of hundreds of tiny fibres. Use tweezers to grab and pull out a single fibre. Examine and describe the fibre.
2 Pinch the cotton wool and gently pull it to stretch it out into a long thin strip. At this point, the fibres should come apart easily.
3 Start to twist the strip of fibres. Do they still come apart easily? What happens when you add more twist?

Extension: Use tweezers to remove single fibres from a range of different yarns. How do they differ?

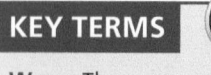

KEY TERMS

Warp: The warp yarn is the stronger yarn in a fabric construction and runs vertically down the length of the fabric. This is also known as the straight grain.

Weft: The weft yarn is woven under and over the warp yarn and runs horizontally, across the length of the fabric. This is also known as the cross grain.

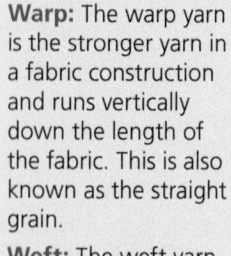

Woven, non-woven and knitted textiles

Fabrics are constructed using a range of different methods, depending on their intended use. The original methods of construction, such as weaving on wooden looms, have been developed and are now mostly completed on industrial machinery, although traditional methods are still favoured by designers for creating small collections or one-off pieces.

Woven fabric construction

Woven fabrics are produced using **warp** and **weft** yarns. The warp yarn is usually the stronger of the two. Warp yarns run vertically down the length of the fabric, which is also known as the straight grain. The straight grain is easily found by using the selvedge, as this always runs parallel to the straight grain. The selvedge is the factory-finished edge of the fabric roll. The selvedge prevents the fabric from fraying or unravelling. Weft yarns run horizontally across the fabric, also known as the cross grain.

The warp yarns are wound on to the loom in parallel rows and are threaded through frames containing plastic, wooden or metal strips. These frames are known as heddles. The heddles lift the warp yarns upwards, allowing a shuttle to be passed through, carrying the weft yarn. The heddles are then lowered, trapping the weft yarn in between the warp yarns. Different heddles are used to lift alternate warp yarns with each pass of the shuttle. This results in the interlocking of the warp and weft yarns, creating a fabric. Weaving patterns are achieved by setting up several heddles to lift different warp yarns at different times, although this is a time-consuming process. Computerised looms allow the creation of complex weave patterns quickly and efficiently.

Figure 12.22 **A traditional loom**

Plain weave

The **plain weave** is the simplest and most widely used weave structure, and it is very versatile. Various weights of fabric can be produced by altering the spacing of the warp and weft yarns or by using coarse or fine yarns. Muslin is a common example of a lightweight plain weave fabric, and calico is a common example of a medium to heavyweight plain weave fabric.

> **KEY TERM**
>
> **Plain weave:** A basic weave construction in which the weft yarn goes under and over alternate warp yarns, giving a strong and flat fabric.

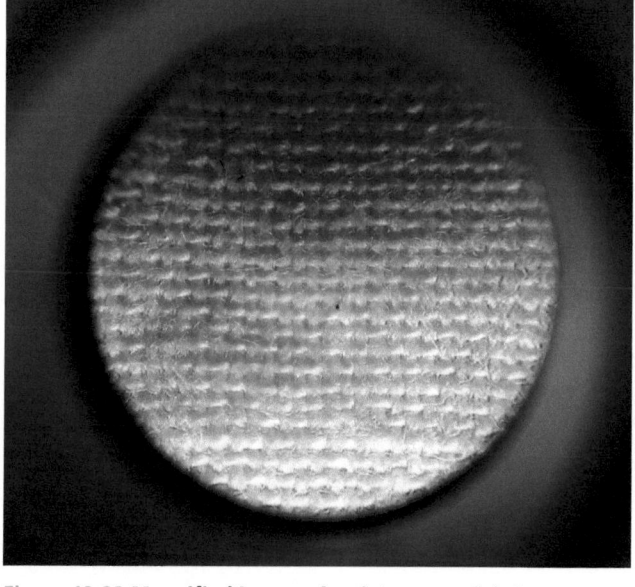

Figure 12.23 **Magnified image of a plain weave fabric**

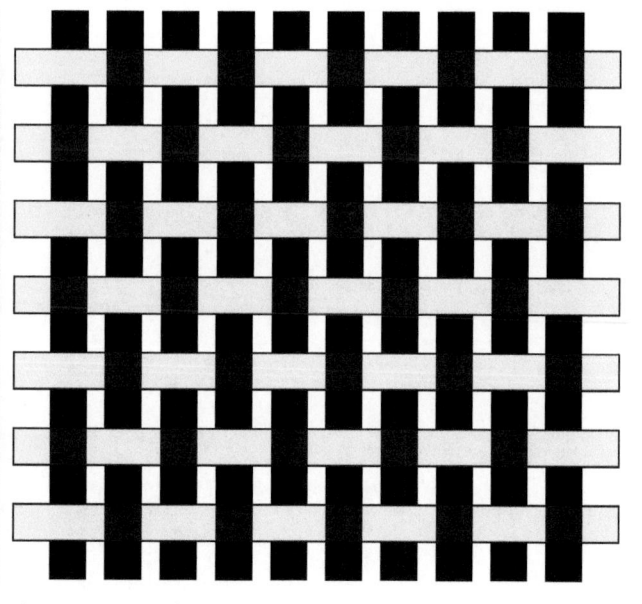

Figure 12.24 **A plain weave structure**

Twill weave

KEY TERM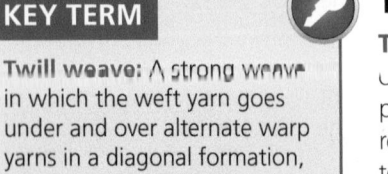

Twill weave: A strong weave in which the weft yarn goes under and over alternate warp yarns in a diagonal formation, giving a textured finish.

Twill weave is easily recognisable due to the diagonal pattern formed by the crossing of the warp and weft yarns. This weave produces a heavier fabric than a plain weave, making twill stronger and more durable than a plain weave. For this reason, twill weaves are often used in products that need to resist heavy wear and tear, such as denim jeans and canvas bags. Twill weaves are also often selected for aesthetic reasons. Herringbone and houndstooth fabrics are popular examples of decorative twill weaves.

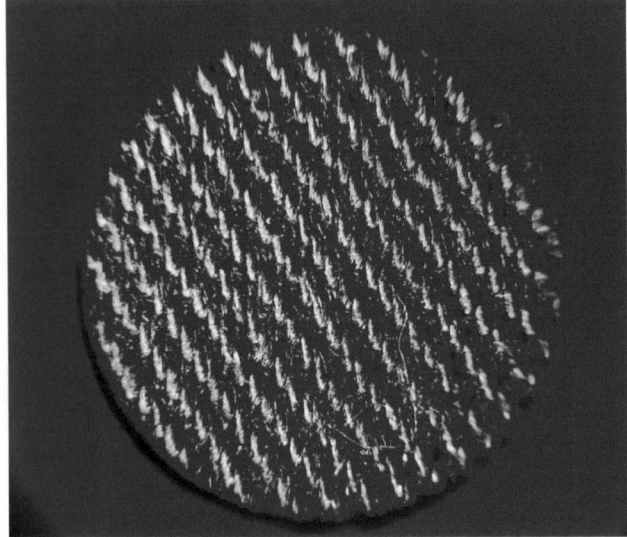

Figure 12.25 **Magnified image of a twill weave fabric**

Figure 12.26 **A twill weave structure**

Satin weave

KEY TERM

Satin weave: A luxurious weave construction in which the weft yarn floats over three or more warp yarns, giving a smooth and shiny finish.

The structure of a **satin weave** produces a smooth, lustrous finish due to the 'floating yarns' on the surface of the fabric – the weft yarns are woven under one warp yarn, then over a minimum of three warp yarns. This results in a larger surface area, which reflects light, giving a shiny finish. Satin weaves are the weakest of the three types of weave and can be snagged easily, due to the structure of the floating yarns.

Figure 12.27 **Magnified image of a satin weave fabric**

Figure 12.28 **A satin weave structure**

Non-woven fabrics

Non-woven fabrics are created using the shape and texture of staple fibres. Staple fibres are short and have crimp (wavy texture). It is the crimped texture that causes the fibres to tangle together.

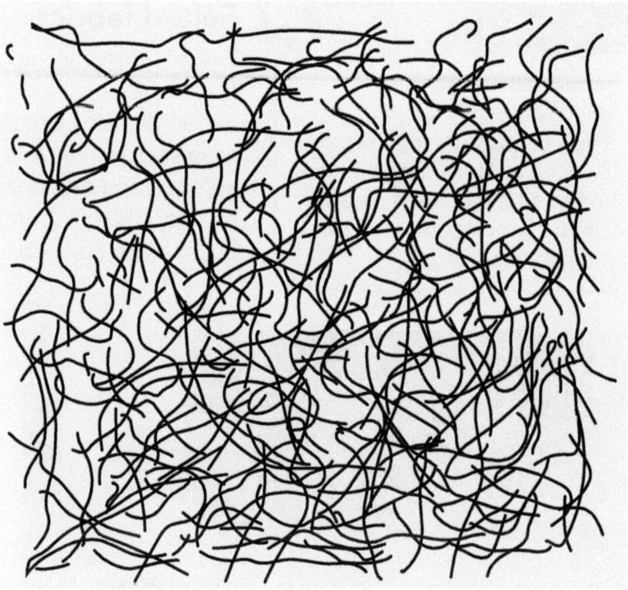

ACTIVITY

You will need: strips of paper (1cm width) in two colours; adhesive tape.
1 Line up ten strips of paper (of one colour) vertically, side by side. Place a strip of tape across the top to secure them.
2 Use the diagrams above as guides to help you create the three types of weave.

Figure 12.29 **A non-woven fabric structure**

Bonded fabrics

Bonded fabrics are manufactured by laying out a 'web' of synthetic fibres, then applying pressure and heat or adhesives to bond the fibres together. Bonded fabrics are often used in **disposable** textiles such as wet wipes, tea bags, surgical masks, dressings and nappies. These fabrics lose their strength and structure once wet, so they are usually only suitable for one use.

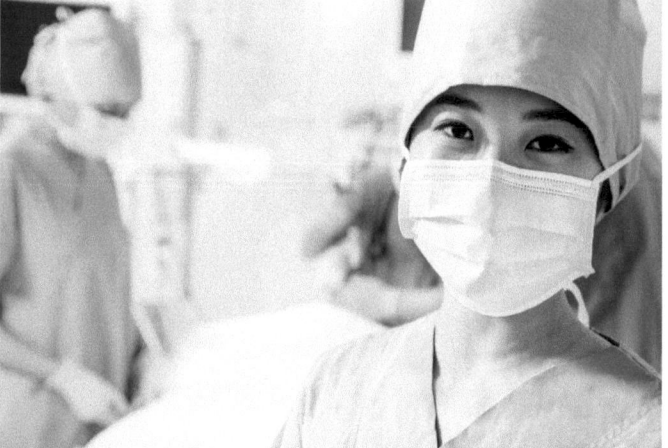

Figure 12.30 **Disposable wet wipes are made from bonded fibres.**

KEY TERMS

Non-woven fabrics: Fabrics made by entangling fibres together using friction, pressure, heat or chemicals. These fabrics are cheaper to produce than woven fabrics and are often used for disposable products.

Bonded fabric: Fabric manufactured by adding pressure, heat, chemicals or adhesives to a web of fibres, causing them to bond together. Examples of bonded fabrics include baby wipes and interfacing.

Disposable: Products intended for a single use for cost, convenience or hygiene reasons. Examples include surgical masks and bandages.

Figure 12.31 **Surgical masks are disposable for hygiene reasons.**

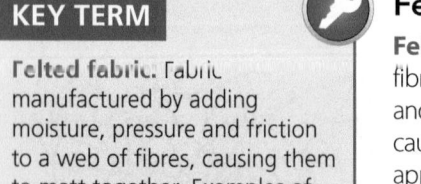

KEY TERM

Felted fabric: Fabric manufactured by adding moisture, pressure and friction to a web of fibres, causing them to matt together. Examples of felted fabrics include craft felt and pool table covers.

Felted fabrics

Felted fabrics are produced by applying moisture, heat and friction to a web of staple fibres, which matt together. The most commonly used fibres in this process are wool and acrylic. The fabric will be denser and stronger the more it is worked – this also causes some shrinkage, however. Felt is often used for decorative purposes such as appliqué, and is historically applied to the surface of pool and snooker tables. Felt is also used for cushioning and insulating various products. Felt is weak and will easily stretch out of shape, particularly when wet.

Figure 12.32 **Traditionally, wool felt was used to top snooker tables.**

Figure 12.33 **Felt doesn't fray, which makes it suitable for a range of crafts.**

Weft knit fabric

Weft knit fabric is comprised of a single continuous yarn, and is constructed in horizontal rows of interlocked loops. The horizontal rows are called **courses** and the vertical columns are called **wales**. Weft knits are often produced using knitting needles, but they can also be manufactured on a larger scale using automated knitting machines. These machines can knit a flat length or a tube of fabric – tubes are very useful in the production of socks. Weft knits may snag and can unravel if part of the yarn is damaged or pulled. Weft knits are used for a wide range of knitwear and home furnishings.

Figure 12.34 **Magnified image of a weft knit fabric**

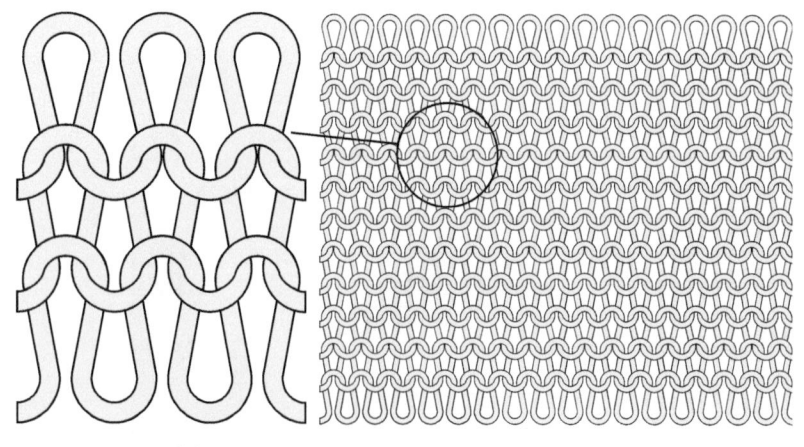

Figure 12.35 **Weft knit structure**

Warp knit fabric

Warp knit fabric has a more complicated structure than weft knit fabric. Warp knits are comprised of multiple yarns. The fabric is constructed in vertical rows of interlocking loops that zig zag from side to side. Warp knits do not run or unravel and are more flexible than weft knits, making them suitable for sportswear and swimwear. Warp knits can only be completed on automated machines.

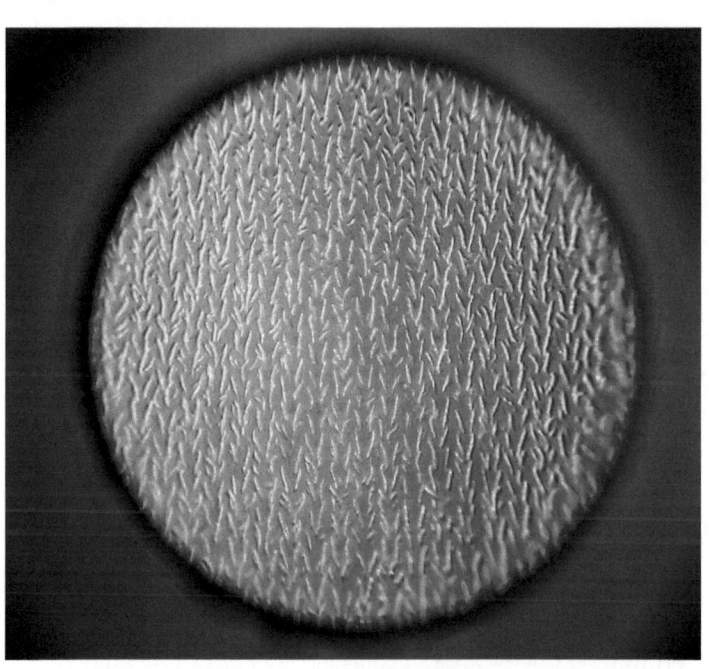

Figure 12.36 **Magnified image of a warp knit structure**

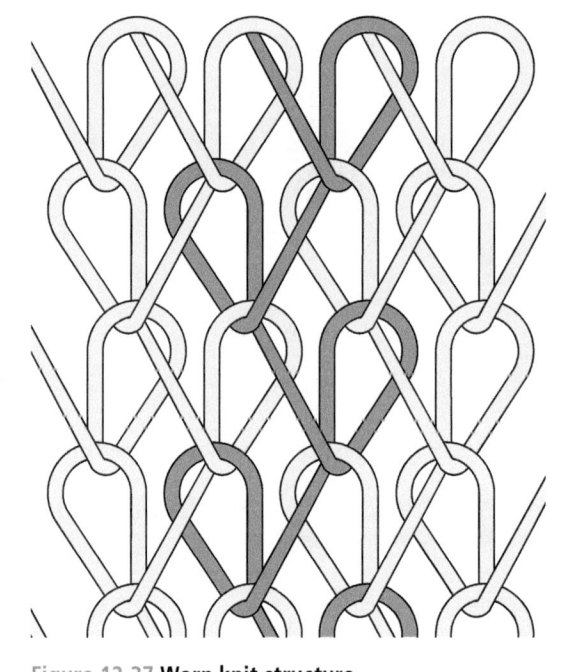

Figure 12.37 **Warp knit structure**

Figure 12.38 **A range of knitted fabrics**

ACTIVITY

Find a fabric sample for each of the following headings:

- Woven – plain
- Woven – twill
- Woven – satin
- Non-woven – bonded
- Non-woven – felted
- Warp knit
- Weft knit

Study the fabrics closely to identify their structure.

Keep your samples to use as a revision aid.

KEY TERMS

Courses: Rows of horizontal loops that run across the width of a knitted fabric.

Weft knit: Fabric constructed with horizontal and vertical rows of interlocking loops. Weft knits can be constructed using knitting needles or on a knitting machine.

Wales: Vertical rows of interlocking loops that run down the length of a knitted fabric.

Warp knit: Fabric constructed with rows of interlocking loops that zig zag down the length of a knitted fabric. Due to their complexity, warp knits can only be constructed on a knitting machine.

Figure 12.39 **Toxic pesticides are heavily used on fibre crops.**

Ecological, social and ethical issues associated with processing

Ecological issues

Ecological impact of natural cellulose fibres

Natural fibre crops are intensively farmed on vast areas of land in order to keep up with global demand. Farmers of crops such as cotton and linen use significant amounts of water as well as toxic pesticides and fertilisers, which damage the ecosystem. Chemicals soak through the soil and into water sources, killing wildlife and contaminating drinking water. These chemicals also strip the soil of nutrients, reducing crop quality and eventually making the land unusable.

Organic farming reduces this impact through the use of natural fertilisers and pesticides – this type of farming, however, still requires a substantial amount of water. It is estimated that approximately 2700 litres of water is required to produce enough cotton for a single t-shirt.

KEY TERM

Recyclable: A material suitable for processing using tertiary recycling methods.

Ecological impact of natural protein fibres

The farming of livestock for wool production has relatively low ecological impact compared to the farming of fibre crops. The land used for the rearing of livestock is usually unsuitable for crops, but livestock can damage the land by overgrazing. On larger wool farms, the number of sheep can exceed 300,000. Chemical dips containing insecticides are used to protect livestock from mites and ticks, and these dips contaminate soil and water sources. Livestock can also produce large amounts of the greenhouse gas methane.

Silk is a highly renewable resource with little ecological impact. The production of silk in captivity, however, is heavily criticised by animal welfare activists. The breeding of silkworms in captivity has led to the demise of the *Bombyx mori* moth in the wild, and captive moths have evolved to be blind and unable to fly. These moths live only for a few days and in this time lay up to 500 eggs.

Transportation and processing

There are few countries where fibre production is local to its manufacturers. Transportation of raw fibres causes air pollution and uses a significant amount of non-renewable resources.

The processing of synthetic fibres uses finite resources such as oil and coal. Extraction and processing of these resources involves the use of high levels of energy and chemicals, and also the production of contaminated waste. Greenhouse gas emissions from factories cause air pollution in local areas and depletion of the ozone layer. Additionally, there is significant water pollution in local areas around factory sites. Synthetic fibres are non-biodegradable: polyester would take up to 450 years to break down in the environment. Synthetic fibres are **recyclable**, however, and this practice is going from strength to strength within the industry.

Figure 12.40 **Large flocks of sheep overgraze land and cause damage to soil.**

The washing, drying, dyeing and printing of fabrics requires lots of energy and also generates large amounts of waste water, which contaminates local water sources and damages habitats. Many factories do not follow rules and regulations on the safe disposal of these waste products due to the cost implications.

Over-packaging of textile products is also creating significant amounts of unnecessary plastic waste each year. When garments are delivered to a retail store, each item is covered by a plastic garment bag that is immediately disposed of. The purpose of these bags is to protect garments from moisture and dust in transit. The garments are also placed inside cardboard boxes, however, which are covered in plastic wrap before shipping.

Figure 12.41 **Drilling for crude oil and coal mining deplete finite resources.**

Social and ethical issues

The treatment of garment workers in the textile industry has been under scrutiny for many years, with garment workers in some countries subjected to poor (and often dangerous) working conditions, long hours and low pay, with little protection from union rights. Children as young as six have been found working in some factories, despite local laws to prevent child labour; and work-related deaths in the textile industry are not uncommon.

Figure 12.42 **Employees work long hours for little money, and often in poor working conditions.**

While steps have been taken globally to improve conditions for employees working within the textile industry, increasing consumer demand for **fast fashion** has led to a **throwaway culture**, with items of clothing being purchased at rock-bottom prices, worn once and quickly disposed of. This has driven down retail prices, meaning that manufacturing costs (including workers' pay and conditions) have also been dropped in order to maintain profits for high street stores. This in turn can have an impact on the pay and conditions of those producing the garments.

KEY TERMS

Fast fashion: A recent trend involving the quick transfer of new collections from the catwalk into stores. Fast fashion is often on-trend, low quality and low in price. The characteristics of fast fashion mean that consumers buy large volumes of clothes more regularly, creating more profit.

Throwaway culture: The rise of fast fashion has made clothing far more affordable for consumers. The incredibly low prices in some high street stores have resulted in a throwaway culture, meaning that consumers don't feel the need to keep clothing that is no longer in fashion, and happily dispose of it.

Those living and working near the textile industry may also be affected by the contamination of the land and water supplies. Skin conditions, lung problems, jaundice and cancers are the possible consequences of exposure to dyes and other toxic chemicals used in the production and processing of textiles. Responsible textile manufacturers should consider the health and safety of their workers, their families and those living near the textile industry, and put suitable measures in place to protect them, for example dealing with waste chemicals properly to prevent contamination of water and land nearby.

KEY TERM

Biodegradable: The ability of a material, substance or object to break down naturally in the environment through the action of micro-organisms, thereby avoiding pollution. Natural and regenerated fibres are biodegradable.

Figure 12.43 **Waste water from factories pollutes local drinking water sources.**

Figure 12.44 **Textiles can be taken to clothing banks for recycling.**

Lifecycle

Natural fibres are **biodegradable** and decompose naturally in the environment within six months; they will release CO_2 during decomposition, however, and natural fibres treated with dyes and chemicals will release these into the ground. Synthetic, polymer-based fibres decompose very slowly, taking between 450 and 1000 years to break down. During this time the fibres will release toxic and greenhouse gases into the atmosphere.

Recycling, reuse and disposal

Most fibres and fabrics can be easily recycled, and it is vital that we do this in order to protect the environment as the textile industry continues to grow. Clothing banks and charity shops play a huge role in ensuring that waste textiles are reused or recycled effectively.

Primary recycling

Primary recycling of textiles involves the reuse of products. Giving clothing to charity shops or handing clothing down to younger siblings are good examples of this practice. It is the responsibility of the consumer to ensure their unwanted items are passed on to prevent them being disposed of prematurely.

Secondary recycling

Secondary recycling is where the materials from a product are used to make something new, for example cutting a pair of jeans to create shorts or cutting an old shirt into strips to be used as cleaning rags. This type of recycling is suitable for old clothing or household textiles that are worn or damaged and therefore cannot be given to charity shops or clothing banks.

Tertiary recycling

Tertiary recycling is when materials are broken down to their original state and made into brand new products. An example of this process is the use of plastic bottles to make polyester fleece. Polyester fibres are thermoplastic and can therefore be cleaned, shredded, melted down and re-formed into new products. Natural fibres can be shredded, bleached and spun into new yarns, or used for padding and insulation applications in the building industry.

Tertiary recycling is beneficial to the environment in many ways. The cleaning, sorting and processing of recycled fibres uses toxic chemicals and energy, however, which must be considered when choosing to use recycled materials.

It is estimated that 350,000 tonnes of used clothing go into landfill in the UK each year. Most of the textile products in landfill could have been used again. Fibres cannot withstand endless recycling, however, and so will eventually need go to landfill sites at the end of their useful life.

Figure 12.45 **Garments can be upcycled to make individual and interesting clothing.**

12.3 Commonly available forms

LEARNING OUTCOMES
By the end of this section you should know about and understand:
→ the range of stock forms available when selecting fabrics
→ the range of standardised components for use with fabrics.

This section will guide you through the stock forms and range of fabrics available and the standardised components used for a variety of textile products.

Stock forms

Fabrics are most commonly bought 'off the roll', in a specialist fabric store or market. There are three standard roll widths:

● 90cm (interfacings or linings)
● 115cm
● 150cm.

Most stores will sell fabric at a minimum length of 50cm. It is important to know how wide the fabric roll is before calculating the number of metres to buy for a specific project.

Figure 12.46 **Fabrics usually come from rolls of 115cm or 150cm.**

Figure 12.47 **Fat quarters are useful for smaller projects or quilting.**

Pre-cut cotton fabrics are also available for quilting or smaller projects. These are called fat quarters and usually come in bundles of complementary colours and patterns. These pieces measure 45 × 55cm. The term 'fat quarter' comes from the historical use of yards when buying fabric, and these pieces measure exactly a quarter of a yard. Felt can also be purchased in squares of various sizes for smaller craft projects.

Fabric weights, textures, handle and drape vary so much that it is vital to go to the store and handle the fabric before selecting the most appropriate one. Most stores will cut off a small sample for you to examine before you make a purchase.

Common fabric names

Fabrics in stores are not usually named just by the fibre content. Many fabrics are made with fibre blends. The table below lists some of the fabric types available.

Table 12.7 **Some of the fabric types available**

Fabric name	Typical fibres used	Example use
Drill	Cotton	Upholstery
Jersey	Cotton/ Wool/ Polyester	T-shirts
Denim	Cotton/ Elastane	Jeans
Voile	Cotton/ Silk	Mosquito nets
Tweed	Wool	Jackets
Gabardine	Wool	Suiting
Felt	Wool/ Acrylic	Applique- crafts
Broadcloth	Cotton	Dresses
Sheeting	Cotton/ Polyester	Bed sheets
Damask	Silk/ Cotton/ Polyester	Decorative napkins
Muslin	Cotton	Sheer curtains
Satin	Silk/ Polyester	Occasion dresses
Crepe	Silk/ Polyester/ Viscose	Blouses
Velvet	Cotton/ Polyester/ Viscose	Evening wear
Corduroy	Cotton/ Polyester	Trousers
Lace	Linen/ Cotton/ Silk	Decorative
Chiffon	Silk/ Polyester	Lingerie
Organza	Silk/ Polyester	Drapes

Figure 12.48 **Voile fabrics are sheer and very lightweight.**

Figure 12.49 **Tweed is a type of twill weave.**

Figure 12.50 **Lace is a very intricate and decorative fabric.**

Figure 12.51 **Various threads**

Standard components

Standard components in textiles are used to join, fasten, shape or stiffen fabrics.

Threads

Threads come in various types, weights and textures. It is important to select the right thread for the job.

- Machine thread – strong thread designed to move smoothly through sewing machine mechanisms.
- Tacking thread – strong, thick thread, designed for hand-sewing seams before machine sewing.
- Embroidery thread – thick, lustrous thread designed for decorative work.
- Monofilament – very strong, usually clear, thread, used for invisible stitching.

Fastenings

Fastenings are selected depending on their aesthetic, function and intended use. It is important to consider safety and ease of use when selecting fastenings, particularly when for children's products where they may be a choking hazard.

- Buttons – used on coats, trousers, shirts.
- Poppers – used on baby clothes, duvet covers, clothing.
- Zips – used on trousers, bags, cushion covers.
- Velcro – used on shoes, outdoor jackets, bags.
- Hooks and eyes – used on bras, dresses, skirts.
- Toggles – used on bags, coats, tents.

Figure 12.52 **Buttons come in an endless range of colours, shapes and materials.**

Figure 12.53 **Hook-and-eye fastenings are strong and can be hidden well.**

Figure 12.54 **Toggles are used to open, close or adjust sizing.**

Structural components

Structural components are used to add support to a fabric or to help shape a garment.

- Boning – a metal or plastic strip that is sewn into reinforced seams to add structure and shape to a garment, such as a corset.
- Petersham – a heavyweight ribbon or band used to stiffen waistbands or reinforce button bands.
- Interfacing – a light- to medium-weight, bonded fabric used to stiffen collars, cuffs or armholes.

Other

There are many other components used in the manufacture of textiles to provide functional or aesthetic effects.

- Bias binding – a strip of fabric, cut on the bias, used to create a finished edge around curves.
- Elastic – various weights, used for waistbands, underwear, swimwear.
- Ribbon, sequins and beads – applied to fabric surfaces for decorative effect.

Figure 12.55 **Bias binding is cut at a 45-degree angle to allow flexibility on curves.**

Figure 12.56 **There are many decorative components available.**

ACTIVITY

Study a range of textile products and fill out the table below.

Product Description	Components used	Purpose of component
Example: Rucksack	Webbing	To provide adjustable straps.
	Clips	To clip pockets shut
	Piping	To add aesthetic value and help maintain the shape of the rucksack.
	Zip	To securely close the rucksack

12.4 Manipulating and joining

LEARNING OUTCOMES

By the end of this section you should know about and understand the use of specialist techniques to shape and fabricate high-quality prototypes, with exemplification of the following processes:
→ Addition
→ Deforming and reforming.

Wastage

Wastage is the process of cutting away material to leave a desired shape.

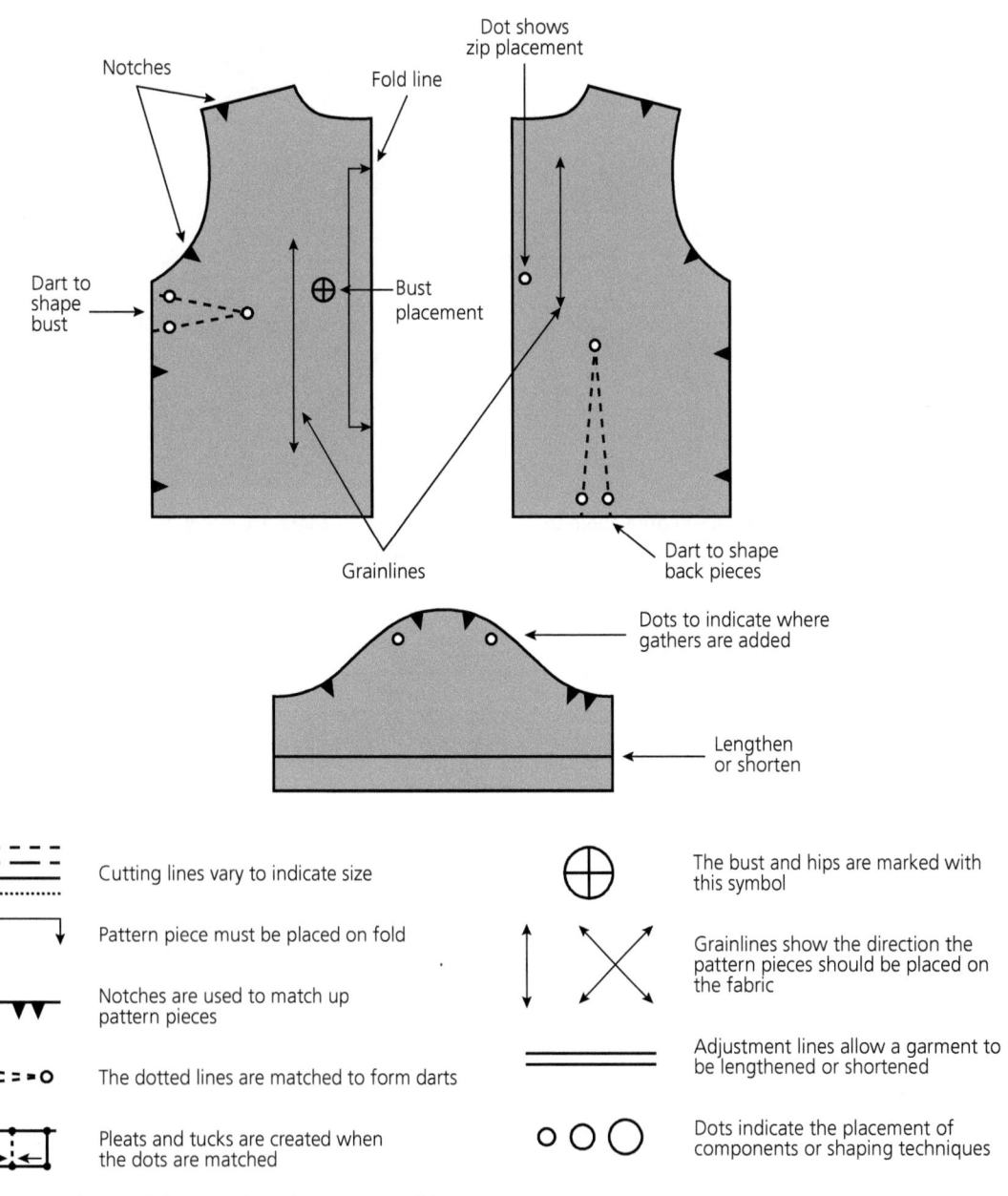

Figure 12.57 Pattern pieces with examples of pattern markings

Pattern marking

Before patterns can be cut, they need to be marked. Commercial pattern pieces and basic blocks come printed with a range of important markings that must be followed accurately to ensure that the finished product is the correct size, shape and quality.

Table 12.8 **Pattern markings that must be followed accurately**

Cutting lines	Lines are patterned to guide the cutting of different-sized pattern pieces.
Fold line	Patterns with this marking should be placed on the edge of a folded fabric. This will give a symmetrical pattern piece when cut.
Balance marks (notches)	These markings are used to match up different pattern pieces, to ensure accuracy. These are cut away from the pattern piece, not into the seam allowance.
Darts	Darts must be accurately marked, then the outer lines brought together and sewn to create shape in a garment.
Pleats and tucks	Pleats and tucks are marked in the same way; the arrows show the direction in which the fabric must be sewn.
Bust/hip placement	This mark helps with fittings; you should copy this to your fabric and check that it aligns with your model's bust or hips, then adjust the pattern to suit them.
Grain line	This shows the direction of the straight grain; pattern pieces must always be placed in the direction of this arrow.
Adjustment lines	These lines indicate where the pattern can be lengthened or shortened.
Dots	These marks show the placement of specific components such as pockets and buttons.

There are several tools that can be used for transferring pattern markings to fabric accurately:

- Tailor's chalk – a small piece of hard chalk, used to temporarily mark fabric.
- Vanishing markers – pens containing vanishing ink that slowly fades on exposure to air, allowing marks to be temporarily made to fabric.

Figure 12.58 **Tailor's chalk**

Figure 12.59 **Vanishing marker**

- Tracing wheel and carbon transfer – in this method of marking out pattern pieces, carbon paper is laid on top of the fabric, with pattern pieces on top. The carbon is transferred by the pressure applied when the tracing wheel is rolled along the surface of the pattern piece.
- Tailor's tacks – this method of marking out uses stitches, instead of marking the surface of the fabric with a pen or chalk. Small loops, which are later removed, are made through the fabric.

Figure 12.60 **Tracing wheel and carbon transfer**

Figure 12.61 **Tailor's tacks**

Pattern cutting

Pattern cutting involves the use of paper or card templates, placed on the surface of the fabric and cut around to produce fabric pieces that are of accurate size and shape for the product being made.

- Commercial patterns contain templates that are printed on large sheets of pattern paper. These often include several sizes and variations.
- Basic blocks are usually made from thin card and are a basic template that can be adapted to produce a range of products.

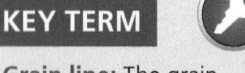

KEY TERM

Grain line: The grain line always follows the direction of the warp yarn, and shows where the pieces should be placed on the fabric prior to cutting.

Commercial pattern pieces and basic blocks are marked with an arrow to show the direction of the **grain line**, and therefore how the pieces should be placed on the fabric for cutting. It is important that these markings are used correctly to ensure that the drape of the fabric is appropriate for the product being made. The straight grain runs vertically down the warp yarn in the fabric, and this grain line is most commonly used in garment production as it gives the best drape. A pattern can also be cut on the bias, which runs diagonally at a 45-degree angle to the straight grain. Fabrics cut on the bias are more flexible and can easily be sewn into curved shapes. Cutting on the bias often causes more wastage, so designers need to purchase more fabric when cutting patterns this way.

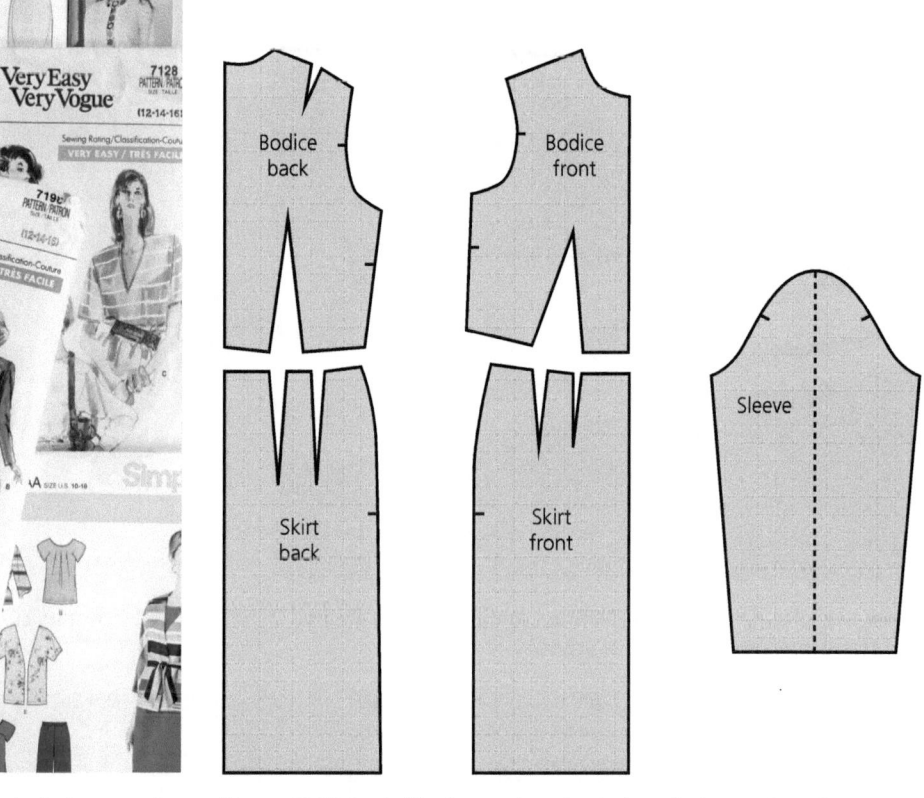

Figure 12.62 Commercial patterns include several styles and sizes.

Figure 12.63 Basic blocks can be adapted easily to create pattern pieces to suit your design.

There are several tools/techniques that can be used for cutting fabric accurately:

- Fabric shears – these are the most commonly used and easiest method for cutting fabric.
- Rotary cutter – these incredibly sharp tools are rolled along the surface of the fabric, following the pattern lines. The surface underneath must be protected with a cutting mat to minimise damage.
- Laser cutting – this can be a quick and efficient way to cut out more complicated pattern pieces if you have access to a laser cutter in school.
- Band saw – this method is used in an industrial setting to cut up to 100 layers at fabric at one time.

Figure 12.64 Fabric shears are the most commonly used method for pattern cutting at home or at school.

Figure 12.65 Rotary cutters are very accurate.

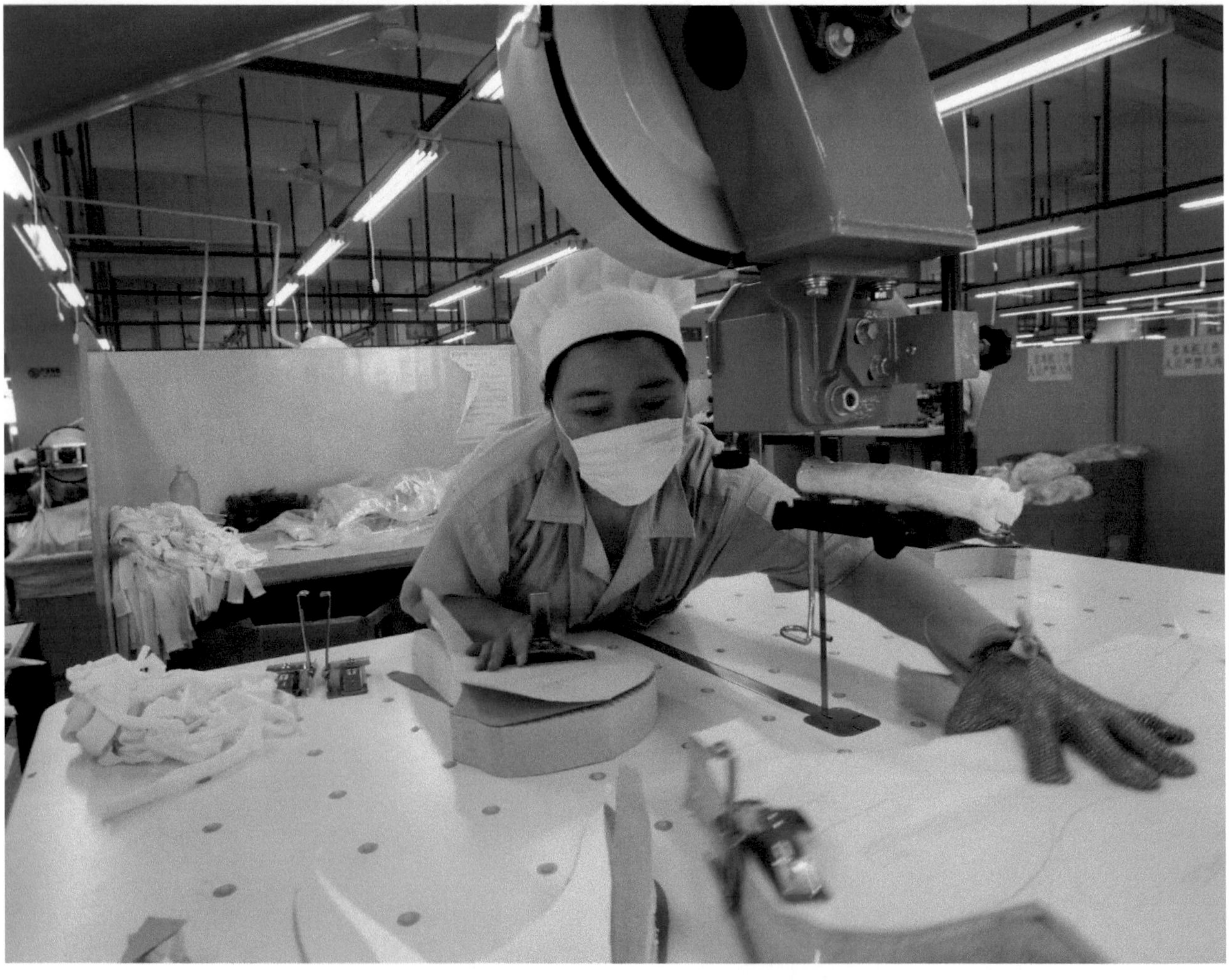

Figure 12.66 Band saws are used in industrial manufacture to cut through several layers of fabric at once.

Addition

This section focuses on the formation of seams that are suitable methods of joining for different fabrics and products. It is important that the correct type of seam is used to ensure the product is high quality and functions as expected.

Plain seams

Plain seams are the most commonly used seam and can be finished neatly using a range of methods, for example with an overlocker, pinking shears or bias binding. Plain seams are suitable for most fabric types.

French seams

French seams are enclosed, hiding any raw edges. This type of seam is suitable for sheer materials, like chiffon, where the seams need to be almost invisible.

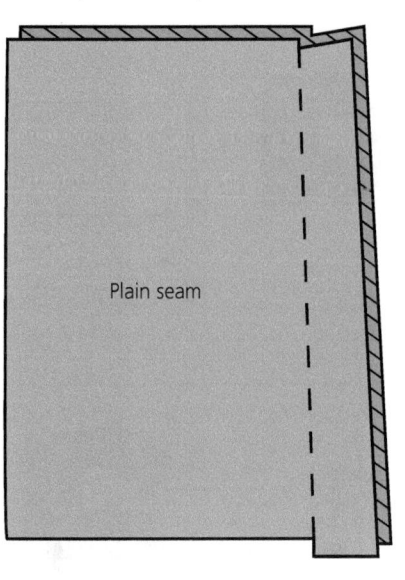

Plain seam

Figure 12.67 **Plain seam**

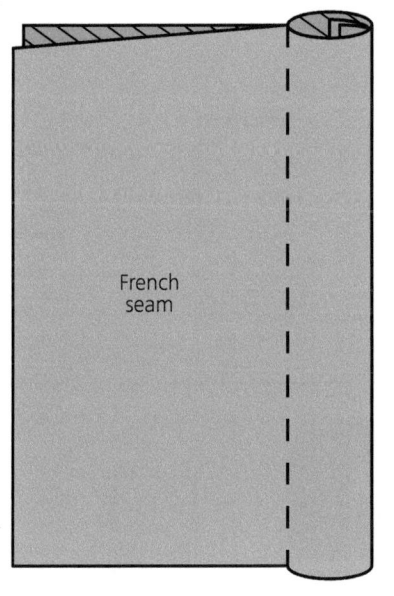

French seam

Figure 12.68 **French seam**

Flat felled seams

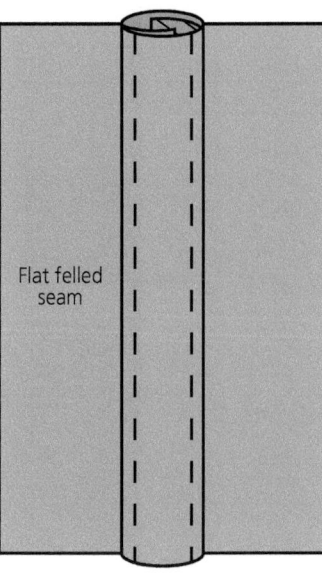

Flat felled seam

Figure 12.69 **Flat felled seam**

Flat felled seams are very strong, enclosed seams. This type of seam is commonly used on denim and sportswear.

KEY TERMS

Plain seam: The most common type of seam formed by placing fabrics right sides together and stitched along the seam allowance.

French seam: A seam used for sheer fabrics where seam finishes such as overlocking may be unsightly. This seam is sewn twice, placing wrong sides together first, then right sides together to enclose the raw edge.

KEY TERM

Flat felled seam: A flat felled seam is an overlapped seam that is topstitched down to hide raw edges and add strength to the seam. This seam is commonly used for denim jeans.

Figure 12.70 **Overlocking encloses the raw edge with stitching.**

ACTIVITY

You will need:
- Fabric shears
- Needle
- Thread
- Sewing machine
- Iron
- Small squares of fabric (approx. 10 × 10cm).

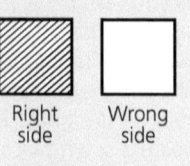

Right side   Wrong side

Plain seam

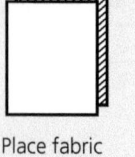 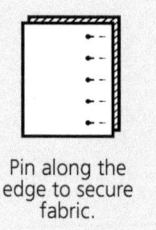 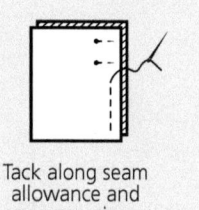 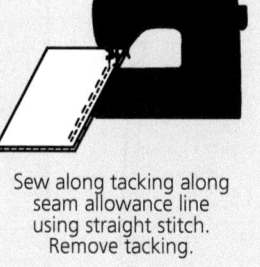 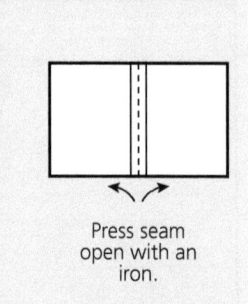

| Place fabric right sides together. | Pin along the edge to secure fabric. | Tack along seam allowance and remove pins. | Sew along tacking along seam allowance line using straight stitch. Remove tacking. | Press seam open with an iron. |

French seam

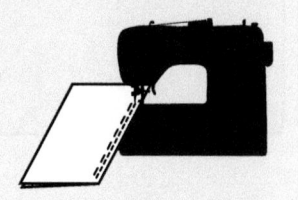

| Place fabric wrong sides together, pin and tack seam allowance line. Remove pins. | Sew along tacking line. Remove tacking. | Trim seam allowance using fabric shears. | Turn fabric right sides together. Pin and tack along seam allowance. Remove pins. | Sew along tacking line, ensuring the raw edges of the original seam are fully enclosed. Remove tacking. |

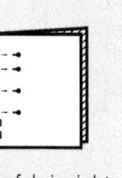 

Flat felled seam

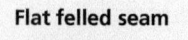 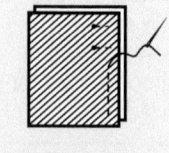 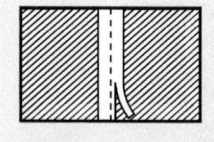 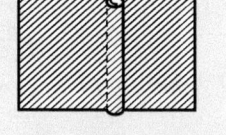 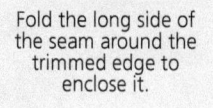

| Place fabric wrong sides together, pin and tack seam allowance line. Remove pins. Sew, using straight stitch. | Trim half of the seam allowance from one side of the seam. | Fold the long side of the seam around the trimmed edge to enclose it. | Press seam flat then pin, tack and sew the opposite edge of the seam to finish it. |

1 Follow the step-by-step diagrams to complete a sample of each of the seams above.
2 Mount each sample and add notes to explain the best uses for each one.
3 Use this page of samples to help you throughout your GCSE course.

Finishing seams

The aim of finishing seams is to neaten them and to prevent fraying. Common methods for finishing seams include:

- Overlocking – the overlocker trims off excess fabric along the seam and stitches around the raw edge, giving a neat and professional finish.
- Pinking shears – these shears produce a cut zig-zag shape along the raw edge of a seam, which prevents fraying.
- Zig zag – if you don't have access to an overlocker, machine sewing a zig-zag stitch along the raw edge of a seam can be just as effective.

KEY TERM

Pleat: A pleat is formed by folding fabric and sewing the fold flat. Pleats add shape and body to garments.

Figure 12.71 **Pinking shears prevent fraying.**

Figure 12.72 **Seam finished with a zig zag**

Deforming and reforming

Fabrics by nature will lie fairly flat, and even knitted and non-woven fabrics will be difficult to shape. The methods in this section are used to add shape or body to fabrics in order to add interest and to improve fit and function.

Pleats

Pleats are formed by folding the fabric back on itself and sewing it into place. Pleating narrows the original width of the material quite significantly, while adding shape or body.

Example pleats

Figure 12.73 **Pleats**

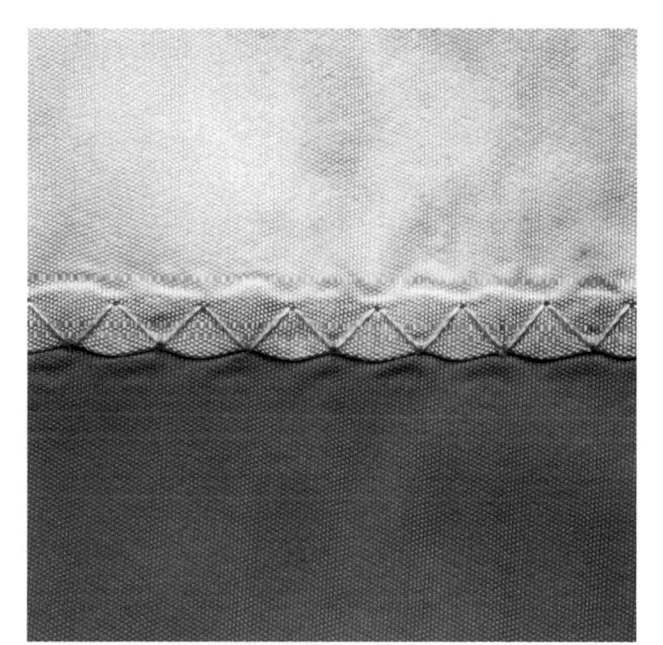

Figure 12.74 **Pleats are often used in skirts.**

Gathers

KEY TERMS

Gather: Gathers are created by sewing along the length of the fabric and pulling the sewing thread, causing the fabric to gather up along its length. Gathers are used to add shape and body to garments.

Dart: Darts are formed by folding and sewing triangular sections of fabric. Darts allow us to shape garments to better fit the body.

Gathers are formed by sewing along the edge of the fabric with a long stitch length. Two rows of stitching are usually used in this process. The thread ends, or tails, are then pulled to create the gathers. This technique narrows the original width of the material and gives fullness to the garment.

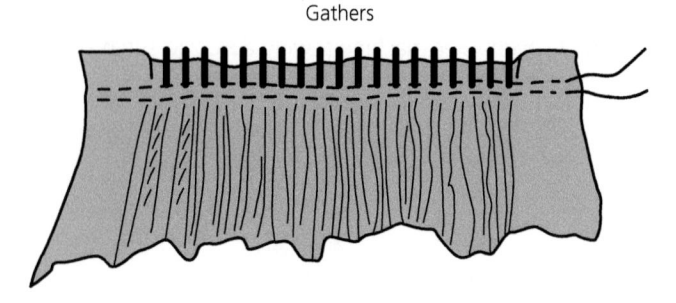

Gathers

Figure 12.75 **Gathers are an easy way to add body and shape to a product.**

Darts

Darts are used to shape fabrics, improving the fit. They are made by creating folds in the fabric that taper to a point. These folds are often sewn into the bust or the back of bodices.

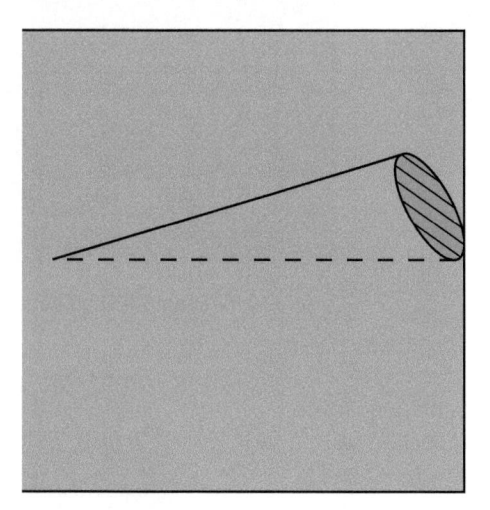

Figure 12.76 **Darts help shape the fabric to fit properly.**

ACTIVITY

You will need:
- Small samples of woven fabric (approx. 20 × 20cm)
- Fabric shears
- Needle and thread
- Sewing machine
- Tailor's chalk.
1 Create a sample of each method of deforming and reforming.
2 Mount each sample and describe how the method has changed the shape of the fabric.

Clipping seams

When a seam is sewn along a curve, particularly with woven fabrics, it will need to be clipped to allow the fabric to lie flat. Clipping involves the use of shears or specialist tools to cut into the seam allowance. This technique is commonly used around necklines and armholes to achieve a neat finish.

12.5 Structural integrity

LEARNING OUTCOMES

By the end of this section you should know about and understand:
→ how and why specific materials need to be reinforced
→ the processes that can be used to ensure structural integrity in fabrics.

KEY TERM

Boning: the process of sewing plastic or metal boning strips into a garment to reinforce seams, support the fabric and prevent it from creasing and buckling.

Most fabrics have natural drape, which is beneficial to designers in many ways. Sometimes, however, we need to add structure to textile products to improve aesthetics and function.

Boning

Boning is a technique that dates back to sixteenth century corsetry, and is named after the whalebones that were originally sewn into garments. Corsets were used for aesthetic or medical purposes. For example, women's corsets typically cinched the waist and exaggerated the bust and hips, giving a more desirable figure – this practice caused many health issues, however, including breathing problems. Medical corsets treat back pain or protect people with spinal or internal injuries by restricting their movement.

Modern-day corsetry uses plastic or metal boning strips, which are sewn into reinforced seams to support the fabric, preventing creasing and buckling.

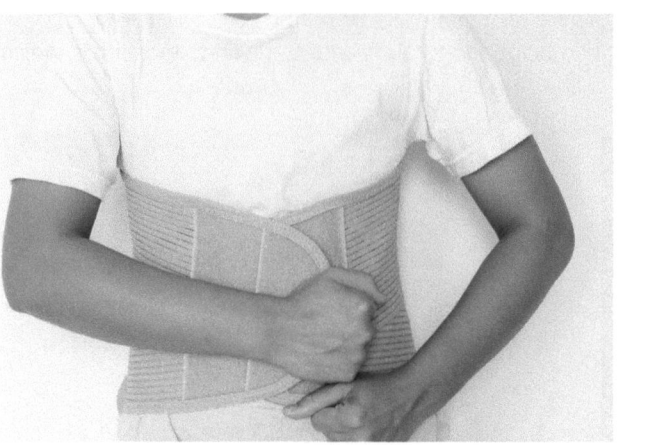

Figure 12.77 A modern torso support, inspired by original medical corsets

Figure 12.78 **Boning is sewn into the seams of the corset to add structure.**

Layering textiles

Layering in textiles can be used for a number of purposes, including structural support, insulation, comfort or to add body.

Interfacing

Interfacing is a non-woven, bonded fabric that can be sewn or ironed on to the outer fabric or lining. It comes in various weights to provide light support or to stiffen fabrics to improve shape. It is most commonly used around necklines, armholes, waistbands, collars and cuffs.

Interlining

Interlining is a layer of fabric added between the fabric and the lining of a garment, most commonly to add a layer of insulation. This is often used in suit jackets and winter coats to provide additional warmth. This layer of fabric can also support the structure of a garment.

KEY TERMS

Interfacing: a non-woven, bonded fabric sewn or ironed onto the outer fabric or lining to provide light support and stiffen fabrics.

Interlining: a layer of fabric added between the fabric and lining of a garment to add insulation and provide warmth.

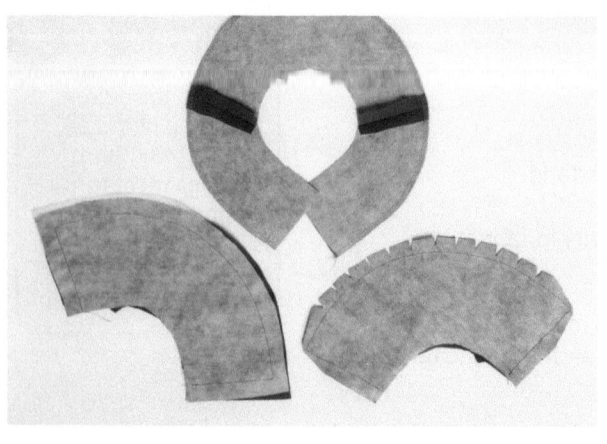

Figure 12.79 **Interfacing can be ironed on (fusible) or sewn into place.**

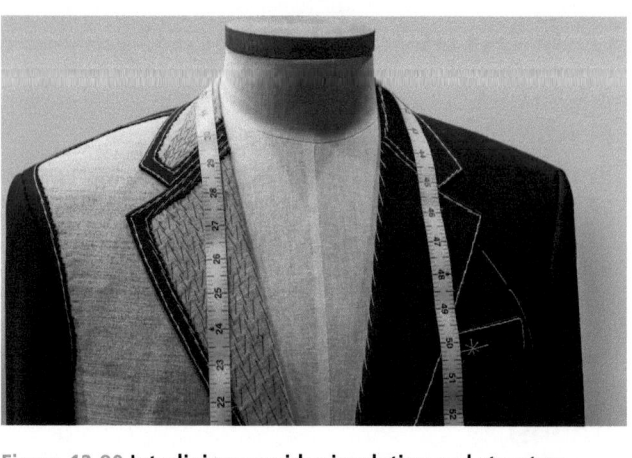

Figure 12.80 **Interlining provides insulation and structure.**

KEY TERMS

Lining: a lightweight fabric used inside a garment to improve comfort and to hide seams.

Underlining: a layer of fabric used underneath sheer fabric to provide opacity or to add extra body to a garment.

Lining

Lining is a lightweight, usually silky, fabric used inside a garment to improve comfort and to hide seams and other construction methods. The lining can also be a design feature – bright or patterned linings are often used inside plain suit jackets.

Underlining

Underlining is a layer of fabric used beneath sheer fabric to provide opacity or to add extra body to a garment. Dresses with lace or chiffon skirts will have an underlining to prevent the skirt being see-through and to add lift to the fabrics, which would otherwise fall flat.

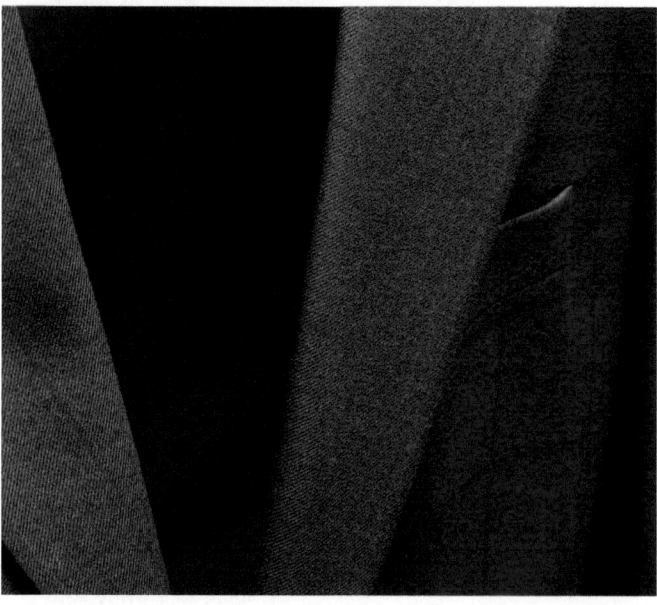

Figure 12.81 **Lining can hide seams, improve comfort and become a design feature**

ACTIVITY

Look through your wardrobe and find products that use each of these types of linings then photograph them and print them to create a revision sheet.

Figure 12.82 **Underlining can add body or provide opacity to sheer fabrics.**

12.6 Making iterative models

LEARNING OUTCOMES

By the end of this section you should know about and understand:
→ the processes and techniques used to produce toiles.
→ How materials, components and manufacturing methods are tested and selected.

Making iterative models is a key part of the design process and allows you to test, manipulate and adapt your design ideas. Realising your designs in 3D form will improve your understanding of the construction methods required to make them, and you will be able to evaluate a range of production techniques.

Toile is a French word meaning fabric, specifically a linen canvas-style material, originally used to make mock-ups of garments and refine them before using more expensive fabrics for the finished product. Over the years, the process of making a garment prototype has adopted the word and it is now referred to as 'making a toile'.

Production of toiles

Toiles are traditionally made in cotton calico fabric, but can also be made from other cheap fabrics or even paper. A basic toile can be made using pattern blocks or by following a simple commercial pattern.

Figure 12.83 A toile made in calico

Figure 12.84 A toile made in paper

ACTIVITY

In a group make a simple calico top, then develop the design by creating several iterations over the next couple of lessons.

Toiles are produced using the same construction methods that will be used on the final product, rather than just pinning and tacking the fabric together. Seams and shaping methods are completed using a long stitch length on the sewing machine, to allow for unpicking and adaptation.

Testing manufacturing methods

When developing a product, the manufacturing methods must be tested to ensure that they will function effectively and fit with the aesthetic of the design. Different types of seam, shaping and decorative methods can be tested by creating small samples which are put to use in a range of contexts to aid decision making. Examples of testing include applying tension to seams to determine the point at which they will split and experimenting with mixing dyes to find the correct colour palette for screen printing.

Testing materials and components

Product development also relies on the use of suitable materials and components to achieve the right function and aesthetic. A designer will collect and test swatches of materials and a range of components to determine the right combination for the end use. Tests include; abrasion resistance, stain resistance, absorbency, strength and insulation. Components are also tested and compared for their size, weight, function and aesthetic.

Design iterations

Once complete, the basic toile is fitted to a model or dressmaker's dummy. It is at this stage that testing and adaptation begins. A designer will draw directly on to the toile to show intended changes in size, shape or structure, then use additional fabrics and techniques to develop the design of the toile. These techniques could include cutting sections away, reshaping or adding sleeves, or creating shape by adding pleats or gathers. Photographs will be taken at each stage to show development, and the designer will often trial several adaptations to explore a range of options for their product.

Results from testing of materials, components and construction methods are used to inform design adaptations. Opinions are also sought from stakeholders as the product is developed to ensure that the product continues to meet their needs. Following this iteration of design, the toile or model can be deconstructed and the pieces used to draft new pattern pieces for the final product.

12.7 Finishes

LEARNING OUTCOMES

By the end of this section you should know about and understand:
→ the processes used for finishing and adding surface treatments to materials for specific purposes.

During the production of fabrics, finishes can be added to improve the aesthetics, comfort or function. These finishes are applied mechanically, chemically or biologically.

Dyeing

The most common method of adding colour to fibres and fabrics is dyeing. Manufacturers can use natural or synthetic dyes, depending on the type of fibre to be dyed. Natural dyes work well on natural fibres and can be made from plants, minerals or insects. Synthetic dyes work on synthetic fibres, but can also give deeper or brighter colours when used with natural fibres. Synthetic fibres require the use of chemicals to enable them to take up the dye.

Fibres and fabrics can be dyed at various stages in the production process:

- **Polymer stage** – this is when the liquid polymer solution is coloured before the extrusion process, which prevents the colour from fading or transferring to other fabrics when washed.
- **Fibre stage** – this is when the raw fibre is dyed before fabric construction to achieve consistent colour.
- **Yarn stage** – yarn dyeing will not give consistent colour in a woven or knitted fabric, but can be useful for creating stripes in woven fabrics like tartan.
- **Piece dyeing** – this is when an entire length of fabric is dyed. This method can produce inconsistencies in colour.
- **Garment dyeing** – this is when a completed garment is dyed. This can be beneficial in t-shirt production, where identical products can be dyed in a range of colours once they have been made.

Mechanical finishes

Brushing

Brushing involves passing wire brushes over the surface of the fabric to raise the fibres to produce a soft, fluffy surface. Brushing is commonly used on fleece and flannel fabrics, but the process can weaken the structure of the weave.

Calendering

Calendering involves pressing the fabric using heated rollers to give it a smoother, more lustrous surface. This is often used on upholstery fabrics to give them a flat surface and sheen.

> **KEY TERMS**
>
> **Brushing:** passing wire brushes over the surface of the fabric to raise the fibres, producing a soft, fluffy surface.
>
> **Calendering:** pressing a fabric using heated rollers to give it a smoother, more lustrous surface.

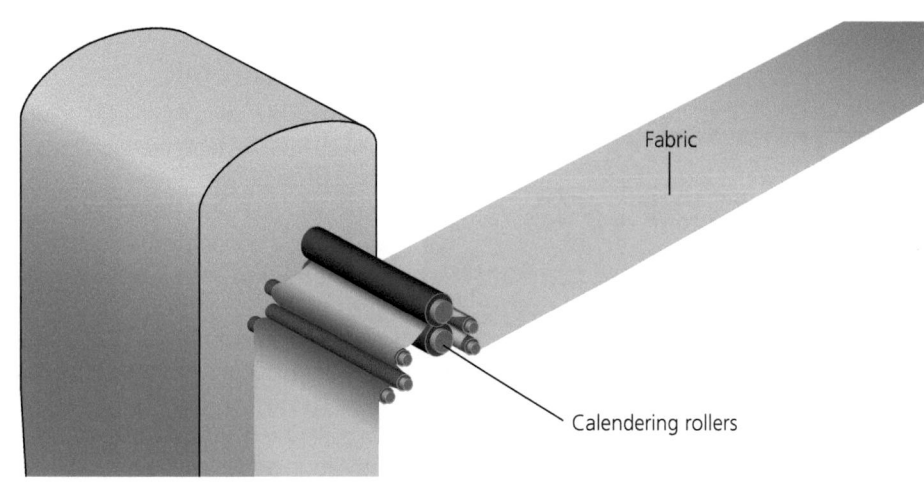

Figure 12.85 **Brushing provides both comfort and warmth.**

Fabric

Calendering rollers

Figure 12.86 **Calendering involves passing the fabric through heated rollers.**

271

KEY TERMS

Mercerising: the process of using caustic soda to cause cellulose fibres in a fabric to swell up, giving a more lustrous, stronger fabric.

Bleaching: the process of using chemicals to make fabrics much lighter or white.

Stonewashing: the process of using large stones to give a newly manufactured cloth a worn appearance.

Thermochromic: products that change colour with changes in temperature.

Chemical finishes

Mercerising

Mercerising uses caustic soda to cause the fibres in the fabric to swell up. The result is a more lustrous, stronger fabric, but it only works for cellulose fibres. It improves the uptake of dye, giving a deeper and more even colour.

Crease-resistance

Crease-resistant finishes are applied in the form of a resin coating. This resin reduces absorbency and stiffens the fibres so the fabric is easier to care for, reducing the need for ironing. Treated fabrics will also dry more quickly.

Flame-resistance

Flame-resistant finishes such as Proban® are applied to the surface of fabrics as a liquid coating; when dry this is durable and long-lasting. These types of finishes are applied to products in high-risk areas such as soft furnishings in hotels, and in public areas such as stage curtains in theatres. They are also widely used in children's sleepwear, bedding and other home furnishings.

Bleaching

Bleaching is the most common chemical finish. It removes any natural colour and prepares fabric well for dyeing and printing. The result of this finish is an even, consistent colour.

Figure 12.87 **Stonewashing gives a faded effect on fabrics like denim.**

Biological finishes

Stonewashing

Stonewashing is most commonly used in the production of denim jeans, giving a popular distressed look. Manufacturers achieve this effect by adding stones to industrial washing machines, along with the jeans. The result is a faded, 'worn out' pair of jeans.

Smart finishes

There are a number of smart finishes used on textiles. These are finishes that can detect and react to changes (for example heat, light or friction) in their environment.

Thermochromic

Thermochromic inks change colour with changes in temperature. These inks can be printed on to the surface of fabrics and have several useful applications, including the use of thermochromic imagery on baby grows. This can alert parents to any changes in body temperature by sight, allowing a quicker response should the baby have a fever. This technology has also been used on cutlery and drinks bottles, to indicate if the food or drink is too hot.

Photochromic

Photochromic inks change colour with changes in natural light. These inks are printed on to the surface of fabrics and come in a range of bright colours. Photochromic technology is commonly used on children's summer clothing to alert parents to move their child to shade, or on hotel sunbeds to alert guests to the most dangerous time of the day for sun exposure.

Microencapsulation

Microencapsulation uses tiny bubbles or capsules that contain chemicals, and these are added to the weave or fibre of fabric. The bubbles burst with friction, releasing the chemical inside. There are many useful applications for this technology, and plenty are already in use in everyday products. Examples of applications of this technology include mosquito-repellent hiking gear, perfumed fabric, moisturising fabric, hypoallergenic fabric and antibacterial fabric (Microban®).

Surface decoration

There are a number of surface decoration techniques that can be used to improve the aesthetics of a product, by adding colour, texture and pattern.
- Appliqué
- Embellishment
- Embroidery
- Patchwork
- Quilting
- Laser etching.

KEY TERMS

Microencapsulation: the process of adding tiny bubbles or capsules that contain chemicals to the weave or fibre of fabric. These burst with friction, releasing the chemical inside.

Photochromic: products that change colour with changes in natural light.

Figure 12.88 **Tiny capsules of odour-controlling chemicals can be added to every day products, such as shoes.**

ACTIVITY

Suggest a suitable product for each decorative method.

Applique	
Embellishment	
Embroidery	
Patchwork	
Quilting	
Laser Etching	
Applique	

12.8 Using digital design tools

LEARNING OUTCOMES

By the end of this section you should know about and understand:
→ the use of 2D and 3D digital technology and tools to present, model, design and manufacture solutions.

Advances in technology mean that many traditional design and manufacturing methods for textiles can now be carried out using specialist software and automated machinery.

Rapid prototyping

The use of rapid prototyping in textiles is still in the early stages of development. This process uses laser cutting to precisely shape individual fabric layers, which are then bonded together to create 3D forms. The development of this kind of technology could see 3D-printed clothing in the future.

Digital manufacture

Digital fabric printing

Digital fabric printing uses a large-scale inkjet printer and specialist dyes to transfer a digital image to the surface of the fabric. This type of printing allows designers to use intricate and detailed images.

Sublimation printing

Sublimation printing uses heat and pressure to transfer dye from specialist printer paper on to the fabric. This is a particularly effective method on fabric, as the process turns the dye into a gas that binds directly to the fibres, leaving a crisp design that is washable.

Figure 12.89 **Digital fabric printers can quickly print imagery on a whole roll of fabric.**

Laser cutting

Laser cutting is controlled by a computer program in which the design is drawn up as a 2D image. The laser strength and speed is set depending on the material to be cut. Once the fabric is placed into the cutter, the laser follows the digital design to quickly and accurately cut the design. Laser cutters can also be set up to etch the surface of a fabric, rather than cut all the way through.

Interpretation of plans

Digital lay planning

Software is available to assist pattern cutters with the creation of digital **lay plans**. These plans show the manufacturer where to place each pattern piece on the fabric so that they follow the grain line and minimise wastage. Digital lay planning software can adapt to different garment sizes and fabric widths. The finalised plans are usually printed on large sheets of pattern paper.

Computer-aided design

There are a number of applications that can be used for image creation and manipulation when designing textile products. Examples include Adobe Illustrator, CorelDraw and Digital Fashion Pro. Hand-drawn sketches can be transferred on screen, where the dimensions, shape and form can be refined and perfected. Various software programs have been developed to give designers a 3D view of their original 2D sketch, allowing them to evaluate the success of a design before **prototyping**. Many of these programs can be linked to company stock data, allowing designers to choose fabrics that are immediately available to them. Development of surface designs for printing can also be completed on similar software and transferred to relevant CAM systems (see below).

Computer-aided manufacturing

Semi- or fully automated machinery is used widely in the manufacture of textiles – this is known as CAM. These systems are expensive, but they speed up manufacturing, improve consistency and reduce the risk of human error.

Figure 12.91 **Computerised embroidery machines complete several designs at once**

Figure 12.90 **Laser cutters can be used to cut out intricate designs in fabric.**

KEY TERMS

Lay plans: Used by pattern cutters like a map to guide them when placing pattern pieces in the correct location and direction before cutting. Lay plans are carefully designed to minimise wastage.

Prototyping: A process that involves the production of a test model, on which the final product is based.

ACTIVITY

Study your school uniform. Make a list of the construction methods used, then suggest the machinery types that might have been used in manufacturing.

KEY TERMS

Bespoke production: Manufacture of 'one off' products that are designed and made for a specific client by an individual or small team of highly skilled workers. Bespoke products are high quality, can be complex and are expensive to make. Examples of bespoke products include wedding dresses, tailored suits and custom fit car seat covers.

Batch production: Batch produced products are manufactured by a large team of workers who each complete a specific stage of the production. These products are usually consistent in quality, available in a range of styles and sizes and fall into the mid to low price range. Typical products include summer dresses, fashion t-shirts and branded school bags.

Mass production: Mass produced products are manufactured mostly on automated machinery, operated by teams of workers. Products are manufactured very quickly, are consistent in quality and the range of available styles is minimal. Mass production costs are much lower than other production methods. Examples of mass produced products include plain socks, plain t-shirts and plain baseball caps.

12.9 Manufacturing methods and scales of production

LEARNING OUTCOMES

By the end of this section you should know about and understand:
→ the methods used for manufacturing at different scales of production
→ the manufacturing processes used for larger scales of production
→ the methods of ensuring accuracy and efficiency when manufacturing at larger scales.

Manufacturing methods vary depending on the production run (the number of items being produced), timescale and budget. The scale of production affects the quality and cost of textile products.

Scales of production

One-off, bespoke production

One-off, or **bespoke**, products are made by highly skilled workers – often an individual or a small team. Products made in this way take a long time to produce, as most of the work is done by hand. The consumer usually has opportunities to attend fittings and make design decisions during the manufacturing process. This type of product is often very high quality, and therefore expensive to purchase.

Batch production

Batch-produced products are made by large teams of workers, working at various stages around the factory. This type of production utilises a mix of semi-automated machinery and hand assembly. Workers are specialised in one element of the construction process, such as collars or hems. Each employee works through a batch of partial products, which are then passed around the production line until they are complete. Batch-produced products are usually of mid- to low quality. Seasonal clothing, such as summer dresses and winter coats, is produced using batch-production methods.

Mass production

Mass production is the largest scale of production available. This method is used for products that are in consistently high demand, such as socks and plain t-shirts. It is because of this demand that many factories run 24 hours per day, in order to maximise output and profit. CAM is used widely in mass production, as consistency and speed are so important. Quality is controlled via computer, so instances of faulty products are low and products are consistent.

Figure 12.92 **Wedding dresses are often bespoke.**

Figure 12.93 **Winter coats are manufactured on batch production lines.**

Figure 12.94 **Everyday items like plain t-shirts are mass-produced.**

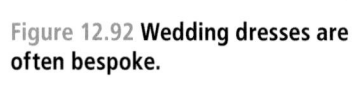

Lean manufacturing and just-in-time methods

These methods of production are used to minimise waste and increase the overall efficiency of a manufacturing system.

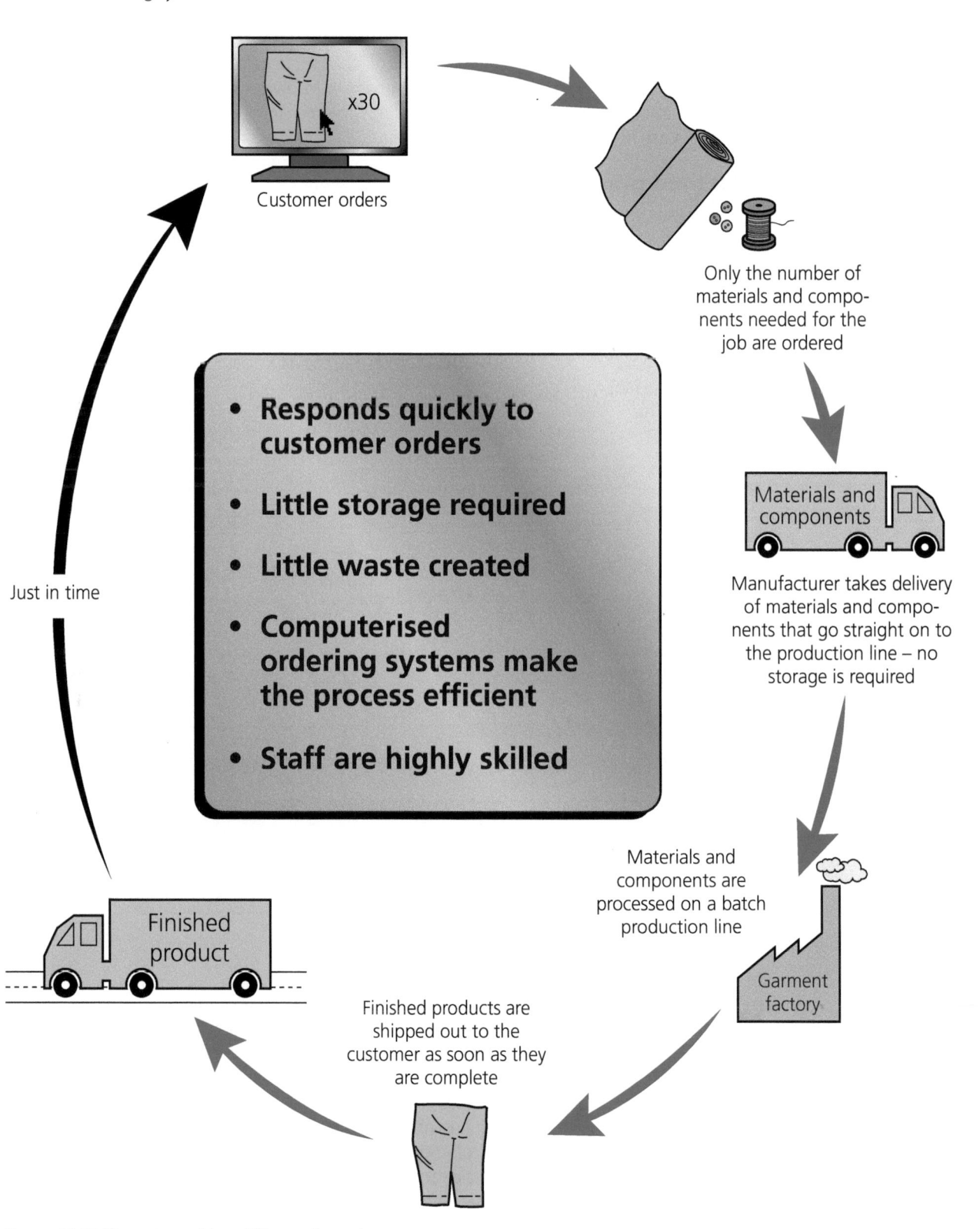

Customer orders

Only the number of materials and components needed for the job are ordered

Materials and components

Manufacturer takes delivery of materials and components that go straight on to the production line – no storage is required

- **Responds quickly to customer orders**
- **Little storage required**
- **Little waste created**
- **Computerised ordering systems make the process efficient**
- **Staff are highly skilled**

Just in time

Materials and components are processed on a batch production line

Garment factory

Finished product

Finished products are shipped out to the customer as soon as they are complete

Figure 12.95 **The stages of lean/JIT manufacturing**

Lean manufacturing:
A manufacturing method designed to minimise waste at each stage of production.

Just in time manufacturing:
A manufacturing method that can quickly respond to changes in trends by ordering materials and component to arrive at the factory 'just in time' for production. This method minimises storage space, wasted materials and left over stock, whilst increasing efficiency.

Manufacturing processes used for larger scales of production

Band saw cutting

For pattern cutting on a large scale, highly skilled workers are employed to use handheld band saws, cutting through up to 100 layers of fabric accurately and quickly. They use card templates to mark out the pattern pieces with chalk, following a digital lay plan to reduce waste. Cutting is quick and efficient, and workers wear chainmail gloves to protect their hands from the sharp blades.

Flatbed printing

Flatbed screen printing is carried out on a conveyor production line. Silk screens as wide as the fabric width are made up and attached to the printing machine in a row. Up to 16 screens can be placed on a single printing run. Each screen applies a different colour and design to the fabric. A conveyor belt moves the white fabric underneath the screens at regular intervals. When the belt stops the screens drop down on to the fabric and an automated squeegee drags printing medium across the screen. The screens then lift up and the fabric moves on. At the end of the printing run, the fabric has a complete design on its surface. The fabric will go on to be fixed, washed and pressed.

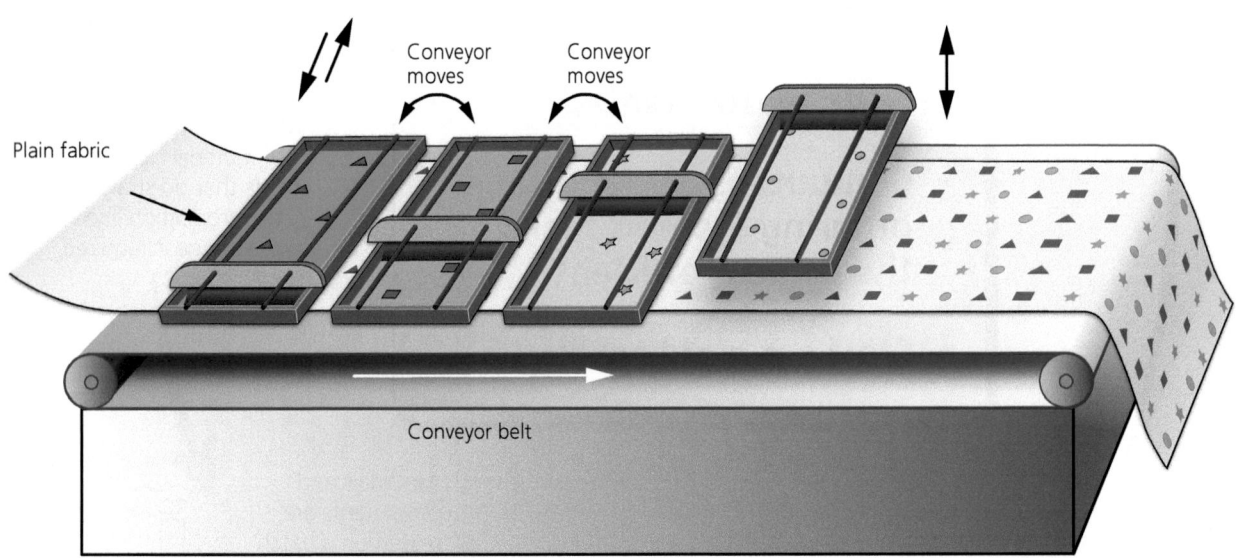

Figure 12.96 **The flatbed printing process**

Rotary screen printing

Similar to flatbed screen printing, rotary screen printing is carried out on a conveyor-driven production line. These production lines are much faster, however, and require less space on the factory floor. This is because this printing process uses printing cylinders instead of flat screens. The fabric design is created on metal sheets, which are then formed into a cylinder. These are mounted on to the printing machine, where the cylinder is filled with printing medium and a static squeegee is placed inside. The cylinders spin as the fabric passes underneath, printing a continuous pattern on to the surface. Each cylinder applies a single colour and design. The continuous motion of this process means printing is much faster and more efficient than flatbed printing. Rotary printing is an expensive option, however, due to the creation of the cylinders.

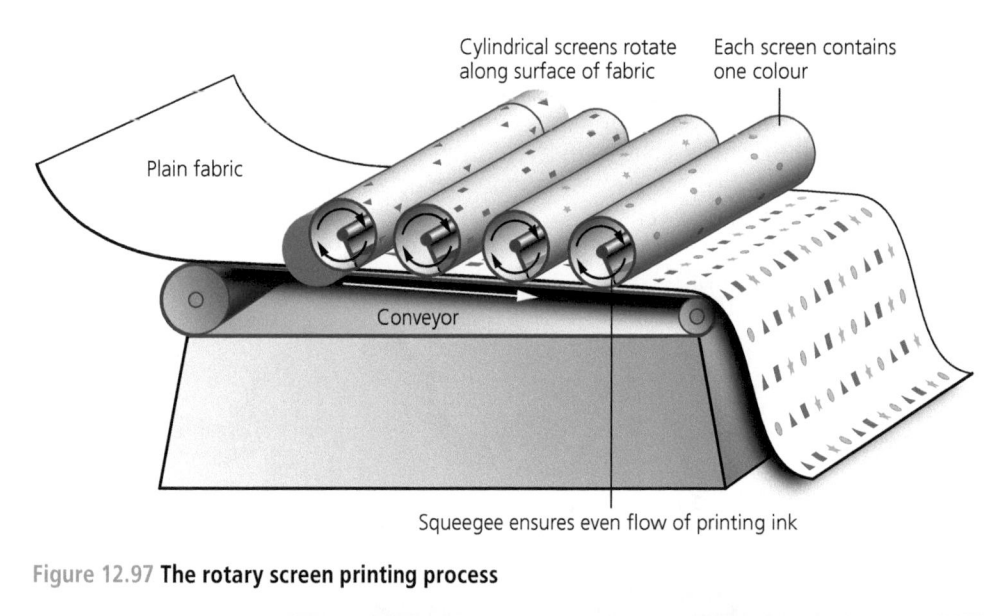

Figure 12.97 **The rotary screen printing process**

Figure 12.98 **Screen printed fabrics are very common and often have a repeat pattern.**

Industrial sewing machines and overlockers

Industrial sewing machines and overlockers are heavy-duty machines, specifically designed to withstand constant use. These machines have large motors that run much faster than domestic machines. They can also take large spools of strong thread to minimise snapping and time lost re-threading machines. These machines are able to sew with ease through heavy or tough materials such as leather and denim.

Automated presses

These presses are used to quickly remove creases and to finish completed garments before packaging. Workers place garments between the pressing plates and use a foot control to start the process. The pressing plates come together, applying pressure, steam and heat to the garment. This process takes only a few seconds per garment, making it much faster and more efficient than traditional ironing.

Figure 12.99 **A worker operates an industrial sewing machine**

Steam dollies

Steam dollies are used to press garments that are unsuitable for the automated presses because of their shape. A steam dolly looks just like a mannequin. The finished garments are placed on to the dolly, which releases jets of steam. The steam fills the garment like a balloon, and this action removes creases almost immediately. This process also helps maintain the shape of the garment.

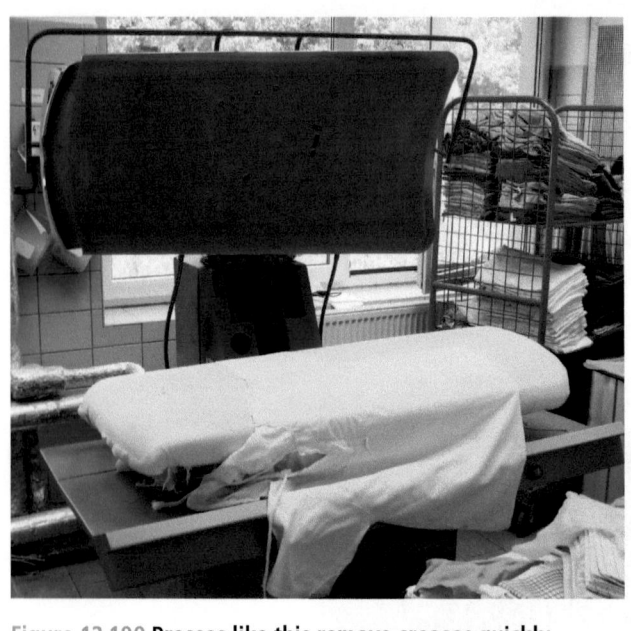

Figure 12.100 **Presses like this remove creases quickly**

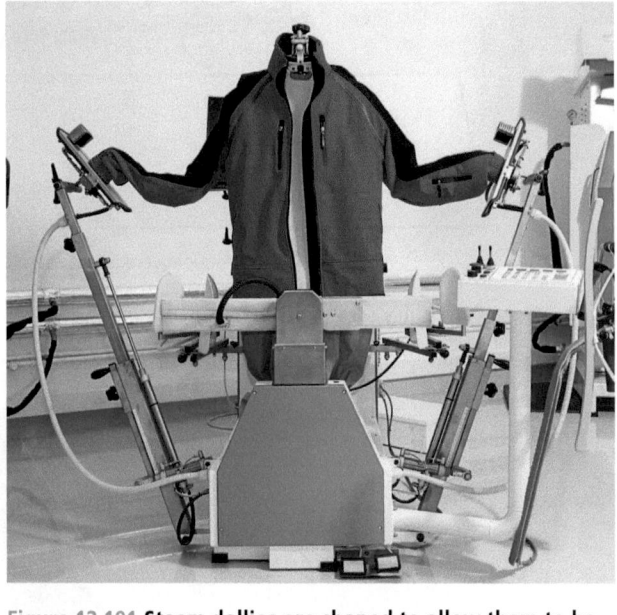

Figure 12.101 **Steam dollies are shaped to allow them to be 'dressed' in the garment to be steamed**

Ensuring accuracy and efficiency

In order to maintain accuracy and efficiency in the manufacture of a new product, a manufacturing specification is used. This is particularly useful in global manufacturing, when the designer and the manufacturer are based in different countries. The manufacturing specification will include a detailed explanation of the materials, components, pattern pieces and construction methods required to make the product. This is followed closely by manufacturers to ensure consistency in the finished products. Failure to follow the manufacturing specification can result in poor quality products, wasted materials and items returned to retailers.

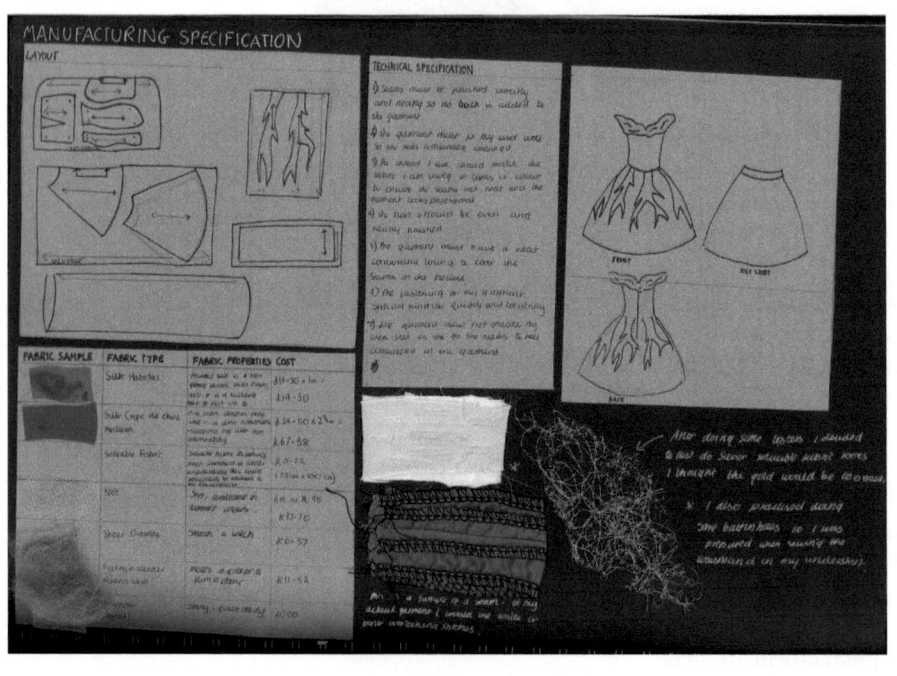

Figure 12.102 **Example of a manufacturing specification**

Quality control checks

In all areas of manufacture, **quality control** checks are carried out at critical control points to ensure consistency in the finished product. These checks are in place to identify faults in the following areas:

- Materials – workers check for faults in the construction of or on the surface of the fabric. These can include snags, misprinted designs, holes or stains.
- Components – workers check components for faults, such as incorrect size/shape, cracked buttons or damaged zips.
- Seams – seams are checked to ensure they are within tolerance (tolerance is explained in the section below), that sewn lines are straight and that there are no holes along the length of the seam.
- Placement and construction – once materials and components are joined, workers check for faults such as incorrect placement of logos or pockets, or incorrect setting of sleeves.

NACERAP

Manufacturers follow various systems to identify and address any faults in production. An example of a quality control system is NACERAP.

Table 12.9 **The stages of the NACERAP system**

	Example: Shirt
N (Name of fault)	Misaligned buttonhole
A (Appearance)	Buttons do not match up with buttonholes
C (Cause)	Error in measurement for button spacing on manufacturing specification
E (Effect)	Hem of shirt does not meet when buttoned up
R (Repair)	Unpick buttons and re-sew using new measurements
A (Action)	Make changes to button spacing measurement on manufacturing specification
P (Prevent)	Check all measurements on manufacturing specification for further errors

Quality assurance

Manufacturers carry out quality control checks consistently in order to provide **quality assurance** to their customers. This means that the customer expects and receives a good level of quality from a particular manufacturer. With good quality products comes a good reputation for the manufacturer, leading to boosted sales.

Tolerances and minimising waste

Tolerances

The processes used in sewing textile products makes it very hard to achieve accuracy. For this reason, tolerances are allowed. This means that any seam could be +/-1cm. Tolerances affect the finished sizing of a product, therefore it is better to use only +1cm tolerances – this means that the finished product will never be too small. Seam allowances are added to pattern pieces to allow enough fabric around the edge of the sewing line for errors. If any seam is out of tolerance, the finished product may be rejected during quality control checks.

Minimising wastage

Wastage is created during pattern cutting, when small amounts of fabric are left around each pattern piece. The size and shape of these pieces means they are difficult to use for other applications. Minimising this waste is the responsibility of lay planners and pattern cutters, who ensure all pattern pieces are placed correctly on the fabric, leaving minimal space between them to reduce these off-cuts. This process ensures the most economical use of fabric rolls, particularly in batch and mass production.

KEY TERM

Quality control: Sample products are taken off the production line at critical control points to check for faults or to test performance.

KEY TERM

Quality assurance: Consistent quality testing at each stage of production means that manufacturers can maintain the required level of quality in their products.

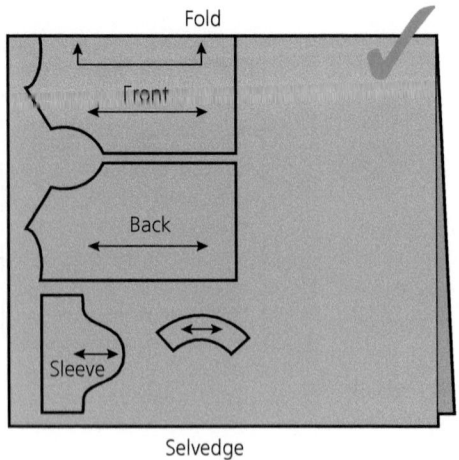

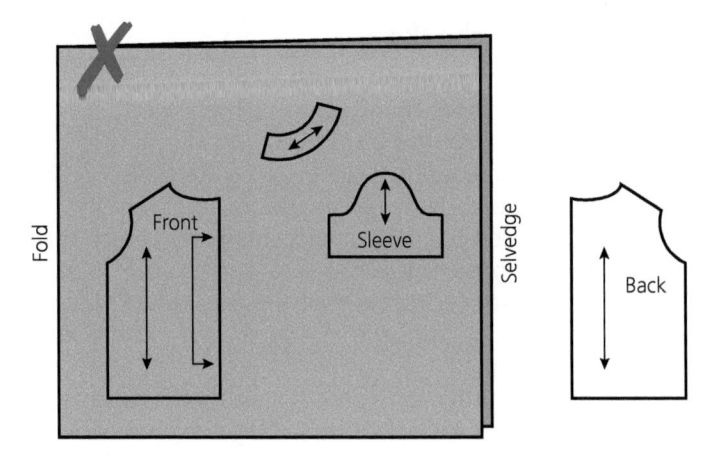

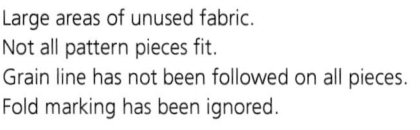

All grain lines are followed correctly.
Minimal wastage between pattern pieces.
Leftover fabric is a useable size and shape.
Fold marking has been followed correctly.

Large areas of unused fabric.
Not all pattern pieces fit.
Grain line has not been followed on all pieces.
Fold marking has been ignored.

Figure 12.103 Pattern pieces must be placed correctly in order to minimise wastage

ACTIVITY

Designing for different scales of production involves the adaptation of designs to make them suitable for production methods. A product designed for bespoke production may include complex construction and finishing methods that aren't suitable for batch or mass production. For this reason, designs may be simplified to make production more time and cost effective.

1. Design a textile product intended for bespoke production inspired by one of the following headings.
   Celebration   Future   Learning   Nature
   Annotate your design to show a range of appropriate production methods.
2. Re-design your product to suit a batch production setting, giving reasons for the changes made to your design.

12.10 Cost and availability

LEARNING OUTCOMES

By the end of this section you should know about and understand:
→ how the cost and availability of specific fabrics and components can affect decisions when designing.

This section will help you to understand how to cost up materials and components for your final product, and how this process differs on an industrial scale.

The significance of cost

The cost of materials and components can affect your final design choices. In some instances it might even be appropriate to scale down or simplify a design due to cost constraints. You must consider stakeholders in these design decisions, to ensure that the final cost of the product is within budget in order to make a profit.

Calculating quantities, costs and sizes of materials

When calculating the amount of fabric required for a prototype, the lay plan, pattern markings and dimensions of the pattern pieces are all required. The example below shows a lay plan with the dimensions added. All the pieces in the lay plan have been placed according to grain lines and fold markings. You need to leave approximately 1cm between pattern pieces to allow space to add balance marks, and these are shown in the diagram.

The fabric will usually be folded in half along the straight grain by placing the selvedges together. This will give two of each pattern piece once the fabric is cut. Where a pattern piece is placed on the fold, a larger symmetrical piece will be produced. Occasionally, a commercial pattern will require the fabric to be unfolded, or folded in different ways. In this case, check the fabric amounts on the back of the packet before purchasing.

To calculate the amount of fabric needed for a basic pattern:
- Measure the height of the lay plan (along the cross grain) and multiply this by 2. This number will indicate the minimum fabric width required.
  Example: Height 64cm × 2 = 128cm. For this you would need to buy fabric from a 150cm roll.
- Measure the width of the lay plan (along the straight grain); this tells you how many metres of fabric you need to buy.
  Example: Width of lay plan = 223cm. For this you would need to buy 2.5 metres of fabric (rounding up to the nearest half metre).

In batch and mass production, fabric and components are ordered in bulk from wholesalers. The price drops as the number of units purchased increases. The table below gives an example of these price differences.

STRETCH AND CHALLENGE

Once you know how many metres you need, you can calculate the cost of this, along with any other fabrics and components.

Visit a website that sells a range of components and fabrics by the metre. Using a commercial pattern, or your own lay plan, calculate the total cost of fabric and components needed for one product.

KEY POINTS

- Remember that fabric is sold on rolls of 115cm or 150cm width.
- Remember that you can buy fabric in 0.5m lengths.

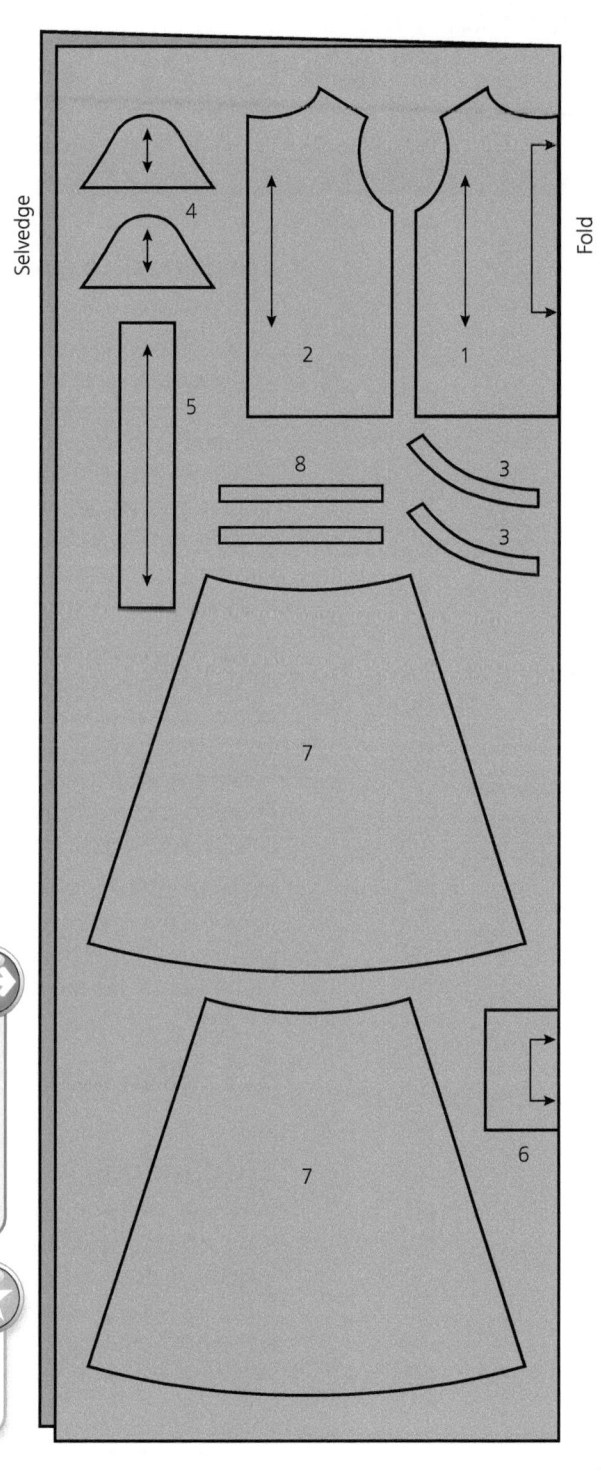

Figure 12.104 **Example lay plan**

Table 12.10 **Price differences as the number of units increases**

Units	Fabric, per metre (£)	Buttons, 12 pack (£)	Interfacing, per metre (£)	Thread, single reel (£)	Ribbon, per metre (£)	Zips, each (£)
1	6.15	2.50	3.55	2.90	1.20	1.75
50	4.40	1.75	2.80	2.10	1.05	0.90
100	3.25	1.05	2.05	1.15	0.75	0.45
500	1.70	0.50	0.65	0.75	0.25	0.15

CHAPTER 13
Design engineering

You should read Chapter 6 before embarking on the material in this chapter.

Chapter 6 introduced some basic mechanical and electronic systems that are frequently found in engineered products. If you wish to specialise in design engineering you will find the following chapter useful in developing your knowledge and understanding of such systems to the point where you will be able to select and use appropriate system components and programmable devices, and to write appropriate programs to benefit your own designs. The ability to develop mechanical and electronic models and solutions during the iterative design process gives you an incredibly powerful approach to creating innovative and interactive outcomes in response to the non-examined assessment Contextual Challenges. It will also be useful to those wishing to focus on the in-depth aspects of design engineering within the written exam.

It is surprisingly straightforward to integrate quite advanced technology into your GCSE projects. As with all new technology, it takes a little time and effort to learn the jargon associated with mechanisms and electronics, and you will need to be patient and logical when trying to work out why your designs do not initially work as you intend. Once you have got over the initial hurdles, however, you will quickly appreciate the advantages of using mechanical and electronic components within your designs. The trick is to keep your early attempts as simple as possible and to get each stage working before you progress on to the next stage.

This chapter contains some necessary mathematics. The beauty of using maths when designing is that you can obtain objective and precise answers to many design questions, which removes the element of guesswork by telling you exactly how big something needs to be, how fast it will move, or what voltage is required to operate it, for example.

13.1 Working with mechanical components

LEARNING OUTCOMES

By the end of this section you should know about and understand:
→ the effect of forces on ease of movement
→ how mechanical devices are used to change the magnitude and direction of forces
→ the working properties of mechanical components
→ the mathematical treatment of mechanical systems.

Chapter 6 introduced some basic mechanical systems without going into much mathematics. The following sections look again at these systems, and some additional systems, and deal with the associated maths.

Mechanical advantage

You will recall that a mechanism can control and change motion, and that it has an input and an output. You should also be familiar with the idea that a mechanism can amplify the input force *or* it can make the output move by a greater distance than the input, but it cannot do both these things at the same time. There is always a trade-off. Mechanisms can either:
● reduce the distance moved but increase the force being exerted, or
● increase the distance moved but reduce the force being exerted.

★ KEY POINT

Mathematically, this trade-off can be written down as follows:

(input force × distance moved by input) = (output force × distance moved by output)

The ratio of the output force to the input force is called the **mechanical advantage (MA)** of the mechanism:

$$\text{MA} = \frac{\text{output force}}{\text{input force}}$$

The mechanical advantage of a mechanism describes the amount by which it amplifies the input force.

★ KEY POINT

For a simple lever, mechanical advantage would be written as:

$$\text{MA} = \frac{\text{load}}{\text{effort}}$$

The effort and load forces are inversely proportional to the lengths of the input arm and output arm of the lever, which means that the mechanical advantage of a lever is also the same as:

$$\text{MA} = \frac{\text{input arm length}}{\text{output arm length}}$$

On some levers, you must look carefully to identify the input and output arm lengths. The arm length is the distance between the force and the fulcrum, and it is always measured at right angles (90°) to the force. Look at the brake lever shown to see what this means in practice.

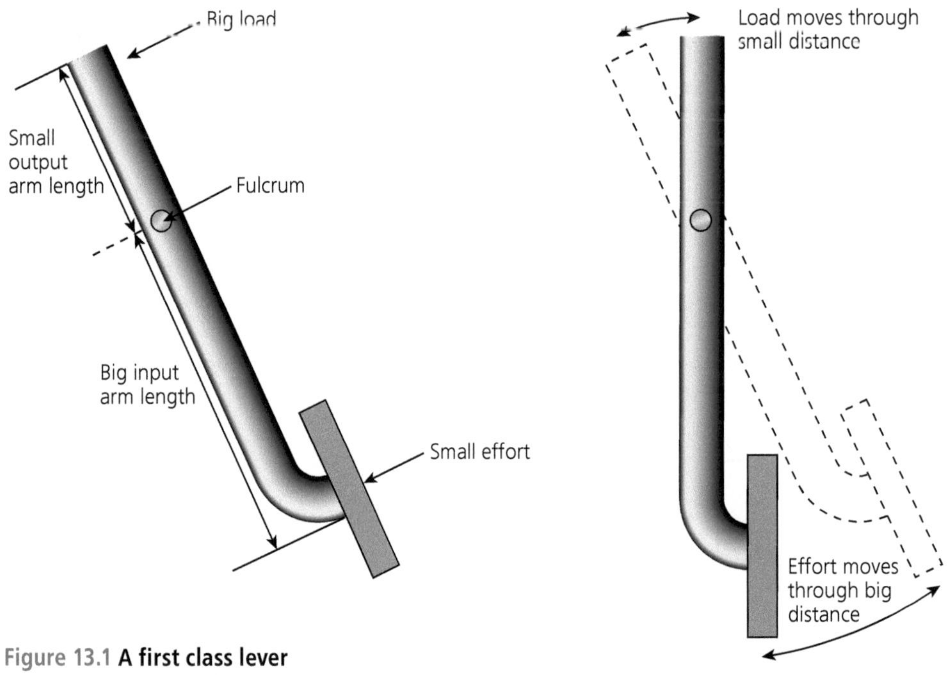

Big load

Small output arm length

Fulcrum

Big input arm length

Small effort

Load moves through small distance

Effort moves through big distance

Figure 13.1 **A first class lever**

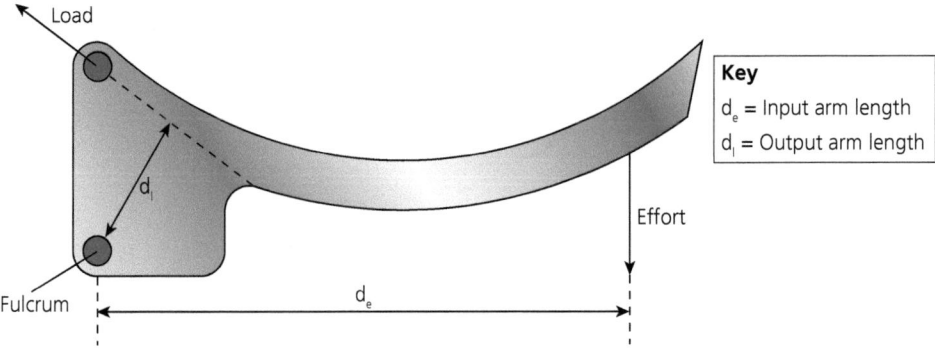

Load

d_i

Fulcrum

d_e

Effort

Key
d_e = Input arm length
d_i = Output arm length

Figure 13.2 **Look carefully to identify the input and output arm lengths on this first class lever**

A pair of scissors is shown below. Use the information in the diagram to calculate the load force generated when an effort of 75N is applied.

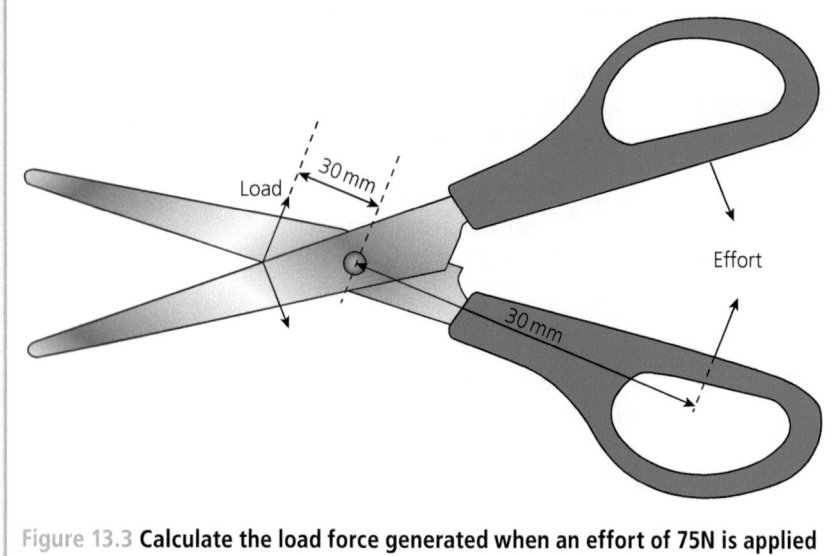

$$MA = \frac{input\ arm\ length}{output\ arm\ length}$$

$$MA = \frac{80}{30} = 2.67$$

$$MA = \frac{load}{effort}$$

So an effort of 75N would be amplified to become a load of:

$$load = MA \times effort$$

$$= 2.67 \times 75$$

$$= \textbf{200N (rounded to 3 significant figures)}$$

Figure 13.3 **Calculate the load force generated when an effort of 75N is applied**

Class of lever

Depending on where the fulcrum is positioned relative to the effort and load, a lever can amplify force or can amplify distance moved, but it cannot do both at the same time – remember, there must be a trade-off. The position of the fulcrum can also result in effort and load moving in the same direction, or in opposite directions. There are three variations, or 'classes', of lever:

- In a first class lever the fulcrum is between the load and the effort. Examples include scissors, and a claw hammer extracting a nail.

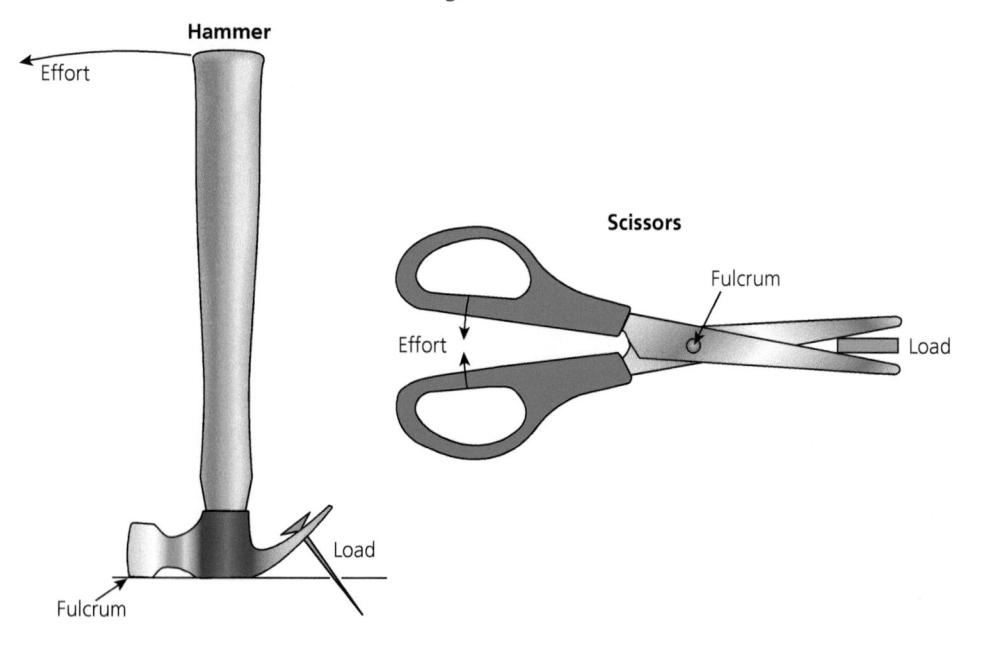

- In a second class lever the load is between the effort and the fulcrum. Examples include a bottle opener and a nutcracker.

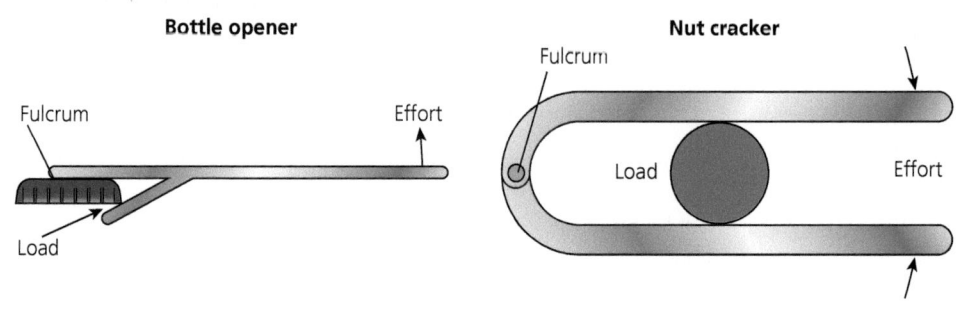

- In a third class lever the effort is between the load and the fulcrum. Examples include tweezers and a staple extractor.

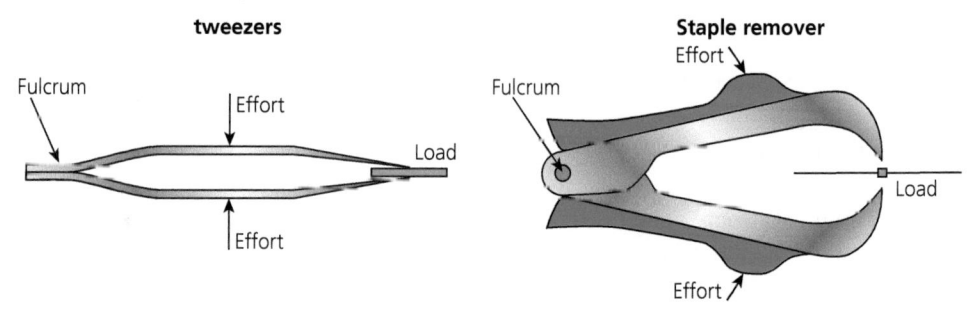

Sometimes it is not obvious exactly where the effort or load acts on a lever. In scissors, for example, the load could be acting anywhere along the length of the blades depending on which part of the blade is being used to snip a material. In these situations, the force arrows on the diagrams will make it clear exactly where the forces are acting.

	Effect on an input force	Mechanical advantage	Effect on the input distance moved	Effect on the input direction moved
1st Class	Depends on input/output arm lengths	Depends on input/output arm lengths	Depends on input/output arm lengths	Reverses the direction
2nd Class	Amplifies	Greater than 1	Reduces the distance moved	Output moves in same direction
3rd Class	Reduces	Less than 1	Increases the distance moved	Output moves in same direction

Do not try to memorise everything in this table; just remember that a lever must trade off force against distance moved, so if the load moves less than the effort, then the load force must be greater than the effort force.

Rotational mechanical systems

Rotating mechanical systems are very common. Most engines and electric motors produce a rotating output, which usually needs to be altered in speed or direction or to be transferred to a different place.

Rotational force or turning effect is called torque.

Rotary mechanical systems still obey the machine trade-off principle, but it is written slightly differently. Rotary mechanical systems can either:
● reduce rotary speed but increase torque, or
● increase rotary speed but reduce torque.

> **KEY POINT**
>
> Mathematically, this trade-off can be written as follows:
>
> $$(\text{input torque} \times \text{input rotational speed}) = (\text{output torque} \times \text{output rotational speed})$$
>
> In a rotational mechanical system, mechanical advantage is defined as:
>
> $$\text{MA} = \frac{\text{output torque}}{\text{input torque}}$$
>
> If you look at the trade-off equation, you might be able to spot that MA is also equal to:
>
> $$\text{MA} = \frac{\text{input rotational speed}}{\text{output rotational speed}}$$

Simple gear train

The diagram shows a simple gear train consisting of a driver gear with N_1 teeth and a driven gear with N_2 teeth.

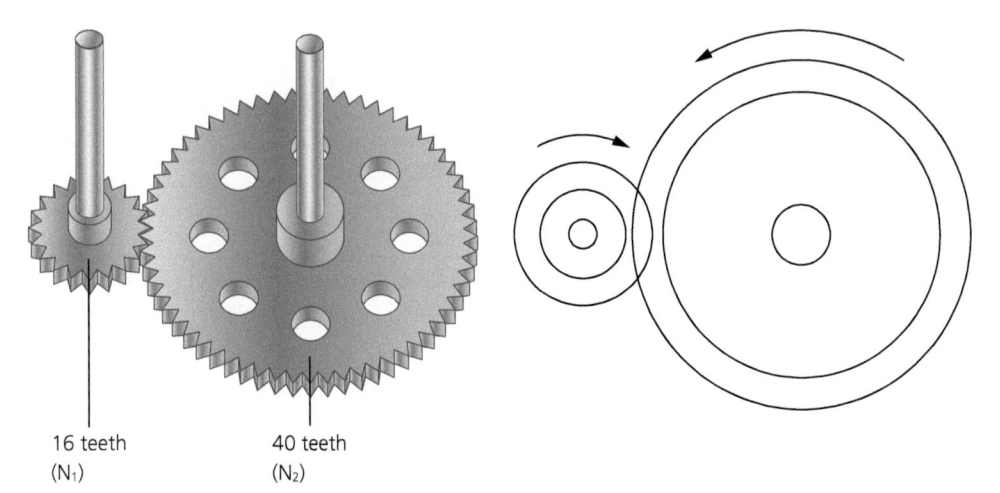

16 teeth
(N_1)

40 teeth
(N_2)

Figure 13.4 **A simple gear train**

★

KEY POINT

The gear ratio is defined as:

$$\text{gear ratio} = \frac{\text{number of teeth on driven gear}}{\text{number of teeth on driver gear}}$$

You will recall that the larger gear rotates more slowly than the smaller gear. The number of teeth around the circumference of the gear is inversely proportional to the rotational speed of the gear, which means that the gear ratio of a simple gear train is also the same as:

$$\text{gear ratio} = \frac{\text{driver gear rotational speed}}{\text{driven gear rotational speed}}$$

The gear ratio of a simple gear train describes the amount by which it reduces the input rotational speed.

If you refer to the mechanical advantage equation, you will see that the gear ratio of a simple gear train is the same as its mechanical advantage, which is the same as the amount by which the torque is increased:

$$\text{gear ratio} = \frac{\text{input rotational speed}}{\text{output rotational speed}} = \text{MA} = \frac{\text{output torque}}{\text{input torque}}$$

In the diagram shown, the gear ratio is:

$$\text{gear ratio} = \frac{\text{number of teeth on driven gear}}{\text{number of teeth on driver gear}} = \frac{40}{16} = 2.5$$

Notice that, although this is called a gear 'ratio', the result is often written simply as a number. Writing 2.5 : 1 would be just as acceptable.

Without worrying too much about equations, now that we know that the gear ratio is 2.5 we know that the mechanical advantage of the system is also 2.5, so the output torque will be 2.5 times bigger than the input torque and, by the trade-off principle, that the output speed must be 2.5 times less than the input speed.

If the input speed is 1200 rpm, then:

$$\text{speed of output} = \frac{1200}{2.5} = 480 \text{ rpm}$$

STRETCH AND CHALLENGE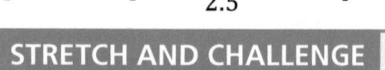

A wind turbine rotates at 90rpm. The turbine drives a 40-tooth gear that meshes with a 12-tooth driven gear.

i   Calculate the rotational speed of the output shaft.

$$\text{gear ratio} = \frac{\text{number of teeth on driven}}{\text{number of teeth on driver}} = \frac{12}{40} = 0.3$$

The smaller (output) gear will rotate faster than the larger gear at a speed of:

$$\text{output speed} = \frac{\text{input speed}}{\text{gear ratio}} = \frac{90}{0.3} = 300 \text{ rpm}$$

ii Describe the torque available at the output shaft.
    There is a trade-off between torque and speed. As the output speed is higher than the input speed, the output torque will be less than the input torque by the same factor:

$$\text{output torque} = 0.3 \times \text{input torque}$$

(Notice that the torque multiplication factor is the same as the mechanical advantage, which is equal to the gear ratio.)

Idler gear

An **idler gear** is a gear that is inserted between two other spur gears, as shown in the diagram.

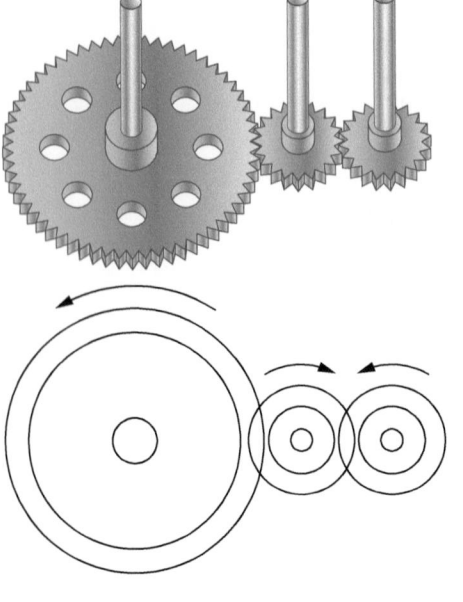

Figure 13.5 **An idler gear between two spur gears**

Figure 13.6 **A simple gear train with several idler gears**

KEY TERMS

Idler gear: A gear that is inserted between two spur gears to change the output direction.

Bevel gear: A gear with teeth cut at a 45° angle to change the direction of the drive shaft by 90°.

The **idler gear** has no effect on the overall gear ratio, but it will reverse the direction of rotation of the output shaft. This can be useful if it is necessary for the input and output shafts to rotate in the same direction. The size of the idler gear does not matter as it has no effect on the gear ratio of the simple gear train.

Another reason for using an idler gear is to increase the separation of the input and output drive shafts. In these circumstances, several idler gears may be used, each of which will have the effect of reversing the direction of rotation. No matter how many idler gears are used, the gear ratio of the system will still depend only on the number of teeth on the first (driver) gear and the final (driven) gear.

Other gear systems

Spur gears are the most commonly used gears, but other types of gears are used for special applications.

- **Bevel gears** have their teeth cut on a 45° angle and they are used when it is necessary to transfer the direction of the drive shaft by 90°.

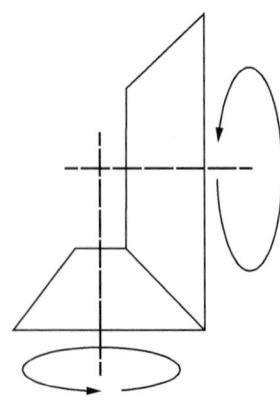

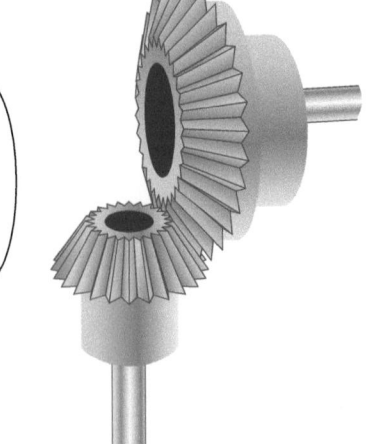

Figure 13.7 **Bevel gears**

- A **worm drive** is a unique type of gear system. The driver is a screw thread called a **worm screw**, which meshes with a **worm wheel** (which is like a spur gear). There are three important things to note about worm drives:
  - They achieve a very high gear ratio, so they can reduce rotational speed by a large factor or increase torque by the same amount. The gear ratio is simply the number of teeth on the worm wheel.
  - They transfer the direction of rotation by 90°.
  - They are self-locking, which means that the input shaft (the worm) can drive the output, but the output cannot drive the input. This feature can be useful in some applications, for example when it is important that the output does not slip if the input drive is turned off, such as in a winch or a lift.
- A **rack and pinion** is a system to change between rotary and linear motion. In the stairlift shown, the rack is stationary and is attached to the wall. The pinion inside the chairlift is driven by a motor and the pinion 'climbs' up the rack.

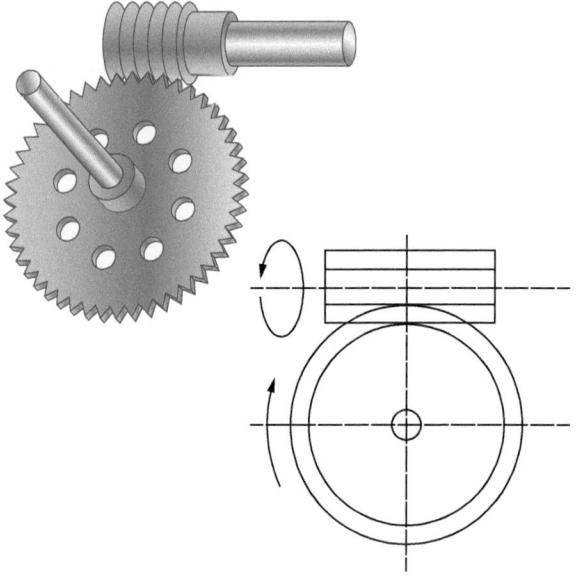

Figure 13.8 **The self-locking action is important in these worm drives.**

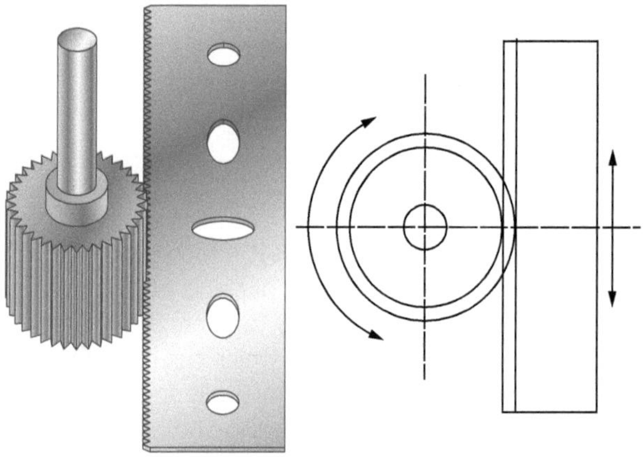

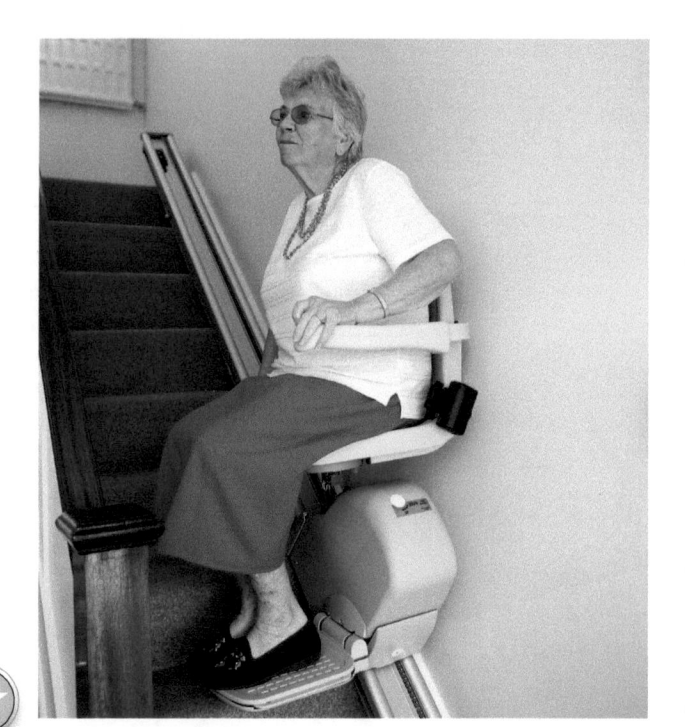

Figure 13.9 **Diagram of a rack and pinion**

Pulley and belt drive

A pulley and belt drive can provide speed reduction or increase, and the 'gear' ratio is calculated in a similar way to a simple gear train.

> **KEY POINT**
>
> For a pulley and belt system:
>
> $$\text{gear ratio} = \frac{\text{diameter of driven pulley}}{\text{diameter of driver pulley}}$$
>
> Just as with a simple gear train, the gear ratio is the same as the ratio of input to output speeds, and is also equivalent to the mechanical advantage of the system.

Figure 13.10 **A stairlift uses a rack and pinion**

> **KEY TERMS**
>
> **Rack and pinion:** A system to change between rotary and linear motion.
>
> **Worm drive:** A self-locking mechanism, achieving a very high reduction ratio.
>
> **Worm screw:** The driver part in a worm drive.
>
> **Worm wheel:** The driven gear in a worm drive.

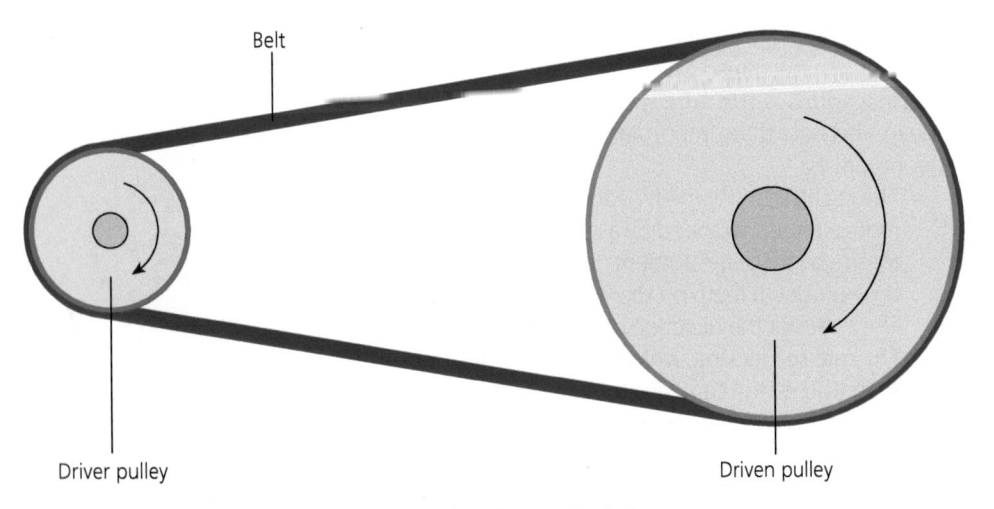

Belt

Driver pulley

Driven pulley

Figure 13.11 Transferring rotary motion with pulleys and a belt

STRETCH AND CHALLENGE

The pulley and belt drive shown in the diagram is being used to reduce the speed of a drive shaft. The input pulley is 60mm in diameter and a 3:1 reduction is required. Calculate the required diameter of the output pulley.

$$\text{Gear ratio} = \frac{\text{diameter of driven (output) pulley}}{\text{diameter of driver (input) pulley}}$$

$$3 = \frac{\text{diameter of output pulley}}{60}$$

Diameter of output pulley $= 3 \times 60 = 180\text{mm}$

The most common type of pulley and belt drive uses a **V-shaped belt**, which grips the pulley side walls to provide a large contact area. This provides lots of grip and means there is less likelihood of the belt slipping on the pulley. In a few applications, belts are deliberately left quite slack so that they can slip if the output shaft gets jammed, which can serve to prevent damage to mechanical systems or the drive motor. This technique is not generally advisable, however, because unless the drive is quickly turned off, the slipping belt will rapidly heat up due to friction and will wear or even catch fire.

Figure 13.12 A V-belt driving a pulley

A pulley and belt drive will be very quiet in operation compared to gears, which tend to 'whine' or rumble. When using belts, provision must be made to keep the tension correct as the belt will stretch with use. If it is too slack there is a risk of the belt slipping, while if it is too tight there will be far too much friction, which causes a loss in efficiency. Sometimes, one of the pulley shafts is designed to be movable to take up the slack in the belt, or **tensioner wheel(s)** are used.

KEY TERMS

Tensioner wheel: A component used to take up the slack in a belt drive system.

V-shaped belt: A belt designed to transfer high loads without slipping.

Cams

Cams were introduced in Chapter 6 as a mechanism for converting rotary to reciprocating motion. They are frequently used in engines and other machines to generate a required kind of motion.

The distance that the follower rises is called the **stroke**. This distance can be calculated from the dimensions of the cam.

STRETCH AND CHALLENGE

Determine the stroke of the following cam system, and describe the motion of the follower through one cycle.

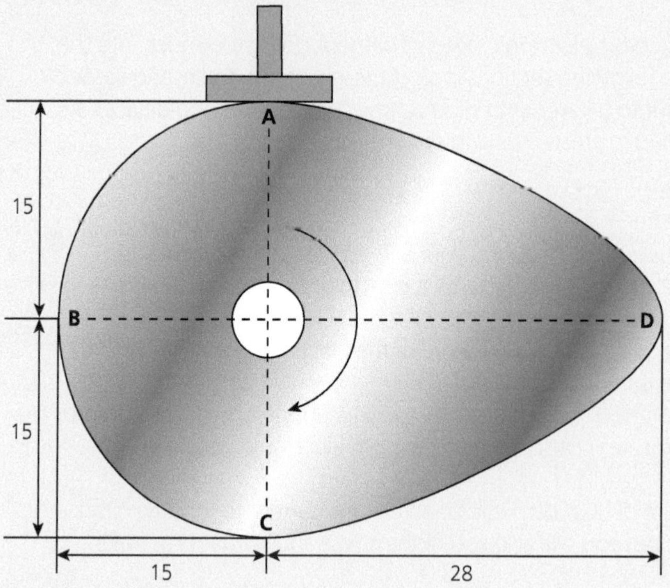

As the cam rotates clockwise, points A, B, C and then D will move past the follower. As points A, B and C are all 15mm from the cam centre, the follower will stay at its lowest level through these points. As point D passes, the follower will rise by an amount called the stroke:

stroke = 28 – 15 = 13mm

As point D passes, the follower will fall through 13mm again until it reaches its lowest point at A.

Figure 13.13 **Two tensioner wheels are used to keep the toothed belt firmly in place in this motorbike engine.**

13.2 Working with electronic components

LEARNING OUTCOMES

By the end of this section you should know about and understand:
→ how sensors respond to a variety of inputs
→ how devices are used to produce a range of outputs
→ the use of programmable components such as microcontrollers to embed functionality into products in order to enhance and customise their operation
→ the working properties of electronic components
→ the mathematical treatment of electronic systems.

KEY TERMS

Voltage: A measure of the electrical 'pressure' causing a current to flow, measured in volts.

Current: The actual electricity flowing, measured in amps.

Resistance: How hard it is for an electric current to flow, measured in ohms.

Chapter 6 introduced some basic electronic systems without going too much into the mathematics involved. The following sections look at these systems again, and some additional systems, and go into greater detail about how components and devices are connected and programmed to create working systems.

Basic electricity

You will have learned about electricity during your Science lessons. In Design and Technology we are interested in applying scientific principles in practical situations to solve design problems.

- The **voltage** at a point in a circuit is a measure of the electrical 'pressure' causing a current to flow. Voltage is measured in volts (V).
- **Current** is a measure of the actual electricity flowing. Current is measured in amps (A), although milliamps (mA) are more commonly used in electronic systems:
  - 1A = 1000mA
- **Resistance** is a measure of how hard it is for an electric current to flow. Resistance is measured in ohms (Ω), but you will also see kilohms (kΩ) and megaohms (MΩ).
  - 1 kΩ = 1000 Ω
  - 1 MΩ = 1000 kΩ

Milli, **kilo** and **mega** are examples of prefixes used to denote powers of ten. The table lists the prefixes commonly used in design engineering.

Prefix	Symbol	Power of 10
tera	T	10^{12}
giga	G	10^{9}
mega	M	10^{6}
kilo	k	10^{3}
milli	m	10^{-3}
micro	µ	10^{-6}
nano	n	10^{-9}

STRETCH AND CHALLENGE

Resistance is linked to voltage and current by the Ohm's law formula:

$$V = IR$$

where V is voltage, I is current, R is resistance.

i   Calculate the current that a 12V power supply unit will need to be able to supply to a 6.8Ω heater. We know the voltage (V) and resistance (R). We will use Ohm's law formula to calculate the current (I).

$$V = IR$$

Rearranging the formula gives:

$$I = \frac{V}{R}$$
$$I = \frac{12}{6.8}$$
$$I = 1.76A$$

ii   Calculate the resistor needed to produce a current of 15mA if the voltage across it is 9.8V.
Ohm's law formula:

$$V = IR$$

$$R = \frac{V}{I}$$

$$R = \frac{9.8}{15 \times 10^{-3}}$$

$$R = 653\Omega$$

iii   Calculate the resistor value which would draw a current of 200mA from a 5V power supply.
iv   Calculate the current through a 47kΩ resistor which has 9V across it.
v   Calculate the voltage you would expect to measure across a 1.5MΩ resistor carrying a current of 4µA

Electronic systems

System diagrams were introduced in Chapter 6 as a way of showing how subsystems are interconnected, and showing the signals that flow between them. We will look at these ideas in more detail in this section.

The electronic systems relevant to the GCSE Design and Technology course consist of one or more inputs, a microcontroller as a process subsystem, and one or more outputs.

Signals in electronic systems are usually voltage levels and it is useful to understand the nature of these signals when you are designing systems. The techniques of connecting subsystems together in a way that allows the signals to pass properly between them is called **interfacing**.

The simple electronic system shown below is an example of how subsystems can be interfaced and made to function in a way that can usefully enhance the function of a product.

The system diagram shown is a design for a bicycle safety lamp that could increase the visibility of cyclists on the road at night. There is a single push-button input and three ultra-bright LED outputs. The microcontroller could be programmed to operate the lamp as follows:

- First push of the button turns on all three LEDs.
- Second push of the button flashes all three LEDs.
- Third push turns off all LEDs.
- The cycle then repeats.

At this point the flexibility of the microcontroller becomes evident, in that it would be a simple matter to reprogramme it to add enhanced features to the product. For example, a fourth push of the button could make the LEDs flash independently in an eye-catching pattern, and a fifth push could change to a different pattern or a faster flash, etc.

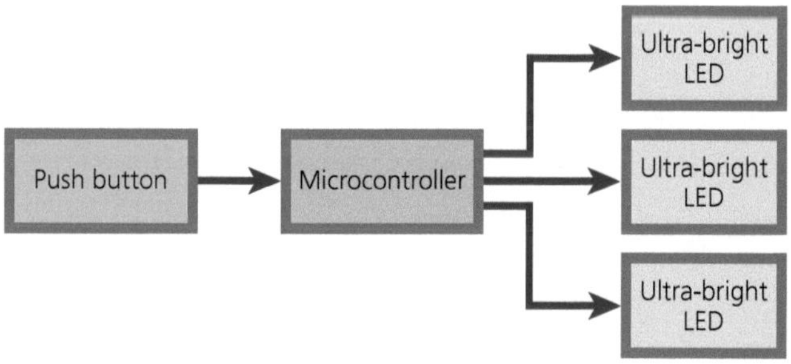

Figure 13.14 **A system diagram for a bicycle safety lamp**

Flowchart programs

The program for a microcontroller is a step-by-step set of instructions that tells the microcontroller what to do. The program should initially be written down as a flowchart, which is a simple way of breaking down the microcontroller task by showing the key steps involved. At a later stage in the design development the flowchart may need to be converted into a specific **programming language**, depending on the microcontroller used. It is beyond the scope of this book to deal with individual programming languages so we will focus on using flowcharts. Some microcontrollers can be programmed directly from a flowchart, making the programming stages much simpler.

A microcontroller can perform only one instruction step at a time, in sequence. When the program is run, each instruction is carried out extremely quickly so the user would not normally notice any delay between the instructions, unless WAIT commands are inserted in the flowchart to deliberately slow things down. Constructing a flowchart helps the designer to 'think like a microcontroller'.

There are five different symbols used for drawing flowcharts:

KEY TERMS

Interfacing: The method of connecting subsystems together to allow signals to pass.

Programming language: The set of instructions and rules used to write a microcontroller program.

Symbol	Name
	Start/end
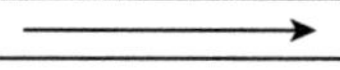 →	Arrows
	Input/Output
	Process
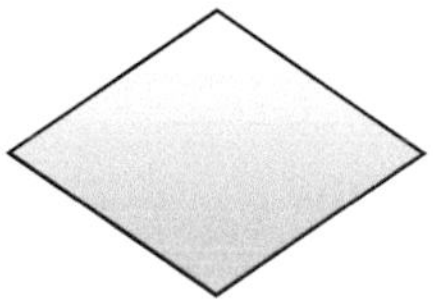	Decision

Figure 13.15 **The symbols used in a flowchart**

295

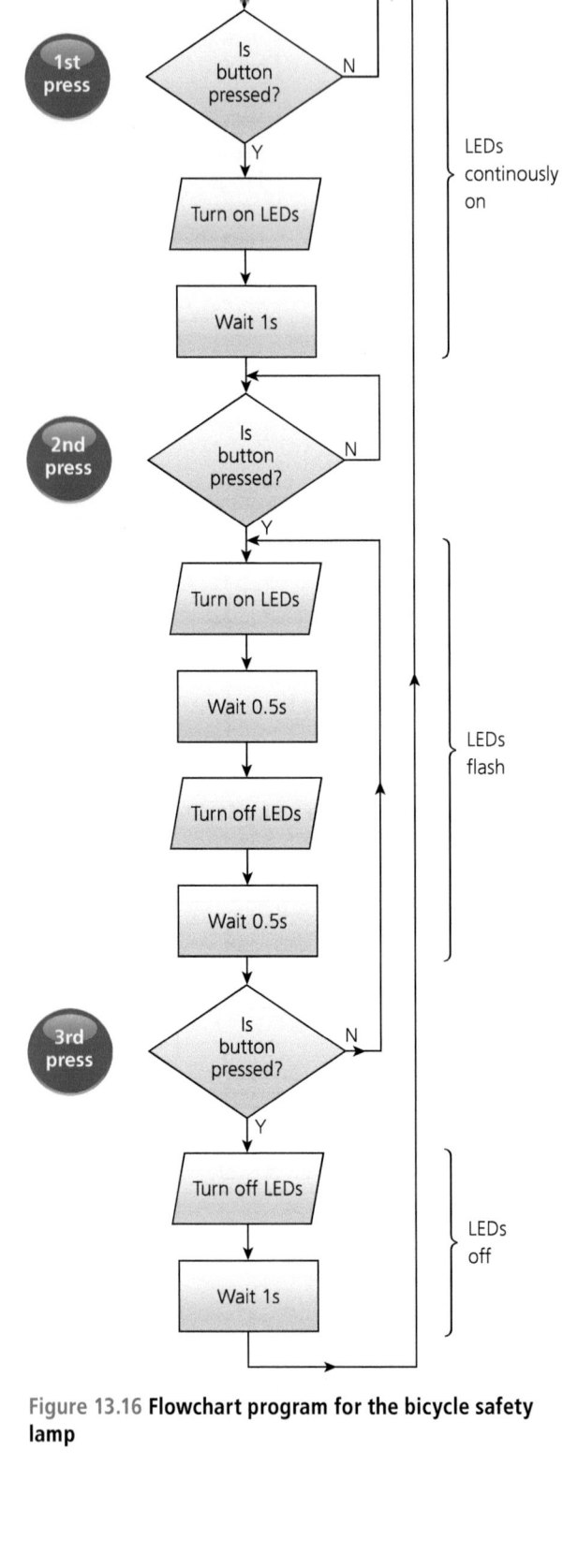

1st press — LEDs continuously on

2nd press — LEDs flash

3rd press — LEDs off

Start

Is button pressed? — N

Turn on LEDs

Wait 1s

Is button pressed? — N

Turn on LEDs

Wait 0.5s

Turn off LEDs

Wait 0.5s

Is button pressed? — N

Turn off LEDs

Wait 1s

Figure 13.16 Flowchart program for the bicycle safety lamp

KEY POINT

It is very important to understand the difference between a system diagram and a flowchart.

● A system diagram shows how subsystems (which are physical parts) are linked together, and the lines on a system diagram show the voltage signals and the direction in which the signal information flows.

● A flowchart shows a sequence of steps (instructions and decisions) that will be carried out by the microcontroller as it performs its task within a system. The lines and arrows on a flowchart show the order in which the instructions are executed.

A program flowchart for the bicycle safety lamp introduced in the previous section is shown below.

Study the flowchart and note the following points:

● The program will begin at the 'START' box.

● The instructions will be executed one after another at very high speed. Consequently, 'WAIT' instructions are used to control the speed through particular sections of the program.

● The two 'WAIT 1s' instructions are needed to give the user time to release the button after pressing it, before the program reaches the next 'Is the button pressed' decision.

● There is no END box because, after the third press of the button, the cycle repeats from the start.

Sensors

Sensors can be broadly classified into two types:

● **Digital sensors** are used for detecting a yes/no situation, for example:
 – Is the button pressed?
 – Is movement detected?
 – Has the product fallen over?

● **Analogue sensors** are used when it is necessary to measure 'how big' a quantity is, for example:
 – How bright a light is.
 – What a temperature is.

The different types of sensor need to be interfaced to the appropriate analogue input or digital input on the microcontroller.

KEY TERMS

Analogue sensor: A sensor to measure 'how big' a physical quantity is.

Digital sensor: A sensor to detect a yes/no or on/off situation.

Switch sensors

All switches are digital sensors, and different switches are available to detect a wide variety of things. A switch can be closed (on) or open (off).

- **Tilt switches** are closed when the switch is upright, and open when inverted.
- **Push-to-make switches** are closed when the push-button is pressed. A momentary switch opens when the button is released, and a latching switch stays closed until the button is pressed a second time.

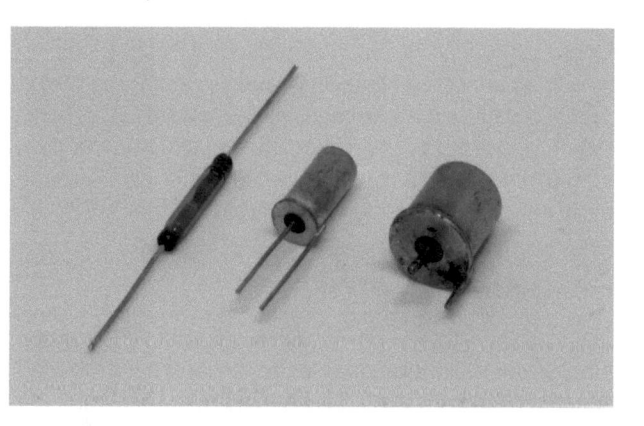

Figure 13.17 **Various types of switch sensors**

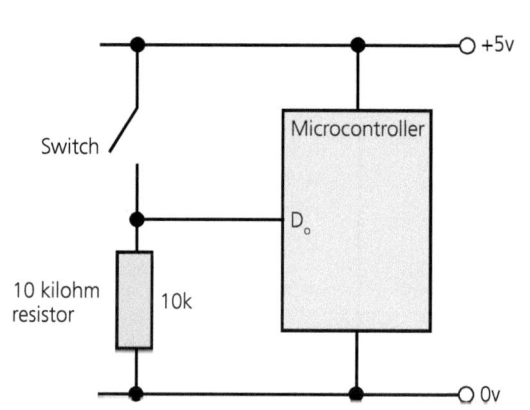

Figure 13.18 **The method of connecting a switch sensor to the input of a microcontroller**

- **Time delay switches** – there are two types, one is like a latching push-to-make switch, which stays closed for a period of time before automatically opening. The other type is programmed to open/close at various times throughout the 24-hour day.
- Another commonly used type of switch is **magnetic reed switches**, which close when a magnet is brought near to them. These are often used to sense a door being opened in a security system. Their benefit is that the magnet does not need to touch the switch, just be close to it.

When a switch is used as an input sensor for a microcontroller, it is connected with another resistor in what is known as a **voltage divider**, as shown in the circuit diagram below. Notice that a digital input pin is used on the microcontroller.

The flowchart program needed for the microcontroller to monitor the switch might look as shown below.

> **KEY TERM**
>
> **Voltage divider:** A method of connecting input sensors to produce a voltage signal.

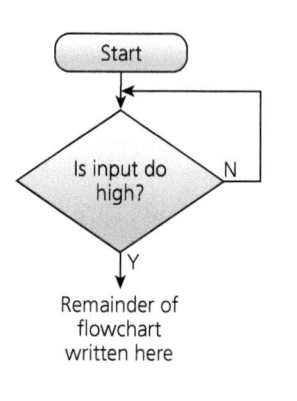

Figure 13.19 **Reading the state of an input switch sensor**

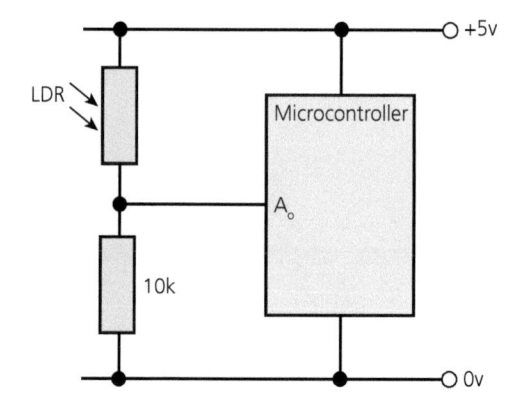

Figure 13.20 **An LDR connected to an analogue input**

KEY TERMS

Force sensitive resistor (FSR): A force-sensing component.

Light dependent resistor (LDR): A light-sensing component.

Multimeter: A device used to take measurements and help find faults in electronic circuits.

Thermistor: A temperature-sensing component.

```
        Start
          |
          v
    Read light
      level
          |
          v
     Is light level ----N-->
        >75?
          |
          Y
          |
          v
```

Figure 13.21 Reading the light level and making a decision

Light sensors

A light sensor is an analogue sensor – the actual light sensing component is a **light dependent resistor (LDR)**, which has a resistance that gets lower as the light level increases. To produce an analogue voltage signal, the LDR is connected in a voltage divider, as shown in the circuit diagram below.

Notice that an analogue input pin is used on the microcontroller.

A **multimeter** can be very useful to help find faults in electronic circuits. The multimeter is set to a voltage range and the black (common) probe is connected to the 0V rail. In this sensing circuit, the red multimeter probe is connected to the analogue input pin. The meter should read a low voltage when the LDR is dark, and the voltage will rise towards 5V as light falls on the LDR. This analogue voltage is converted by the microcontroller to a number that will be somewhere between 0 (in total darkness) and a maximum value that depends on the microcontroller (the maximum value, in extremely bright light, is likely to be either 255 or 1023, but check this for the microcontroller that you use).

A flowchart program to read and respond to the light level is shown below.

This flowchart would follow the YES path if the light level is high (above number 75) and the NO path if the light level is low. It is important to understand that the sensor is not calibrated, so the number 75 in this flowchart does not have any units, it is just a number that would be determined experimentally during the iterative designing stages. If the system responds 'YES' at too low a light level, then the value 75 would be increased until a suitable value is found.

The LDR can easily be changed for a different resistive component to sense other physical quantities. Replacing it with a **thermistor** will produce a temperature sensor. Replacing the LDR with a **force sensitive resistor (FSR)** will produce a force (or a weight) sensor, and so on.

Infra-red sensors

Several different types of infra-red (IR) sensors are available for various applications, including:
- IR sensors that detect the presence of a warm object, e.g. a hand sensor for an automatic tap; these have a digital output.
- Passive infra-red (PIR) sensors that detect a moving warm object, e.g. an intruder detector for an alarm system.
- IR distance sensors that measure the distance to a nearby object; these produce an analogue output.
- IR receivers that pick up data signals from an IR transmitter, e.g. a TV remote control system.

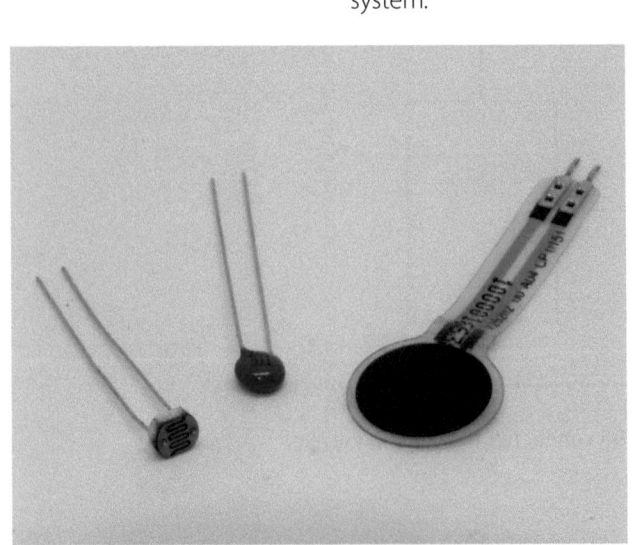

Figure 13.22 An LDR, thermistor and FSR

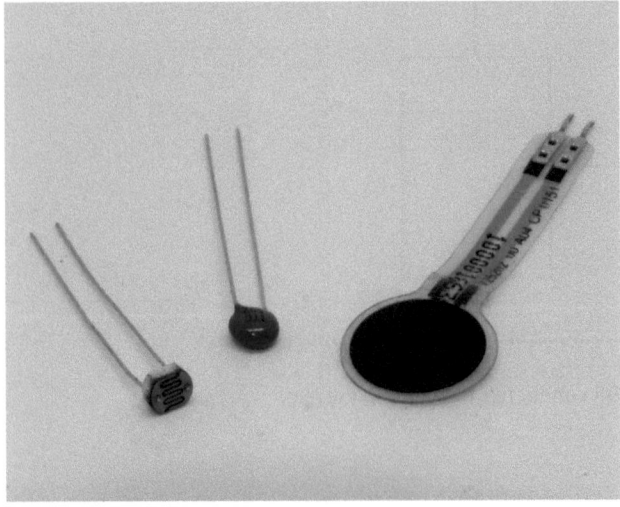

Figure 13.23 A range of IR sensors

Output devices

Outputs consist of components to produce light (for information and illumination), or to produce sound or motion, etc. Many outputs can also be digital or analogue, for example a light that is either on or off is digital, but the same light could be used in an analogue way by making it brighter or dimmer. An electric motor can be varied in speed (analogue) or simply turned on or off (digital). Many microcontrollers have both digital and analogue outputs. Make sure you select the appropriate ones for the output device you intend to use.

Light emitting diodes

Light emitting diodes (LEDs) are available in an enormous range of sizes, colours, brightnesses and shapes. They need a resistor to be placed in series with them to limit the current flowing, otherwise the LED will burn out.

Most small LEDs can be connected (through the resistor) directly to the digital output pin of a microcontroller. LEDs are **polarised**, meaning they have a positive lead (**anode**) and a negative lead (**cathode**), and they must be connected the correct way round or they will not work. The circuit diagram shows how an LED would be connected to the output pin of a microcontroller.

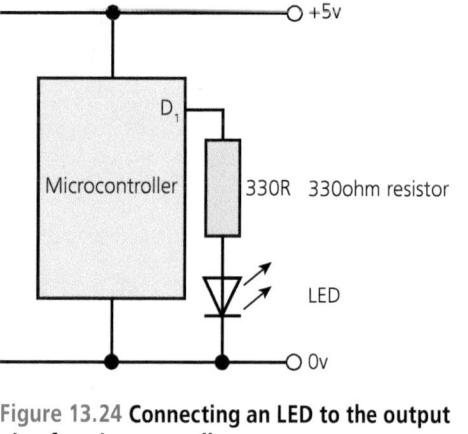

Figure 13.24 Connecting an LED to the output pin of a microcontroller

KEY TERMS

Anode: The positive connection on a component.

Cathode: The negative connection on a component.

Driver: A component to boost the output current.

Piezo-electric sounder: A miniature speaker for producing sounds from a microcontroller.

Polarised: A component that must be connected the correct way round in a circuit.

KEY POINT

The series resistor for an LED can be calculated using Ohm's law:

$$R = \frac{V_s - V_{led}}{I}$$

where V_s is the power supply voltage, V_{led} is the voltage drop across the LED, and I is the current the LED needs to light up.

The values for V_s and I can be looked up for the LED you intend to use.

Example:

For a 5V power supply, and a red LED for which $V_s = 2.0V$ and I = 10mA:

$$R = \frac{5 - 2}{0.01}$$

$$R = 300\Omega$$

In practice, a 330Ω resistor would be used.

More powerful LEDs are used when it is necessary to provide illumination, such as in a lamp. These LEDs require a higher current and it would be necessary to use a **driver** to boost the current. Drivers are explained later in this section.

Speakers and buzzers

The difference between a speaker and a buzzer was explained in Chapter 6.

The speaker component in most projects will be a **piezo-electric sounder**. These are small and cheap components that can generate surprisingly loud tones, especially at higher frequencies. A piezo sounder would be connected to the output pin of a microcontroller as shown in the diagram.

Small buzzers may also work connected directly to an output pin, but louder sirens would need to use a driver, as explained below.

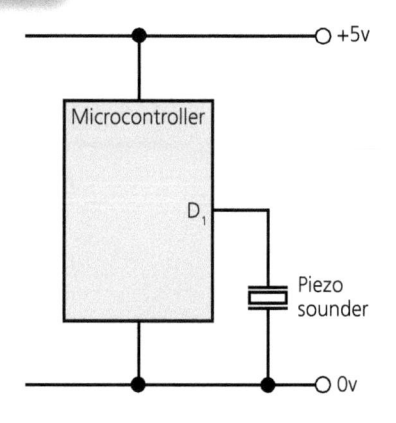

Figure 13.25 Connecting a piezo sounder to an output pin

DC motors

Electric motors are the prime source of motion in mechanical systems, and they were discussed in Chapter 6. DC motors are the most common type and their speed and direction can both be controlled using a microcontroller. A driver is always necessary to provide sufficient current to operate the motor.

Drivers

Some output devices need a **driver** between the microcontroller and the output device to boost the current, otherwise the output device will not work properly.

The driver component is called a **MOSFET**. It has three leads, named drain (d), gate (g) and source (s), and it is essential to connect these correctly as shown in the following circuit.

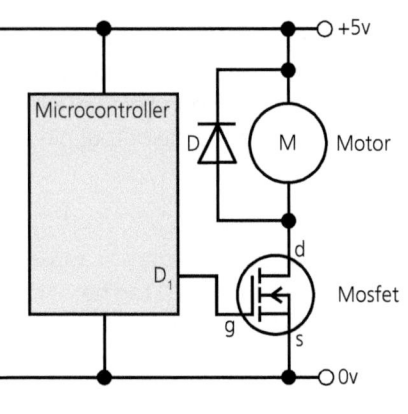

This example shows a MOSFET driving a DC motor, but the motor can be replaced with a buzzer or an LED (don't forget the series resistor with an LED). The **diode** (D) is included in this circuit as a protection device for the MOSFET. The diode is needed only if an electric motor is used, as motors generate a lot of electrical 'spikes' that can damage the MOSFET. If LEDs or buzzers are being used, the diode can be omitted.

Figure 13.26 **Using a MOSFET as a driver**

> **KEY POINT**
>
> MOSFETs are available with different current ratings. Two common types are shown in the table below.
>
> Table 10.1 **MOSFET information**
>
> | MOSFET type | ZVN2106A | IRF510 |
> |---|---|---|
> | Pinout | 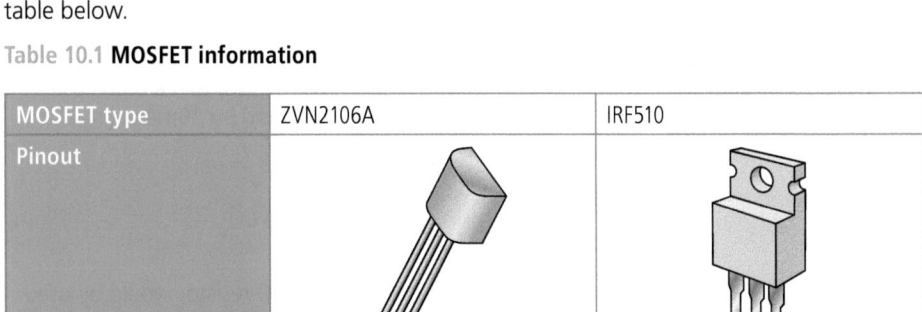 | |
> | Maximum drain current | 0.45A | 4.0A |

Choosing the right microcontroller

There are many microcontrollers available, and the range is constantly changing as new technologies are developed. Each microcontroller has benefits and drawbacks. The choice of microcontroller for a specific application will depend on:

- the technical requirements of the application, such as the number of inputs/outputs needed
- how easy or challenging the microcontroller is to programme
- the choice of power supply (e.g. which batteries you intend to use)
- the range of dedicated accessories available, such as sensors or output devices, which make the interfacing much easier.

KEY TERMS

Diode: A component that allows current to flow in one direction only.

MOSFET: A type of transistor used as a driver.

Designers tend to become familiar with a particular family of microcontrollers, so you may find that your teachers prefer you to use a particular type of microcontroller as they have the resources and experience to guide you through its use in a project.

Some microcontrollers are designed for designers with no previous knowledge, and these should be the easiest to use. Other microcontrollers are aimed at applications in textile or wearable projects. Some are intended to be connected temporarily with crocodile clip leads, others with conductive thread, and some microcontrollers will need you to design your own **printed circuit board (PCB)** and solder them in place.

The programming language will have a big influence on your success with a microcontroller. The more complex languages – while very powerful – are best avoided if you are new to using microcontrollers, because you will spend a lot of time learning the language and **debugging** your program. For all microcontrollers, there are active online communities offering tutorials, advice and project ideas, so you can draw on these for support. With all projects, start simple and gradually build up the complexity of the system and the program as you iteratively develop your design.

The range is constantly evolving. At the time of writing, some of the most popular microcontrollers for GCSE work are:

- **Crumble** can be connected with crocodile clips or conductive thread or it can be soldered. Crumble has a range of expansion boards called Crumbs and full-colour LEDs called Sparkles. The programming language is easy to learn.
- The **BBC micro:bit** was developed to help children learn to 'code', but it is also very useful as a controller within simple Design and Technology projects. It can be programmed using a range of programming languages, so you can learn the language you feel most comfortable with.
- **PICAXE and GENIE** are two similar types of microcontroller **integrated circuits (ICs)**. The easiest way of using them in a project is to purchase them on a 'development board', which provides connections for input sensors and output devices. Or, for those users with a particular interest in electronics and who want to design their own printed circuit boards (PCBs), PICAXEs and GENIEs can be used as standalone ICs. They are powerful and can be programmed using flowcharts or a language called BASIC.
- **Arduino** is a very popular family of microcontrollers but their potential for advanced applications can make them tricky for beginners to use in a simple project. There is a very active online support community for Arduino, so for the keen design engineer who is prepared to find answers to technical problems, Arduino would make a good choice. Different models are available; the Arduino Uno or Nano would be popular choices for design engineers, while the Flora is a round, sewable device intended for wearable products. Several accessories are available including NeoPixels, which are tiny full-colour LEDs available in strips, rings and a pixel matrix.

> **KEY TERMS**
>
> **Debugging:** Finding and removing errors in a microcontroller program.
> **Integrated circuit (IC):** A microchip.
> **Printed circuit board (PCB):** The support and connections for the electronic components in a product.

Figure 13.27 **Microcontrollers and NeoPixels**

Success with microcontrollers

Using microcontrollers is great fun and you should be able to get a simple project running without much trouble – lights, colours, sounds and simple movement should be quickly achievable even by beginners. More advanced projects require you to gain some experience and to spend time learning some more advanced programming skills. Do not try to run before you can walk – remember, as with many aspects of Design and Technology, you do not need to learn everything at once. Master the basics and draw upon deeper technical information as and when you need it.

KEY POINT

Tips for success with microcontrollers:
- Start every project with the most basic function, e.g. flashing a single LED. This makes sure that the microcontroller is receiving power correctly and that the program is being successfully downloaded from the computer.
- As you develop your project, make only one change at a time, and always test the microcontroller after each change. If it stops working, undo the change and get it working again in its previous state.
- There is plenty of support online and someone has probably already done the thing you are trying to do, so search for answers to your problems – this is all part of the iterative designing process. Remember to acknowledge all sources of help.

ACTIVITY

1   Experiment with a microcontroller. If you have never used one before, choose a microcontroller that can be connected using crocodile clips. Using the IDE software on a PC, start by programming the microcontroller to flash a single LED. Then flash a few LEDs in a repeating pattern. Programme it to play a simple tune or sound effect. Then try detecting when an input button has been pressed. Then use analogue commands to sense the light level from a light sensor input and respond by turning on LEDs or making sounds as the light level gets progressively darker.

2   Design and build electronic systems, based on a microcontroller, to achieve the following functions:
   a   A system to warn if a door has been left open for more than 30 seconds (retail stores use similar devices for security reasons, to warn if the stock delivery back door has been left open). Use a magnetic reed switch sensor to detect the door being open. The output device will be a buzzer.
   b   A system to control a cooling fan so that it runs at a speed dependent on the room temperature – the hotter the room, the faster the fan. For this application, use a temperature sensor and a low voltage electric motor to drive the fan. Remember that you will need a driver between the microcontroller and the fan.

3   Search for online tutorials for ideas to help you develop your programming skills and increase your electronics knowledge.

13.3 Sources and origins

LEARNING OUTCOMES

By the end of this section you should know about and understand:
→ the sources and origins of system components
→ consideration of the ecological, social and ethical issues associated with processing system components
→ the lifecycle of system components when used in products
→ recycling, reuse and disposal of system components.

Sources of system components and ecological, social and ethical issues

System components include electronic components and mechanical components. These components are manufactured from a range of materials using a range of processes, and the manufacture may take place in different factories, perhaps in different countries. System components are manufactured on a global scale. Most electronic components are now manufactured in Asia, with China, Japan, India and Taiwan being the big players. The United States leads the world in manufacturing microchips (integrated circuits), and certain other high-value parts are made in Europe.

Electronic components frequently require a range of chemicals and raw materials that are often difficult to source, difficult to use and to process, and often toxic. Also, extracting these raw materials from the earth requires energy and depletes natural resources, and requires energy for transportation. Companies that deal with such materials need to invest large amounts of money in equipment and staff training, and they will be extremely specialised manufacturers. As such, they often produce components on a very large scale at very low unit cost.

It is predicted that, as the complexity of microchips increases, the number of companies with the technology to manufacture them will drop to perhaps only four companies worldwide. Problems may then arise because of the domination of these companies over the industry: lack of competition can lead to rising prices and a lack of manufacturing capacity can lead to 'chip famine', when demand for certain microchips exceeds supply. This happened in 2011 when the damage caused by the Japanese earthquake and tsunami led to a world shortage of computer memory chips and display screens.

Fortunately, system components that are manufactured across the globe are readily available to buy from UK-based retailers, which stock huge ranges of components and can also provide the technical data needed to use them. In fact, electronic components have never been easier to buy and use.

Lifecycle of system components

Electronic components are extremely reliable in use. Quality assurance and quality control during manufacture mean that it is almost unheard-of nowadays for a new component to be faulty. Provided a component is operated within its design ratings, it is also very rare for electronic components to fail during use. The faults that do occur in electronic products are often due to failed solder joints (perhaps due to physical shock such as the product being dropped) or other problems like water ingress.

The **rating** of a component is the maximum value of a specified quantity the component can handle. If this value is exceeded, the component will be damaged, perhaps instantly, or the component may continue to work but its life expectancy will be drastically reduced. Sometimes, the manufacturer will state that it is permissible to exceed the rating briefly.

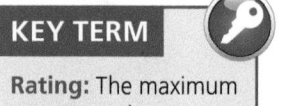

KEY TERM

Rating: The maximum parameter that a component can reliably handle.

Table **13.2 Some examples of component ratings**

Component	Rating
Resistor	Maximum power: 0.25 watt
Microcontroller	Maximum supply voltage: 5.5V
Battery	Minimum temperature: −15°C
Motor	Maximum speed: 12,000 rpm

Some electronic components, however, do suffer from a limited lifespan. Rechargeable batteries are one example; after several hundred charge–discharge cycles the battery's capacity reduces to the point where it may not be able to hold enough charge to usefully power the product for a long period. You may have noticed such problems with your mobile phone or laptop battery as it ages.

Some system components are deliberately disposable, such as non-rechargeable batteries. The manufacturer of a product that uses disposable batteries should demonstrate a responsible attitude by balancing the energy needs of the product and the user's needs to reliably use the product, with the environmental impact caused by manufacturing and disposing of batteries during the product's life.

Mechanical components generally have shorter lifespans because moving parts always involve some degree of friction, which causes wear. Higher-quality mechanical components will use harder-wearing materials that last longer, and provision may be made for servicing mechanical parts, such as renewing lubrication or changing individual parts that are known to wear quickly. In the same way as for electronic components, mechanical components that are operated within their design ratings should have a long and predictable lifecycle.

KEY TERMS

Obsolescence: Becoming outdated or no longer wanted.

WEEE directive: A sustainability scheme to reduce the amount of waste electrical products sent to landfill.

Recycling, reuse and disposal

Figure 13.28 Waste printed circuit boards ready to be recycled

The principles of **obsolescence** and the problems caused by our 'throwaway society' are covered in Section 1.2.

The point at which it is better for the environment to buy a new product rather than keep using an old one is called the breakeven point. The typical breakeven point for a mobile phone is 7 years, but the average user will exchange their phone for a new model after just 11 months.

Electronic and mechanical products may contain hundreds of different components and a wide range of raw materials, many of which will be toxic. Some of the hazardous materials used in electronic components include lead, cadmium, mercury, sulphuric acid and radioactive substances. If products are disposed of in normal waste they will go straight to landfill, where they will decompose and the hazardous materials will leak into the environment. This means they will get into the water system and potentially cause serious health problems for humans.

The average UK citizen will dispose of over 3 tonnes of electrical and electronic products in their lifetime. The WEEE Man is a thought-provoking sculpture based at the Eden Project in Cornwall, constructed from the typical quantity of electronic products an average person will throw away in their lifetime.

The **WEEE (Waste Electrical and Electronic Equipment) directive** now requires all manufacturers and producers to take responsibility for what happens to the products they sell at the end of their lives. In practical terms, this means that retailers of electronic products must now provide a free take-back service for customers to hand in the unwanted product they are replacing. The retailer must then dispose of the products at an approved treatment facility. You may have spotted in your local supermarket the collection bin for old batteries – all battery retailers must now provide this take-back facility.

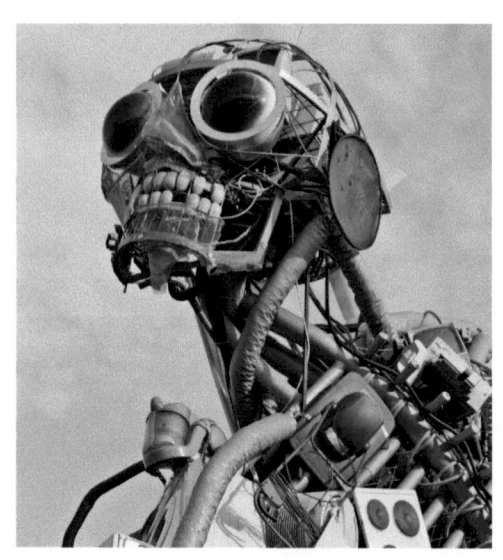

Figure 13.29 The WEEE Man sculpture

13.4 Manipulating and joining

LEARNING OUTCOMES
By the end of this section you should know about and understand:
→ The specialist manipulating and joining techniques used in Design Engineering

When developing iterative designs, a Design Engineer will need to consider electronic and/or mechanical systems to achieve a desired function, but they will also need to consider the use of materials to provide support, structural integrity or protection for the system. Consequently, a working knowledge of polymers, metals, timbers and textiles is required to successfully produce a Design Engineering solution. Core knowledge will provide students with a broad understanding of what can be achieved with these materials. As a design is developed, it will be necessary to learn some in-depth knowledge in a few specific areas of material properties and how to manipulate and join them to achieve a desired outcome. For this, the reader is encouraged to 'dip in' to other sections of this book to extract the precise information required to develop their specific project.

Printed circuit board manufacture

In an electronic system, the printed circuit board holds the electronic components and makes the circuit connections between them. The board is made from an insulating material (often glass-reinforced plastic, GRP) and the connections are formed by copper tracks, to which the components are soldered.

In a school environment, printed circuit boards (PCBs) are usually made by one of two methods:
- Photo-etch method
- Isolation routing

Photo-etch method

The PCB begins as a sheet of GRP onto which is bonded a thin laminate of copper with a layer of photo sensitive film applied. The PCB artwork (the pattern of tracks) is printed in black ink onto a transparent film. During processing, ultra-violet (UV) light is shone through the artwork film onto the PCB and then the board is developed in a chemical tank. This produces an image of the track pattern on the surface of the copper laminate. The board is then immersed into an etchant which chemically removes all the copper except in the areas where the track pattern has been formed. At the end of the process the PCB is drilled to allow the component wires to be inserted and then soldered. School-made PCBs are usually single-sided, having tracks on only one side of the board.

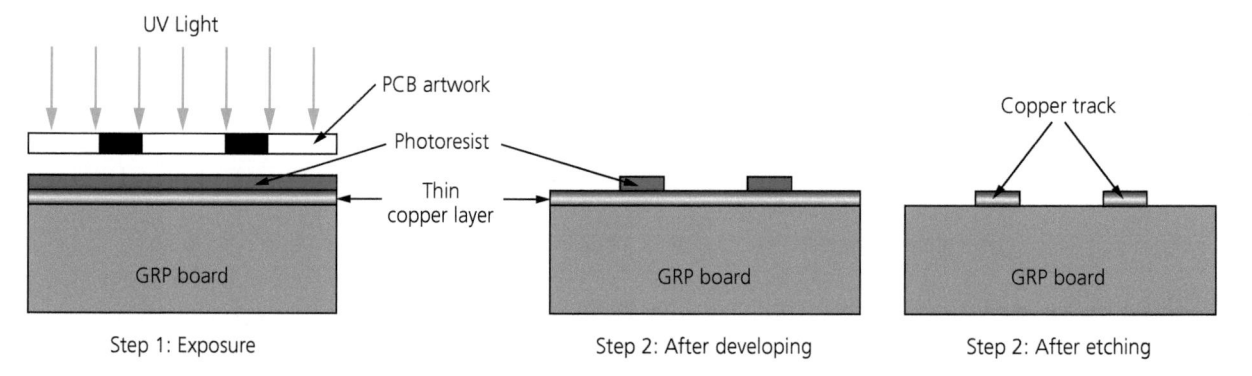

Figure 13.30 **Photo-Etching**

The photo-etch method allows high quality PCBs to be made, but it does require the use of quite hazardous chemicals so safety precautions have to be taken when using these. There are also a large number of variables which can affect the outcome such as UV exposure time, chemical temperature, developer and etching times, age of chemicals etc, so it can be difficult to guarantee a good outcome at the first attempt, especially for an inexperienced user. The pre-sensitised PCB sheet and the chemicals are also quite expensive.

PCB isolation routing

In this method (which is also known as PCB engraving), the PCB begins as a sheet of GRP onto which is bonded a thin laminate of copper. A CNC engraving machine is used to remove copper from the board to leave isolated areas which are then the connecting tracks between components. The CNC machine is controlled from an output file generated by the PCB design software. The same machine will often be able to drill the holes in the PCB pads.

Isolation routing has some advantages over the photo-etch method in that no unpleasant chemicals are used which makes the process somewhat safer, and the running costs are relatively low. The processing time for a single circuit board is quicker by the routing method. However, the initial equipment is expensive and a skilled technician is needed to set up the CNC machine, ensure the PCB is loaded in exactly the right place and securely clamped. It is important that the engraving tool is kept sharp otherwise the cut edges of the copper will be burred and soldering will then be very difficult. Incorrect insertion of PCB or tools can result in expensive damage to the CNC machine.

Figure 13.31 PCB engraving

Soldering

Soldering is of paramount importance in electronic systems as it is the main method used to make the physical and electrical connections between electronic components. Poor soldering, or failure of a solder joint is by far the most common cause of faults in electronic systems.

Solder is traditionally an alloy of tin and lead but, in 2006, the Restriction of Hazardous Substances (RoHS) directive prohibited the use of significant quantities of lead in consumer products produced in the European Union. This resulted in most solder now being lead-free, and many schools will also use lead-free solder as 'good practice'. Lead-free solder is predominantly made from tin, with small amounts of copper and, possibly, silver added.

Soldering for small scale electronics construction will be carried out by hand using a soldering iron. The hot end of the soldering iron is called the bit and this will be shaped down to a small size in order to direct the heat to precisely where it is needed. The solder will be in the form of a thin wire with a core of flux, which is a cleaning chemical needed to ensure that the solder binds effectively to both parts of the joint. Soldering by hand is a skill which can be developed and it must be mastered before you can expect to produce reliable electronic systems. Poor solder joints can occur for a number of reasons, including:
- The joint not being hot enough for the solder to flow completely around the joint
- Dirt or oxidisation on the metals, or a dirty soldering iron bit
- The joint moving before the solder solidifies
- Not feeding in fresh solder wire – trying to transfer hot solder from the iron onto the joint.

A careful visual inspection must be made after every joint is made. Examples of good soldering are shown in the photograph. Bad joints must be corrected or they are likely to cause problems later on during testing. In some cases, a poor joint can simply be reheated and fresh solder applied. If the joint already contains too much solder then this will need to be removed with a solder-sucker tool, or with de-soldering wick, before fresh solder is then used. As with all manufacturing, it is better to get it right first time!

Attaching to a rotary system

It is frequently necessary for one part of a rotating system to transfer its motion to another part on the same axis. Examples are explained below.

Attaching a wheel or a gear to a shaft

Drive wheels need to be attached to their drive shaft so that the rotation of the shaft is transferred to the wheel without slipping. For low torque systems using plastic wheels this could be achieved by drilling a hole in the wheel which is slightly smaller than the diameter of the metal shaft so that the wheel is a tight fit onto the shaft, which results in the rotation being transferred through friction.

This method would not be satisfactory in higher-load systems where a more positive method of drive is called for.

The diagram illustrates two other popular methods of attaching wheels to drive shafts in school projects. Both methods allow for the wheel to be removed. The grub screw method is probably the simplest and most popular choice; notice that a section of the shaft needs to be filed flat so that the grub screw has a surface to 'bite' on.

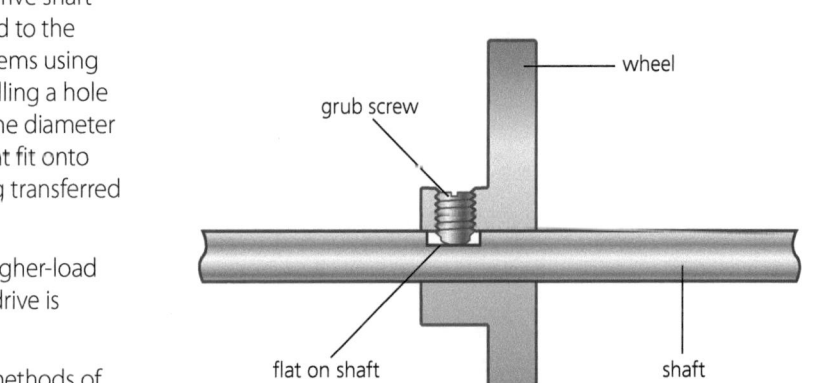

Figure 13.32 **Different ways of attaching drive wheels to a shaft**

Attaching an electric motor to a drive shaft

Electric motors usually have an output shaft which is just a few millimetres long. In some applications, a pinion will be attached to this shaft (possibly using the grub screw method described above). In other applications it is necessary to couple the motor output directly to a drive shaft in order to transfer the rotation to a different place.

The diagram shows how a drive shaft coupler can be used to join two drive shafts end-to-end. The shafts may be of different diameters. As it is very difficult to precisely align two shafts, especially if they are quite long, flexible couplings are available which allow for slight misalignment.

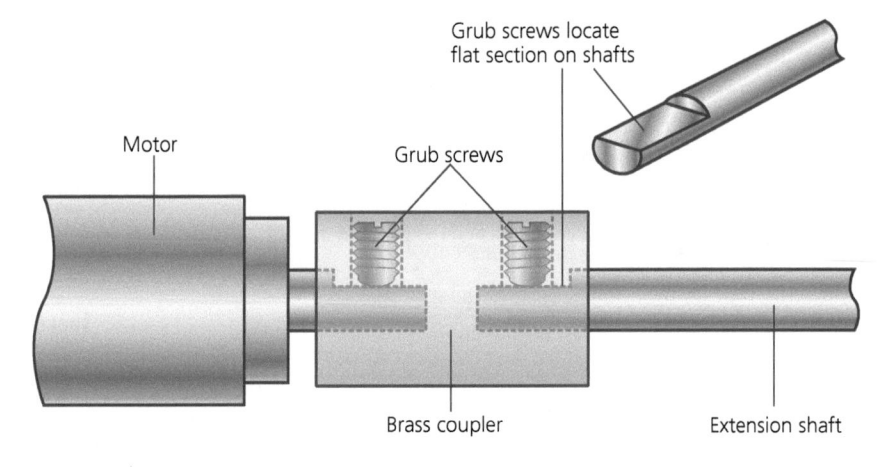

Figure 13.33 **Extending a motor drive shaft**

Linkages, mounts and special components

In a Design Engineering project, most of the components will be purchased ready to use. However, it is quite common, especially in mechanical systems, for certain parts to be needed which cannot be bought and which need to be designed and manufactured specially to perform a particular job.

Examples include a linkage to connect one part of a mechanism to another, or a motor mounting bracket, or a holder for a sensor in an electronic system.

Such components can be made using a variety of materials and workshop tools and machines, referring to other sections in this book. Polymorph (see section 5.2) can be a useful material for quickly manufacturing special parts, especially in the development stages. Once it has fully hardened, polymorph can be drilled and machined to accept screws and other fasteners.

Special components can also be designed using 3D CAD and then produced on a 3D printer to produce accurate and durable parts which can quickly be modified and re-printed if necessary.

Fasteners and fixings

Section 10.7 of this book describes various temporary fasteners such as screws and nuts and bolts. These fasteners are frequently used in Design Engineering to assemble mechanical and electronic components. In mechanical systems, screw fasteners can sometimes work loose due to vibration and care must be taken to prevent this from happening.

Figure 13.34 Nyloc nuts

Figure 13.35 **Shakeproof washers**

Nyloc nuts contain a ring of nylon which grips the screw threads and prevents the nut from vibrating loose. Shakeproof washers are placed under a conventional nut and, when the nut is tightened, the teeth on the washer bite into the nut and prevent it from turning easily.

Threadlock is a liquid substance which is applied to a screw thread before the screw is inserted. Once it dries, the threadlock effectively glues the thread in place preventing it from accidentally loosening. Threadlocked screws can still be loosened if required using tools.

13.5 Structural integrity

LEARNING OUTCOMES

By the end of this section you should know about and understand:

→ how and why system components need to be reinforced to withstand forces and stresses

→ processes that can be used to ensure the structural integrity of a product.

How system components can be reinforced to withstand forces

The individual parts of a mechanical system must be held firmly in place while the mechanism is in motion and while it is subject to various forces and torques. A **structure** is a collection of parts that work together to provide support. The parts in a structure are called **members**.

The support structure for a mechanical system is often called the **chassis**. If the chassis is not rigid enough to withstand the loads without flexing, there is a risk that gears, shafts, belts, etc., will move out of alignment, which could cause problems and possible damage to the mechanical components. Achieving structural rigidity is very important in a mechanical system.

A structural system is said to be **rigid** if it can withstand forces and stresses without bending or flexing. The diagram shows a thin bar of material supported between two posts. A force applied to the centre of the bar will cause it to bend. The amount it bends depends on:
- the size of the force
- the material the bar is made from
- the distance between the supports
- the thickness of the bar
- the cross-sectional shape of the bar.

> **KEY TERMS**
>
> **Chassis:** The support structure for a mechanical system.
>
> **Member:** The individual components in a structure.
>
> **Rigid:** The ability to withstand forces without flexing.
>
> **Structure:** A collection of parts that provide support.

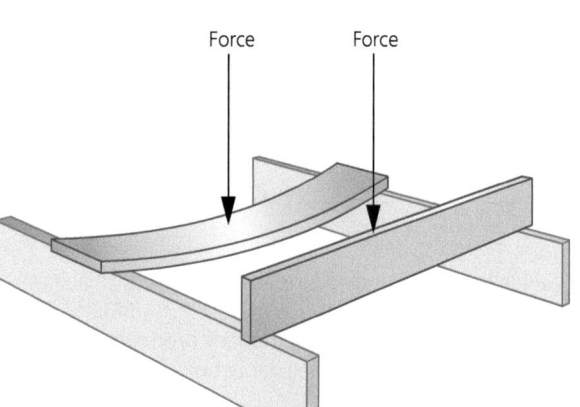

Figure 13.36 **A bar bending under load**

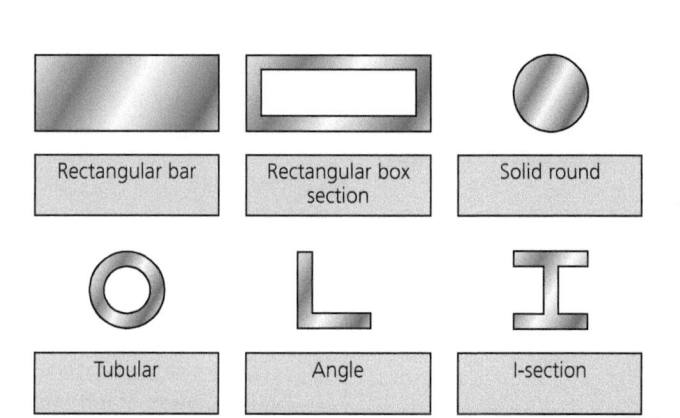

Figure 13.37 **Examples of material cross-sections**

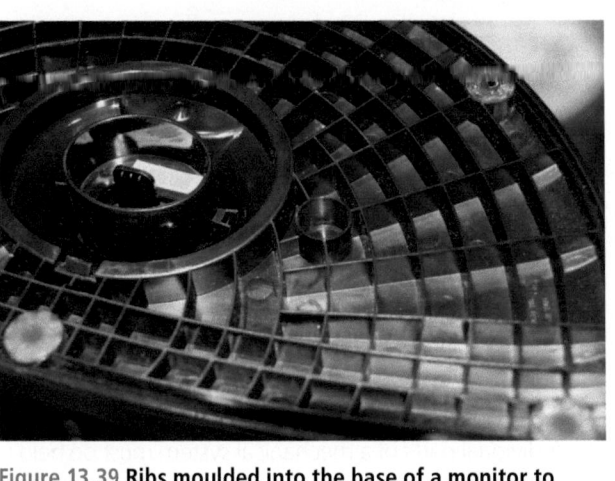

Figure 13.38 Corrugated metal sheet and a metal panel with indents for rigidity

Figure 13.39 Ribs moulded into the base of a monitor to increase rigidity

Placing the bar vertically as shown in the second diagram will reduce the bending because the effective thickness has been increased.

One way the bar could be made more rigid is by making it thicker, but a more efficient method is to change the cross-sectional shape. Stock lengths of metal and plastic are available in a range of different cross-sections. Some examples are shown in the diagram below.

The use of hollow tubular or box section materials allows designers to achieve almost the same rigidity as a solid material with the same dimensions, but with reduced weight and reduced cost. This works because when a material is bent, most of the bending forces are concentrated in the outer surfaces. This means that removing material from the centre has little effect on the rigidity. Having less material in the middle section and more material on the outer surfaces makes for a more efficient and rigid design.

Large sheets of flat material are made more rigid by including folds and indents in the material, as shown in the photographs.

 **KEY TERM**

Rib: A thin support added to increase rigidity.

Ribs can also be added to increase rigidity. A rib is a thin support that runs at right angles to the surface of the sheet and that is attached to the sheet. In plastic parts, ribs are moulded at the point of manufacture. For metal parts, ribs are often welded or riveted on to the sheet during assembly of the system. An example is plastic ribs being used to reinforce the plastic base of a monitor, as shown in the photo.

ACTIVITY

1 Find five examples of products that use hollow or angled cross-section pieces to achieve rigidity. Photograph the examples and annotate the photos to explain your findings. You may need to look carefully to find examples; remember that, in many products, hollow tubes or box sections will have plastic end caps to finish the product nicely – these end caps are a clue that the material is hollow!
2 Find and photograph five examples of products that feature indents or ribs to achieve rigidity. You may need to look inside or underneath the products, as ribs are often hidden because they spoil the aesthetics of a smooth, flat surface.

Ensuring the structural integrity of a product

Even if the individual components are made rigid, a complete chassis structure will not be rigid unless it has been designed properly. A rectangular framework is naturally unstable and it can collapse when a sideways force is applied.

The only way to keep a rectangular framework rigid is to make the corner joints very strong, for example by using large screws, strong adhesive or welds. This is the only practical method in some designs, but a better approach is to design the frame using **triangulation**, which is where the frame is composed of triangles rather than rectangles.

A triangular frame cannot change its shape – it is naturally rigid and doesn't rely on the stiffness of the joints to keep its shape. Adding a diagonal **cross-member** across a rectangular frame creates two triangles and this makes it become a naturally stable framework.

If a design prevents a full cross-member being used, then **gusset plates** can be used to achieve a partial approach to triangulation and, therefore, increase the rigidity of the framework.

KEY TERMS

Cross-member: A member added to achieve triangulation.

Gusset plate: A corner reinforcement to increase structural rigidity.

Triangulation: Achieving rigidity by producing triangular structures.

Force →

Figure 13.40 A rectangular frame is naturally unstable

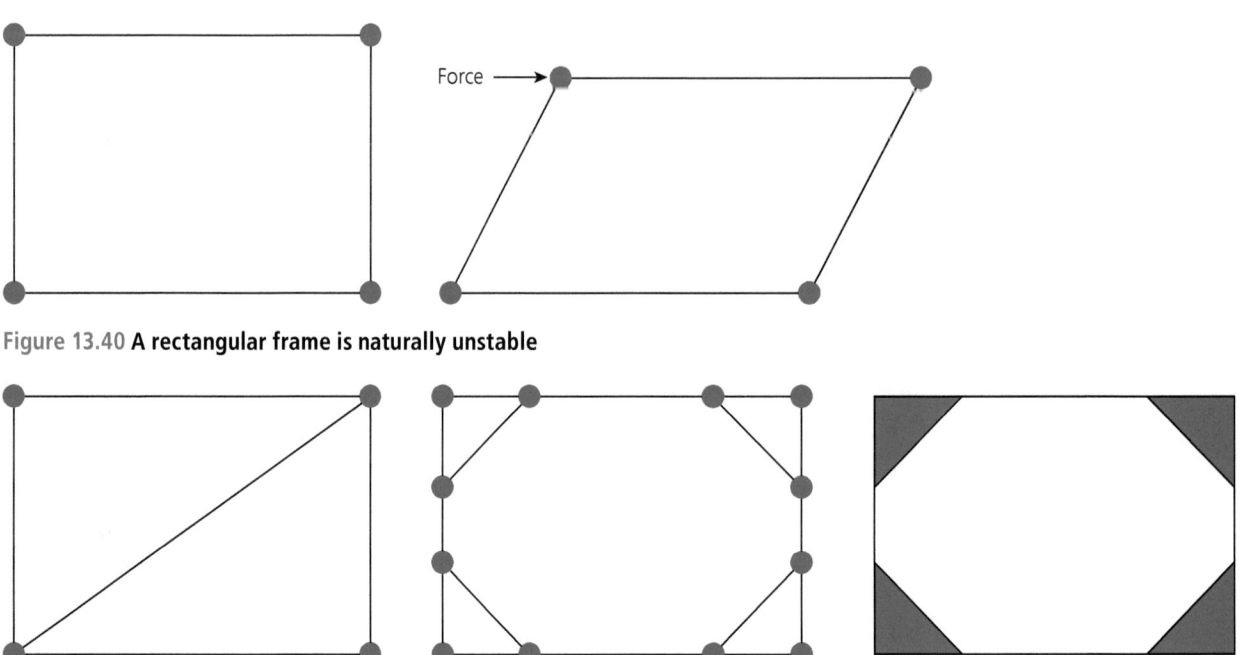

Adding a cross-member   Corner-braces   Gusset plates

Figure 13.41 Methods of triangulating a framework to increase its rigidity

ACTIVITY

1. Find five examples where cross-members have been used to achieve triangulation in a framework. Photograph and annotate your findings. Look for large-scale frameworks in towers and buildings, but also for smaller examples around your house or school.
2. Take five photographs of examples where gusset plates or other methods of corner reinforcement have been used in products.

13.6 Making iterative models

LEARNING OUTCOMES

By the end of this section you should know about and understand:
→ the processes and techniques used to produce early models to support iterative designing.

Other sections in this book explain how physical and aesthetic models can be made from a range of materials. This chapter focuses on models used to test and develop mechanical and electronic systems.

Iterative designing relies on models being created to test ideas, and the results of those tests being used to improve the design. Further models may then need to be made, until the design achieves the desired outcome.

Models should be made only because they are needed to answer a question about a design, or to provide some parameters (numbers) or limitations for the design. Never make a model just because you think you are expected to; before making any model you should know exactly what it is you are setting out to investigate. Also, think carefully before building a model to find out data that you could easily look up from an existing source.

Broadly speaking, models can be:
- **physical models**, made using real materials and components
- **computer-aided models**, such as 3D computer-aided design (CAD) or electronic circuit simulation using computer-aided engineering (CAE)
- **software models**, which test flowchart programs for a microcontroller
- **mathematical models**, where a real situation is explained through a mathematical or scientific equation in order to make predictions about the situation.

Remember that models are useful for helping to solve small problems within a larger design. When designing a functioning system, the most useful models may be the ones that explore, for example, the different ways of attaching a wheel to a drive shaft, or the optimum number of LEDs to use to illuminate a given area.

Making mechanical models

Mechanical systems are quite difficult to design without actually building and testing them. Forces, mechanical advantages and gear ratios can be calculated, but it is often not clear whether a system will actually work until it is built.

For example, when using electric motors and gearboxes it is difficult to predict how fast the motor will turn when it is under load. Some of the uncertainty can be reduced by trying out different motors/gearboxes, and this gives a better idea whether the system is likely to be too fast or whether it will produce sufficient torque to keep it turning under load. Some motor/gearbox kits allow you to vary the gear ratio by using different combinations of gears, so there are opportunities here to model various combinations in order to decide upon an optimum ratio for a given application.

While some mechanical components will be purchased, parts such as linkages, levers, cams or chassis components can be designed on CAD and then laser cut out of a suitable material. The modelling stage is unlikely to involve cutting these parts out of card in the first instance. This would allow you to check that the dimensions are correct, including features such as bearing mounting hole sizes and separation of shafts to ensure the correct meshing of gears. Corrections would then be made before committing to cut out in the final material. 3D printing provides an alternative method of modelling mechanical components, or even of producing the final work part. During the modelling stages the designer may start to experiment with bought-in components such as motors/gearboxes/sensors etc. and the fixed sizes of these parts will begin to determine certain dimensions of the overall design. The final scale of the prototype will need to be kept under tight control during the iteration stages before presentation to stakeholders.

Foam board (see Chapter 11) is another useful sheet material for modelling mechanical components, and also has the advantage of being easy to cut using a modelling knife.

A useful method for testing the movement of linkages is to make a 2D model out of foam board (or card) and to use paper fasteners as pivots. Such models can be used to test that the full range of required motion can be achieved, and to prove that parts don't collide or lock-up. Foam board is easy to cut, so iterations of the system can be quickly developed and tested.

CAE software has some use for modelling mechanical systems but, as mechanical systems tend to be bespoke designs for specific applications, the best results are often obtained using 3D CAD software. It can, however, take a long time and a lot of user skill to develop virtual mechanical systems using this software.

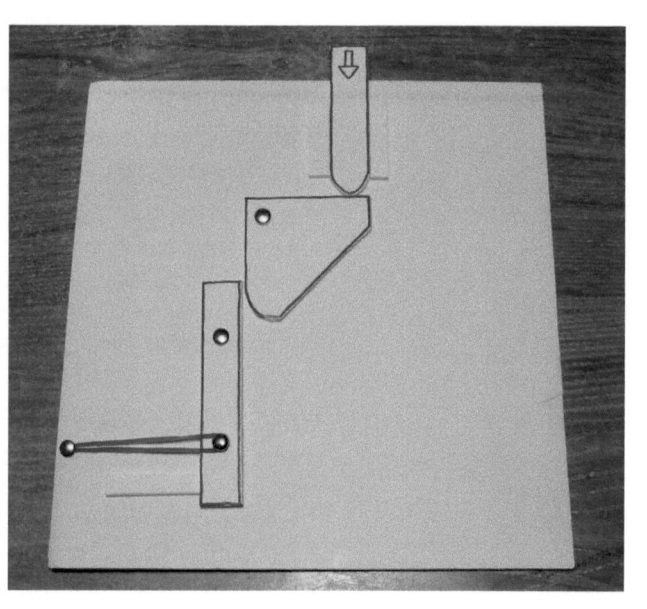

Figure 13.42 A linkage mechanism being modelled in foam board

Making electronic models

Electronic models are usually constructed in order to:
- check that the interfacing works between the input/output devices and the microcontroller
- check that the input/output devices achieve the desired effect
- check the functioning of the flowchart program.

> **KEY TERM**
>
> **Breadboard:** A prototyping tool for building electronic circuits.

In Section 10.2, it was explained that some microcontrollers are designed for experimental work and that these types allow for electrical connections to be quickly made using crocodile clip leads. This means that input and output devices can very easily be connected, disconnected and swapped. For more advanced designs, which require the use of other components such as resistors or driver ICs, a **breadboard** may be needed. A breadboard (or 'prototyping board') consists of holes into which component wires can be inserted. The holes are electrically joined in rows of five, and wires sharing the same row will be connected. This allows electronic circuits to be quickly built without the need for any soldering. Breadboards take a little practice to use effectively, but they are by far the most effective method of modelling electronic circuits.

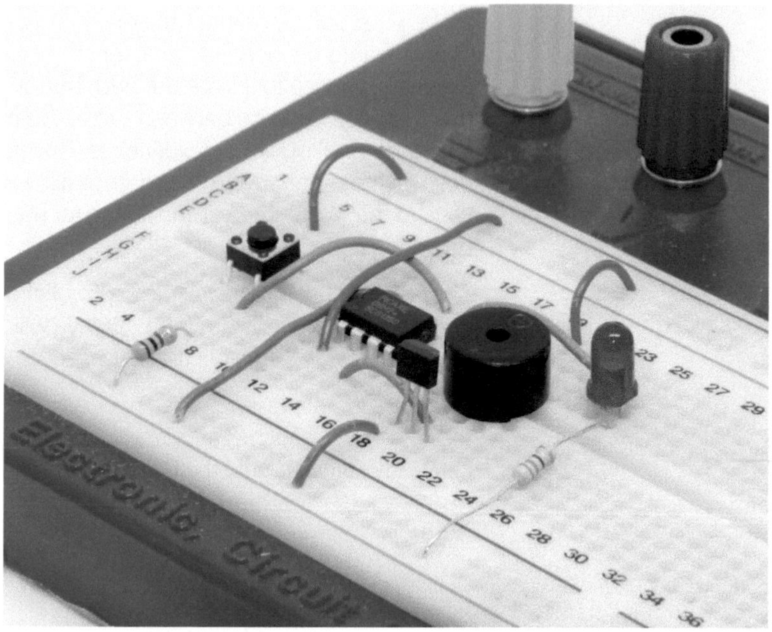

Figure 13.43 An electronic system modelled using a breadboard

CAE software is available for modelling electronic circuits and this is generally very effective and can save hours of breadboard prototyping. Components can be quickly connected and disconnected and the designer can easily modify a component's value (e.g. the resistance of a resistor) and observe the effect. This provides a rapid method of optimising component values by trial and error.

Integrating the full programmable functionality of microcontrollers is a problem for some circuit simulation software. Certain software packages will simulate only specific microcontrollers, while others can't simulate a microcontroller program at all. It is likely that this functionality will improve as the software develops, but it is important to understand the limitations of CAE circuit simulation for a particular design situation.

Making microcontroller software models

A very important part of a successful functioning system in design engineering is the iterative development of the microcontroller program. It is very tempting to rush this aspect, to leave it to the end, or to assume that the first version is going to be satisfactory. The microcontroller program is crucial to the successful operation of the whole system, however, and it must play a continuing part in the iterative design process.

The most obvious reason to iteratively test the microcontroller program is to remove bugs – errors that cause the program to behave in an unexpected or undesirable way. Some software packages for programming microcontrollers allow the program to be run and simulated before downloading it to the microcontroller, and this can be a useful way to iron out obvious bugs. Running the program on a PC also allows system parameters to be monitored; the program can usually be run one step at a time, giving you a chance to identify exactly where a problem is arising.

Many products have a user interface (UI) that is directly influenced by the microcontroller, and a poor UI will make the product difficult or even impossible to operate properly. The microcontroller program, therefore, can directly affect the ergonomics of a product.

For example, refer back to the bicycle safety lamp in Figure 13.14 and Figure 13.16. Each press of the push button changes the operating mode of the lamp. It was explained that the 'WAIT 1s' commands are needed to prevent the program skipping through the modes before the user has a chance to release the push button. If the 'WAIT' interval is too long, however, the user will have to pause between successive pushes of the button, which gives an awkward (and annoying) operating experience. Therefore, the duration of the 'WAIT' interval is best determined by iterative trial-and-error, allowing real operators to use the prototype system, varying the 'WAIT' interval and asking for their feedback on which value gives the best operating experience.

Allowing real users to test a microcontroller program is essential, because they are likely to use the product or system in an intuitive way, and your first attempts at writing the software may not provide a suitable user experience.

Iterative software testing is also needed to fine tune program parameters, or to calibrate the system. One such example would be in a cooling fan system, where program values are iteratively adjusted until the fan switches on at the desired temperature.

13.7 Finishes

LEARNING OUTCOMES

By the end of this section you should know and understand:
→ The processes used to finish design engineered products for specific purposes including:
→ Function, such as durability and ability to overcome environmental factors
→ Aesthetics

Durability and environmental factors

Section 13.5 dealt with the issues of designing products to have structural integrity so that they can withstand the forces they will be expected to experience during use.

In the broader sense, durability can be thought of as the requirement for the product to continue to function when it is exposed to a wide range of environmental factors such as dirt, water, extremes of temperature, knocks and bumps etc.

The solar-powered lamp in the photograph will be left outside all year long. The outside environment is very harsh and the product will be exposed to rain, ice, long periods of sunlight (which can cause plastics to fade in colour and to become brittle), and temperature ranges from perhaps -10°C to +30°C in the UK.

Figure 13.44 A solar-powered garden lamp

ACTIVITY

Research and consider the ways in which the solar-powered garden lamp has been designed to be durable and to withstand the outdoor environment.

Water ingress can be extremely damaging to electronic and mechanical products, and it is very difficult to make a product fully waterproof. However, by careful design, it may be possible to make the product 'splashproof' and more durable at withstanding accidental splashes such as may occur in a kitchen, for example. The photograph shows the control panel of PCB developing and etching tanks. Notice the flexible plastic covers over the rocker switches and the clear plastic splash-screen over the entire control panel.

It may be tempting to try to completely seal the casing of a product using adhesive so that water cannot enter, but this needs balancing with the need to open the case to service the device, change batteries, lubricate the mechanism etc. A better solution would be to use a rubber seal at the point where the case parts join. A suitable seal in a school project would be the self-adhesive rubber strip used to seal against draughts around doors.

Figure 13.45 Splashproof covers on a control panel

Figure 13.46 **Waterproof seal on a sports camera case**

Figure 13.47 **Weatherproof box with cable glands**

For very demanding, fully waterproof solutions it will probably be necessary to purchase a waterproof 'project case' and place the system inside this. Remember, however, that whenever a hole is drilled in the case for cables, switches etc, then the waterproofness can be compromised. Cables should exit through a 'cable gland' which maintains the seal of the case.

In addition to water, dirt ingress can be very damaging, especially to mechanical systems. The lubricant used in some mechanisms can be very sticky and can pick up grit which then acts as an abrasive which can cause serious wear to moving parts. Therefore, the choice of lubricant is important. For example, a bike chain is fully exposed to road dirt, so cyclists often use a 'dry' lubricant on the chain rather than conventional grease, as the dry lubricant does not collect road dirt like grease would.

When buying mechanical components, it is possible to buy 'sealed' bearings which are pre-lubricated and have covers on their sides to keep out dirt and moisture.

Figure 13.48 **Sealed and open bearings**

Aesthetic finishes

Much is written elsewhere in this book about how to achieve a quality finish on a range of materials, and how to protect the materials from environmental factors.

In a Design Engineering context, the aesthetic finish sometimes extends to labelling: there is frequently a need to mount controls on a panel which also requires labels to identify the controls. Prior thought about how to label the panel can result in a high quality labelling outcome. A poor outcome results when, at the last minute, paper labels are simply stuck on to the panel!

Figure 13.49 Simple labelling on an aluminium panel

Various options exist for labelling prototype panels. Sheets of transfer letters are available which can be placed directly onto the panel; these would require a clear lacquer finish to protect them from being scratched off.

A CNC vinyl cutter can be used to cut out bespoke labels which can then be stuck onto an existing panel.

By using a laser cutter, a complete panel can be cut out, including holes and engraved labels in one go. Careful choice of panel material for laser cutting can result in some excellent, professional-looking panels.

Also available are a wide range of 'engraving laminates' which are plastic sheets available in a wide range of finishes, including aluminium and wood grain effects. When engraved by a CNC machine (or a laser cutter), a contrasting colour shows through, allowing creative panel designs to be produced.

13.7 Using digital design tools

> **LEARNING OUTCOMES**
>
> By the end of this section you should know about and understand:
> → the use of 2D and 3D digital technology and tools to present, model, design and manufacture solutions.

Rapid prototyping

Through the iterative designing stages of a design engineering project, it sometimes becomes apparent that a custom-designed part is needed. This often occurs in a mechanical system when a particular size or shape of linkage is required, or a special bracket is needed to mount a motor or a sensor.

Such parts can be manufactured by hand in the workshop using machine tools and hand tools, but the advent of rapid prototyping has meant that a part can be designed using CAD software and then produced in a matter of minutes or hours using CAM machinery. 3D printers are now able to produce resilient and strong complex-shaped parts that, in many cases, can be used directly in the final prototype system.

Conventional machining of the same part would require a high level of skill and be time-consuming (and, therefore, expensive). Rapid prototyping has given birth to the process of iterative designing, where designers can now build first generation rapid prototypes, test and improve them, before building second generation prototypes, and so on.

In addition to the actual rapid prototyped part, the designer also has a CAD model of the part that can then be assembled into the CAD model of the entire system.

CAD, CAM, CAE, digital manufacture and interpretation of plans, circuit diagrams, PCB layouts

Design engineers also make use of other digital technologies that can link to various CAM machines to aid the manufacture of systems.

2D CAD software can output cutting information to a laser cutter, allowing sheet materials to be cut to accurate dimensions and intricate shapes. This can be useful for cutting out functioning parts such as a chassis for a mechanical system. The chassis may need to support various components such as motors, bearings or pivots, and it is crucial that these are all mounted in exactly the right positions. Producing parts on a laser cutter ensures accuracy and repeatability.

Other industrial sheet material cutting machines include plasma cutters, CNC routers and vinyl cutters. All perform a similar job but each one is suited to machining different materials and different sheet sizes.

CAD software can also be useful for printing out full-size templates that can be used to assist the manufacture of parts or help ensure that parts are aligned properly within an assembly. Sticking a template on to a sheet of material allows the material to be cut out and drilled accurately.

The electronic system in many prototypes is built on a printed circuit board (PCB). A PCB is a custom-designed board with the components soldered to copper tracks, which complete the circuit connections. Most CAE circuit simulation software will link to PCB design software so that a functioning circuit can be developed as a real PCB design.

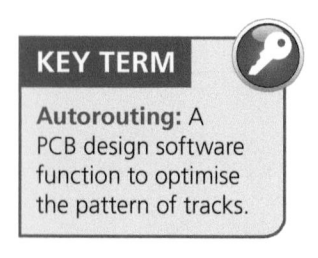

KEY TERM

Autorouting: A PCB design software function to optimise the pattern of tracks.

Designing the PCB copper track pattern can be quite tricky as the components need to be laid out so that the tracks do not cross over each other. PCB design software uses an **autorouting** function to achieve this. The PCBs in modern electronic appliances such as computers and mobile phones are so complex that they could not possibly be designed without the aid of PCB design software.

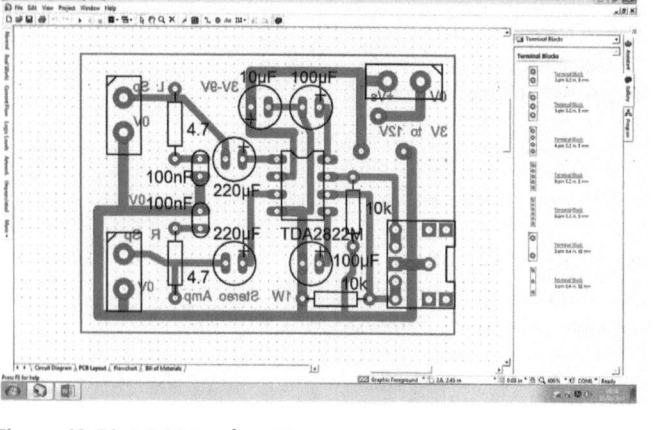

Figure 13.50 **A PCB track pattern**

Once the PCB design is complete, the software will print out the track pattern artwork on to a translucent film, which is then used to produce the actual PCB.

13.9 Manufacturing methods and scales of production

> **LEARNING OUTCOMES**
>
> By the end of this section you should know about and understand:
> → the methods used for manufacturing at different scales of production
> → the manufacturing processes used for larger scales of production.

Scales of production

A project's design brief and initial market research will determine how many products need to be manufactured. The method of manufacture will be directly influenced by the scale and speed of production required. The design of the product will also be influenced by the manufacturing scale, as products that were originally intended to be manufactured on a small scale are not likely to be suitable for mass production unless they are significantly re-designed.

One-off, bespoke production

One-off prototypes are quite common in design engineering. For example, many industrial manufacturing processes are unique to a factory, and a design engineer may be asked to produce a bespoke control system for part of a manufacturing plant.

A one-off design is usually expensive because the designer needs to receive payment for their time spent developing the prototype as there will not be any future product sales to bring in a long-term income. The designer may spend significant time on the iterative development of the product until it functions to the client's satisfaction.

Bespoke production usually involves a high degree of manufacturing skill, much of which may be carried out manually, for example a PCB may be soldered by hand. In some cases, manufacturing may be contracted out to specialists, adding to the costs. Rapid prototyping may be useful for manufacturing bespoke parts, and software for the microcontroller will need to be written and developed.

Batch production

For a scale of production that is beyond a few items, a manufacturer will organise the production in a more efficient way. The exact manufacturing process will depend on the product and the manufacturing facilities. Generally, the manufacturer will focus on producing an entire batch of products in one go. Once the batch is complete, the manufacturer may then switch production to an entirely different product.

The production will be organised to make the most efficient use of the machinery available and the skills of the workers. In the case of manufacturing a small batch of a simple mechanical product, this might be:
- Day 1: The entire factory manufactures chassis parts.
- Day 2: The chassis are assembled.
- Day 3: The final components are added.
- Day 4: Product testing.
- Day 5: Packaging and dispatch.

You should be able to see that this approach requires every member of staff to be skilled in carrying out all the processes, because every worker must be kept busy every day. This might be the case in a smaller factory, but in a larger manufacturing plant staff tend to have skills in certain areas so the manufacturing would be organised differently.

Re-organise the five-day manufacture of the mechanical product described above for a factory in which some workers are skilled only at manufacturing, some are skilled only at assembling, and others are skilled at testing and packaging.

As the product complexity increases, the manufacturer will contract out the specialist tasks. For example, it is common for a product manufacturer to leave the moulding of plastic parts and PCB manufacture to specialist companies.

Batch production describes the way the manufacturer organises the production and the fact that a target quantity is agreed before manufacturing starts. There is no limit on batch size. A larger, more complex product may be manufactured in a batch of ten, while a simpler product may be produced in a batch of 10,000.

Mass production

As the name suggests, mass production involves the manufacture of very large numbers of products. Such manufacture is typical for commonly used components such as screws, connectors, batteries, etc. Mass producers are usually very specialist manufacturers that have invested large sums in machinery capable of producing large volumes of parts, with repeatability and reliability being important. Once the production is underway it is often most cost-effective to leave machinery running continuously, with staff working shifts to monitor the process. The sheer scale of production and economies of scale mean that mass production is the cheapest method of manufacture.

Lean manufacturing and just-in-time methods

A manufacturer that is producing a product batch will need to order materials and components that are specific to that product. Consequently, for production to start on day one, the manufacturer relies on receiving delivery of the materials just before manufacturing starts. They will not want to take delivery of materials too early because this causes problems storing them. The materials are ordered to arrive just in time for manufacturing to commence.

Just-in-time (JIT) manufacturing depends on reliable suppliers that will deliver on time, because a missed delivery will hold up the entire production. In practice, the process works surprisingly well as every part of the supply chain realises that their reputation depends on meeting production promises.

Manufacturers are usually keen to ship out the products as soon as they are completed as this clears the factory ready for the next production run to commence. The whole JIT process is aimed at producing efficient 'flow' through the factory.

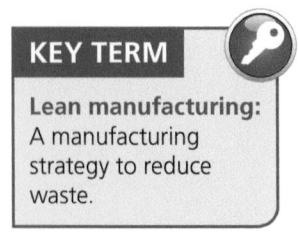

KEY TERM

Lean manufacturing: A manufacturing strategy to reduce waste.

The phrase **lean manufacturing** refers to a management strategy of reducing waste at every stage of the production process. JIT is one example of lean manufacturing. Apart from the obvious aims of reducing material waste by efficient cutting, the general aim in lean manufacturing is to remove any activity that does not directly add value from the customer's viewpoint. Lean manufacturing aims to produce higher quality products at lower cost.

Manufacturing processes used for larger scales of production

The manufacturing methods used to produce a prototype in a school workshop are usually different from those used at larger scales of production.

PCB manufacture

The PCBs produced industrially for modern products differ from school made PCBs in a number of ways. Industrially produced PCBs are also manufactured by a photo-etching process, but they are usually double-sided and have tracks on both the top and bottom of the board. This allows more complex designs to be achieved without the problem of tracks crossing each other. In fact, many modern PCBs are multi-layered, and these can be thought of as several thin single-sided boards laminated together. Multi-layered boards allow highly complex circuits to be constructed in a very compact space, and they are invariably used in products such as mobile phones.

Pick-and-place PCB assembly

Prototypes made in a school workshop will generally use through-hole components, where the components sit on one side of the PCB and their wires pass through the board to be soldered on the reverse side. Industrially produced electronic products use **surface mount technology (SMT)**, where the components do not have wires and are placed directly on and soldered to the surface of the PCB.

Some of the resistors and capacitors used are less than a millimetre in size and they are placed on to the PCB by a **pick-and-place machine**. The components are temporarily held in place by a sticky solder paste. These machines operate at astonishing speed, typically placing five or more components per second. The PCB is then passed through a reflow solder oven, in which the board is raised to a temperature high enough to melt the solder paste.

Ensuring accuracy and efficiency

The PCB manufacturing and assembling process is controlled by computers, as the very small component sizes demand very close tolerances when placing components.

Optical recognition technology is used to visually inspect the PCB after manufacture. A computer checks images of the PCB against a library image to identify if any components have been misaligned during placement.

The final stage is an electrical check, during which power is applied to the PCB to ensure it is functioning correctly.

Mechanical assembly

Mechanical sub-assemblies within products can be assembled by robotic pick-and-place arms, in a similar way to electronic products. Large scale production systems would probably make use of such expensive technology. However, it is not uncommon to find that many simple mechanical products are still assembled by humans, perhaps with the aid of power tools such as electric screwdrivers, or presses to insert gears onto shafts.

The final assembly stages of nearly all products are carried out by human workers. Plugging together electronic sub-assemblies, applying lubrication to gear systems, carrying out quality checks and inserting the final case screws is usually done by hand. People are far better at spotting problems and carrying out final checks than a machine.

> **KEY TERMS**
>
> **Pick-and-place machine:** The machine used to assemble SMT circuit boards.
>
> **Surface mount technology (SMT):** Miniature electronic components assembled by machines.

PRACTICE QUESTIONS: In-depth principles of design and technology

Timbers

1 This figure shows a sandwich board, used to advertise outside shops and cafés.

The sides and legs are made from a hardwood timber.

a Materials need to be sourced and processed in order to be used to make products. For one specific material from the sandwich board:
   – State the source of the material.
   – Describe how it is processed into a workable form.
   – Discuss how the selection of that material is influenced by social and ethical issues.

2 This figure shows a side table.

Designers make prototypes to show their designs to key stakeholders.

Study and use the images and technical information in the figure.

a Produce a step-by-step plan to explain the stages that you would take if you were making a final prototype of your chosen product in a school workshop.
   You must include details of:
   – specific materials and components you would use to make the prototype
   – the processes, techniques or skills you would use
   – tools you would use, including digital technology as appropriate
   – how you would ensure accuracy when making the prototype
   – how you would finish it to present it to stakeholders.

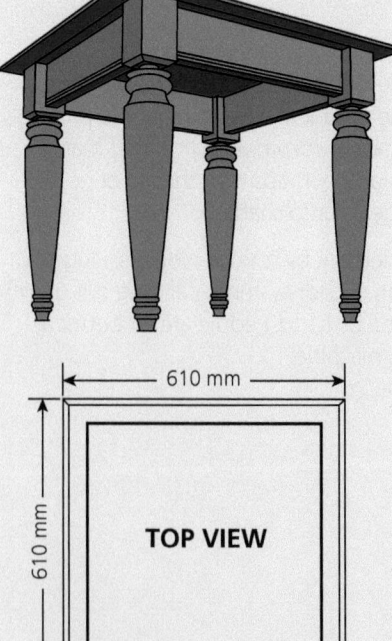

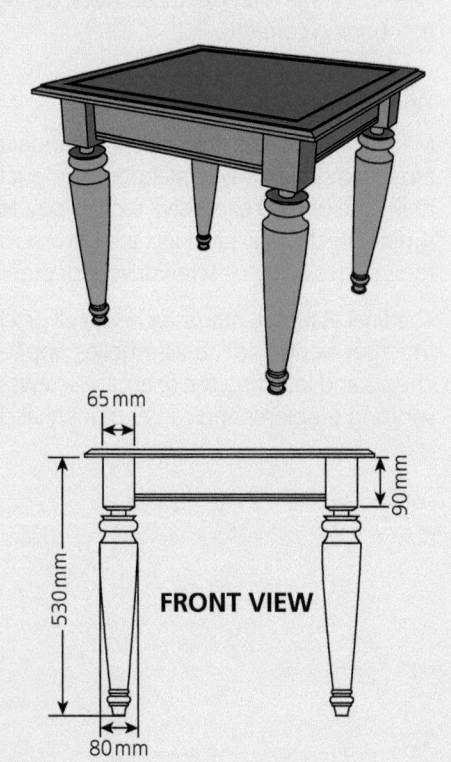

Metals

3 a Chemical reaction is one method of extracting metals.

Name two other methods.

b Chemical reaction uses a blast furnace. A diagram of a blast furnace is shown below.

Label the parts identified with arrows.

4 The structural integrity of metals can be altered by different processes such as annealing.

Describe the purpose of annealing metals

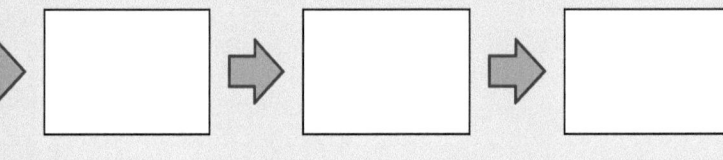

5 Case hardening is another treatment done to metals.

Complete the diagram of the case hardening process below.

Stage 1: Heat metal until it glows red ⇨ ☐ ⇨ ☐ ⇨ ☐

Paper and boards

6 Paper and boards are versatile manufacturing materials that are used for a wide variety of products.

Discuss, using relevant examples, the environmental impact of using paper

7 Describe the embossing process.

8 A cardboard leaflet holder is shown below.

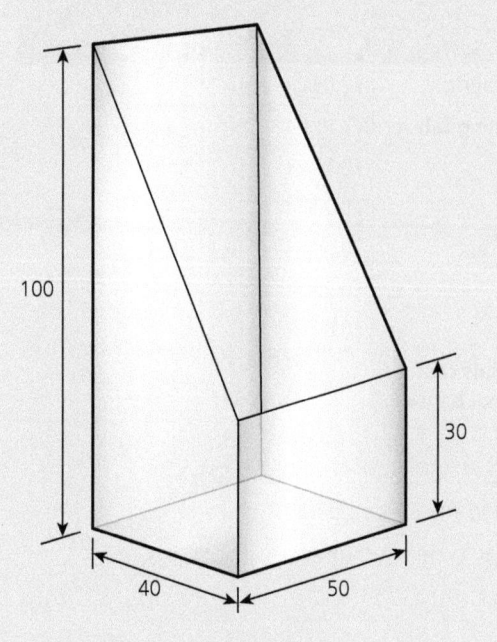

a Draw the development (net) of the leaflet holder including all glue flaps.

b Using notes and sketches show how you would create a 90° fold in foamboard.

Polymers

9 This figure shows a child's sit-on toy. The body is made from a thermopolymer.

a Name a suitable thermopolymer that could be used for the main body.

b Give two properties of thermopolymers that make them suitable for toys like this.

c Name a suitable manufacturing process that could be used for the main body.

d Explain why the process named in (c) would be suitable.

e Discuss how digital technologies have provided opportunities for designers and manufacturers of children's toys.

Fibres and fabrics

10 The image below shows a school bag made from a plain weave polyamide fabric. The bag includes a polyester lining and a number of components.

a A retailer has made an order for 285 bags. Complete the tables below to show the final material costing for this order.

Materials and components required.	Quantity for 1 school bag	Quantity for 285 school bags
Polyamide fabric	1.0m	m
Polyester lining fabric	0.5m	m
Piping	1.0m	m
Webbing	0.25m	m
Strap Adjuster	2	
Zip 22"	1	
Zip 10"	2	

Materials and components	Price per unit/m	Cost for 285 school bags
Polyamide fabric	£2.05	
Polyester lining fabric	£1.10	
Piping	£0.18	
Webbing	£0.20	
Strap Adjuster	£0.05	
Zip 22"	£0.20	
Zip 10"	£0.09	
Total materials cost for 285 school bags=		

b What is the total cost of materials for 1 school bag?

c Explain why a lining fabric is used inside the school bag.

d Name the scale of production that would be used for the school bags.

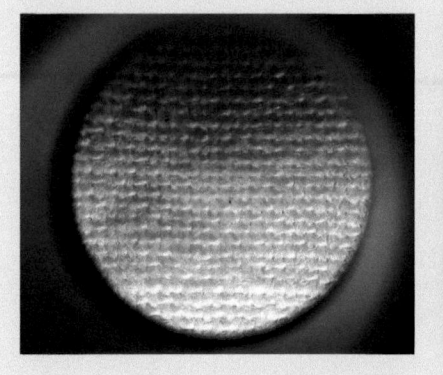 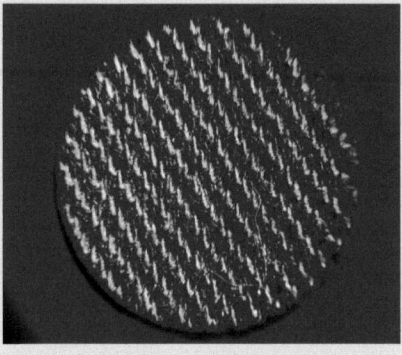 

11 a Identify the twill weave structure from the images below.

b Denim is a cotton twill weave fabric. Explain the benefits and drawbacks of using a cotton twill weave to make denim jeans.

c Eco-Denim is a company specialising in creating jeans that are carbon neutral. They wish to add a logo to the back pocket of their new line of jeans to help build brand awareness.

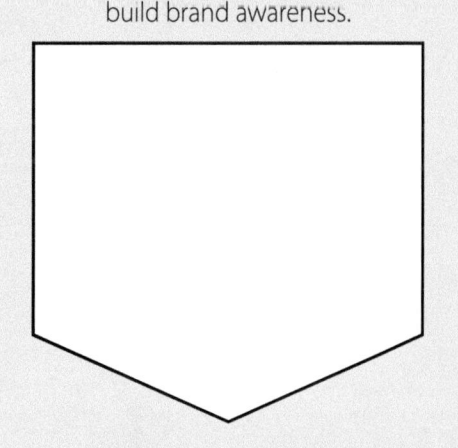

d Describe a suitable decorative method to add a logo design to the pocket.

e Describe the environmental impacts of cotton production.

Design engineering

12 a Explain why electronic products should not be disposed of in normal domestic waste.

b Explain why electronic products can be difficult to recycle.

c Give one example of how the RoHS directive has changed the design of electronic products.

d Describe one way in which the WEEE directive impacts on retailers of electronic products.

13 a Use sketches and notes to describe the stages involved in producing a printed circuit board in school.

b Identify ways in which industrial PCB assembly differs from the methods used in school.

14 Give three reasons why a designer may choose to use a pulley and belt drive rather than spur gears to transfer rotary motion.

Preparing for assessment

This section looks at the ways in which your knowledge and understanding of the principles of design and technology will be assessed.

All students will be assessed by an examination on Principles of Design and Technology and a non-exam assessment in which you will receive an Iterative Design Challenge. This section explains how each part of the assessment will work.

The two parts of the assessment are covered in the following chapters:

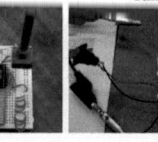
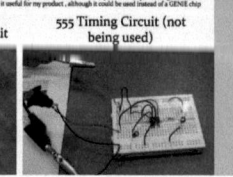
CHAPTER 14
Non-exam assessment: Iterative Design Challenge

LEARNING OUTCOMES
By the end of this section you should know about and understand:
➜ the structure and format of the NEA
➜ the requirements of your chronological portfolio
➜ the requirements for the mark scheme
➜ a range of approaches for presenting the evidence for the NEA.

Introduction

The **Iterative Design Challenge** is the non-exam assessment (NEA) component of your GCSE, and it accounts for 50 per cent of your total GCSE marks. It gives you the opportunity to demonstrate the knowledge, understanding and skills you have developed throughout the course in order to create a final prototype or group of prototypes that reflects a real-world design consideration.

What is iterative design?

Iterative design is a way of designing that is based on a cyclic process of prototyping, testing, analysing and refining a design solution: **Explore** – **Create** – **Evaluate**. Based on the results of testing, the most recent iteration (or version) of a design is refined to more closely meet the needs of the primary user and the stakeholders. The process followed may appear unconventional, but be assured that it closely matches what happens in the real world and your assessment is more concerned with the wider **process of design** than with the practical outcomes you produce.

How long should I spend on the NEA?

Although there is no maximum or minimum time requirement, as a guideline you will have approximately **forty hours** to complete the whole Iterative Design Challenge. You should be wary of spending too long on the NEA as it could cause you to lose focus and so affect the level of your work.

When can I start?

You won't be able to start your challenge until 1 June in the year before your final year. This is when OCR releases the **contextual challenges** for you to use as a starting point.

These will be open-ended, real-world challenges for you to interpret and respond to.

KEY POINT

Contextual challenges are released by OCR on 1 June each year.

Example contextual challenges

- **Public spaces** – The sensitive design of public spaces can enhance users' experiences and interactions with that space. Explore a space in your locality with the view to enhancing the users' experiences.
- **Security** – The theft of people's possessions is a problem in modern society. Explore the role that design can play in securing people's belongings.
- **Dining** – Dining can be a wonderful social and cultural experience that not only focuses on the eating of food. Explore the ways in which design can enhance the experiences for any of the stakeholders involved.

As you can see, the challenges are open to interpretation and are not restricted to a specific material or process. They offer an authentic starting point for you to explore and to consider in relation to your subject interests and the problems and opportunities that you can identify within the context(s).

It is important before you start the challenge that you have experienced a broad range of learning and practical activities to prepare you for demonstrating your ability to undertake an iterative challenge. You need to be confident in your decision making when dealing with a given context and you must avoid having pre-conceived ideas of what you could design and make.

What format will the NEA take?

You should produce a **chronological portfolio** supported by **real-time evidence** that demonstrates your complete response to the challenge, along with a **final prototype(s)**.

A chronological portfolio is a portfolio of evidence that records the whole process of designing as it happens. This includes exploring the context, analysing products and components, communicating with stakeholders, and sketching, modelling and testing your designs. It should outline your design thinking as it occurs in **real time**. For example, instead of presenting all your research at the start of your portfolio in a single section, you would present it *throughout* your portfolio, as and when it happened. You would present sketching and modelling as they occurred, instead of in separate sections headed 'sketches' and 'models'.

Evidence of your final prototype(s) should be clearly documented in your portfolio through the use of photography and video as appropriate. For example, if a key aspect of your prototype(s) has a function that includes movement or sound, then a video of it in action must be presented.

All portfolio evidence for this assessment should be **easy to follow and understand.** You could use presentation software such as **PowerPoint** to present your portfolio as this allows information to be quickly added to the slides to record and present your real-time evidence as it happens. If a paper portfolio is submitted you can enhance it by supplying audio/video files separately. Whatever system you use the chronology must be made clear.

The majority of your visual evidence can be captured using a still digital image (suitably annotated), which can then easily be inserted into your PowerPoint portfolio. Short video clips, however, can be used to great effect when conducting product analysis, user interviews/feedback, model testing, and so on. It is very important that any video clips used are relevant and to the point. They must also be inserted into your portfolio correctly for them to work properly. The file size of videos can cause problems when opening your portfolio, so make sure you compress your videos before inserting them using a suitable app on your phone or computer.

What will I need evidence of?

What you need evidence of will depend on the nature of your project, but evidence must be **relevant** and **may** include some of the examples shown in Table 14.1.

Table 14.1 **Examples of evidence that could be used**

Actions	Examples
Visits	To fashion shows, trade fairs, factories, workplaces, etc.
Interviews	To establish the needs of your primary user and other stakeholders
Observations	To study users in their environment; for user interaction
Gathering data	Anthropometrics, surveys, legislation, other measurements
Product analysis	Disassembly, use, inspiration, reverse engineering
Written reports	Write-ups, analysis of research, testing, etc.
Collaborative work	To generate ideas with others, peer testing, feedback
Sketching	Hand drawn, digital, over photos of models, etc.
Modelling	CAD and physical models, to scale and full size
Materials testing	Desirable properties, sample processes
Prototype testing	For viability, user feedback
Component testing	Using destructive and non-destructive methods
User feedback	Observations, interviews, etc.
Stakeholder feedback	Emails, video chats, etc.
Manufacturing	To explore potential processes, actual processes used, CAM
Final prototype	Clear images, close-up views, moving/working parts, testing in situ
Visualisation	Photoshop, image manipulation, etc.

Remember that evidence alone is not enough. You must **summarise** and **analyse** what you find and you must guide your reader. It must be clear **what** you are doing, **why** you are doing it and **how** it is helping your project.

How will I be assessed?

You will be assessed on your thinking and on your creative and practical skills and abilities through the designing and making of a prototype(s).

There are four Assessment Objectives in OCR GCSE Design and Technology, which are detailed in the table below. AO1, AO2 and AO3 are assessed through the NEA. (AO3 is also assessed in the written exam, along with all of AO4.)

The Assessment Objectives relate directly to the iterative processes of **Explore**, **Create** and **Evaluate**, as shown below. Those three processes are spread across five strands, also shown in the table.

Your chronological portfolio will be marked out of 100 and those marks are spread across the iterative processes.

Table 14.2 **The four Assessment Objectives in OCR GCSE Design and Technology**

Assessment Objective		Strand	Number of marks
AO1 (Explore)	Identify, investigate and outline design possibilities to address needs and wants.	Strand 1 – Explore	20
AO2 (Create)	Design and make prototypes that are fit for purpose.	Strand 2 – Create: Design Thinking	24
		Strand 3 – Create: Design Communication	16
		Strand 4 – Create: Final Prototype(s)	20
AO3 (Evaluate)	Analyse and evaluate: • design decisions and outcomes, including for prototypes made by you and by others • wider issues in design and technology.	Strand 5 – Evaluate	20
		Total:	**100**

The number of marks allocated to each strand gives you a rough guide to how you should divide up your time on the challenge. Remember, though, that you are **not** just exploring for eight hours at the start of your project – you must explore whenever it is relevant **throughout** the design process.

Within each strand your teachers will be looking for evidence to assess so they can award you a 'best fit' mark for each strand. We will look more closely at what that evidence might look like later on in this chapter.

An outline of the Iterative Design Challenge

The Iterative Design Challenge requires you to design and make a prototype (or prototypes) through iterations of exploring, creating and evaluating that constantly respond to stakeholder needs, wants and interests.

Everybody will start with an **exploration** of the contextual challenge and will end with an **evaluation** of their final prototype(s).

However, the path from that initial exploration to the final evaluation will vary from student to to student as they repeatedly Explore, Create and Evaluate throughout the challenge.

The process

This process should be followed and evidenced to form an accurate account of your progress, and will involve the following:

1 **Developing a brief** – Before you begin writing your design brief, it is important that you fully explore the contextual challenges. You are then required to write your own unique design brief as a response to your exploration of one of the contextual challenges (set by OCR).

2 **Outlining requirements** – As you follow the iterative design process, you will identify various stakeholder and technical requirements. Some will be established early on; others will be identified and refined as you develop your ideas. It is important that any requirements you identify are clearly outlined and presented in a way that supports the design process. You will need to review these as you progress, but they should be evident as and when they occur.

3 **Generating initial ideas** – There are various approaches that can be taken to conceive initial ideas. A minimum of ten should be delivered appropriately within the design process. Your initial ideas should be innovative, challenging and focused on responding to the problems and requirements you identified in the earlier sections of your portfolio. Initial ideas can be seen as ways of getting those ideas in your head out as quickly and appropriately as possible. Presentation is not the most important focus here, but rather sharing with others what is in your mind.

4 **Design developments** – When developing your designs, the focus must be on narrowing down and improving your ideas through more detailed iterations that gradually resolve your stakeholder requirements. The quality of these iterations and how well they meet the technical requirements is important throughout this section.

5 **Developing a final design solution** – When developing your designs, you will carry out additional investigations, sketching, modelling and testing, as appropriate to the direction of your thinking. It is important that this development process and the thinking behind it is clearly documented in your portfolio. It must be obvious how your designs have progressed towards your final design solution. Your final design solution is the conclusion to your development prior to making a final prototype(s); it must present your design as it would look and function if developed as a saleable product.

6 **Delivering a technical specification** – Following the presentation of your final design solution will you will need to deliver a technical specification that shows how your design meets your stakeholder requirements. It will include specific written and graphical information (such as working drawings) that is sufficient for a third party to understand your intentions.

7 **Producing a final prototype** – A plan of how to make your final prototype(s) will be required first, then evidence of the specialist tools and processes you used to make it. Clear photos and videos of the finished prototype will be needed to provide evidence of the level of skill involved in its realisation. The making of your final prototype(s) must be completed at school under the direct supervision of your teachers. In order to make a final prototype(s) in the school workshop, it may be necessary to use different materials and processes than your design solution would be manufactured with. You must plan for these possible changes in order to make a final prototype(s) that best presents your intentions to a third party.

8 **Analysing the validity of the final prototype** – In order to make an appropriate evaluation of the final prototype(s), you will need to analyse your stakeholder's opinions. You will also need to effectively test the prototype to determine its strengths and weaknesses. Suggestions for modifications based on this testing must be presented. This is similar to the on-going process review you will have made throughout the development of your design solution, but will be more conclusive.

NEA marking criteria

Your teacher will use the marking criteria to assess your work once you have completed your Iterative Design Challenge. Each criterion has four mark bands that get progressively more difficult to achieve.

Table 14.3 **Example marking criteria**

	Mark Band 1 (1–5)	Mark Band 2 (6–10)	Mark Band 3 (11–15)	Mark Band 4 (16–20)
Quality of planning for making the final prototype(s)	Offers little or no support to the making process.	Generally supports the management of the making process with some relevant requirements identified from the technical specification.	Good level of detail and relevance, covering most requirements identified from the technical specification to manage the making process.	Comprehensive and relevant, covering all requirements identified from the technical specification to effectively manage the making process.

In this section, we will look at what you should be thinking about and the evidence you could provide to your teacher to enable them to confidently assess you against each of the marking criteria.

● The **What?** sections suggest questions you should ask yourself as you work through the challenge.
● The **How?** sections suggest the possible techniques and approaches you could use.

There is then an explanation of some of the techniques you may use to provide evidence and how this might look in your e-portfolio.

Strand 1: Explore (A01) – 20 marks

The work being assessed in this strand will be evidenced from your complete portfolio. This assessment focuses on all of the exploration you undertake and the needs, opportunities and facts that you uncover as part of these investigations.

	Mark Band 1	Mark Band 2	Mark Band 3	Mark Band 4
Investigations of the context	Superficial investigations identify few or no problems and/or opportunities for further consideration.	Investigations are of sufficient quality to identify some problems and/or opportunities for further consideration.	Investigations offer a good level of detail and identify a breadth of problems and opportunities for further consideration.	Comprehensive investigations identify a breadth of challenging problems and opportunities for further consideration.

What?

How well have you explored your chosen context? Have you used a variety of effective methods to explore the context? Have you identified a broad range of problems?

How?

Moodboards, concept maps, observations, interviews, surveys

How you start your initial investigation is very important. The contextual challenges are starting points for your own investigations.

A **moodboard** is a quick collection of images and text that can help you start to get a feel for and understanding of your context. It is important to analyse the images you find by explaining how they might identify a potential problem or need. They are useful in the early stages of a project and can give you feedback on potential ideas before you have invested too much time. They can also be used to start identifying possible stakeholders. Any potential project ideas should be clearly highlighted.

A **concept map** is another useful tool for exploring a given context. It allows you to start recording your initial thoughts. Although mainly textual, images can be used to reinforce certain key areas. Start with a central idea or theme (the context), then branch out into key themes/areas. These branches can be expanded further as you explore the context. Use a range of different colours to group together related ideas and to highlight potential project ideas. A concept map can also be used to begin identifying possible stakeholders.

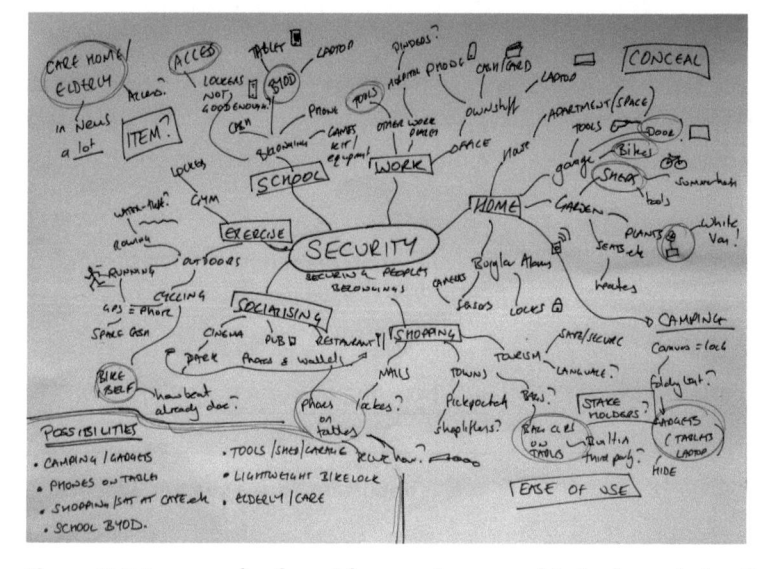

Figure 14.1 **An example of a quick concept map used to begin exploring the context of security**

Observing people in their natural environment provides a good sense of the difficulties they experience within a real-world context. Studying what people do – rather than what they say they do – provides a more realistic overview of their actions. Observation may involve shadowing a person over a period of time as they progress through various activities related to your chosen context. Specific problems should be clearly highlighted and linked to potential project ideas.

Interviewing people within a context is another useful tool for establishing needs and identifying potential project ideas. Interviewing basically consists of asking a series of questions directly to an individual. You could ask how they feel when using a particular product, how easy it is to use, whether they find using it enjoyable or frustrating. Your analysis of their responses is important and will guide your thinking. Any potential ideas for suitable design briefs should be clearly highlighted.

Questionnaires and surveys are simple and effective ways to gather information from a broad range of people. Response rates can vary depending on the methods you use. A printed questionnaire given out by hand could be time-consuming and there are several online alternatives that can be used to quickly generate and send out survey questions. These should be targeted at groups of people relevant to the context being investigated. Surveys are a useful method to establish the particular traits and values of large numbers of users relatively quickly. Online surveys can be sent as links and posted on relevant social networking sites. Responses should highlight potential project ideas.

Survey (sent to cycling forums)

1. Age
The majority of people who responded were between 25 and 24. *This means I will have to work hard to design a product that appeals this age range. I might have to explore some image/style boards as they are older than me.*

2. Please select from the list below, what best describes your circumstances.(Student/full-time/part-time/unemployed/retired)
The majority were employed full time. *This should hopefully give me scope to design an effective yet affordable product. I wont have to cut back on features or materials.*

3. What is the value of your bicycle?
Most had bikes between £500-£1000 and several had bikes over £1000. *This means security is definitely an issue and I will have to make sure my product works and will give peace of mind to these cyclists.*

4. Do you currently own an Electric Bicycle (E-Bike)?
Only 5 of the respondents had an E-bike and these were both retired from work. *E-bikes range from £2000-£4000 so I'm quite relieved I don't have to focus on them. However, their popularity is increasing so it should not be ignored.*

5. Why do you cycle?
The majority of cyclists used their bikes for fitness/leisure, cycling at weekends or holidays mainly. *So product life is not an essential (although making it durable enough for rugged use is) perhaps this could affect battery life if an alarm is used?*

6. How often do you cycle?
The majority cycle at the weekends, although some did the odd evening run (which increases in the summer months). *Some did commute with their bikes – so maybe I should ensure its suitable for urban use as well?*

7. Where do you store your bicycle(s) whilst at home?
The majority stored their bikes in a garage *so adapting my product for additional situations would probably not be required*

8. If you answered (a-g in question 6) do you also use a bicycle lock or equivalent whilst at home?
This was a surprising split – roughly half did use a lock whilst it was in the garage but I was worried that the other half just relied on their locking garage for security. *Maybe this is another marketing opportunity?*

9. When away from home, where do you store your bicycle(s)? Please select all that apply.
This answer varied a lot. Some brought their bikes into work during the week. At weekends, most simply locked it to a post/fence or something similar. *Several didn't lock them at all as they were always with their bike whilst out at weekends. Does this mean my product is not needed?*

10. When away from home, do you use a bicycle lock or equivalent?
This was mainly yes – but a few nos. I suspect this is the bunch who take their bikes out for a leisure ride and return to home. *However, this means that I have to pitch my idea just right to appeal to people who already have a bike lock*

11. Have you ever had a bicycle stolen from home?
Only 3 said that this had happened. *So maybe my idea for an adaptable lock is overkill?*

12. Have you ever had your bicycle stolen away from home?
Sadly this has happened to quite a few people. Mainly in town centres, *but some had had theirs taken from campsites whilst they were out sightseeing – which supports my argument*

You can see my original survey here:
https://www.surveymonkey.co.uk/r/2MKZ62

I sent it to these Forums:
- Mountain bike rider Forums
- Mountain biking Forum
- Singletrack Forum
- Mtbr Forum
- Cycling UK Forum

And the following facebook groups:
- Cycling UK
- Sherwood pines Cycling club
- Bad brains Mountain bike club
- Leeds mountain bikers
- Wakefield Disctrict Cycling club

RESEARCH ACTIVITY

1 Observe a classmate performing a specific task. Record the stages involved. Can you identify the problems they experience? Present your findings as an annotated photo storyboard.

2 Write a survey using a suitable online survey tool.

Wine storage system survey

1. Are you male or female?
O Male
O Female

2. How old are you?
O Under 20
O 21–40
O Over 40

3. How often do you drink wine?
O Often
O Regularly
O Sometimes
O Never

4. How many wine bottles do you have in your house?
O 1–3   O 4–6   O 7 or more

5. How often do you have friends round for drinks?
O Often   O Sometimes
O Regularly   O Never

6. Where do you store your wine now?
[   ]

7. Is it important to see the wine labels?
O Yes   O No

Figure 14.2 **Examples of how you could present interview and survey analysis**

	Mark Band 1	Mark Band 2	Mark Band 3	Mark Band 4
Design brief	Limited relevance to the context and little or no identification of a primary user or other stakeholders.	Some relevance to the context and identification of a primary user and/or other stakeholders.	Mostly has relevance to the context, offering scope for challenge and identification of a primary user and other stakeholders.	Clear and full relevance to the context, offering scope for challenge and a focused identification of a primary user and other stakeholders.

What?

Have you set yourself a challenging design brief? Is it relevant to the chosen contextual challenge? Have you identified a primary user and a range of other stakeholders?

How?

Written work, contextual images, links to contextual investigation, concept maps.

Throughout your initial investigations into the context, you should have highlighted any potential project ideas. You must now set yourself a suitably challenging design brief, which must respond to the context and must give you the opportunity to develop a suitably creative and innovative design solution. It should offer significant scope for challenge and consider who the stakeholders are who could have an interest in your potential outcome.

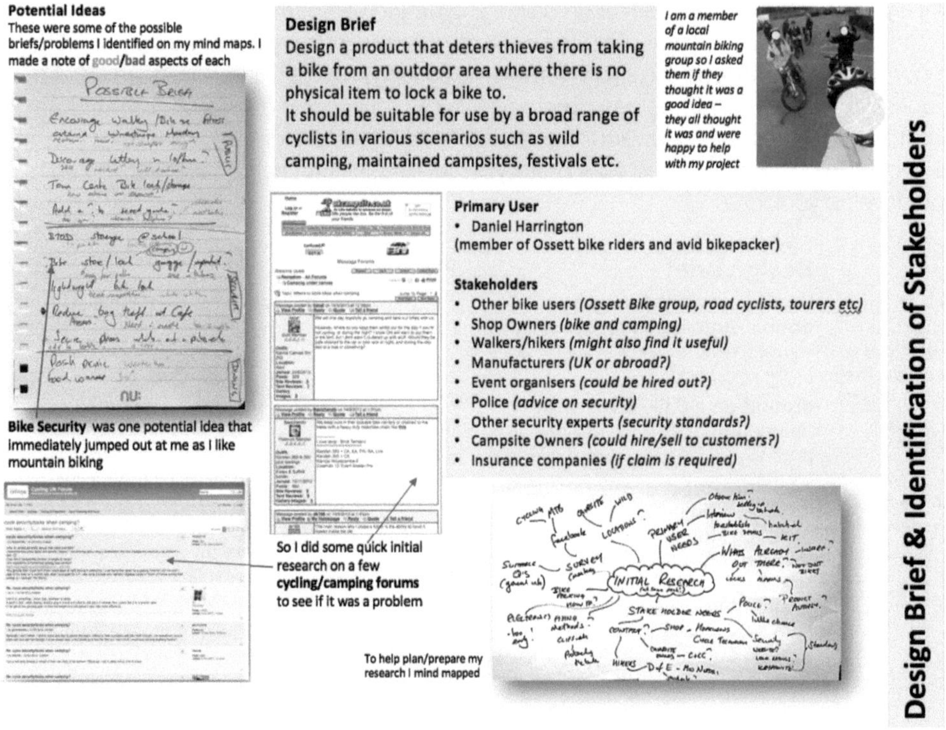

Figure 14.3 **Example of a range of design briefs**

KEY POINT

When writing a design brief, you must avoid design fixation – this can happen when you focus on a product rather than on a problem. For example, 'Design a lunchbox' immediately locks you into a certain way of thinking, whereas 'Design a method of transporting a packed lunch from home to school' opens up several design possibilities.

CLASSROOM DISCUSSION

You have been asked to 'Design a chair'.

- In what ways, might this 'design an object' focus encourage design fixation?
- How could this 'design an object' be reworded into a 'problem to be tackled' in order to avoid potential design fixation?

Discuss these questions in groups.

	Mark Band 1	Mark Band 2	Mark Band 3	Mark Band 4
Investigations of user and stakeholder needs and wants and the outlining of stakeholder requirements (non-technical specification)	Superficial consideration of primary user's needs and wants, with little or no consideration of other stakeholders. Few or no requirements have been identified and these are outlined with limited scope to support the future design process.	Some relevant consideration of primary user's needs and wants, and some consideration of other stakeholders. Some requirements are identified, which offer some scope to support the design process.	Informed consideration of primary user's needs and wants, and those of other stakeholders. A range of requirements is identified with a good level of detail, which offers scope to support the design process.	Full and objective consideration of primary user's needs and wants, and those of other stakeholders. A range of comprehensive requirements is identified, which offers scope to support the design process.

What?

Have you gathered information from your primary user and other stakeholders? Have you analysed and presented this information clearly? Have you summarised their requirements (in the form of a non-technical specification)?

How?

Interviews, surveys, observations.

Once you have set yourself a suitably challenging design brief, it may now be appropriate to conduct some investigations into the needs and wants of your **primary user** and other important **stakeholders**. The aim of these investigations is to generate a detailed list of their requirements. The techniques described earlier in this section can once again be employed, only this time with more focus. Your stakeholder requirements might well change as your project progresses and you test your prototypes. Make sure any changes are highlighted as you develop your ideas.

The need

Public transport is used by **hundreds** every single day. Aeroplanes are used by many too. If you have luggage with you, whether you're a businessman/woman and need a small case for work, or are going on holiday, then mostly everyone has had difficulty getting places quickly if you're in a rush and you have luggage.

Suitcases can be **very big so it can be a struggle getting them through crowds of people**. They can also delay you… If you're in an airport and are late, then running with your suitcase can be a pain, especially if the airport is busy. Carrying a suitcase can be **very heavy and very large**.

If the user is carrying other travel bags as well as the suitcase then it can be very inconvenient. There are different types of suitcases that can make things easier, such as lightweight ones, but none has been made that can help you get around a lot **more quickly and easily**.

Design brief

I will design a **lightweight, double-functioned** product that will transport you **quickly** around places.

It will be aimed at people who **travel around on a daily basis with heavy luggage**. However, the product will be a very **good idea for tourists** too.

It will be used by people who will be travelling around a lot and use public transport most days. It will be mostly **sold in airports**. It will also be sold in **all public transport stations** such as railway stations and bus stations.

Primary user

This is Richard (35). He lives in the middle of London with his family. He is a very successful businessman who earns £160,000 per year. The company he works for is very good. He likes to spend almost all of his free time **travelling** to new places.

Richard has worked for the same company for many years and is getting **tired of having to rush every day to work**. He gets a **train at a certain time every day with his suitcase**. But his suitcase is **hard to get through people** and is heavy. He is always almost late for work due to people not moving out of the way for his suitcase. His job requires a suitcase because he has his own laptop and has to take a lot of paperwork home every night, which can be heavy. He decided not to use a bag to carry his laptop and paperwork because he didn't want to put so much strain on his shoulder.

So he thought a suitcase would be better. The suitcase is **better for his health** but is very **difficult to cart around every day**, especially on public transport in London, where it's very busy. His hobbies are travelling, swimming and drawing.

Figure 14.4 **Example of a design brief with identification of stakeholders**

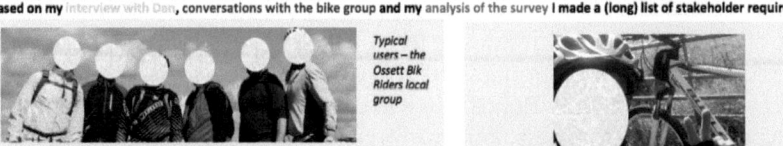

Based on my interview with Dan, conversations with the bike group and my analysis of the survey I made a (long) list of stakeholder requirements

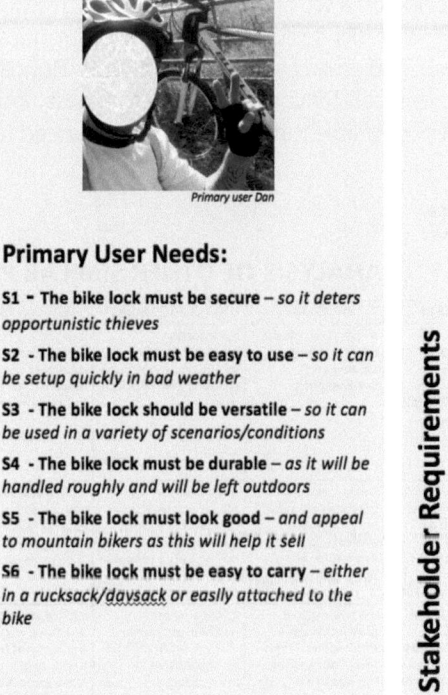

Typical users – the Ossett Bik Riders local group

Primary user Dan

Stakeholder Needs & Wants

1. **The bike lock must be secure**
2. The bike lock must not be cut through by bolt-cutters
3. The bike lock should look strong and unbreakable
4. The bike lock must deter would-be thieves from attempting to steal the bike
5. **The bike lock must be easy to use**
6. The bike lock should be long and flexible enough to be able to pass through spokes, frame, etc...
7. The bike lock should have no keys so there's less things to carry
8. The bike lock should not be too heavy
9. **The bike lock should be flexible enough to use in multiple situations**
10. The bike lock should be an appropriate length
11. The bike lock should be able to lock more than one bike together
12. The bike lock must lock the frame to an anchor
13. The bike lock could make use of natural anchors such as trees and light poles
14. **The bike lock must be durable**
15. The bike lock must work in all weathers - freezing/wet/hot
16. The bike lock must not rust
17. The bike lock should withstand misuse
18. **The bike lock must look good**
19. The bike lock could match what the user wears
20. The bike lock could matches the bike colour
21. The bike lock could be made from recycled bike parts
22. **The bike lock must be easy to carry**
23. The bike lock could be stored without the need of a special fitting
24. The bike lock should not be heavy in the backpack
25. The bike lock should fit easily into a bag
26. The bike lock should not scratch the bike frame / rider
27. The bike lock should be compact
28. The bike lock should offers an additional function (say a built-in light)

Primary User Needs:

S1 – The bike lock must be secure – *so it deters opportunistic thieves*

S2 – The bike lock must be easy to use – *so it can be setup quickly in bad weather*

S3 – The bike lock should be versatile – *so it can be used in a variety of scenarios/conditions*

S4 – The bike lock must be durable – *as it will be handled roughly and will be left outdoors*

S5 – The bike lock must look good – *and appeal to mountain bikers as this will help it sell*

S6 – The bike lock must be easy to carry – *either in a rucksack/daysack or easily attached to the bike*

Stakeholder Requirements

Figure 14.5 **An example of stakeholder requirements**

STRETCH ACTIVITY

Toy store Toys R Us wants to introduce a new range of children's cycles.
1 List as many stakeholders as you can think of.
2 Do they fit into categories that might be helpful? (Use sticky notes so each stakeholder can be sorted into possible categories.)
3 Indicate how important you think each stakeholder's 'stake' is. (Their levels of influence and interest will vary.)

	Mark Band 1	Mark Band 2	Mark Band 3	Mark Band 4
Investigations of existing products and design practices	Little or no information or sources of inspiration are identified to offer support to design iterations and thinking.	Some information and/or sources of inspiration are identified, which may not always be relevant but do offer some influence on design iterations and thinking.	A good amount of relevant information and sources of inspiration are identified to influence design iterations and thinking when required throughout the design process.	Comprehensive and relevant information and sources of inspiration are identified to influence on design iterations and thinking when required throughout the design process.

What?

Have you examined other products while designing? Have you studied any other design practices? Is it clear how they have influenced your design thinking?

How?

Product analysis, using a recognised design principle such as product disassembly or mimicry.

It can be useful at the start of a project to quickly explore what solutions are already available. Hands-on **product analysis** can help you understand why certain products perform better than others. When done at the start of the process, this can help you evaluate competing products. You should select a reasonable range of products to analyse and decide on a set of criteria against which to judge them. Put each product through a series of typical tasks. If you record how each product performs you will be able to evaluate each in turn. You could ask questions such as 'How easy is it to use?', 'How reliable is it?', 'How much does it cost?' If you can summarise these findings, this can be used to generate a list of issues or of desirable functions that you will need to include in your product.

The stars mean how good the product is compared to the desk tidy. Blue is the same, red is worse and green is better. The smiley faces show if the product is overall better (green) or worse (red).

ANALYSIS OF OTHER SIMILAR PRODUCTS

PRODUCT	DESCRIPTION	FUNCTION	MATERIALS	MANUFACTURING	ERGONOMICS	AESTHETICS	COST	OVERALL CONCLUSION
Desk tidy number 1	This is another desk tidy, it is designed to hold variety of stationary also note pads and sticky notes etc.	It can **hold more** office desk items and it is **easy to access** these items. However, it is **slightly obtrusive** on the desk. ★	It is made of **plastic**, made in **many colours** and is lightweight. ★	It is **injection moulded**, required a mould and is **not very cheap** or **cost efficient** to make. ★	It can hold a **larger variety** of items, they are **easier** to get out. However, **not easy to hold and move** around. ★	It looks **more aesthetically pleasing** and is **not too obtrusive**. ★	The **cost** is very **reasonable** for what it is and how much it holds compared to other products. ★	In conclusion, I think this is a better product as it can hold more items, it is not too obtrusive and can easily access items. Although the product is not very cost efficient and can be expensive to manufacture and transport. ☺
Bathroom tidy	This is a bathroom tidy, it is designed to hold bathroom items like a face towel, toothpaste, toothbrushes.	It **holds few items** and is not designed to hold much else. It does what it is supposed to but not a lot ★	This is made of **plastic** in a **similar way** and **more intricate** so it **takes more time** and is **more difficult**. ★	It will be made out of **plastic** and **injection moulded**, it will **not be a cheap mould** and **not be very profitable** at first for manufacturers. ★	It can **not hold many items**, it is **not easy to hold** and **cannot be moved** as it is **screwed** to the wall. ★	It does **not** look aesthetically pleasing and is obtrusive in a bathroom and looks to childish and **not very modern**. ★	It costs £9 for one of these which is fairly **expensive** and too much for such a simple product that only has one purpose. ★	In conclusion, I think this is a **worse** product as it is not very appealing, obtrusive and looks tacky in a modern bathroom. It can not be easily moved. ☹
Desk tidy number 2 (rotational)	This is another desk tidy, it is designed to hold a variety of stationary	It **holds a lot of items** and it can **rotate** to make items **easily accessible**. It **is not too obtrusive** on the desk but can be **difficult to get items out**. ★	It is made out of **plastic** and has a **mechanism**. This is fairly **lightweight**, but **heavier** than my product. ★	It uses a **rotary mechanism, plastic**, so it will **more difficult** and **expensive** to manufacture. ★	It is **not easy to hold** and **move** around and does **not hold a variety** of items. It is also very **difficult** to get **small items** out of the **smaller compartments**. ★	It is **more aesthetically pleasing** than my product as it **rotates** and looks more **sleek** and **professional**. ★	It costs arouns £8 whicch is relatively **cheap** for what it does as there is a **mechanism**. ★	In conclusion, I think this is a product as it can rotate in a 360 degree circle and can easily access the products. Although the smaller compartments are more difficult to get items out of as they are too small. ☺
Scissors holder	This is a scissors holder, it is designed to hold many scissors that protects the blade with holes the same size in the peice of wood.	It does what is suppose to, it protects the blades from cutting people. It is **not multifunctional**. ★	It is made out of **wood** and has a large number of holes drilled into it all the same depth. It is **lightweight** but can give you **splinters**. ★	It is very **easy** and **cheap** to manufacture as it is just **machine** based job to create these. ★	It is **not easy** to **hold** or **move** around and cannot be **used** for other items. ★	It is **not** very aesthetically pleasing as it is just a block of wood but it does its job and is **not obtrusive**. ★	It is very **cheap** and **easy** to **manufacture** and **transport** so makes a high **profit**. ★	In conclusion, I think this is a **worse** product as it can not hold anything but scissors and is not aesthetically pleasing, it can also harm the user by giving them splinters. ☹
Cutlery tray	This is a kitchen utensil tody, it is designed to cutlery and utensils found in the kitchen to help keep draws tidy and organised.	It organises the kitchen cutlery and is **not obtrusive** and **easy to move** around. ★	It is made out of **wood**, depending how filled it gets it can get **heavy and difficult to move**. ★	It is **made by machinery** and is **easy** to make as it is a **simple mould**. ★	It is **not very easy** to **move** around and is **not ergonomically** designed but is **not too obtrusive**. ★	It is **aesthetically pleasing** as long as the wood is finished well. It does its job. ★	It is a moderate price for what it is, it look **appealing** and is **easy to manufacture**. ★	In colclusion, I think this is a **better** product as it is aesthetically pleasing, easy to manufacture, not very difficuly to transport and can be multifunctional. ☺

Figure 14.6 **An example of comparative product analysis**

It is also useful to seek out sources of **inspiration** while you are designing. Existing products can help inspire your own ideas and provide clarity in your own design thinking if an aspect of your design is similar. Inspiration may however also come from visual sources, such as nature, from historical design movements or thinking what might exist further in the future.

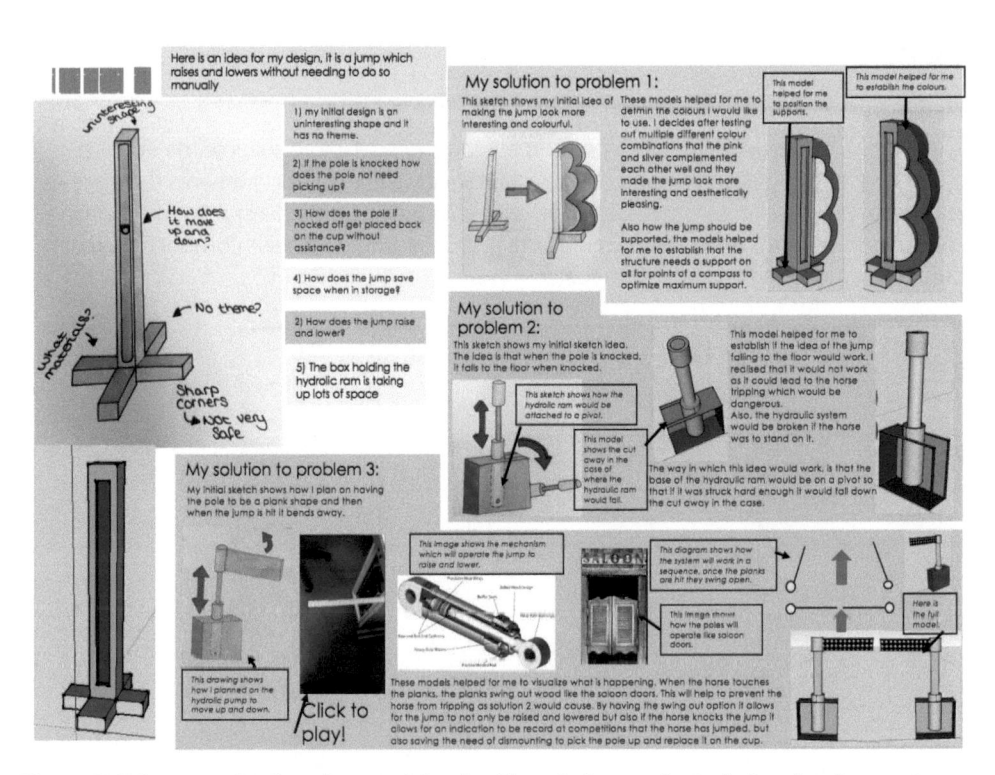

Figure 14.7 An example of student work inspired by existing products during development

Product disassembly can be used to get a better understanding of how a product has been designed and manufactured. It is done in three stages:
- Taking the product apart.
- Recording through photographs and sketches how the parts go together and how they relate to each other.
- Evaluating the disassembled product.

There is more on disassembly in Chapter 1.

Mimicry is the process of copying pre-existing solutions to problems. This could be copying the way things look, the way things act or the way things work. Your inspiration may come from a wide variety of source materials and should be built into your development.

STRETCH ACTIVITY

Find out what inspired the following products:
- Velcro
- The Shinkansen bullet train
- The Fastskin swimsuit.

	Mark Band 1	Mark Band 2	Mark Band 3	Mark Band 4
Exploration of materials and possible technical requirements	Superficial consideration of materials and/ or possible technical requirements.	Some relevant consideration of materials and possible technical requirements.	Informed consideration of materials and possible technical requirements when required throughout the design process.	Full and objective consideration of materials and possible technical requirements when required throughout the design process.

What?

Have you thought about materials and their working properties throughout the design process? Have you looked at specific technical requirements (such as finishing, sustainability, flexibility, toughness, etc.)?

How?

Gathering data, materials testing, component testing.

As you develop ideas and test your prototypes, you should explore suitable materials to further clarify your design thinking and to establish any technical requirements that may be required of your final prototype(s). This could include specific tests to establish the working properties of potential materials. It could also include testing of specific components that could be included in the final prototype(s).

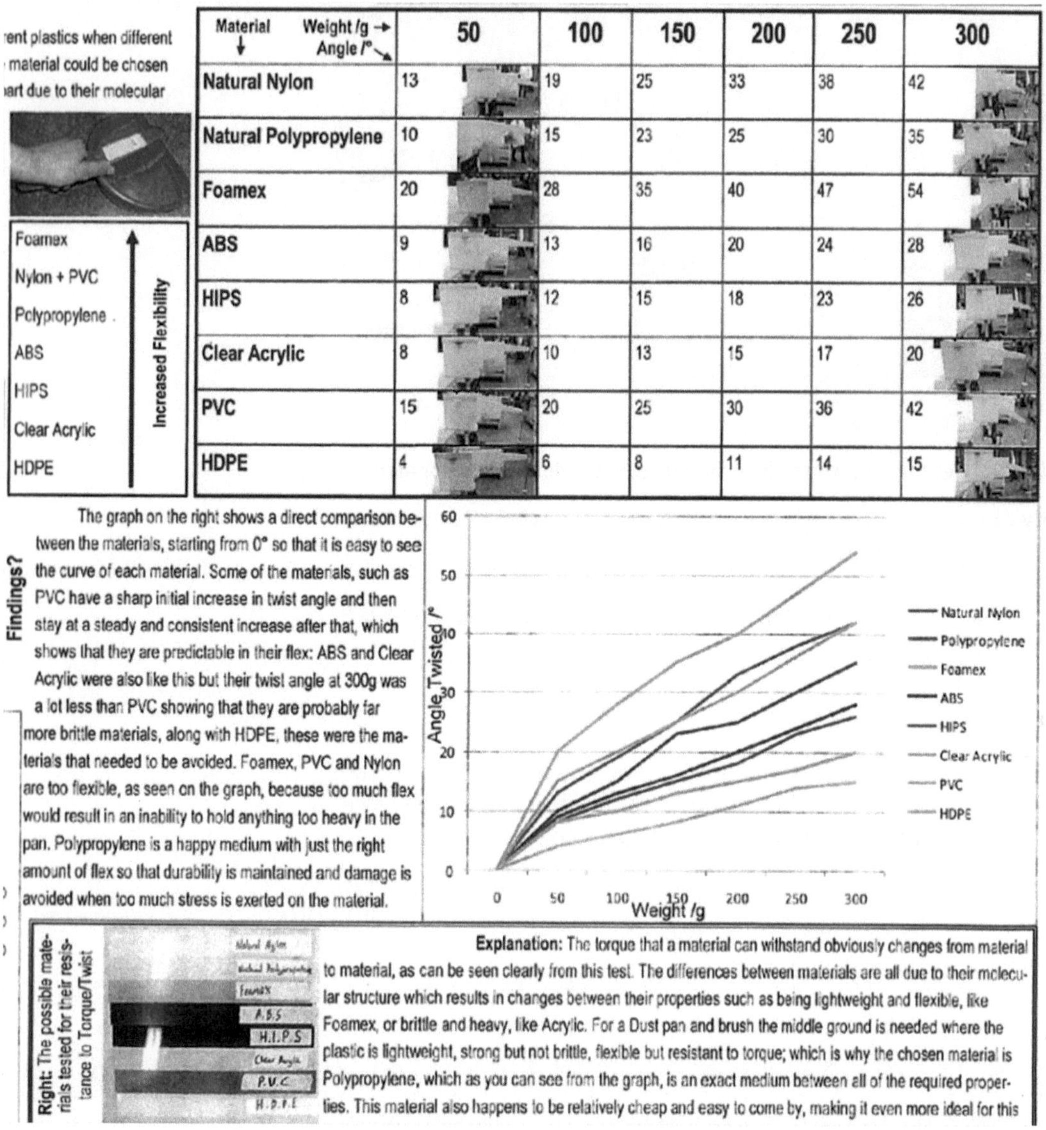

...ent plastics when different ...material could be chosen ...art due to their molecular

Increased Flexibility:

Foamex
Nylon + PVC
Polypropylene
ABS
HIPS
Clear Acrylic
HDPE

Material ↓ / Weight /g → Angle /° ↘	50	100	150	200	250	300
Natural Nylon	13	19	25	33	38	42
Natural Polypropylene	10	15	23	25	30	35
Foamex	20	28	35	40	47	54
ABS	9	13	16	20	24	28
HIPS	8	12	15	18	23	26
Clear Acrylic	8	10	13	15	17	20
PVC	15	20	25	30	36	42
HDPE	4	6	8	11	14	15

Findings? The graph on the right shows a direct comparison between the materials, starting from 0° so that it is easy to see the curve of each material. Some of the materials, such as PVC have a sharp initial increase in twist angle and then stay at a steady and consistent increase after that, which shows that they are predictable in their flex: ABS and Clear Acrylic were also like this but their twist angle at 300g was a lot less than PVC showing that they are probably far more brittle materials, along with HDPE, these were the materials that needed to be avoided. Foamex, PVC and Nylon are too flexible, as seen on the graph, because too much flex would result in an inability to hold anything too heavy in the pan. Polypropylene is a happy medium with just the right amount of flex so that durability is maintained and damage is avoided when too much stress is exerted on the material.

Graph: Angle Twisted /° vs Weight /g (0 to 300) — Natural Nylon, Polypropylene, Foamex, ABS, HIPS, Clear Acrylic, PVC, HDPE

Right: The possible materials tested for their resistance to Torque/Twist

Natural Nylon
Natural Polypropylene
Foamex
A.B.S
H.I.P.S
Clear Acrylic
P.V.C
H.D.P.E

Explanation: The torque that a material can withstand obviously changes from material to material, as can be seen clearly from this test. The differences between materials are all due to their molecular structure which results in changes between their properties such as being lightweight and flexible, like Foamex, or brittle and heavy, like Acrylic. For a Dust pan and brush the middle ground is needed where the plastic is lightweight, strong but not brittle, flexible but resistant to torque; which is why the chosen material is Polypropylene, which as you can see from the graph, is an exact medium between all of the required properties. This material also happens to be relatively cheap and easy to come by, making it even more ideal for this

Figure 14.8 An example of materials testing during development to establish technical requirements such as flexibility

RESEARCH ACTIVITY

Design and make a test rig capable of testing samples of different materials relevant to your NEA activity.

You might test for toughness, durability, flexibility, resistance to tearing, etc.

	Mark Band 1	Mark Band 2	Mark Band 3	Mark Band 4
Technical specification	Inaccurate, outlines basic details and/or is incomplete, making it difficult for a third party to understand.	Generally accurate, outlines details that communicate some requirements to a third party.	Good levels of accuracy, outlines details that communicate most requirements to a third party.	High levels of accuracy, outlines details that clearly communicate all requirements to a third party.

What?

Have you provided specific technical information? Are your working drawings accurate? Have you specified real-world materials? Do you have a cutting list? Have you documented your final design/prototype(s) features?

How?

Appropriate working orthographic, and exploded drawings, CAD drawings, parts list, features list.

Before you make your final prototype(s) you must present a technical specification that outlines the details of your design solution to a third party. These are the details required for if the design solution were manufactured, NOT for making the final prototype(s). Relevant, accurate **working drawings** will be your priority. These could be drawn using a suitable CAD software package.

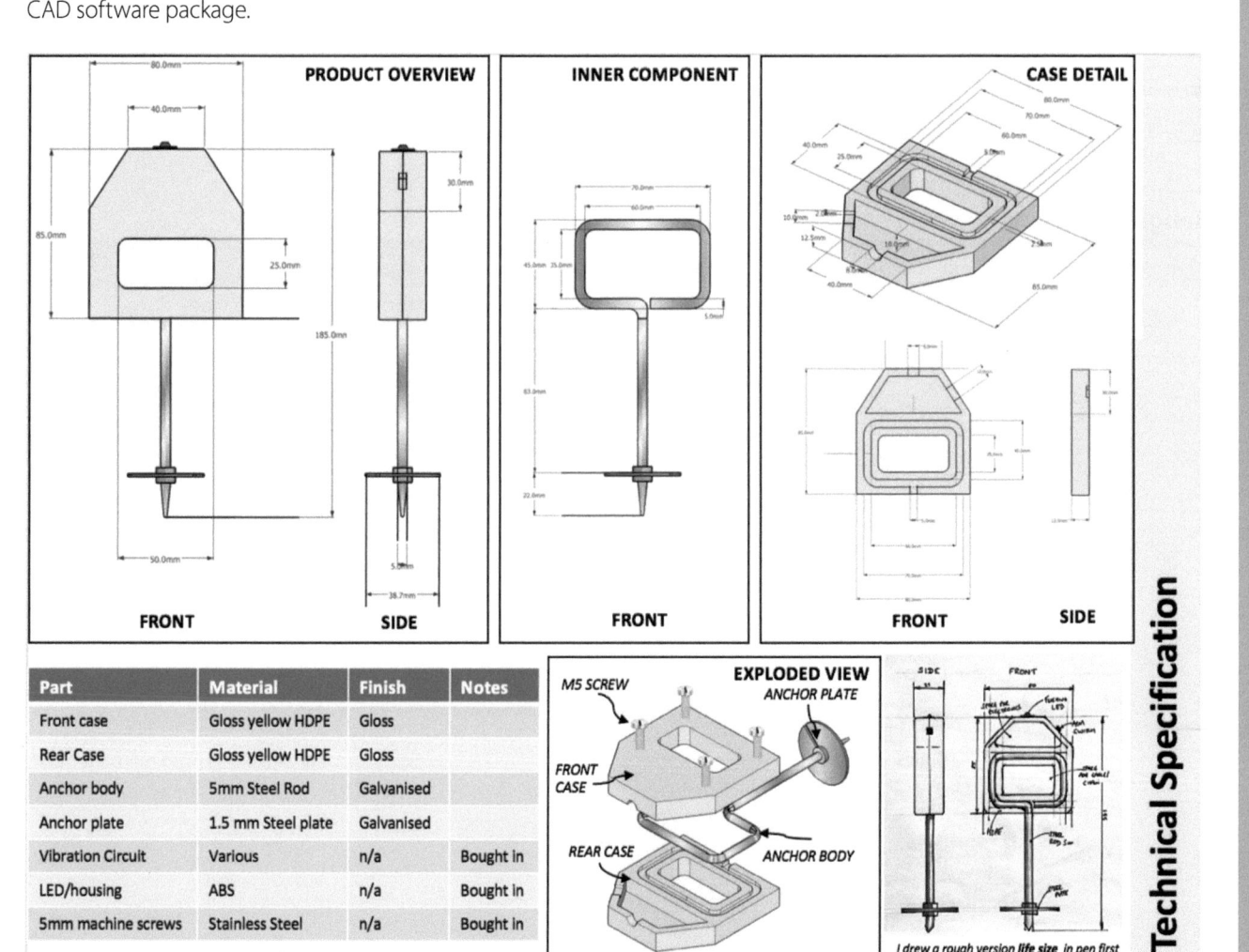

Figure 14.9 **An example of a suitable technical specification for a bike lock/ground anchor concept**

If your design solution has a particular detail, feature or function that may need clarity to explain it to a third party, then this should also be explained. This might involve the use of an exploded view, details of any bought-in components required, a circuit diagram or a cutting list, for example.

Cutting list and bought-in materials

Cutting list

Number	Name of part	Number of	Sizes (mm)	Material
1	Main frame	2	Length – 800mm Height – 85 up to 170mm	Sheet aluminium (2mm)
2	Aluminium spacers	2	Height – 19mm Width – 19mm Length – 500mm	Sheet aluminium (2mm)
3	Handles	2	Length – 1200mm Width – 19mm Thickness – 1mm	Thin-walled (1mm) Box-section steel
4	Bottom support panel that battery sits on	1	Length – 400mm Height – 450mm	Sheet aluminium (2mm)
5	Joining cylinder that connects brushes to frame	2	Length – 50mm Diameter – 20mm	Aluminium rod
6	Ball ramp	1	Length – 500mm Width – 400mm	3mm acrylic
7	Hooks that hold basket in place on main panel	2	Height – 200mm Width – 2mm Length – 50mm	Sheet aluminium (2mm)
8	Battery cover	1	Length – 420mm Width – 3mm	3mm acrylic
9	Aluminium hooks that hold basket on handles	4	Height – 40mm Width – 50mm Thickness – 2mm	Sheet aluminium (2mm)

Bought-in materials

Number	Name of part	Number of	Sizes (mm)	Material
1	Door brush	2	Length – 35mm Width – 2mm	Nylon brush
2	12v motor	1	Length – 90mm Width – 50mm	
3	Ball basket	1	Length – Width – Height –	Galvanised steel mesh

Figure 14.10 **An example of a cutting list for a tennis ball collector**

Strand 2: Create: Design Thinking (A02) – 24 marks

The work being assessed in this strand will be evidenced from your complete portfolio. This assessment focuses on standard of thinking, problem solving and development that you have undertaken and understood.

	Mark Band 1	Mark Band 2	Mark Band 3	Mark Band 4
Generation of initial ideas	Limited use of different design approaches, which leads to ideas that do not always reflect the requirements and may appear stereotypical.	Some different design approaches, which lead to some ideas that avoid design fixation and generally reflect the requirements.	Different and relevant design approaches, which lead to ideas that mostly avoid design fixation, offer scope for challenge and reflect requirements.	Different and relevant design approaches, which lead to ideas that fully avoid design fixation, offer scope for challenge and fully reflect requirements.

What?

Have you used a range of design methods to generate your initial ideas? Have you generated a broad enough range of ideas? Are they all quite different? Do they offer sufficient scope and challenge? Do they avoid stereotypical solutions?

How?

Collaboration, sketching/ideation, sketch modelling, use of a design principle such as SCAMPER.

It is important that you start designing as soon as possible; there are a number of ways that can help you generate a wide range of quick early ideas.

Collaboration is when you and your classmates work together to help generate each other's ideas. It is a useful technique to avoid fixation at an early stage and it can generate a range of ideas in a relatively short space of time. The quality of the sketching is not important so long as you understand the idea. A combination of descriptive writing and quick, annotated doodles works best.

Collaboration may also come from ideas that stakeholders have suggested. It is important in all cases of collaboration that you acknowledge where the ideas came from and how you are going to use or refuse them.

Ideation is a combination of the words 'idea' and 'creation'. It is the term used when designers quickly generate successive ideas in a short space of time. These are often quick explorations of form but can also be individual ideas.

Initial Ideas

WHY THIS IS A GOOD IDEA:
- Because would be suitable for sleeping/relaxing and comfortable.
- Easy to listen to music as near ears.
- If you have got headache, can play soothing music in the dark.
- Good for travel, would help block out light and sound.

WHY THIS IS A GOOD IDEA:
- Could be put over a pillow
- Could use it in travel, therefore would be well used.
- Would have an aux cord connection and pocket to hold phone/MP3 device.

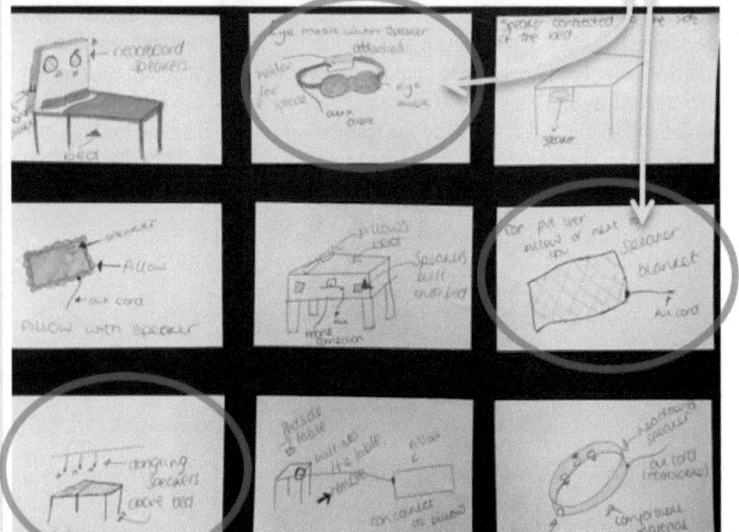

EXPLANATION:
We worked together in **small groups** of 4 to create **innovative ideas** for each others products. We talked about each others product by telling our **specification** points to each other, and then our **brief**.
- Firstly, we sketched quick designs onto **post it notes** which related to a persons product.
- We then **annotated** the ideas with points of what the idea showed/presented.
- We then stuck the 9 post it notes to an A3 card, then photographed them.

By using this method, we quickly created plenty of ideas as to which I could develop further, or incorporate into my own product. It was good that we did it in groups as if working on my own, the ideas would not have been as diverse and the activity would of taken a lot longer.

There were no problems with the activity, other than if people ran out of ideas and then drifted off, but this only happened when people had run out of creative ideas towards a few peoples briefs.

WHY THIS IS A GOOD IDEA:
- Safe as there are no wires near the bed that could get damaged and cause electrocution.
- Appealing appearance which would look great for relaxation and sleep.
- Easy access to it, no complication with the device.
- Multifunctional as well as serving sleeping purposes, therefore wider target market.

Figure 14.11 **An example of collaboration to generate initial ideas**

STRETCH ACTIVITY

As an example of an ideation task, take a sheet with 30 roughly drawn circles. Transform each circle into a recognisable object, such as a wheel, a ball, a planet, etc. Transform all 30 circles in two minutes. You can draw outside of the lines.

Design is a three-dimensional subject and it is essential that you model your ideas physically at the earliest opportunity so that you can test them in the real world. **Sketch models** are full-size or scale models that capture your ideas as they occur. They tend to use readily available materials depending on the idea you have in mind. You could use card, paper, foam, bin bags – whatever is at hand and can be manipulated quickly. Sketch models can be used to make a quick evaluation of your idea's ergonomics, aesthetics, functionality and usability. They can be used to communicate your ideas to your stakeholders and to gauge their reactions.

SCAMPER is an idea-generation technique that uses action verbs as stimuli. It is a well-known kind of checklist that can help when coming up with ideas, either for modifications to an existing product or for making a new product. SCAMPER is an acronym:

- **Substitute** – What materials or shapes can you change? Can you change the purpose of the design?
- **Combine** – Can you combine this product with another, to create something new?
- **Adapt** – Which parts of the product could be adapted to change the nature of the product?

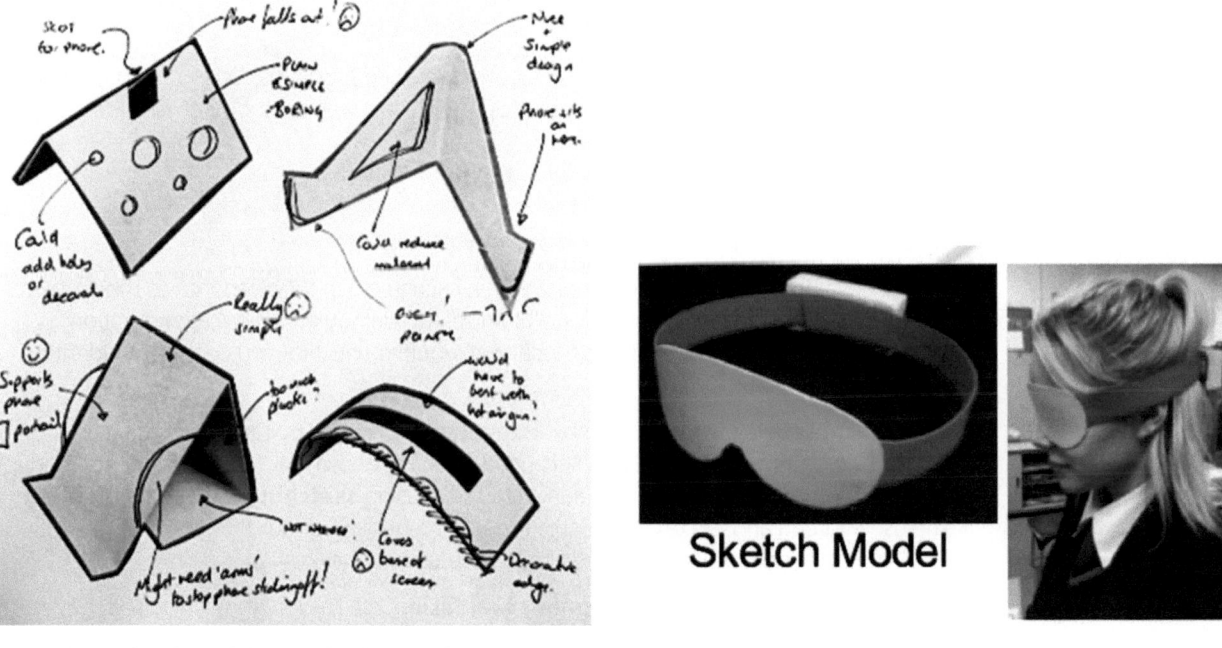

Figure 14.12 **Sketch model examples for a mobile phone stand and a sleep mask**

- **Modify** – Can you change the shape of the product? What could you add to modify this product?
- **Put to other use** – Could this product be used for something else, other than the original intention?
- **Eliminate** – How could you simplify this product? What features or parts could you eliminate?
- **Reverse** – How could you reorganise this product? What roles could you reverse or swap?

	Mark Band 1	Mark Band 2	Mark Band 3	Mark Band 4
Design developments	Limited developments are superficial and/or not iterative.	Iterative developments are generally progressive and respond to some identified next steps of development.	Iterative developments are progressive, incorporate technical requirements and respond to most identified next steps of development.	Iterative developments are comprehensive and progressive, incorporate all technical requirements and fully respond to identified next steps of development.

What?

How well does each development address any perceived problems? Does it build on what came before it? How detailed is each development? How well does it meet your identified stakeholder and technical requirements?

How?

Sketching, modelling, annotation, ongoing evaluation.

As you develop your design it is important that each iteration fully responds to the problems you identified in the previous iteration. You must also demonstrate how each iteration incorporates the stakeholder and the technical requirements that you continue to identify throughout your portfolio.

In the example below you can see how the student has clearly pointed out the initial problems with the current iteration (left-hand side). They have then methodically outlined a possible solution to each problem (right-hand side) using **sketching** and CAD **modelling**. The use of green and red text highlights the advantages and disadvantages of each solution in relation to certain stakeholder and technical requirements.

As the design progresses, using **ongoing evaluation**, the student has concluded this section of their development by formally assessing their iteration against the stakeholder requirements. This could be further enhanced by evidence of stakeholder testing.

5) In order to make sure the wine keeps cool, I need to change my design.

The first idea is to add a fan on the top of the wine rack. This will ensure it keeps the wine cool and ready to drink. A problem is that it makes the design look less attractive.

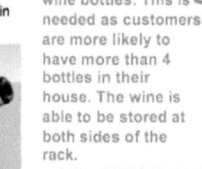

My second idea is to use a triangular shape wine rack. This will allow you to store wine at both sides so more can be stored. The wine bottles slot in the holes.

Having the wine rack in a triangle shape also makes it less like idea 1, which is one of my bad points in idea 2.

Final developed iteration

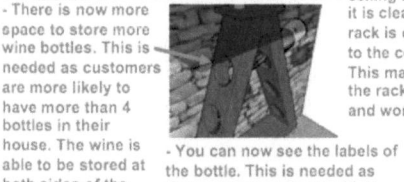

- There is now more space to store more wine bottles. This is needed as customers are more likely to have more than 4 bottles in their house. The wine is able to be stored at both sides of the rack.

- You can now see the labels of the bottle. This is needed as then customers know which wine bottle they are picking out.

- I have added a ceiling mount so it is clear how the rack is connected to the ceiling. This makes sure the rack is secure and wont fall.

- I have changed my design to allow wine glasses to be stored at the bottom. This is good as then everything is stored together and easier to find. I added a rack so they are secure and wont fall off and break.

(6) In order for the rack to store more wine bottles, I will need to change my idea as it only holds three/four.

My first idea is to use a large thin piece of wood with holes in to allow the wine to fit in. This will be hung from the ceiling. A disadvantage to this is that it would take a lot space up and if too low, hit the customers head.

My first idea is to add a ceiling mount so the poles can be connected to it by using screws.

(8) In order to make the poles adjustable so they are the right height and customers wont hit their head, I need to change my idea.

(7) In order to make sure it is clear how to connect the rack to the ceiling, I will need to change my design.

My second idea is to add a thin rounded piece of wire and for it to be connected at both sides. This will allow the pole to be connected to the middle.

My first idea is to use rope that you can pull to adjust.

Stakeholder Requirements	Why?	Overall Score – 44.5/80
Aesthetics = 7.5/10	My product looks attractive so it will stand out from other wine racks. Also it has a unique look which will appeal to people age 20+, which is my target market. It also is a modern design. However, it would not look in a range of different styled houses.	
Customer – 8/10	My redesigned product is easy to use and is suitable for 20+ age people. It is also suitable for both sexes. However, is not easy to store in cupboards if required to put away but this design is good quality.	
Cost – 10/10	My product will be able to cost £60 or plus. The price is this because the materials used would be high quality. It will also be able to be made for no more than £40.	
Ergonomics – 5/10	My product is not easy to move around as it is heavy and would be secured to the ceiling. However, the places to put the wine bottle in are big enough to fit them in securely and therefore they wont fall out and break.	
Safety – 5/10	My product has sharp edges, which could be a danger hazard to people around the rack. But it has no loose or small parts which therefore is safe to be around children as it is not dangerous for them.	
Size – 0/10	My product is not small enough to store away because it is meant to be secured on the wall, and is not lightweight because heavy and durable material would be used which will increase the weight of it.	
Function – 5/10	My product does not hold more than 7 bottles of wine. However, it does keep the wine bottles well organised and easy to access. But it does not keep the wine cool.	
Manufacturing – 4/10	My product is strong but is not made from environmentally friendly and sustainably materials and is made out of a heavy material.	

Figure 14.13 An example of progressive iteration with ongoing evaluation

The following activity shows how physical modelling can be used to check how well a design incorporates the technical requirements.

STRETCH ACTIVITY

Give yourself two minutes to make a sketch model of an idea for an innovative shelf/wall display for small collectables.

Use card, scissors and sticky tape.

	Mark Band 1	Mark Band 2	Mark Band 3	Mark Band 4
Development of final design solution(s)	Little or no progression from earlier developments, and few or none of the identified opportunities and requirements met.	Some progression from earlier developments, and some of the identified opportunities and requirements met.	Clear progression from earlier developments, and most of the identified opportunities and requirements met.	Clear and comprehensive progression from earlier developments, and all of the identified opportunities and requirements met.

What?

How clearly does the final design solution conclude your development work? Is there a clear path from initial development to the final design solution? Is it clear how well it meets the needs of your primary user and other stakeholders?

How?

Sketching, modelling, visualisation, annotation, evaluation.

To score highly in this section it is vital that you **communicate** how your idea has developed from its earlier stages through to the final design. This will take place over several pages so it is important that you guide the reader through your development using clear headings to enable them to clearly follow how your design evolved over time. Each development must be detailed and you must make clear how each development meets both the technical and stakeholder requirements as the design evolves. You will **sketch**, **model**, **test** and **evaluate** as you go and you must make sure that any **annotation** you use helps explain and justify your decisions.

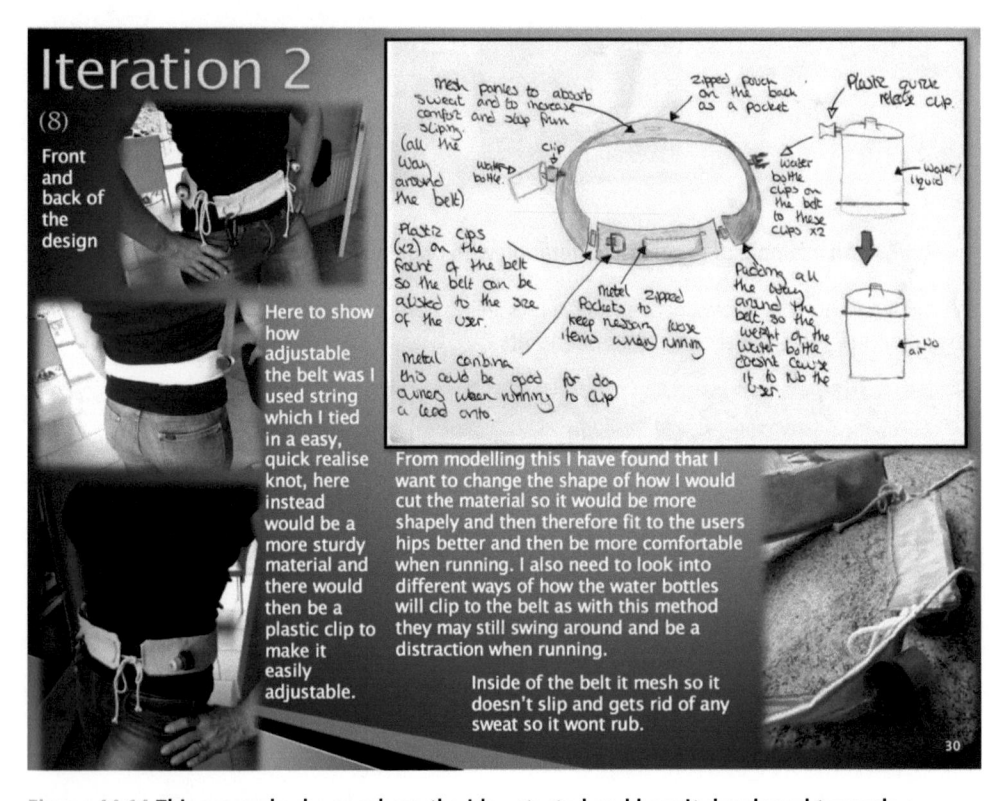

Figure 14.14 **This example shows where the idea started and how it developed towards a further iteration. This iteration was then developed further before being presented as the final design solution. The final design has been annotated with details of how it meets the requirements that were established at the start of the process.**

	Mark Band 1	Mark Band 2	Mark Band 3	Mark Band 4
Critical thinking	Superficial responses when problems are identified. Little or no evidence of innovation throughout the design process.	Effective responses to some identified problems. Some evidence of innovation throughout the design process.	Effective responses to most identified problems. Clear evidence of innovation throughout the design process.	Systematic and effective responses to all identified problems. Clear and systematic evidence of innovation throughout the design process.

What?

Have you been clear in identifying problems as they arise? How well have you responded to problems during the development? How methodical and logical has your approach been? Have you shown innovative thinking through your design developments?

How?

Sketching, modelling, presentation, annotation.

There are two requirements for this criterion. The first relates to how well you have responded to problems throughout your development. As mentioned earlier, it is important that your development work is clearly laid out and easy to follow using relevant headings. Key to achieving this will be clearly identifying problems as you develop your ideas. 'Systematic evidence of innovation' relates to you having some sort of plan or format to help you respond to problems. Whatever method you decide on, stick with it throughout your development. It is important here that you don't see any problems as mistakes to hide, but rather an opportunity for further improvement.

STRETCH ACTIVITY

To practise your critical thinking, choose a product you are familiar with, and:
- choose three things you like and explain why you think they work in the design
- share three things you think need improvement and how you would improve them.

The second requirement of this criterion is **innovation** – whether you have considered new methods or ideas to improve and refine your design solutions. For something to be considered innovative you should demonstrate how it offers something new to the intended market and/or primary user. Try to respond to problems with unexpected solutions or new approaches. Get radical – try weird, wild and absurd solutions. It is easier to tone down a solution than it is to pump one up. Is there clear evidence of innovation throughout your development?

Strand 3: Create: Design Communication (A02) – 16 marks

The assessment of this strand relates to the appropriate quality of communicating your design ideas, in order that a third party would be able to understand you intentions.

	Mark Band 1	Mark Band 2	Mark Band 3	Mark Band 4
Quality of chronological progression	Design iterations are not always clear and/or chronological, with little or no support from real-time evidence.	Design iterations are sometimes clear and predominantly chronological, with some support from real-time evidence.	Design iterations are clear and chronological, and mostly supported by real-time evidence.	Design iterations are clear, systematic and chronological, and fully supported by real-time evidence.

What?

How well have you documented your development? Is there a clear sense of progression through iterations? Have you recorded it in real time?

How?

Layout and presentation, videos, photographs.

This criterion is similar to earlier ones but requires you to make sure that your development truly reflects a real-time approach to design. This means recording everything **as it happens**. If, for example, you explored some existing products halfway through iteration 2, then this must be clear in your portfolio. Although this will be based on your teacher's judgment, how well you document and communicate this will be crucial to your success.

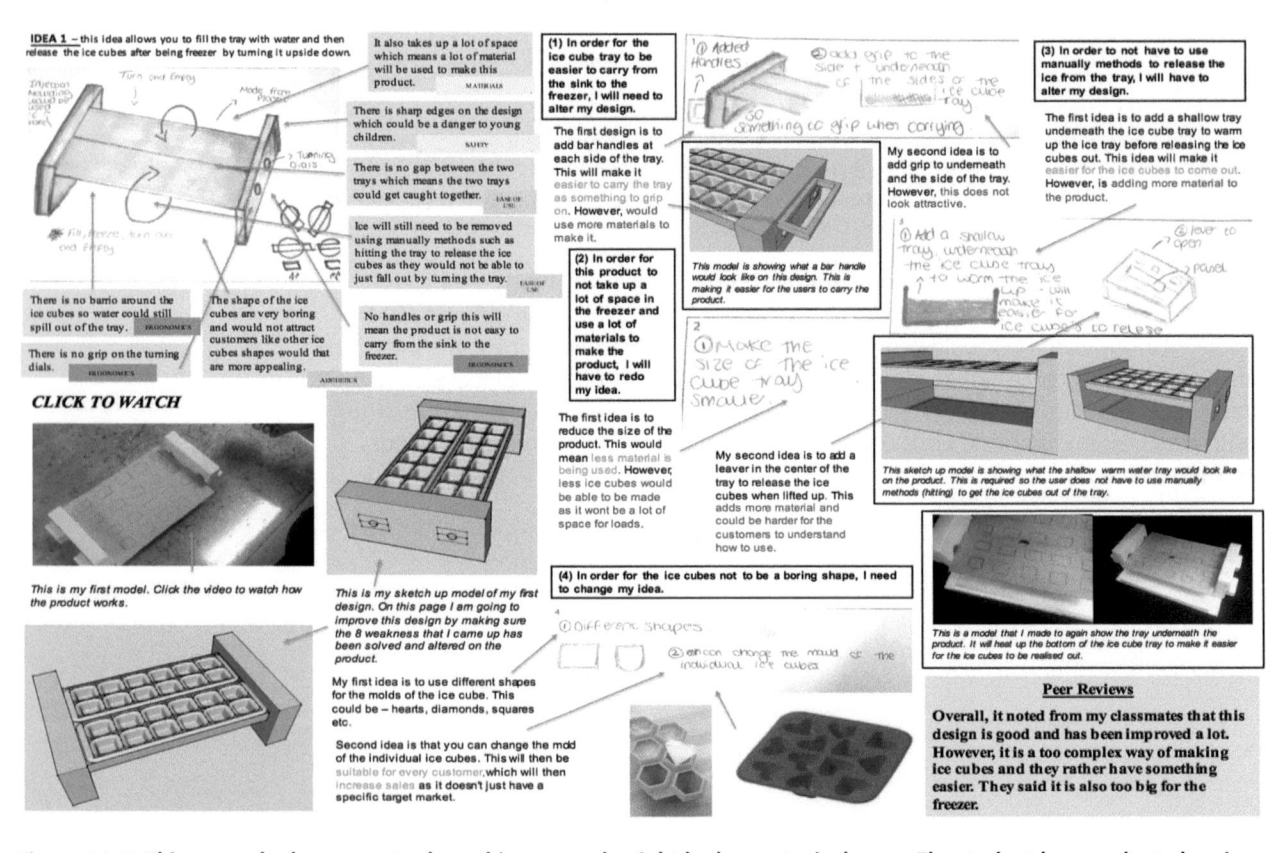

Figure 14.15 **This example demonstrates how this approach might look on a typical page. The student has conducted various pieces of research at appropriate stages and this has been recorded as it happened in their portfolio. From an investigation into potential components in solution 5, to further exemplification of an idea in solutions 6 and 7, to some material investigation in solution 8.**

	Mark Band 1	Mark Band 2	Mark Band 3	Mark Band 4
Quality of initial ideas	Informal graphical and modelling skills are limited, and rarely clear enough to appropriately communicate initial thinking.	Informal graphical and modelling skills are sufficient, but are not consistent in appropriately communicating initial thinking.	Informal graphical and modelling skills are good, and are consistent in appropriately communicating initial thinking.	Informal graphical and modelling skills are excellent, and are effective and consistent in appropriately communicating initial thinking.

What?

How good is your initial sketching and modelling? Does it effectively communicate your initial ideas? Have you used a range of suitable techniques?

How?

Ideation, sketching, sketch models.

This criterion assesses how well your sketching and modelling communicates your initial ideas. Early sketches could vary from simple doodle-type sketches, to more accurate 3D sketches. The key requirement is that they communicate your ideas effectively. Simple sketches will require more explanation. Try to show any moving parts using a sequence of drawings and arrows to show movement. Simple exploded and close-up views could also be used at this stage.

Figure 14.16 **Sticky note approach to generating ideas – note the sequence sketch at the bottom left**

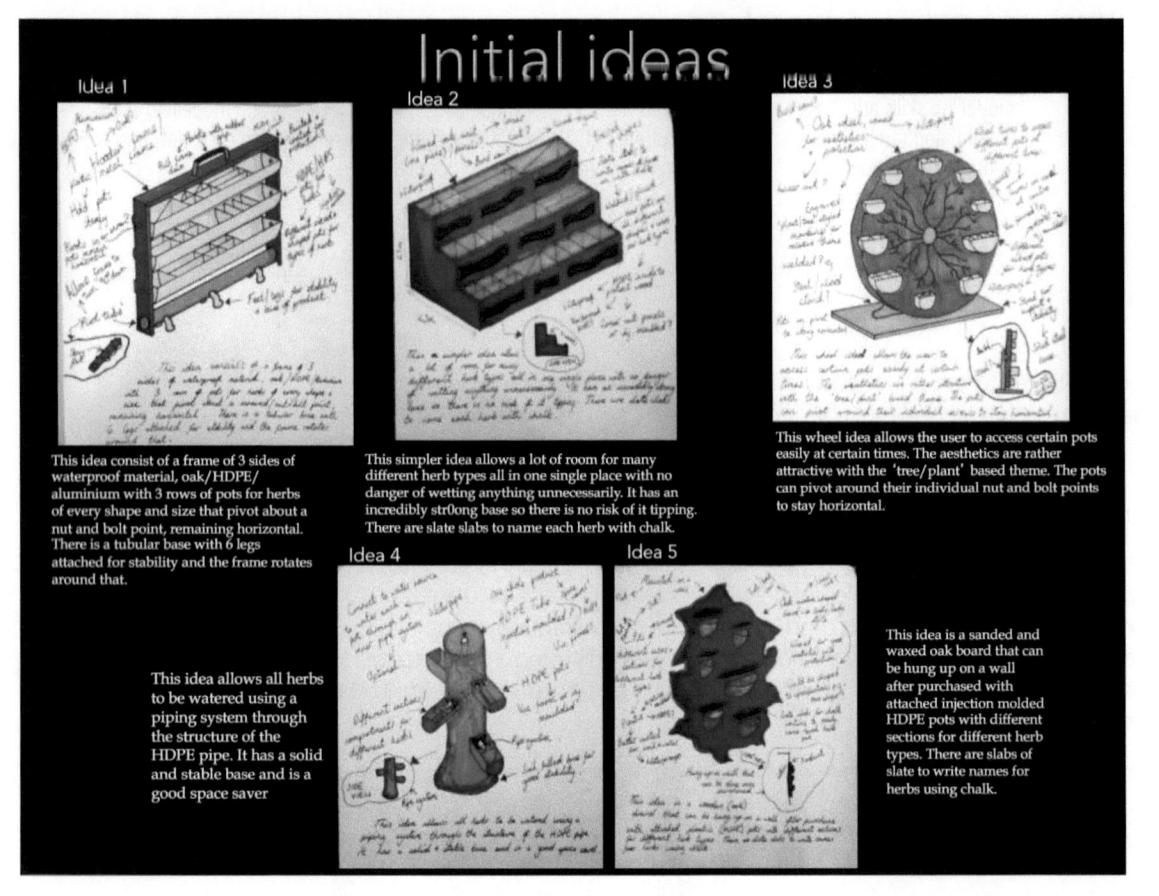

Figure 14.17 **More refined initial ideas that rely heavily on detailed annotation to communicate**

The following text accompanies the ideas:

This idea consist of a frame of 3 sides of waterproof material, oak/HDPE/aluminium with 3 rows of pots for herbs of every shape and size that pivot about a nut and bolt point, remaining horizontal. There is a tubular base with 6 legs attached for stability and the frame rotates around that.

This simpler idea allows a lot of room for many different herb types all in one single place with no danger of wetting anything unnecessarily. It has an incredibly str0ong base so there is no risk of it tipping. There are slate slabs to name each herb with chalk.

This wheel idea allows the user to access certain pots easily at certain times. The aesthetics are rather attractive with the 'tree/plant' based theme. The pots can pivot around their individual nut and bolt points to stay horizontal.

This idea allows all herbs to be watered using a piping system through the structure of the HDPE pipe. It has a solid and stable base and is a good space saver

This idea is a sanded and waxed oak board that can be hung up on a wall after purchased with attached injection molded HDPE pots with different sections for different herb types. There are slabs of slate to write names for herbs using chalk.

CLASSROOM DISCUSSION

Working in groups of four, take it in turns to discuss each other's design briefs. Spend 10 minutes coming up with quick ideas for each other's projects. Use the Tell me/Show me approach by first describing the idea and then presenting it as a simple doodle-type sketch. Collate the ideas on to a large sheet of cardboard and discuss further to identify the ideas with the most potential.

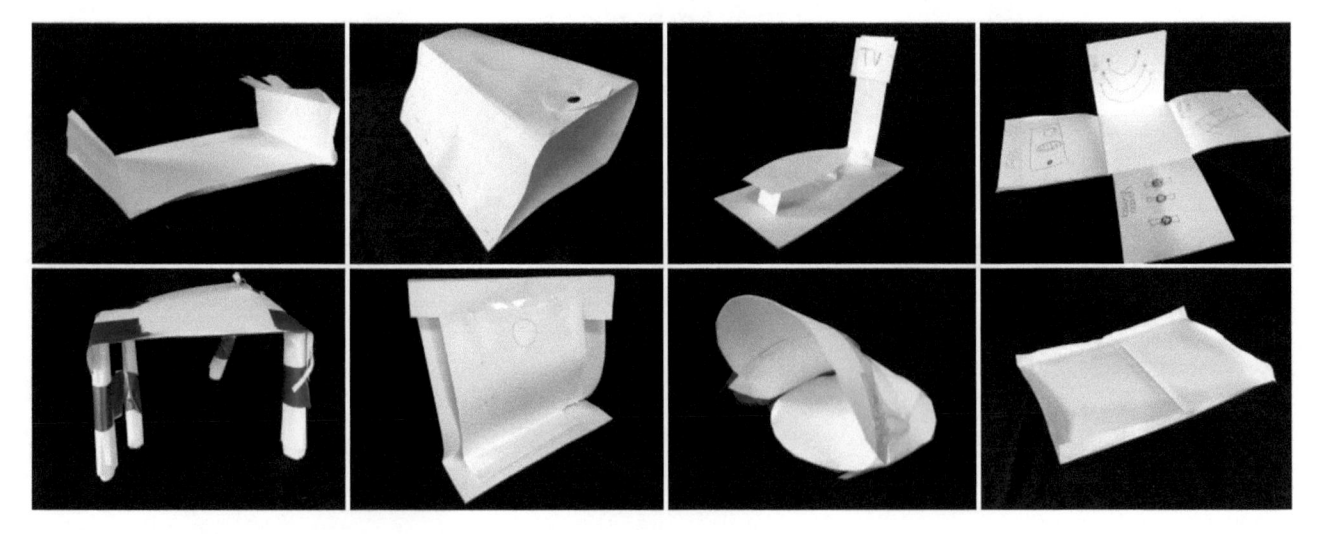

Figure 14.18 **Quick paper models can be used to generate initial ideas**

	Mark Band 1	Mark Band 2	Mark Band 3	Mark Band 4
Quality of design developments	The range of communication techniques used is limited, and is rarely clear enough to appropriately develop or communicate design concepts.	The range of communication techniques used is sufficient, but is not consistent in appropriately developing or communicating design concepts.	The range of communication techniques used is good, and is consistent in appropriately developing or communicating design concepts.	The range of communication techniques used is excellent, and is effective and consistent in appropriately developing or communicating design concepts.

What?

Have you used a variety of methods to communicate your developments? How well do they communicate your ideas? How appropriate are they?

How?

Sketching, modelling, prototypes, visualisation, video.

This criterion focuses on the range and quality of your sketching and modelling throughout the development. In terms of your sketching this would involve such techniques as 2D diagrams, sequence drawings, exploded and cut-away views, close-ups, 3D sketching and rendering. The key point to remember is that you should use the appropriate technique at the appropriate time. For example, producing a fully rendered 3D exploded view sketch at the start of the process when ideas are just forming would be time-consuming and not especially useful to your development.

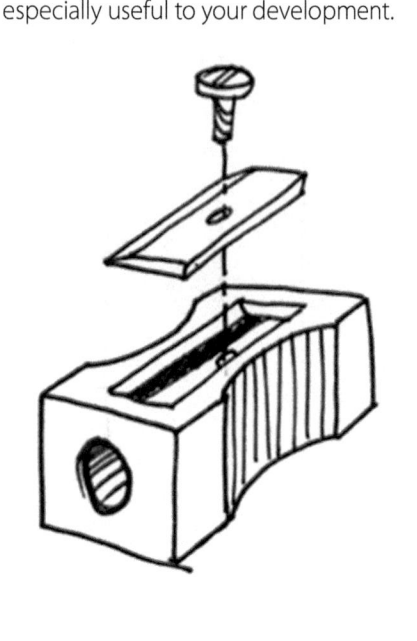

Figure 14.19 **A simple exploded view sketch**

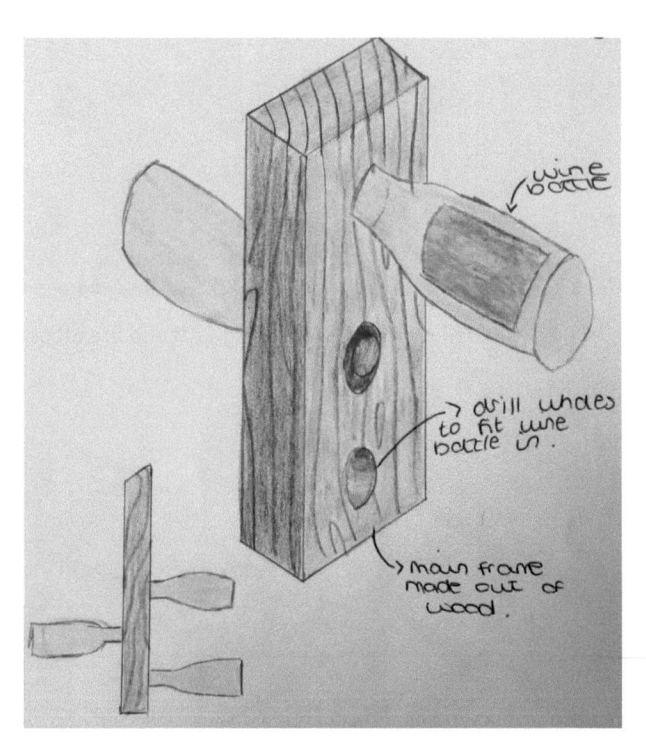

Figure 14.20 **A three-dimensional sketch with a supporting 2D diagram**

As with the sketching, your modelling must be appropriate and relevant throughout your development. You should use a range of modelling techniques to communicate your ideas. These should include techniques such as: quick paper models to establish initial ideas; full-size mock-ups in card, foam and fabric; breadboard prototypes; samples of techniques to consider incorporating; CAD models; through to more complex materials as you develop working prototypes.

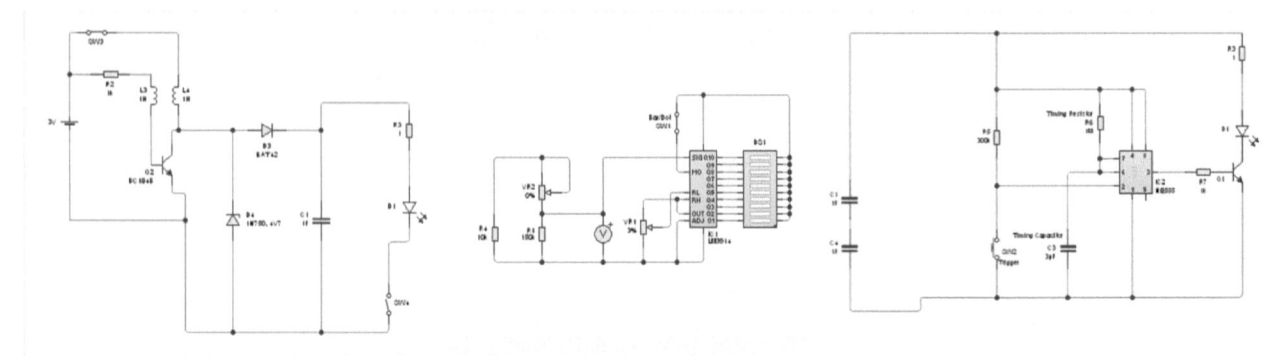

Here you can see that I have gone from virtual to physical designing. This has been done by me firstly designing the circuit on the computer on Circuit Wizard and then after testing the circuit, moving on to bread boarding, this has a major advantage as I have found out that sometimes the circuit needs adjusting on the breadboard, instead of going straight into production the PCB. The reason for choosing a 555 timing circuit is due to circuit being easy to construct , and also is cheap (cost of components would be around 40p, when bulk purchasing is taken into account), plus it was just showing that the circuit is viable and then it does not have enough features to make it useful for my product , although it could be used instead of a GENIE chip for a lower end market product.

Charging Circuit   Measuring Voltage Circuit   555 Timing Circuit (not being used)

Figure 14.21 **Modelling circuit ideas using CAD and breadboard prototyping**

	Mark Band 1	Mark Band 2	Mark Band 3	Mark Band 4
Quality of final design solution(s)	Formal presentation of the final design solution(s) is limited, making it difficult for a third party to understand.	Formal presentation of the final design solution(s) is sufficient and provides some clarity to a third party.	Formal presentation of the final design solution(s) is good and provides appropriate clarity to a third party.	Formal presentation of the final design solution(s) is excellent and provides impact and appropriate clarity to a third party.

What?

How well does the quality of your final design presentation communicate the final idea? Is it clear enough for a third party to understand it?

How?

Sketching, rendering, modelling, visualisation.

As your development concludes, you are expected to present a final design solution. This should be formally presented in a way that is appropriate to the product. It should provide impact to your stakeholders and help them to clearly understand your design solution. This can be done using: a 3D rendered drawing, and/or a 3D CAD model.

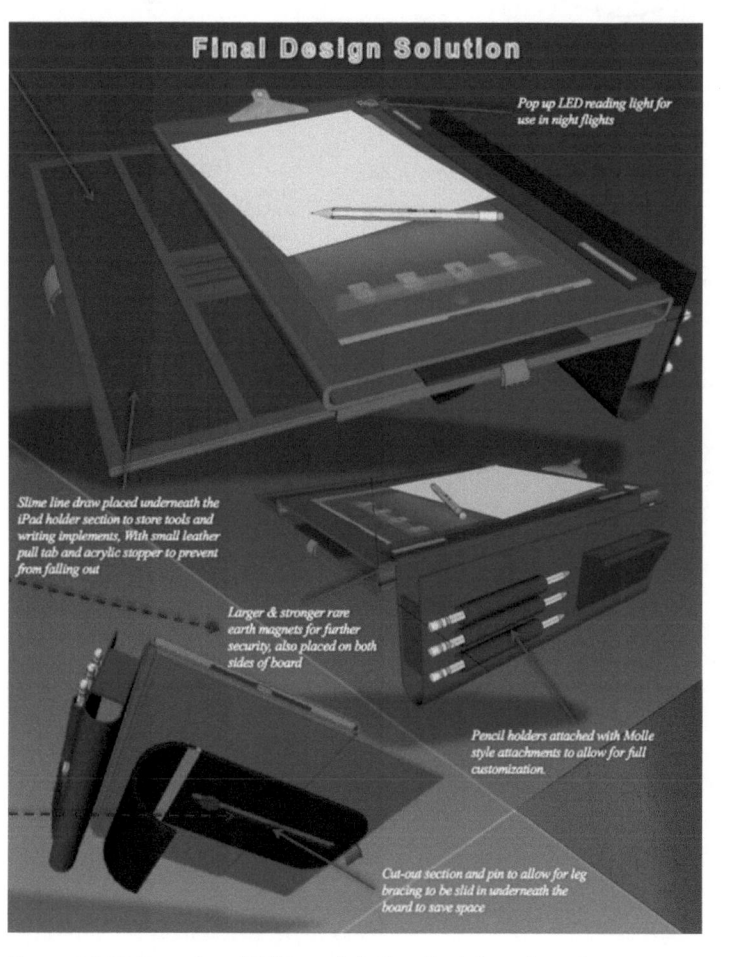

Figure 14.22 **A rendered CAD model of a pilot's kneeboard**

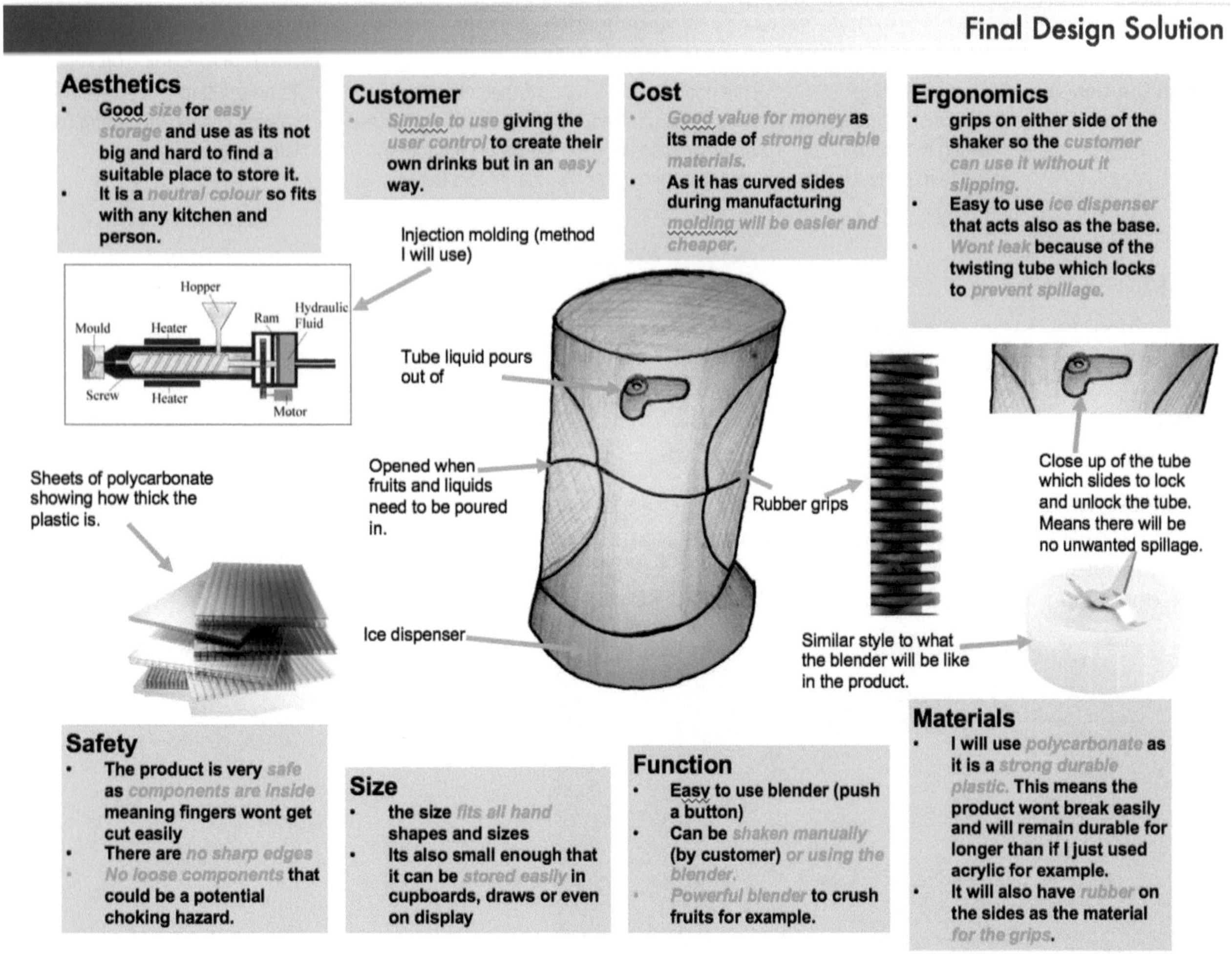

Aesthetics

- **Good** *size* for *easy storage* **and use as its not big and hard to find a suitable place to store it.**
- **It is a** *neutral colour* **so fits with any kitchen and person.**

Customer

- *Simple to use* **giving the** *user control* **to create their own drinks but in an** *easy* **way.**

Cost

- *Good value for money* **as its made of** *strong durable materials.*
- **As it has curved sides during manufacturing** *molding will be easier and cheaper.*

Ergonomics

- **grips on either side of the shaker so the** *customer can use it without it slipping.*
- **Easy to use** *ice dispenser* **that acts also as the base.**
- *Wont leak* **because of the twisting tube which locks to** *prevent spillage.*

Injection molding (method I will use)

Hopper

Mould   Heater   Ram   Hydraulic Fluid

Screw   Heater   Motor

Sheets of polycarbonate showing how thick the plastic is.

Tube liquid pours out of

Opened when fruits and liquids need to be poured in.

Rubber grips

Close up of the tube which slides to lock and unlock the tube. Means there will be no unwanted spillage.

Ice dispenser

Similar style to what the blender will be like in the product.

Safety

- **The product is very** *safe* **as** *components are inside* **meaning fingers wont get cut easily**
- **There are** *no sharp edges*
- *No loose components* **that could be a potential choking hazard.**

Size

- **the size** *fits all hand* **shapes and sizes**
- **Its also small enough that it can be** *stored easily* **in cupboards, draws or even on display**

Function

- **Easy to use blender (push a button)**
- **Can be** *shaken manually* **(by customer)** *or using the blender.*
- *Powerful blender* **to crush fruits for example.**

Materials

- **I will use** *polycarbonate* **as it is a** *strong durable plastic.* **This means the product wont break easily and will remain durable for longer than if I just used acrylic for example.**
- **It will also have** *rubber* **on the sides as the material** *for the grips.*

Figure 14.23 **A hand-drawn final solution with additional written details**

KEY POINT

Practise presenting your ideas to other people as often as you can. Build these presentations into your portfolio and work up to regular consultation with your key stakeholders.

Figure 14.24 **A rendered CAD model placed into a photograph to help visualise the final design**

Strand 4: Create: Final Prototype(s) (A02) – 20 marks

The assessment of this strand relates to the impact and quality of your final prototype(s), In order that a third party would be able to understand your design intentions.

	Mark Band 1	Mark Band 2	Mark Band 3	Mark Band 4
Quality of planning for making the final prototype(s)	Offers little or no support to the making process.	Generally supports the management of the making process with some relevant requirements identified from the technical specification.	Good level of detail and relevance, covering most requirements identified from the technical specification to manage the making process.	Comprehensive and relevant, covering all requirements identified from the technical specification to effectively manage the making process.

What?

How well have you planned the making of your final prototype(s)? Will it meet the requirements of your technical specification? Do you need to consider alternative methods and materials suitable for the prototype(s)? Have you outlined the tools, machinery and processes you will need? Have you estimated the time it will take? Have you used a useful and relevant planning method? Have you built in specific Quality Assurance and Control measures?

How?

Flowcharts, Gantt charts, task lists.

This criterion assesses how well you have planned the making of your final prototype(s). Your making plan should detail how each component is to be made in the workshop. It should take into account the tools, machinery and processes that you intend you use. You must make it clear how the plan allows you to meet the technical specification that came before it. Any techniques must be relevant to the product you have designed and demonstrate that you have managed the making process effectively.

Production Plan

Backboards
- Cut all holes- 2 in thicker back board, 3 in front layer of layered back **board**- Drill, centre punch and Piercing Saw
- Guillotine 2nd layer for tiers to hook onto
- **File and round edges**- Flat file and Shaped file
- **Glue together**- between 2nd and front layer glue material to hook tier on
- **Round all corner edges once all layers are together**- Flat file

Tiers
- Cut up using guillotine
- **Cut slits for pods to hang onto**- Piercing Saw
- **Round edges which will stick out furthest**- Flat file and Half-round file
- **Glue material between layers x2**

Pods
- Complete on Creo
- Transfer into Techsoft
- Colour line for engraving and where to cut out
- Laser cut
- **Build any necessary formers- to help with alignment**- Coping saw, flat file and Sand paper
- Glue together
- **Drill in holes for hooks**- Drill
- **Make hooks**- line bender
- Attach hooks
- **Finish- remove excess glue**- Sand paper, wet and dry paper, small file if necessary

KEY:
(T)- theory lesson
(P)- ,manufacturing lesson
(GR)- Graphics room- allows me to work on Creo or in the metal room to use the piercing saw or file
(WK)- workshop

To the right is a rough production plan for lessons and after school manufacturing sessions depending on where I can manufacture and how much time I have so I will finish my manufacturing before the end of term, allowing me to market and evaluate the final product

STRETCH ACTIVITY

Plan the stages involved in making a cup of tea. Outline the equipment needed and the time you estimate each step to take. This could be done as a table or flowchart.

Figure 14.25 **An example of an approach to planning**

	Mark Band 1	Mark Band 2	Mark Band 3	Mark Band 4
Quality of final prototype(s)	Inaccurate and/or basic standards demonstrated. Finishing may not be appropriate and/or the outcome would not present well to a stakeholder.	Sufficient standard demonstrated through a generally accurate outcome. Finishing is appropriate but the outcome could be better presented to a stakeholder.	Good standard and levels of accuracy demonstrated. Finishing is appropriate and the outcome will present well to a stakeholder.	Excellent standard, demonstrating high levels of accuracy. Finishing is appropriate and the outcome will present well and provide impact to a stakeholder.

What?

How accurate is your final prototype? What is the level of finish? Does it reflect the final design? How effectively will it communicate to your stakeholders?

How?

Photographs, video.

This criterion focuses entirely on the quality of your final prototype(s). You will need a range of clear photographs that fully demonstrate your attention to detail and the level of accuracy in your work. If there is any movement or sound in your prototype(s) then this must be clearly shown in the form of a video. The level of finish must be appropriate to clearly represent your design solution. Your prototype will also be judged on the level of impact that it will provide if presented to a third party.

FINAL PROTOTYPE

This shows the desk tidy in use, in a work place and with stationary and items in the product to show how it works and what it will look like as an example.

This picture shows the development of the product, it shows how the product compartments are **stackable** therefore **easy to store away** and **less obtrusive** in a work place. The image below shows the **divided sections** so that it is **easily organised** and does **not** get **clustered** and smaller items can be **separated easily**. The tube also shows a section for pens and pencils and other larger items.

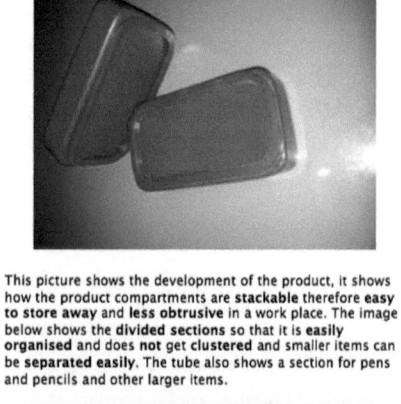

The picture above shows the **stack ability** and **trays/compartments** of the larger and smaller compartments. The image below shows the desk tidy not in use but in the intended environment. It also shows another way it can be arranged.

Figure 14.26 **Examples of final prototype presentations – Desk tidy**

	Mark Band 1	Mark Band 2	Mark Band 3	Mark Band 4
Use of specialist techniques and processes	Limited and rarely appropriate to materials/components used.	Sufficient, but not consistently appropriate to materials/components used.	Good and consistently appropriate to materials/components used.	Excellent and effective and consistently appropriate to materials/components used.

What?

What techniques and processes have you used? How specialised are they? Are they appropriate to the materials you have chosen? Have you documented this in real time?

How?

Photographs, videos.

This criterion (and the next) rely on you successfully documenting the making processes undertaken to fully realise your final prototype(s). This will be wholly dependent on a range of photographs and videos that clearly show the main processes involved. In addition your teacher and technician will make observations on your ability in the workshop.

	Mark Band 1	Mark Band 2	Mark Band 3	Mark Band 4
Use of specialist tools and equipment	Use and selection of hand tools and/or machinery is limited and rarely appropriate. Digital design and/or manufacture is limited and demonstrates little or no skills or knowledge.	Use and selection of hand tools and machinery is sufficient, but not always consistently appropriate. Digital design and manufacture is not always used appropriately, but demonstrates sufficient skills and knowledge.	Use and selection of hand tools and machinery is good and consistently appropriate. Digital design and manufacture is used appropriately to demonstrate good skills and knowledge.	Use and selection of hand tools and machinery is effective and consistently appropriate. Digital design and manufacture is used effectively and appropriately to demonstrate excellent skills and knowledge.

What?

What tools and equipment have you used? Are they appropriate to the materials you have chosen? How have you used digital design and manufacture techniques? Have you documented this in real time?

How?

Photographs, videos.

As with the previous criterion, a fully documented 'making diary' is required to provide evidence of the tools and machinery you have used to make your final prototype(s). Note that you are required to demonstrate use of hand tools, machinery, digital design and digital manufacture. This should not detract from the most appropriate use of each item. This could involve use of CAD/CAM to make your final prototype(s) or to make a specific element/component. If digital design has not been appropriate to the making of your final prototype(s), however, it will be assessed on its use in your development modelling. This could equally relate to use of hand tools and machinery through earlier developments if the final prototype(s) are exclusively made through the use of CAD/CAM. You are not required to show every individual step of the making; but it must be sufficient to support the demonstration of your skills. You must be selective about the images and video you use to show the main stages of making.

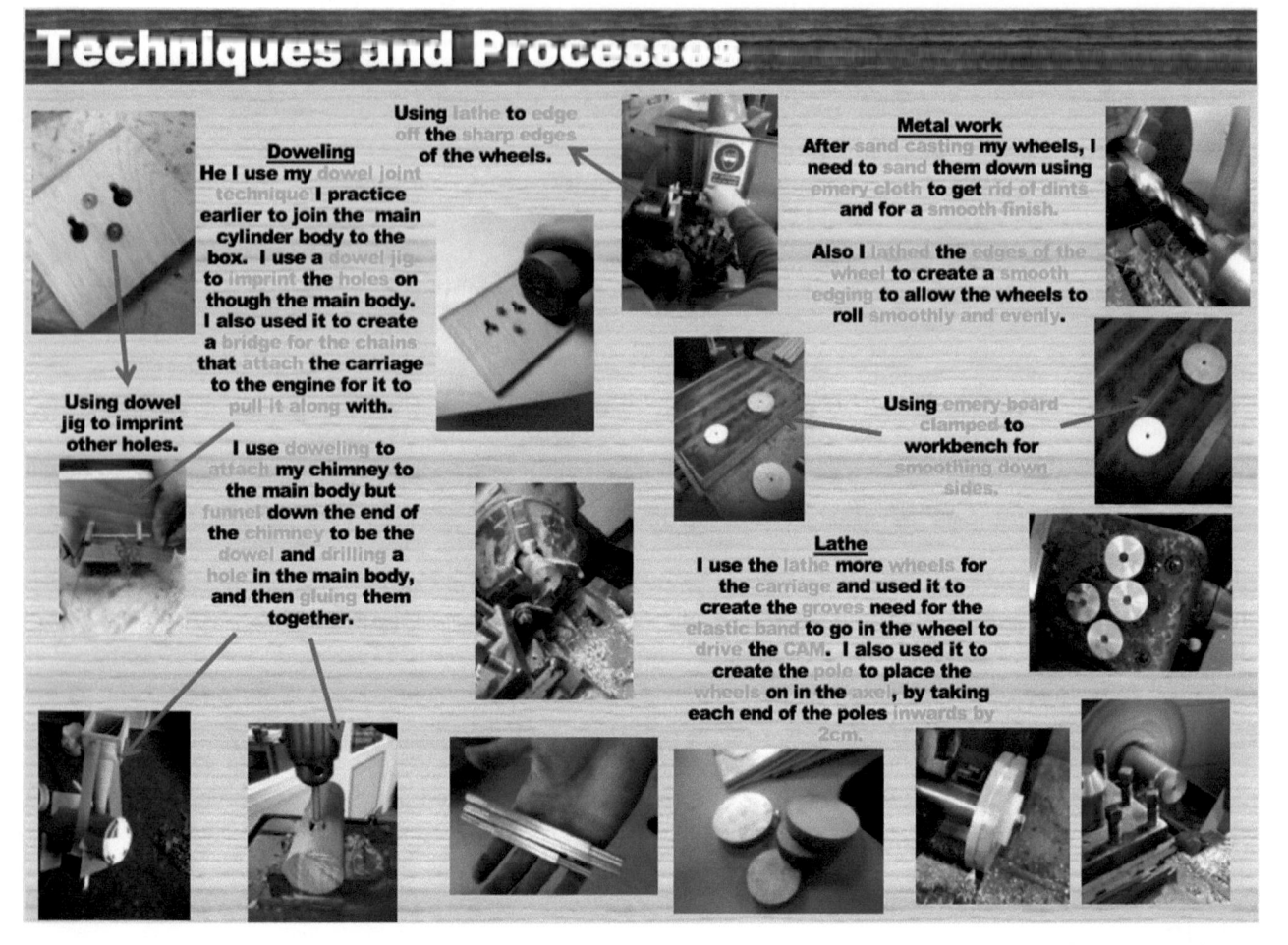

Techniques and Processes

Using lathe **to** edge off **the** sharp edges **of the wheels.**

Doweling
He I use my dowel joint technique **I practice earlier to join the main cylinder body to the box. I use a** dowel jig **to** imprint **the** holes **on though the main body. I also used it to create a** bridge for the chains **that** attach **the carriage to the engine for it to** pull it along **with.**

Using dowel jig to imprint other holes.

I use doweling **to** attach **my chimney to the main body but** funnel **down the end of the** chimney **to be the** dowel **and** drilling **a** hole **in the main body, and then** gluing **them together.**

Metal work
After sand casting **my wheels, I need to** sand **them down using** emery cloth **to get** rid of dints **and for a** smooth finish.

Also I lathed **the** edges of the wheel **to create a** smooth edging **to allow the wheels to roll** smoothly and evenly.

Using emery-board clamped **to workbench for** smoothing down sides.

Lathe
I use the lathe **more** wheels **for the** carriage **and used it to create the** groves **need for the** elastic band **to go in the wheel to** drive **the** CAM. **I also used it to create the** pole **to place the** wheels **on in the** axel, **by taking each end of the poles** inwards by 2cm.

Figure 14.27 **An example of documenting making**

	Mark Band 1	Mark Band 2	Mark Band 3	Mark Band 4
Viability of the final prototype(s)	Few or no links to the technical specification, demonstrating limited potential to become a marketable product.	Meets some of the technical specification, demonstrating some potential to become a marketable product.	Meets most of the technical specification, demonstrating good potential to become a marketable product.	Meets all of the technical specification, demonstrating excellent potential to become a marketable product.

What?

How well does the prototype(s) meet your technical specification? Can you justify its marketability?

How?

Photographs, videos.

This criterion requires you to show how your final prototype(s) meets your technical specification. This could be done by comparing your final prototype(s) against your working drawings and describing how it meets each of your technical requirements. A justification of why you consider your prototype(s) to be marketable is also required.

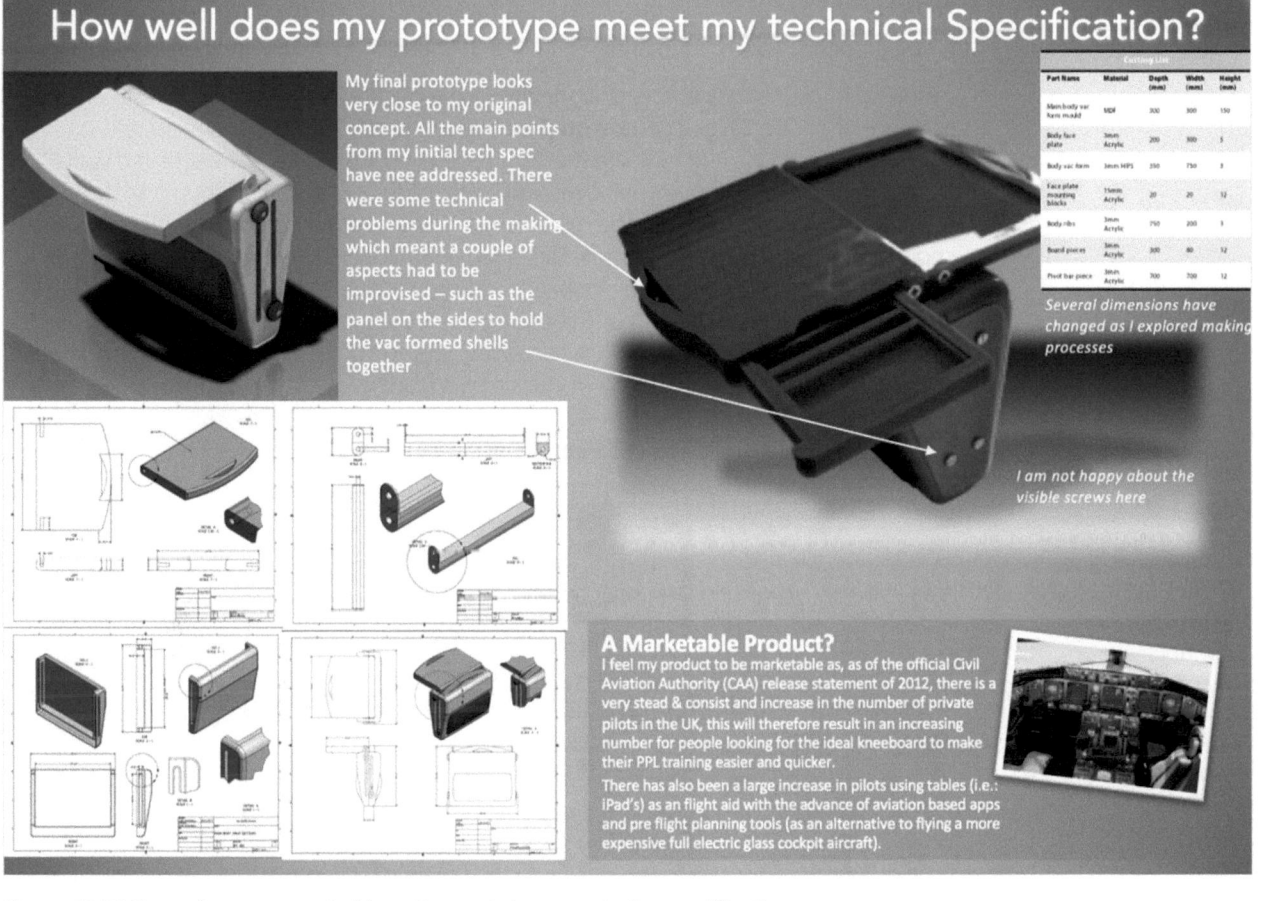

Figure 14.28 **Example assessment of how the prototype meets the specification**

Strand 5: Evaluate (A03) – 20 marks

The work being assessed in this strand will be evidenced from your complete portfolio. This assessment focuses on the quality of analysis and evaluation you have undertaken and how well this relates and reflects on the context, brief and requirements of the developments you have gone through.

	Mark Band 1	Mark Band 2	Mark Band 3	Mark Band 4
Analysis and evaluation of primary and/or secondary sources	Limited analysis and evaluation of investigated sources of information from stakeholders, existing products and/or wider issues, offering little or no support to inform the design process.	Sufficient analysis and evaluation of investigated sources of information from stakeholders, existing products and wider issues, offering some support to inform the design process.	Good level of analysis and evaluation of investigated sources of information from stakeholders, existing products and wider issues, offering clear support to inform the design process.	Comprehensive and systematic analysis and evaluation of investigated sources of information from stakeholders, existing products and wider issues, offering clear and focused support to inform the design process.

What?

How well have you analysed the information you found? Have you clearly explained how it supports the design process? Have you evaluated its usefulness?

How?

Written reports, videos, annotations, graphs and charts.

Despite this criterion appearing near the end of the mark scheme, you will need to provide evidence of detailed evaluation **throughout** your e-portfolio. This criterion focuses on your ability to analyse and evaluate the information you found as you investigated your stakeholders and developed your ideas. As you develop your ideas you should analyse existing products and examine the relevant wider issues appropriate to your ideas. It is also important that you make it clear how the information you analyse affects the direction of your development. A primary source is information you have found yourself. This would include hands-on product analysis, interviews, etc. A secondary source of information is someone else's research that you have analysed. This could be a document form the internet or one you have scanned in from a magazine or textbook.

KEY POINT

Make it clear how any ongoing research you carry out has influenced the development of your ideas. Clear headings and explanations will help you with this. Colour highlighting might also help to identify specific stakeholder requirements that are being met.

Figure 14.29 **An example of hands-on analysis of a product – this is primary research. Note the use of colour and bold to highlight key information.**

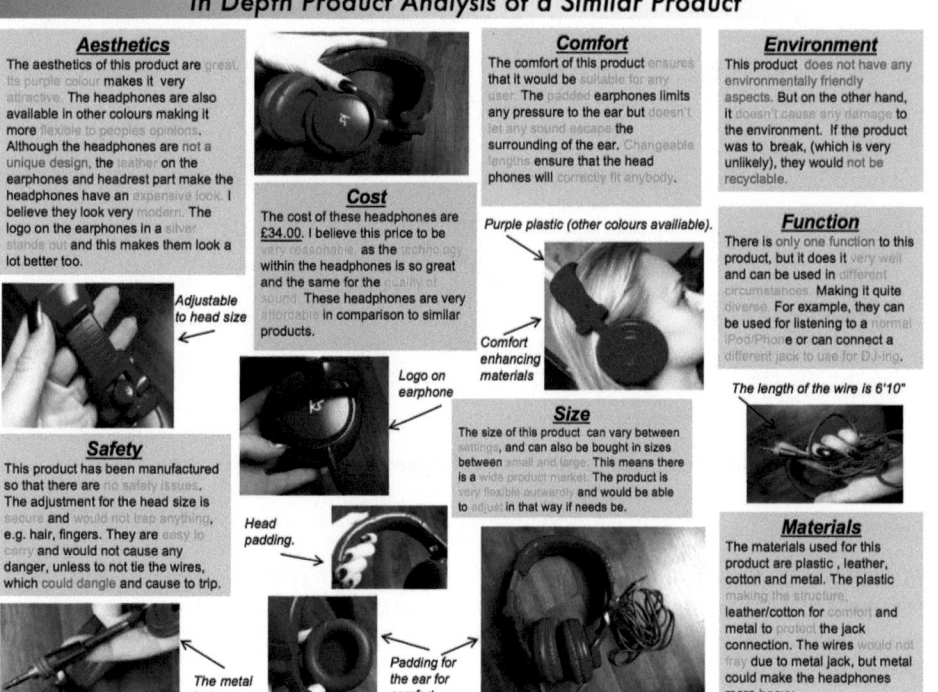

In Depth Product Analysis of a Similar Product

Aesthetics
The aesthetics of this product are great, its purple colour makes it very attractive. The headphones are also available in other colours making it more flexible to peoples opinions. Although the headphones are not a unique design, the leather on the earphones and headrest part make the headphones have an expensive look. I believe they look very modern. The logo on the earphones in a silver stands out and this makes them look a lot better too.

Adjustable to head size

Safety
This product has been manufactured so that there are no safety issues. The adjustment for the head size is secure and would not trap anything, e.g. hair, fingers. They are easy to carry and would not cause any danger, unless to not tie the wires, which could dangle and cause to trip.

The metal jack.

Cost
The cost of these headphones are £34.00. I believe this price to be very reasonable, as the technology within the headphones is so great and the same for the quality of sound. These headphones are very affordable in comparison to similar products.

Logo on earphone

Head padding.

Comfort
The comfort of this product ensures that it would be suitable for any user. The padded earphones limits any pressure to the ear but doesn't let any sound escape the surrounding of the ear. Changeable lengths ensure that the head phones will correctly fit anybody.

Purple plastic (other colours availiable).

Comfort enhancing materials

Size
The size of this product can vary between settings, and can also be bought in sizes between small and large. This means there is a wide product market. The product is very flexible outwardly and would be able to adjust in that way if needs be.

Padding for the ear for comfort.

Environment
This product does not have any environmentally friendly aspects. But on the other hand, it doesn't cause any damage to the environment. If the product was to break, (which is very unlikely), they would not be recyclable.

Function
There is only one function to this product, but it does it very well and can be used in different circumstances. Making it quite diverse. For example, they can be used for listening to a normal iPod/Phone or can connect a different jack to use for DJ-ing.

The length of the wire is 6'10"

Materials
The materials used for this product are plastic , leather, cotton and metal. The plastic making the structure, leather/cotton for comfort and metal to protect the jack connection. The wires would not fray due to metal jack, but metal could make the headphones more heavy.

	Mark Band 1	Mark Band 2	Mark Band 3	Mark Band 4
Ongoing evaluation to manage design progression	Superficial evaluations with little or no reflection on requirements or feedback.	Some critical evaluations with sufficient reflection on requirements and feedback.	Mostly critical evaluations with good reflection on requirements and feedback.	Full and critical evaluations with focused reflection on requirements and feedback.
	Few or no reviews to identify any problems and/ or next steps for future iterations, resulting in limited support for design progression.	Infrequent reviews to identify some problems and/or next steps for future iterations, which are not always consistent in supporting design progression.	Ongoing and clear reviews to identify problems and next steps for future iterations to consistently support design progression.	Ongoing, clear and comprehensive reviews to identify problems and next steps for future iterations to effectively and consistently support design progression.

What?

How well have you evaluated each iteration? Have you focused on technical and stakeholder requirements? Have you used feedback from your stakeholders? Have these reviews identified problems? Have you used this information to plan what to consider in the next iteration? Has the process been ongoing?

How?

Annotations, videos, written reports.

This criterion looks at how well you make use of **ongoing evaluation** while you develop your ideas. Any evaluation must be full and must focus on the critical details that will help direct subsequent iterations. You must make it clear how each solution you suggest clearly addresses any problems you have discovered through model testing and stakeholder feedback.

> **KEY POINT**
>
> Make it clear how the regular feedback from your stakeholders is influencing and directing your development. This could be done using colour coding or reference numbers that relate to your non-technical specification.

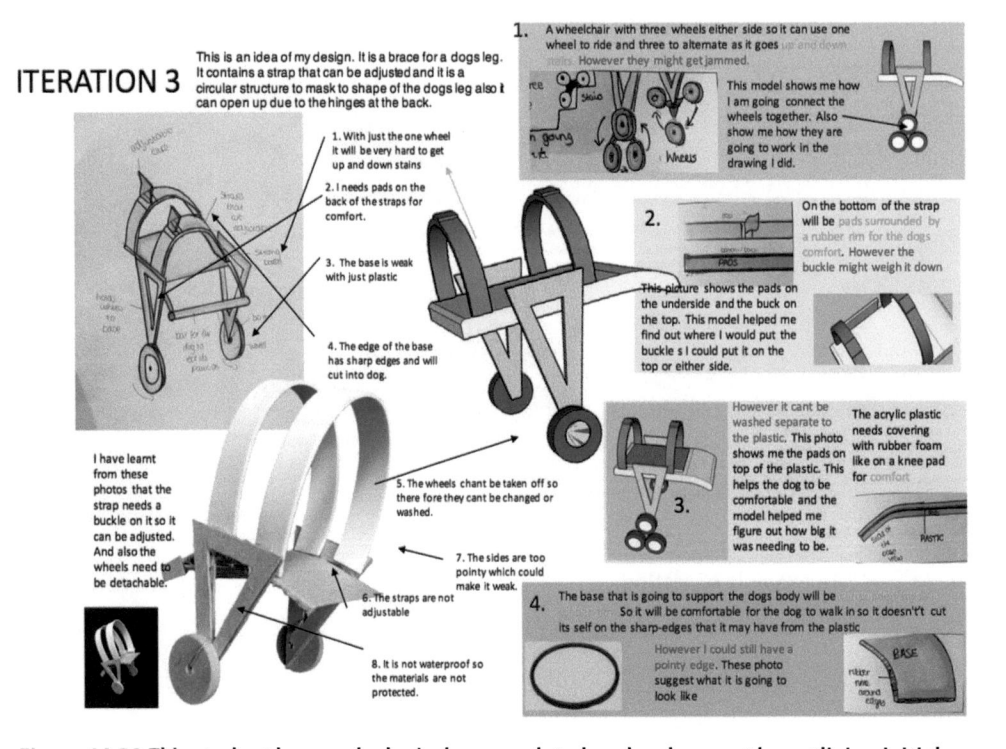

Figure 14.30 **This student has used a logical approach to her development by outlining initial problems with the design, then focusing on a solution that deals with each one**

	Mark Band 1	Mark Band 2	Mark Band 3	Mark Band 4
Feasibility of the design solution	Limited with few or no methods used to appropriately analyse and test whether the design solution is fit for purpose.	Sufficient with some appropriate methods used to analyse and test whether the design solution is fit for purpose.	Good level of detail with mostly appropriate methods used to analyse and test whether the design solution is fit for purpose.	Comprehensive with fully appropriate methods used to analyse and test whether the design solution is fit for purpose.

What?

How effectively have you tested your design solution? Have you used a variety of appropriate methods? Have you effectively analysed the results? Is your design fit for purpose?

How?

Photographs, videos, annotations and written reports.

How well you analyse and test your design solution will determine whether or not it is fit for purpose. This criterion will assess the appropriateness of the methods you use to fully test your design solution. This will depend on the type of design solution that you make. Most design solutions, however, can be successfully tested in a number of ways.

Usability can be evaluated by testing the design solution on its intended user group. This should include your primary user, but it should also include other relevant stakeholders. It involves asking the user to test it in a realistic environment, meaning that the school workshop will not be suitable. It could also include more general stakeholder surveys and opinions.

Material tests could be carried to determine if the design solution will cope with anticipated use in a variety of environments. A **test rig** might be built to help you test out specific features, such as stability. Photographs and videos will be essential to record the specific tests that you carry out.

KEY POINT

Ensure your testing is 'in situ'. This means testing it in the environment where it is intended to be used. For example, a tea bag disposal system would best be tested in a kitchen or café.

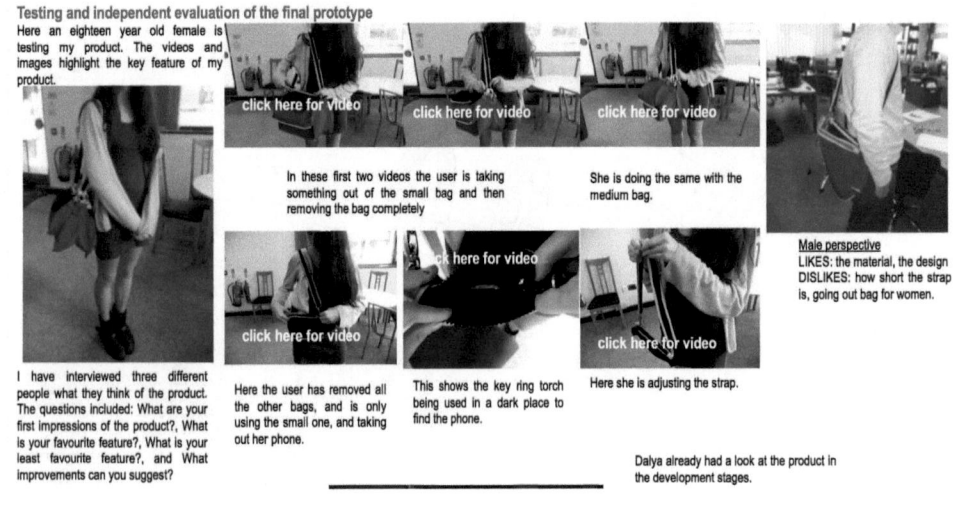

Figure 14.31 **An example of testing**

	Mark Band 1	Mark Band 2	Mark Band 3	Mark Band 4
Evaluation of the final prototype(s)	Superficial evaluation of strengths and/or weaknesses, with few or no suggestions for modification and/or consideration of possible design optimisation presented.	Sufficient critical evaluation of strengths and/or weaknesses, with some suggestions for modification and/or consideration of possible design optimisation presented.	Good critical evaluation of strengths and weaknesses, with detailed suggestions for modification and consideration of possible design optimisation presented.	Full and critical evaluation of strengths and weaknesses, with comprehensive suggestions for modification and consideration of possible design optimisation presented.

What?

Have you clearly identified your design's strengths and weaknesses? Have you based these on your testing? Have you suggested modifications to address any weaknesses? Have you suggested any modifications that would optimise the design?

How?

Written work, sketching, modelling.

This criterion assesses how well you have analysed your testing to arrive at a set of detailed strengths and weaknesses. You must then suggest comprehensive modifications that will address the weaknesses you have found.

Design optimisation is about balancing the trade-off between your prototype's cost, weight and manufacturability. Any modifications you suggest will have an impact on one or all of these. You must demonstrate an awareness of these issues and demonstrate how they might begin to be resolved.

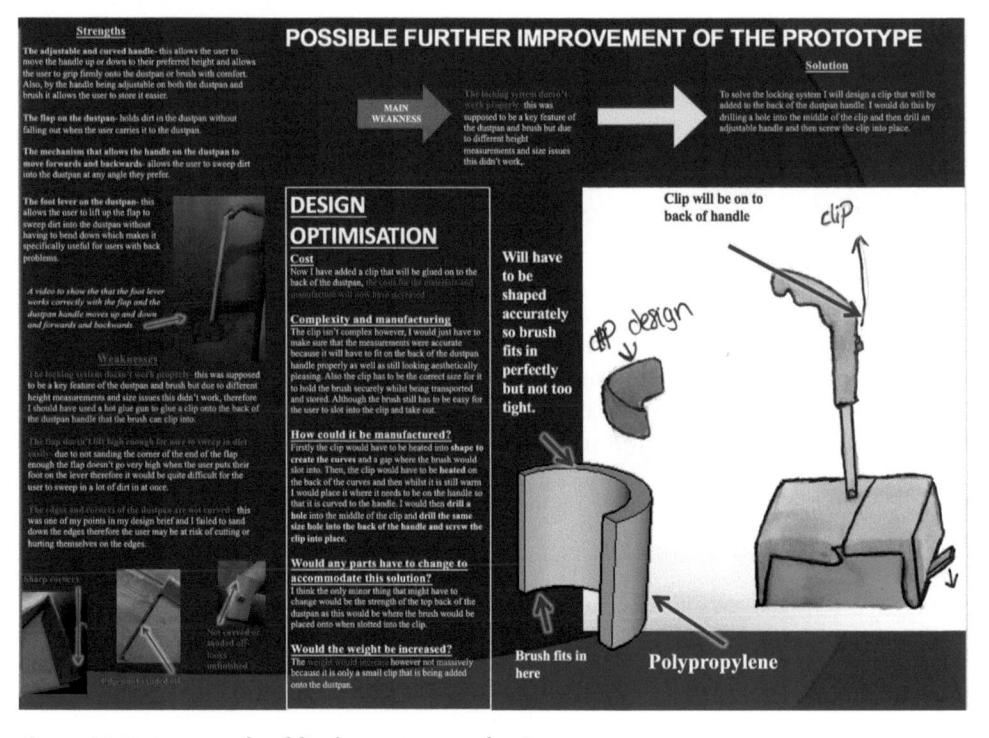

Figure 14.32 **An example of final prototype evaluation**

The written paper

Principles of Design and Technology is a single written exam with questions covering both 'core' and 'in-depth' content. It covers the specification content and can draw on any of the following areas:

- Identifying requirements
- Learning from existing products and practice
- Implications of wider issues
- Design thinking and communication
- Material considerations
- Technical understanding
- Manufacturing processes and techniques
- Viability of design solutions

When will the written paper be taken?

You will sit the paper at the end of the GCSE course in the May or June of the second year of the course, which is usually in Year 11.

How long will I have?

This examination will be 2 hours long. The paper is divided into Sections A and B (see below for more information), and while both sections are available for the full 2 hours, you should use more time to answer the in-depth questions in Section B of the paper.

What format will the written paper take?

The paper will be split into two sections:

Section A (55 marks)

This section of the paper will consist of three sets of wider questions that mainly require you to demonstrate your 'core' knowledge (although there may be some questions that rely on you drawing on your 'in-depth' toolkit of knowledge. In no particular order:

- There will be a product analysis question, and this will be a multi-material product.
- There will be questions that test your mathematical skills and can relate to different types of products.
- There will be a question that requires technical understanding.

These questions will also cover wider issues and other core topics.

There will be a mixture of different levels of questions, these will expect you to demonstrate a comprehensive and thorough knowledge of the topic and one extended response question that will have an * next to it. This question will look for an in-depth understanding of the topic, a detailed and logical answer that is supported by examples.

Section B (45 marks)

This section of the paper will assess your 'in-depth' knowledge. There is an insert document that will support the questions given in Section B. The insert document will provide a choice of products within a situational context that relate to the six main categories of in-depth learning.

The beginning of this section of the paper is likely to test further 'core' knowledge within the situation outlined in the insert document. The remaining question will require you to choose a product in order to demonstrate your in-depth understanding in relation to any **one** of the main categories of materials and/or systems, including the sources, development and manufacture of prototypes and products using the chosen categories of materials and/or systems.

There will be a mixture of different levels of questions, that examine both core and in-depth knowledge as laid out in the earlier chapters. As in Section A, there will be a mixture of different levels of questions, these will expect you to demonstrate a comprehensive and thorough knowledge of the topic and one extended response question that will have an asterisk * next to it. This question will look for an in-depth understanding of the topic, a detailed and logical answer that is supported by examples.

How will I be assessed?

In the written paper you will be assessed against the two assessment criteria shown in the table below. (AO1 and AO2 are not assessed in the written paper.)

Assessment objective		% of GCSE (9-1) Design and Technology
AO3	Analyse and evaluate: • design decisions and outcomes, including for prototypes made by themselves and others • wider issues in design and technology	10%
AO4	Demonstrate and apply knowledge and understanding of: • technical principles • designing and making principles	40%

Fifteen per cent of the paper will assess mathematical skills within a design and technology context. Mathematical skills that may be assessed include arithmetic and numerical computation, handling data, graphs, and geometry and trigonometry. See 'Questions assessing your mathematical skills' on page 376 for more information on this.

General advice on answering exam questions

Your exam paper will be scanned and marked online, so you need to take care not to write outside the L-shaped brackets at the corner of the paper, otherwise the examiner may not be able to see your answer. Practise keeping your answers to the booklet spaces provided – these have been carefully thought-out. Additional paper can be used, which will be scanned along with your paper.

Section A (55 marks)

Typically questions in this section of the paper will start with an image of a product and you will be asked questions about the materials, properties and features of that product. The questions in Section A tend to cover core principles, although extended discussions and explanations may require the use of examples to support your answers from the in-depth areas of learning that you have covered.

Some questions will draw on core design principles, contexts and material properties. They may involve product analysis or problem solving and you may not always be familiar with the product that is presented, but you will still need to apply your knowledge and understanding. Avoid relying solely on the photograph, and think about the context in which the product would be used. Practise this by looking at existing products that are made from multiple materials and by thinking about materials and components – why they have been chosen and what properties they have to make them suitable for this purpose.

Figure 15.1 Some products made from multiple materials

Some products to start you thinking are shown below. Think about the different components that have been used in the design of these objects: the materials they are made from, the properties they need, and the user and the contexts they are used in. Consider how this affects their design, thinking about the users and their needs. Are the products aesthetically pleasing? Are they ergonomically sound? What anthropometric data would have been needed to design them? What features of the products are successful, and why?

You can also consider the wider issues involved in sourcing materials and their manufacture. Looking at existing products provides a great starting point for learning how to approach these types of questions.

Questions assessing your mathematical skills

Some questions will require you to demonstrate your mathematical skills, but this will be within a realistic and meaningful Design and Technology context.

Questions assessing your mathematical skills may ask you to:
- calculate quantities of materials, costs and sizes
- calculate percentage profits or waste saving
- calculate surface areas or volumes
- apply tolerances
- analyse or interpret data in tables, graphs, charts and diagrams
- extract information from technical specifications and graphical sources
- calculate angles to support accurate marking out
- understand symmetry and tessellation when minimising waste
- calculate areas to determine quantities of materials to apply scale factors.

To help you prepare to answer questions that assess your mathematical skills you can calculate the costs of materials in your NEA.

ACTIVITY

A manufacturer of desk top storage solutions is looking to increase their product range. The figure below shows an acrylic magazine storage unit, it retails at £10 and measures H305 × W95 × D265mm
Sketch and work out the area of sheet material one storage unit would need.

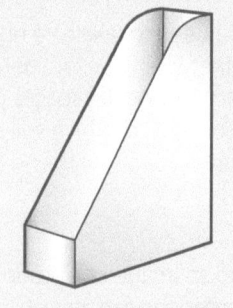

The images in Figure 15.2 show a range of Year 9 design work calculating areas and the cost of prototype products. They include the use of trigonometry to find lengths of sides, calculation of circumferences and areas of shapes such as circles, triangles and polygons. This provides a grounding for the Maths questions at GCSE.

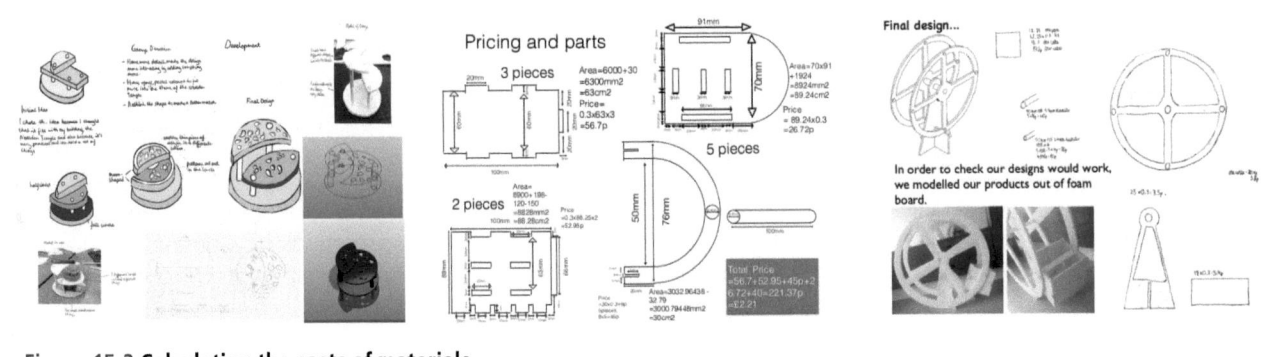

Figure 15.2 **Calculating the costs of materials**

Technical understanding questions

These questions will ask you to apply your knowledge of new technologies, digital technology, mechanisms, motion, forces, electronics systems including the use of programmable components. To practise answering technical understanding questions, it may be useful to find examples of products that are multi-material and/or multi-functional in order to think about the systems, structures and materials they use. You should also consider how the design of the product benefits the user. Handheld devices that make tasks easier in the kitchen can be a good starting point for example products.

Figure 15.3 **An example of a product that uses mechanisms**

Section B (45 marks)

A situational context will be provided through an insert booklet, along with products related to the context that are associated with different areas of in-depth learning. There may be questions that still cover core knowledge and you will be able to choose a product within an area to demonstrate your in-depth knowledge on materials, processes, technical understanding and wider issues.

To practise this section you should look at real-life products, for example in a public space such as a park, and also select some made from different materials within your subject area. Think about how they are made.

Wider issues or technical understanding may also be tested, for example the issues that face designers in the selection of materials and manufacturing methods: how designs need to be easy to use, how prototypes can achieve their functional requirements; what will happen to products after they are no longer needed, and how designers pitch and sell ideas, and present them to users and stakeholders within the given context.

Figure 15.4 **You need to consider the situational context**

GLOSSARY

Absorbency: The ability of a material to suck up moisture.

Additives: Substances added to polymers to improve their mechanical properties.

Aesthetically pleasing: Attractive to the senses.

Alloy: A metal made by combining two or more metals to give greater strength or resistance to corrosion.

Annealing: Reducing the hardness of a metal and making it more ductile.

Anthropometrics: The study of the sizes of the human body.

Asymmetry: The absence of symmetry of any kind.

Batch production: Making a set number of identical products.

Bevel gear: A gear with teeth cut at a 45° angle to change the direction of the drive shaft by 90°.

Biobased: A product made from a renewable resource.

Biodegradable: The ability of a material, substance or object to break down naturally in the environment through the action of micro-organisms, thereby avoiding pollution. Natural and regenerated fibres are biodegradable.

Blending fibres: Mixing fibres of different origins together in order to improve the properties of the finished yarn.

Block model: An informal model that captures the form of an idea – tends to have no moving parts.

Bonded fabric: Fabric manufactured by adding pressure, heat, chemicals or adhesives to a web of fibres, causing them to bond together. Examples of bonded fabrics include baby wipes and interfacing.

Brazing: Soldering at very high temperatures.

Breadboard: A prototyping tool for building electronic circuits.

Brittle: Likely to snap, crack or break when bent or hit with an impact.

Bulk discount: Reduced price of items for buying a greater number at a time.

CAD: Computer-aided design.

CAE: Computer-aided engineering.

Cam and follower: A mechanism to convert rotary motion into reciprocating motion.

CAM: Computer-aided manufacture.

Case hardening: Hardening the outer surface of a metal.

Catalysts: Chemicals that cause a reaction to happen.

Circular economy: An alternative to a traditional linear economy (manufacture, use, dispose), in which we use resources for as long as possible then reuse and regenerate products and materials; a cradle-to-cradle model.

Climate change: A change in global climate apparent from the mid- to late twentieth century onwards.

Cloud computing: A network of online servers that store and manage data.

CNC: A computer-controlled machine used in industry.

CNC: Computer numeric control.

Co-injection moulding: An injection-moulding process that uses two different polymers.

Collaboration: Working with others to achieve a shared goal. There is a shared expectation which is critical in understanding collaboration.

Commodity: A raw material that can be bought and sold, such as coffee or gold.

Compressive strength: The resistance of a material to breaking as a result of compression/squashing.

Context: The setting or surroundings for a design solution.

Crease-resistance: A fibre's ability to recover after being wrinkled.

Culture: The ideas and activities of groups of people; the way that people behave and relate to one another; the beliefs and aspirations of a group of people.

Current: The actual electricity flowing, measured in amps.

Debugging: Finding and removing errors in a microcontroller program.

Diode: A component that allows current to flow in one direction only.

Downcycling: Recycling into a lower-quality product.

Drape: The way a fabric hangs under its own weight.

Driver: A component to boost the output current.

Ductile: Able to be bent or deformed without losing toughness.

Durability/hardness: The ability to resist cutting, wear or abrasion.

Eco-material: An environmentally friendly material.

Economies of scale: The cost advantages that manufacturers obtain due to the size, output or scale of their production.

Effort: The input force in a mechanism.

Electrolysis: Extraction of metals by melting and passing electric currents through it.

Emerging technologies: New technologies that are currently being developed, or will be developed within the next five to ten years.

Ergonomics: The study of how we use and interact with a product or system.

Ethical: Correct, fitting, good or honourable.

Fabricating: Creating products by assembling parts and components together.

Fast fashion: A trend involving the quick transfer of new collections from the catwalk into stores.

FEA: Finite element analysis.

Felted fabric: Fabric manufactured by adding moisture, pressure and friction to a web of fibres, causing them to matt together.

Ferrous metal: A metal that contains iron.

Filament fibre: A long, smooth, fibre of synthetic or natural protein origin.

Finite resources: Non-renewable sources that cannot be replaced in a sufficient timeframe to allow further human consumption. Examples include crude oil and natural gas and coal, as these resources have taken millions of years to form.

Focus group: An organised discussion led by a moderator, where a group of people are asked about their views and experiences, perceptions of and attitudes towards a product, brand, service, idea, advertisement or packaging.

Force: A push, a pull or a twist.

Former: A block made to hold material in the shape required.

Fulcrum: The pivot around which a lever turns.

Function: What a product will do and how it will work.

Fused deposition modelling (FDM): Building up a 3D shape by laying down material in layers.

Galvanising: Coating steel with zinc to stop it corroding.

Globalisation: Businesses and organisations operating globally and developing international influence.

Gloss: A shiny, reflective surface finish.

Golden ratio: A common mathematical ratio found in nature that can be used to create pleasing, natural-looking compositions in your design work; also known as the Golden Mean, the Golden Section, or by the Greek letter Phi.

GRP: Glass-reinforced plastic.

gsm: Grams per square metre. Used to classify the weights of paper and card.

Idler gear: A gear that is inserted between two spur gears to change the output direction.

Import: To bring goods in from another country.

Inclusive design: Designing for the widest possible audience.

Input: The type of motion or force put in to a mechanism.

Insulator: Resists the passage of heat or sound, keeps heat inside/blocks sound.

Integrated circuit (IC): A microchip.

Internet of Things (IoT): Where electronic devices connect within the existing internet infrastructure, to send and receive data without human intervention.

Iterative design: A design process based on a cyclic process of prototyping, testing, reflecting, analysing, evaluating and refining a product or process. The processes are repeated with a focus of achieving the optimum design.

Iterative modelling: Repeated modelling to develop an idea.

Laminating: Building up a shape in thin layers.

Lathe: A machine used for making cylindrical-shaped objects by spinning.

Lay plans: Used by fabric pattern cutters to guide them when placing pattern pieces in the correct location and direction before cutting.

Lean manufacturing: A systematic method for the elimination of waste within a manufacturing system.

Lever: A rigid bar that turns around a fulcrum.

Light dependent resistor (LDR): A light-sensing component.

Lightweight: Weighs very little.

Line bending: Bending a polymer sheet after softening a narrow strip.

Linkage: A component used to direct forces and movement to where they are needed.

Load: The output force in a mechanism.

Lubrication: A substance applied to reduce friction between moving parts.

Malleable: Able to be hammered or pressed into shape without breaking or cracking.

Market pull: Refers to the market (users/customers) identifying and requiring the need for a new product or solution to a problem.

Marketing: The business of promoting and selling a product; can include advertising and promotion, and market research.

Mass/high volume production: Producing very large numbers of an item or product.

Mathematical modelling: The representation of a real situation using mathematical concepts and language.

Matt: A dull, non-reflective surface finish.

Mechanism: A series of parts that work together to control forces and motion.

Microcontroller: A programmable electronic component that adds functionality to a product.

Micron: One-thousandth of a millimetre. Used to classify the thickness of paper and card.

Monomer: A molecule that can be bonded to other identical molecules to form a polymer.

MOSFET: A type of transistor used as a driver.

Mould: has various references in design and manufacture; it can be a hollow container used to give shape to a molten or hot material while it cools and hardens; or used to describe the forming of an object from a malleable material

Moulding: can describe the process of forming a shape from a mould; or be used to describe a specific form, e.g. a decorative moulding.

Nanotechnology: Technology on a microscopic scale.

Natural cellulose fibres: Fibres that come from plant-based sources, for example cotton and linen.

Natural polymers: Chains of protein or cellulose molecules (monomers), such as keratin, glucose or fibroin. These chains are the basis of all natural fibres.

Natural protein fibres: Fibres that come from animal-based sources. These include hair, fur or silk fibres.

Non-ferrous metal: A metal that does not contain iron.

Non-renewable energy: Energy derived from resources that come out of the ground as liquids, gases and solids and cannot be quickly replenished.

Non-woven fabrics: Fabrics made by entangling fibres together using friction, pressure, heat or chemicals. These fabrics are cheaper to produce than woven fabrics and are often used for disposable products.

Obsolescence: Becoming outdated or no longer wanted.

One-off production: Making only one or a small number of products.

Organic: Derived from living matter.

Output: The type of motion or force a mechanism produces.

Oxidation: Discolouring, tarnishing and/or rusting of metal through reaction with air or water.

Physical and working properties: The properties of a material that affect the way it is used.

Piezo-electric sounder: A miniature speaker for producing sounds from a microcontroller.

Pinion: A small driver gear (smaller than the driven gear).

Plain weave: A basic weave construction in which the weft yarn goes under and over alternate warp yarns, giving a strong and flat fabric.

Planned obsolescence: The business practice of deliberately outdating an item before the end of its useful life.

Plating: Coating one type of metal with another to improve appearance or corrosion-resistance.

Polarised: A component in electrical systems that must be connected the correct way round in a circuit.

Polymer: A substance that has a molecular structure built up from a large number of similar units (monomers) bonded together.

Polymerisation: The process of joining small molecules to form polymers (long chains).

Polymer memory: The ability of thermo polymers to return to their original state after reheating.

Powder coating: A coating of electrically charged powder that is baked on to achieve a tough finish.

Press moulding: Forming a hollow shape from a softened polymer sheet.

Primary user: The person or group of people who will use a product or system.

Printed circuit board (PCB): The support and connections for the electronic components in a product.

Program: A set of instructions to tell a microcontroller how to carry out a task.

Programming language: The set of instructions and rules used to write a microcontroller program.

Proportion: The relative size and scale of the various elements in a design.

Prototyping: A process that involves the production of a test model, on which the final product is based.

Pulley and belt drive: A method of transferring rotary motion between two shafts.

Quenching: Rapidly cooling hot metal by immersing in cold water.

Rack and pinion: A system to change between rotary and linear motion.

Rapid prototyping: The process of making a 3D shape from a digital file.

Recyclable: A material or product suitable for re-processing back to raw forms.

Recycled paper: Paper made from used paper products.

Regenerated fibres: Fibres derived from natural cellulose sources, using synthetic chemical processes. These include viscose and bamboo fibre.

Reinforce: To strengthen or support an object or substance with additional material.

Renewable: A natural resource that is not depleted by use.

Resistance: How hard it is for an electric current to flow, measured in ohms.

Rigid: The ability to withstand forces without flexing.

Satin: An in-between gloss and matt finish.

Satin weave: A luxurious weave in fabric construction in which the weft yarn floats over three or more warp yarns, giving a smooth and shiny finish.

Scale of production: The number of products being produced in one go.

Seam allowance: Margin added in textiles to pattern pieces to allow enough fabric around the edge of the sewing line for errors.

Seasoning: Adjusting the moisture content of timber to make it more suitable for use.

Self-finishing: A material that requires no further coatings or finishing processes.

Sensor: A component that produces a signal in response to a specific physical quantity.

Shaft: A rod that transfers the rotation through a mechanism.

Signal: An electrical voltage that is used to represent information.

Simple gear train: A pair of gears consisting of a driver gear and a driven gear.

Sketch models: Quick models, often of just parts of a design, made from easy-to-work and low-cost materials such as cardboard or foam.

Smelting: Melting down metals into molten liquid.

Soldering: Melting solder around two or more metals to create a joint.

Spur gear: A wheel with teeth around its edge.

Stabiliser: An additive added to polymers to help them withstand degradation.

Stakeholder: A person, group or organisation with an interest in a product/system, for example parents/schools when designing products for children.

Standardised component: An individual part or component, manufactured in thousands or millions, to the same specification.

Staple fibre: A short fibre with crimp (wavy texture); most natural fibres are staple fibres, although synthetic fibres can be manufactured to form the staple shape.

Steam bending: Softening the fibres of wood with steam to allow it to bend.

Stock form: Commonly available forms of a material that can be bought.

Stove enamelling: A paint coating that is baked on to achieve a tough, durable finish.

Stroke: The distance the follower (of a cam) rises.

Structure: A collection of parts that provide support.

Subsystem: A section with a specific role within a system.

Sustainable: Refers to the consideration of whether something is able to be maintained at a certain rate or level without impacting on future considerations.

Sustainable economic growth: A rate of growth that can be maintained for future generations without causing significant economic problems.

Symmetry: When elements are arranged in the same way on both sides of an axis or when rotated around a point.

Synthesised: Made by combining parts of, e.g. combining different carbon compounds.

Synthetic: A manufactured substance that imitates a natural product.

System: The general name for a set of mechanical or electronic parts that work together to produce a desired output.

System diagram: A diagram of the interconnections and flow of signals in an electronic system.

Systems thinking: The understanding of a product or component as part of a larger system of other products and systems. In the iterative design process, consideration of the role of all components and sub-systems of the product or system, including the user experience and the marketing of the object being designed, ensures all aspects of the product are given the required attention to detail.

Technical textiles: Textiles manufactured specifically for their performance properties instead of their aesthetic value. Examples of technical textiles include Kevlar and Stomatex®.

Technology push: When research and development of new technology drives new product development, e.g. touchscreen and fingerprint technology in smartphones.

Tempering: Using heat to make a metal less brittle.

Tensile strength: The resistance of a material to breaking under tension/stretching.

Thermistor: A temperature-sensing component.

Throwaway society: A society influenced by consumerism and excessive consumption of products.

Tolerance: An allowable amount of variation of a specified quantity, especially in the dimensions of a part.

Torque: A turning or twisting force.

Toughness: Ability to resist breaking, bending or snapping.

Triangulation: Achieving rigidity by producing triangular structures.

Twill weave: A strong weave in fabric construction in which the weft yarn goes under and over alternate warp yarns in a diagonal formation, giving a textured finish.

Upcycling: Reusing and transforming products into a higher-quality product.

Usability: How easy a product is to use, how clear and obvious the functions are.

User-centred design (UCD): Sometimes called 'human-centred design', user-centred design is a design strategy, or design approach, with the aim of making products and systems usable. It focuses on the user interface and how the user interacts with and relates to the product.

UV degradation: The weakening of polymers when exposed to the ultraviolet light in sunlight.

Vacuum forming: Producing thin hollow items over a shaped mould using a heated polymer sheet and vacuum.

Viability: The ability to work successfully.

Virgin fibre paper: Paper made from 'new', unused wood fibres.

Voltage: A measure of the electrical 'pressure' causing a current to flow, measured in volts.

Wastage: Removing material to leave a desired shape.

Water-resistance: The ability of a material to resist water droplets for a period of time.

WEEE directive: A sustainability scheme to reduce the amount of waste electrical products sent to landfill.

Weft knit: Fabric constructed with horizontal and vertical rows of interlocking loops.

Welding: A fusion of materials caused by intense heat.

INDEX

Terms in **bold** indicate pages where key term definitions can be found.

ACKNOWLEDGEMENTS

The authors and publishers would like to thank the following schools and students:

Silcoates School in Wakefield; Kenilworth School in Kenilworth; Philip Robinson from Thrybergh Academy and Sports College and Terry Bream.

Picture credits

All photos not listed below are copyright or the authors.

p.iv © Brain light/Alamy Stock Photo **p.1** © mrivserg - 123RF; **p.2** top© ullstein bild/Getty Images, bottom © Suricoma/123RF; **p.3** top © Nicola Ferrari/Alamy Stock Photo, bottom © Chris Harris/Alamy Stock Photo; **p.4** left © Monkey Business Images/Shutterstock.com, right © Alexandre zveiger/Shutterstock.com; **p.5** © Fiphoto/Shutterstock.com; **p.6** top © Han Myung-Gu/Getty Images ©Entertainment/Getty images, left© Beat Bühler, right ©Byelikova Oksana/Shutterstock.com; **p.7** top © Ratmaner/123RF, bottom © Alexander Davidyuk/Shutterstock.com; **p.8** © Coprid/Shutterstock.com; **p.10** left © ullstein bild/Getty Images, right © Duncan Snow/Alamy Stock Photo; **p.11** left © Hulton Archive/Getty Images, right © Justin Sullivan/Getty Images News/Getty Images; **p.12** right © Maxim Aksutin/Shutterstock.com, left © suns07butterfly/Shutterstock.com; **p.14** © lightmoon - Fotolia; **p.16** top row, l-r: © Notachai plugjaiseua /123RF, © Phumeth Nithikulprecha/123RF, ©Viktorija Reuta/123RF, bottom row, l-r: © Nito500/123RF, © Kchung/123RF, © Belchonock/123RF, bottom © Dorling Kindersley ltd/Alamy Stock Photo; **p.17** © Nick Moore/Alamy Stock Photo; **p.18** left © FoodStocker/Shutterstock.com, right ©Hugh Threlfall/Alamy Stock Photo; **p.20** © Mark Boulton/Alamy Stock Photo; **p.22** © lunamarina - Fotolia.com; **p.23** © Ken226/123RF; **p.25** © Jeffery Kent/Alamy Stock Photo; **p.27** © Jezper/Shutterstock.com; **p.28** top © Scanrail/123RF, bottom © SOKO EPZ Ltd; **p.29** © Julio Etchart/Alamy Stock Photo; **p.30** © Suricoma/123RF; **p.33** top ©Illustrart/123RF, bottom ©Victoroancea/123RF; **p.34** left © Alri/123RF, right © Viktor Bonda/123RF; **p.38** top © B.A.E. Inc./Alamy, bottom © Tom Mc Nemar/istockphoto.com; **p.39** © Wichien Tepsuttinun/Shutterstock.com; **p.40** LEFT © batchimages/Alamy Stock Photo, © kingan - Fotolia, right © auremar - Fotolia.com; **p.41** top left © slydgo1111 – Fotolia, top right © Anton Starikov/123RF, bottom left © Jari Hindstrom/123RF, bottom right © Fotolia; **p.42** © Blackshark66/Shutterstock.com; **p.43** top left © Nattakit Jeerapatmaitree/Shutterstock.com, top right © Winai Tepsuttinun/Shutterstock.com, bottom left © Lucky Dragon - Fotolia, bottom right © Yurchello108/Shutterstock.com; **p.44** top left © Lebedinski Vladislav/Shutterstock.com, middle © Anton Khegay/Shutterstock.com, bottom left © Vitalii Tiahunov/123RF; **p.45** left © dzmitri mikhaltsow/123RF, right © Realimage/Alamy Stock Photo, bottom © hskoken/123RF; **p.46** top left © Udom Jinama/123RF, top right © lightmoon – Fotolia, bottom left © P Wei/istockphoto.com, bottom right © Berna Namoglu/Shutterstock.com; **p.47** top © Leonid Shcheglov - Fotolia, middle © vitaliy_73/Shutterstock.com, bottom © Cardaf/Shutterstock.com; **p.48** left © XiXinXing/Shutterstock.com, right © Noomcpkstic/Shutterstock.com, bottom © Julia Lototskaya/Shutterstock.com; **p.49** © tlorna – Fotolia; **p.50** © Antonius Egurnov/Shutterstock.com, bottom © RGB Ventures/SuperStock/Alamy Stock Photo; **p.51** top © PASCAL GOETGHELUCK/SCIENCE PHOTO LIBRARY, middle © Phil Degginger/Alamy Stock Photo, bottom © Ekaterina Garyuk - Fotolia.com; **p.52** top © Duncan Astbury/Alamy Stock Photo, middle left © Richard Heyes/Alamy Stock Photo, middle right © lasha/Shutterstock.com, bottom © diter/Fotolia; **p.53** left © freedomnaruk/Shutterstock.com, right © luissantos84/iStock/Getty Images Plus/Getty images; **p.56** © Matt Blythe//Shutterstock.com; **p.57** ©Murat Baysan/Shutterstock.com; **p.60** top © Westend61/Getty Images, left © Kostic Dusan/123RF, right © Kitch Bain/123RF; **p.61** left © Westend61/Getty Images, right © Trevor Walker/Alamy Stock Photo; **p.62** left © Anton Samsonov/123RF, right © Olegsam/123RF; **p.67** © Tiago Zegur/Alamy Stock Photo; **p.68** top © Olegsam/123RF, bottom © Cheskyw/123RF; **p.69** l-r © Ian Allenden/123RF, © Artur Marciniec/Alamy Stock Photo, © Cultura Creative (RF)/Alamy Stock Photo; **p.74** © Shutterstock/lucadp; **p.76** © Clearjade/123RF **p.77** top ©Iaroslav Danylchenko/123RF, bottom ©Victor W. Adams/Alamy Stock Photo; **p.78** left © C Squared Studios/Photodisc/Getty Images/Travel Vacation Icons OS23, right © Stockbyte/Getty Images/Entertainment & Leisure CD35; **p.79** top ©R MACKAY/Fotolia, bottom © Brad Pict/Fotolia; **p.81** © Mikael Damkier - Fotolia.com; **p.82** © Moodboard/Alamy Stock Photo; **p.83** © latham & holmes/Alamy Stock Photo; **p.84** top © Gts/Shuttetstock.com, bottom © Wichien Tepsuttinun/Shutterstock.com; **p.88** © Yuri Tuchkov/123RF; **p.90** left ©Kitch Bain/Shuttetstock.com, right © Paul Smith/123RF; **p.97** left © Peter Gudella/Shutterstock.com, right © Adisak Rungjaruchai/Shuttetstock.com; **p.98** © Quang Ho/Shutterstock.com; **p.99** top © Robert Ashton/Massive Pixels/Alamy Stock Photo, middle © B Christopher/Alamy Stock Photo; **p.101** top © Reamolko/Shutterstock.com, bottom ©le Moal Olivier/Alamy Stock Photo; **p.102** ©Zoonar GmbH/Alamy Stock Photo; **p.104** ©jurra8/Shutterstock.com; **p.105** top © FotoLibre Studio/Shutterstock.com, bottom ©dpa picture alliance/Alamy Stock Photo; **p.106** ©Maksym Dykha/Shutterstock.com; **p.108** ©Jiří Zuzánek/123RF; **p.112** © stocksolutions – Fotolia **p.115** ©Carolyn Jenkins/Alamy Stock Photo; **p.117** © Jack_photo/Shutterstock.com, © Roman Babakin/Shutterstock.com; **p.126** top left © Olegsam/123RF, top right © 895_The_Studio/Shutterstock.com, bottom left © Rashid Valitov/Shutterstock.com, bottom right © Gina Vescovi/Shutterstock.com; **p.130** top left © Oleksandr Chub/Shutterstock.com, top right © m.bonotto/Shutterstock.com, bottom left © OlegSam/Shutterstock.com, bottom right © Stockphoto-graf/Shutterstock.com; **p.131** left © ESB Essentials/Shutterstock.com, right © Carlos andre Santos/Shutterstock.com; **p.143** top © Optimarc/Shutterstock.com; bottom ©Marc Tielemans/Alamy Stock Photo; **p.144** © Steven Heap/123RF; **p.145** ©Marc Tielemans/Alamy Stock Photo; **p.147** © Guchici/Shutterstock.com; **p.148** left © Art Directors & TRIP/Helene Rogers/Alamy Stock Photo, right ©Cultura Creative (RF)/Alamy Stock Photo; **p.150** © ZUMA Press, Inc./Alamy Stock Photo; **p.152** © DmyTo/Shutterstock.com; **p.153** © Arterra Picture Library/Arndt Sven-Erik/Alamy Stock Photo, bottom © Maxim Tupikov/Shutterstock.com; **p.154** © Photographee.eu/Shutterstock.com; **p.156** © Imagestate Media (John Foxx)/Vibrant Backgrounds SS; **p.163** © 赵 建康/123RF; **p.164** © Sashkin/Shutterstock.com; **p.165** © Dan Hughes; **p.166** top © Yanas/Shutterstock.com, bottom © David J Green/Alamy Stock Photo; **p.167** top © Luckyraccoon/Shutterstock.com, left © Ppn2047/Shutterstock.com, right © Difydave/E+/Getty Images; **p.168** © Alacatr/E+/Getty Images; **p.169** © Alfred Hofer/123RF; **p.172**© Zoltan Major/Shutterstock.com; **p.173** left © Anton Khegay/Shutterstock.com, right © Homonstock/Shutterstock.com; **p.174** © Kbwills/E+/Getty Images; **p.175** © Asharkyu/Shutterstock.com; **p.179** right © Aon168/Shutterstock.com, bottom © Alex Cao/Photodisc/Getty Images; **p.180** © Scanrail1/Shutterstock.com; **p.181** © Rocharibeiro/Shutterstock.com; **182** top © Danielle Nichol/Alamy Stock Photo, bottom © Maya Kruchankova/Shutterstock.com; **p.184** © Kues/Shutterstock.com; **p.185** top © Randall Schwanke/Shutterstock.com; **p.187** © John99/Shutterstock.com; **p.188** © Supergenijalac/Shutterstock.com; **p.190** © greenbird-photography/Alamy Stock Photo; **p.192** © Srebrina Yaneva/iStockphoto.com; **p.195** top © Mnoor/Shutterstock.com, bottom © V&A Images/Alamy Stock Photo; **p.197** © Huguette Roe/Shutterstock.com; **p.198** © MediaWorldImages/Alamy Stock Photo; **p.199** left © MIHAI ANDRITOIU/Alamy Stock Photo, right © Jirawatfoto/Shutterstock.com; **p.200** left © daniella christoforou/Alamy Stock Photo, right © Oramstock/Alamy Stock Photo; **p.201** © Digital Genetics/Shutterstock.com, © Unkas Photo/Shutterstock.com, © Coprid/Shutterstock.com, © KRIANGKRAI APKARAT/123RF, © Achim Prill/123 RF, © Elena Abduramanova/Shutterstock.com; **p.202** © pzAxe/Shutterstock.com; **p.204** © Chris

Fertnig/iStockphoto.com; **p.205** © Adrian brockwell/Alamy Stock Photo; p.208 © Mark Boulton/Alamy Stock Photo; **p.206** © Duncan Astbury/Alamy Stock Photo; **p.210** left © Happy Stock Photo/Shutterstock.com, right © Nor Gal/Shutterstock; **p.212** © Dan Hughes; **p.213** © Jo ingate/Alamy Stock Photo; **p.214** © Skoropadska Maruna/Shutterstock.com; **p.216** top © Artem Loskutnikov/Shutterstock.com, bottom © Improve Your Injection Molding; **p.217** author supplied; **p.219** left © mariakraynova/Fotolia, right © Skoropadska Maruna/Shutterstock.com; **p.220** © Dan Hughes; **p.222** © Yilmaz Uslu/Shutterstock.com; **p.223** bottom © Giftzaa5069/Shutterstock.com; **p.224** © Zoonar/Marko Beric/Alamy Stock Photo; **p.225** © Baloncici/Stock/Getty Images Plus/Getty images; **p.228** © Cai Liang/123 RF; **p.230** © Hero Images Inc./Alamy Stock Photo; **p.231** © Lourens Smak/Alamy Stock Photo; **p.232** top © lunamarina - Fotolia.com, Bottom ©JGI/Jamie Grill/Blend Images/Getty Images; **p.235** ©Walter Bibikow/AWL Images/Getty Images; **p.236** top © Ibreakstock/Shutterstock.com, bottom ©LAGUNA DESIGN/SCIENCE PHOTO LIBRARY; **p.237** top left © Berna Namoglu/Shutterstock.com, top right © Spline_x/Shutterstock.com, bottom left © Doubleo44/Shutterstock.com, bottom right © Tinnko/Shutterstock.com; **p.238** left © Cl2004lhy/Shutterstock.com, right ©LAGUNA DESIGN/SCIENCE PHOTO LIBRARY; **p.239** left © JoobheadiStock/Getty Images Plus, right ©Design Pics Inc/Alamy Stock Photo; **p.240** top left © Photogolfer/Shutterstock.com, top right © Coprid/Shutterstock.com, bottom © Galyna Andrushko/Shutterstock.com; **p.241** top © Ross Gordon Henry/Shutterstock.com, left © Christopher May/Shutterstock.com, right © Phoompiphat Phoomkaew/Shutterstock.com; **p.245** © Dja65/Shutterstock.com; **p.247** top © Sergiy Kuzmin/Shutterstock.com, bottom © XiXinXing/Shutterstock.com; **p.248** left © Noomcpkstic/Shutterstock.com, right © Sergey Chayko/Shutterstock.com; **p.249** © Julia Lototskaya/Shutterstock.com; **p.250** top ©Design Pics Inc/Alamy Stock Photo, bottom ©Art Directors & TRIP/Alamy Stock Photo; **p251** top © Noomcpkstic/Shutterstock.com, bottom © Mamunur Rashid/Alamy Stock Photo; **p.252** top © China Photos/Stringer/Getty Images News, bottom © Kevin Britland/Alamy Stock Photo; **p.253** ©Steven Scott Taylor/Alamy Stock Photo; **p.254** left © Shutter Top/Shutterstock.com, right © Catherine Jones/Shutterstock.com; **p.255** top right © Feng Yu/123 RF, left © Marcoventuriniautieri/E+/Getty Images, right © Maly Designer/Shutterstock.com; **p.256** top © Peepo/E+/Getty Images, bottom left © Evgeny_Popov/Shutterstock.com, top right ©SCIENCE PHOTO LIBRARY, bottom right © Tom Gowanlock/Shutterstock.com; **p.257** left © Pete Anderson/Dorling Kindersley/Getty Images, right © Fmua/Shutterstock.com; **p.259** left © Pressmaster/Shutterstock.com, © Monika Wisniewska/Shutterstock.com; **p.260** left © Hill Street Studios/Hill Street Studios, right ©Maarigard/Dorling Kindersley/Getty Images; **p.261** © UrbanImages/Alamy Stock Photo; **p.262** top left © Golubovy/Shutterstock.com, top right ©Mint Images Limited/Alamy Stock Photo, bottom © Natalie Behring/Aurora Photos/Alamy Stock Photo; **p.263** © Levent Konuk/Shutterstock.com; **p.265** left ©Siede Preis/Stockbyte/Getty Images, top right © Dorling Kindersley/Getty Images, bottom right © Vitalii Gorbatiuk/Shutterstock.com; **p.267** left © Toeytoey/Shutterstock.com, right © Indigolotos/123RF; **p.268** top left ©Maarigard/Dorling Kindersley/Getty Images, top right © gridenko/123RF, bottom left © Michael Harvey/Alamy Stock Photo, bottom right © Punchallt Chorlkasatian/Shutterstock.com; **p.271** © DJ Srki/Shutterstock.com; **p.272** © Audy39/Shutterstock.com; **p.273** © Dorling Kindersley/Shutterstock.com; **p.274** © dpa picture alliance/dpa/Alamy Stock Photo; **p.275** top ©Dario Cantatore/Stringer/Getty Images Entertainment/Getty Images, bottom © David Litschel/Alamy Stock Photo; **p.276** left © creative labe/Shutterstock.com, middle © Roman Sigaev/Shutterstock.com, right © Tom Wang/Shutterstock.com; **p.279** left © Helmut Meyer zur Capellen/imageBROKER/Alamy Stock Photo, right © Kzenon/Shutterstock.com; **p.280** left © Xtrekx/Shutterstock.com, right © Levent Konuk/Shutterstock.com; **p.284** © Bancha atsawatawon/123RF; **p.290** © Zigzag Mountain Art/Alamy Stock Photo; **p.291** © John Hopkins/Alamy Stock Photo; **p.292** © Bancha atsawatawon/123RF; **p.293** © Frank Kletschkus/Alamy Stock Photo; **p.304** top © Powered by Light/Alan Spencer/Alamy Stock Photo, bottom © DBURKE/Alamy Stock Photo; **p.308** left © Colin moore/123RF, right © AlexLMX/Shutterstock.com; **p.310** © alexlmx/123RF; **p.315** © Grigory_bruev/123RF; **p.316** top right © D J Myford/Alamy Stock Photo, bottom left © David Willman/123RF, bottom right © Sergey Soldatov/123RF; **p.317** © Raymond McLean/123RF; **p.322** © Brad Pict/Fotolia; **p.324** © Coprid/Fotolia.com; **p.327** © picsfive - Fotolia **p.367** top © Viktorija Reuta/123RF, bottom © Vladimir Tarasov/123RF; **p.369** bottom © Stockbyte/Getty Images/Child's Play SD113; **p.370** © Cathy Yeulet/123RF